Air Pollution

Other books in Clay's Library of Health and the Environment:

E Coli: Environmental Health Issues of VTEC 0157 – *Sharon Parry and Stephen Palmer* 0415235952

Environmental Health and Housing – *Jill Stewart* 041525129X
Air Quality Assessment and Management – *Owen Harrop* 0415234115

Also available from Spon Press:

Clay's Handbook of Environmental Health 18th Edition – *edited by W. H. Bassett* 0419229604

Decision-making in Environmental Health – *edited by C. Corvalán, D. Briggs and G. Zielhuis* 0419259406 HB, 0419259503 PB

Environmental Health Procedures 5th Edition – *W. H. Bassett* 0419229701

Groundwater Quality Monitoring – *S. Foster, P. Chilton and R. Helmer* 0419258809 HB, 0419258906 PB

Legal Competence in Environmental Health – *edited by Terence Moran* 0419230009

Monitoring Bathing Waters – *edited by Jamie Bartram and Gareth Rees* 0419243704 HB, 0419243801 PB

Statistical Methods in Environmental Health – *edited by J. Pearson and A. Turton* 0412484501

Toxic Cyanobacteria in Water – *edited by Ingrid Chorus and Jamie Bartram* 0419239308

Upgrading Water Treatment Plants – *Glen Wagner and Renato Pinheiro* 0419260404 HB, 0419260501 PB

Water Pollution Control – *Richard Helmer and Ivanildo Hespanhol* 0419229108

Water Quality Assessments – *Deborah Chapman* 0419215905 HB, 0419216006 PB

Water Quality Monitoring – *Jamie Bartram and Richard Balance* 0419223207 HB, 0419217304 PB

Urban Traffic Pollution – *D. Schwela and O. Zali* 0419237208

Air Pollution

Second Edition

Jeremy Colls

SPON PRESS

Taylor & Francis Group

London and New York

First published 2002
by Spon Press
11 New Fetter Lane, London EC4P 4EE

Simultaneously published in the USA and Canada
by Spon Press
29 West 35th Street, New York, NY 10001

Spon Press is an imprint of the Taylor & Francis Group

© 2002 Jeremy Colls

Typeset in Times by
Integra Software Services Pvt. Ltd, Pondicherry, India
Printed and bound in Great Britain by
Biddles Ltd, Guildford and King's Lynn

British Library Cataloguing in Publication Data
A catalogue record for this book is available
from the British Library

Library of Congress Cataloging in Publication Data
A catalog record has been requested

ISBN 0–415–25564–3 (hbk)
ISBN 0–415–25565–1 (pbk)

Clay's Library of Health and the Environment

An increasing breadth and depth of knowledge is required to tackle the health threats of the environment in the twenty-first century, and to accommodate the increasing sophistication and globalisation of policies and practices.

Clay's Library of Health and the Environment provides a focus for the publication of leading-edge knowledge in this field, tackling broad and detailed issues. The flagship publication *Clay's Handbook of Environmental Health*, now in its 18th edition, continues to serve environmental health officers and other professionals in over 30 countries.

Series Editor:

Bill Bassett: Honorary Fellow, School of Postgraduate Medicine and Health Sciences, University of Exeter, and formerly Director of Environmental Health and Housing, Exeter City Council, UK

Editorial Board:

Xavier Bonnefoy: Regional Adviser, European Centre for Environment and Health, World Health Organization, Bonn, Germany
Don Boon: Director of Environmental Health and Trading Standards, London Borough of Croydon, UK
David Chambers: Head of Law School, University of Greenwich, UK
Michael Cooke: Environmental Health and Sustainable Development consultant, UK, formerly Chief Executive of the CIEH

To Mum and Dad

Contents

12   Ozone depletion and ultraviolet radiation   472

13   Standards and legislation   511

Acknowledgements

I would like to thank Deborah Tasker for her help in organising the Figures; Anil Namdeo for the results from his Leicester modelling work; Robert Storeton-West and David Fowler for the 1993 dataset from CEH Bush and my students pursuing Environmental Science for their constructive criticisms. There are a few figures for which I could not trace the source – I gratefully acknowledge their use.

Introduction

Air pollution has been with us since the first fire was lit, although different aspects have been important at different times. On the small scale, point source releases of individual pollutants can cause localised responses ranging from annoyance to physical injury. In urban areas, high concentrations of gases and particles from coal combustion and, more recently, motor vehicles have produced severe loss of air quality and significant health effects. On a regional scale, tropospheric ozone formation and acid deposition have been the major threats. Finally, emissions of carbon dioxide and other radiatively active gases, together with stratospheric ozone depletion, represent planet-scale assaults on the quality of our atmospheric environment.

This book is designed to cover the whole gamut of air pollution issues from a quantitative standpoint. In Chapters 1 and 2, the major sources of gaseous and particulate air pollution, together with an outline of possible control measures, are described. Mobile sources, which have taken over from stationary ones as the major threat to local air quality, get their own space in Chapter 3. Chapter 4 describes the most widely-used methods for measuring these pollutants. The temporal and geographical variations of concentrations and deposition on a national and international scale are outlined in Chapter 5. Once released, the effects of these pollutants depend critically on their dilution during dispersion, a process which is covered in Chapter 6. Chapter 7 gives an extended example of the data processing techniques that can be used to extract different types of information from a set of air pollution measurements. Although people tend to associate air quality, or the lack of it, with the outdoors, most of us spend most of our lives indoors, and specific aspects of this specialised environment are highlighted in Chapter 8. The effects of air pollution on plants, animals, materials and visual range are described in Chapters 9 and 10, and the recent issues of climate change and ozone depletion in Chapters 11 and 12. Finally, the effects of pollutants on the environment have led to a wide variety of standards and legislation for their control, and these are reviewed in Chapter 13.

With such a broad spectrum of topics, there is inevitably a considerable variation in the depth of coverage that reflects my own professional interests and experience. I have used as much original research material as possible, since that is where the knowledge comes from in the first place and it is important for the reader to keep in touch with that process. I have also given relevant equations and tables of data to support the statements made in the text; there is little purpose in writing out the content of these in longhand as well, so they will usually repay more detailed examination. Although air pollution is a wholly international issue, my own access to data has resulted in a UK bias, followed in order of emphasis by Europe, the USA and the world at large. Readers are encouraged to pursue other sources for greater depth of coverage on any particular issue. Some suggestions are given as 'Further Reading' at the end of each chapter. These are not only useful documents in their own right, but also contain references to many more specialist research papers. Similarly, the figure captions cite many books, reports and research papers from which Figures for this book have been taken. If further information is required on a particular topic, then simply entering that phrase into a good Web search engine will usually provide some leads.

This book is aimed at a wide target audience – much of the material has been taught on both undergraduate and taught Masters programmes at Nottingham to students from a wide range of academic backgrounds. I hope it will be useful to as wide a range of readers elsewhere.

Acronyms and abbreviations

AA	ambient air – usually refers to plants growing in the open for comparison with those in chambers
AAS	atomic absorption spectroscopy
ACE	aerosol characterisation experiment
ACH	air changes per hour – an estimator of building ventilation rate
AES	atomic emission spectroscopy
AGCIH	American Conference of Government Industrial Hygienists
AMIS	air management information system
ANC	acid neutralising capacity
AOT40	accumulation over threshold – the measure currently favoured by UNECE for estimating ozone impact on plants
APHEA	air pollution and health – european approach
AQCD	air quality criteria document (US)
AQI	air quality index
AQMA	air quality management area (UK)
AQRV	air quality related value (US)
AQS	air quality standards (US)
ARN	automated rural network
AUN	automated urban network
BAF	biological amplification factor – used to describe the overall response of biological systems to ozone changes
BaP	benzo[a]pyrene
BART	best available retrofit technology
BATNEEC	best available techniques (or technology) not entailing excessive cost
BC	black carbon
BCC	basal cell carcinoma
BFO	bunker fuel oil
BPEO	best practicable environmental option
BPM	best practicable means – the long-established UK philosophy for pollution control
BS	British Standard

BTX	benzene, toluene and xylene
BUN	basic urban network – urban sites in the UK national network of eight-port 24-h samplers for black smoke and SO_2
CAA	Clean Air Act (US)
CAAA	Clean Air Act Amendments (US)
CALINE	California line source model – one of the most widely used dispersion models for vehicle emissions
CARB	California Air Resources Board
CCN	cloud condensation nuclei – the particles on which condensation initially occurs to form cloud droplets
CEC	Commission of the European Communities
CEH	Centre for Ecology and Hydrology (UK)
CF	charcoal-filtered – an OTC supplied with cleaned air
CFC	chlorofluorocarbon – family of chemicals responsible for depleting ozone in the stratosphere
CHESS	Community Health and Surveillance System (US)
CLRTAP	Convention on the Long Range Transport of Air Pollutants
CNC	condensation nucleus counter
COH	coefficient of haze
COHb	carboxyhaemoglobin – produced when blood haemoglobin absorbs CO
COMEAP	Committee on the Medical Effects of Air Pollutants (UK)
COP	Conference of Parties (for UNFCCC)
COPD	chronic obstructive pulmonary disease
CORINAIR	The EU programme to collect and map emissions data for all significant sources of eight gaseous pollutants
CPB	Canyon Plume Box model for calculating dispersion in urban areas
CPC	condensation particle counter
CPF	clothing protection factor
CRT	continuously regenerating trap
DALR	dry adiabatic lapse rate – the rate of decrease of temperature with height in the atmosphere applicable to a parcel of air that contains no liquid water. Value 9.8 °C km^{-1}
DEFRA	Department for the Environment, Food and Rural Affairs (UK)
DEP	diesel exhaust particles
DI	direct injection
DIAL	differential absorption lidar
DMA	differential mobility analyser
DME	dimethylether
DMS	dimethyl sulphide – organic sulphur compound released from marine phytoplankton that is eventually oxidised to sulphur dioxide and particulate sulphate in the atmosphere
DOAS	differential optical absorption spectroscopy

DOC	diesel oxidation catalyst
DoE	Department of the Environment (UK)
DOM	dissolved organic matter
DRAM	direct reading aerosol monitor
DTLR	Department for Transport, Local Government and the Regions (UK)
DU	dobson unit – for the column depth of ozone in the atmosphere
DVI	dust veil index
EA	Environment Agency (UK)
EAA	electrical aerosol analyser
EC	European Community
ECD	electron capture detector
ECE	Economic Commission for Europe (same as UNECE)
EDAX	energy dispersive analysis of X-rays
EDU	ethylenediurea – a chemical that protects plants from ozone
EEA	European Environment Agency
EEC	European Economic Community
EER	erythemally effective radiation – sunburning potential of a given radiation environment
EESC	equivalent effective stratospheric chlorine
EF	emission factor – e.g. g km^{-1}
EGR	exhaust gas recirculation
EIONET	European Environmental Information and Observation Network
ELISA	enzyme-linked immunosorbent assay
ELPI	electrostatic low pressure impactor
ELR	environmental lapse rate – the vertical profile of temperature in the atmosphere
EMEP	European Monitoring and Evaluation Programme
ENSO	El Niño southern oscillation
EPA	Environmental Protection Act (UK)
EPA	Environmental Protection Agency (US)
EPAQS	Expert Panel on Air Quality Standards (UK)
ERBS	Earth Radiation Budget Satellite
ESP	electrostatic precipitator
ETC/AQ	European Topic Centre on Air Quality
ETS	environmental tobacco smoke – the combination of MTS and STS that makes up the atmospheric load
EU	European Union
EUDC	extra-urban drive cycle (EC)
EUROAIRNET	European Air Quality Monitoring Network
FACE	free-air carbon dioxide enrichment – the system developed in the US for elevating the CO_2 concentration above field crops

FAR	First Assessment Report (by IPCC on climate change)
FEAT	fuel efficiency automobile test – an optical gas sensor that scans across the road width
FEV	forced expiratory volume – a measure of lung response to air pollutants
FGD	Flue gas desulphurisation – a range of chemical process plant that strips sulphur dioxide from flue gases before they are released to atmosphere
FID	flame ionisation detector
FTIR	Fourier transform infrared
FTP	Federal Test Program (US)
GC	gas chromatography
GCM	general circulation model
GCTE	Global Change and Terrestrial Ecosystems project
GHG	greenhouse gas
GWP	global warming potential
HAP	Hazardous Air Pollutants (US)
Hb	haemoglobin
HCFC	hydrochlorofluorocarbon – substitute for CFCs
HDV	heavy duty vehicle – such as a truck
HEPA	high efficiency particulate air
HFC	hydrofluorocarbon – substitute for CFCs
HGV	heavy goods vehicle
HMIP	Her Majesty's Inspectorate of Pollution (UK)
HPLC	high pressure liquid chromatography
HVAC	heating, ventilating and air-conditioning
ICAO	International Civil Aviation Organisation
ICP	inductively coupled plasma
IDI	indirect injection
IGAC	International Global Atmospheric Chemistry project
IPC	integrated pollution control
IPCC	intergovernmental panel on climate change
IPPC	integrated pollution prevention and control
IR	infrared
ISO	International Standards Organisation
ITE	Institute of Terrestrial Ecology (UK)
LA	local authority (UK)
LAPC	local air pollution control
LAQM	local air quality management (UK)
LCPD	large combustion plant directive (EC)
LDV	light duty vehicle – such as a van
LEV	low emission vehicle (US)
LGV	light goods vehicle
LIDAR	light detection and ranging

LNG	liquified natural gas
LRTAP	long-range transboundary air pollution
LTO	landing and take-off
MACT	maximum achievable control technology (US)
MATES	Multiple Air Toxics Exposure Study
MDO	marine diesel oil
MEET	methodology for calculating transport emissions and energy consumption
MOUDI	micro-orifice uniform deposit impactor
MRGR	mean relative growth rate – a measure of plant or animal vitality
MSW	municipal solid waste
MTBE	methyl tertiary butyl ether
MTS	mainstream tobacco smoke – drawn from the cigarette during puffing
NAA	neutron activation analysis
NAAQS	National Ambient Air Quality Standards (US)
NAEI	National Atmospheric Emissions Inventory (UK)
NADP	National Atmospheric Deposition Program (US)
NAPAP	National Acid Precipitation Assessment Program – the major coordinated programme in the US to understand the processes of, and responses to, acid rain
NAQS	National Air Quality Strategy (UK)
NCLAN	National Crop Loss Assessment Network – the US experimental programme on plant responses to air pollutants
NETCen	National Environmental Technology Centre – performs a variety of air pollution services for the UK Government, including management and data processing for the AUN
NF	non-filtered – an OTC supplied with ambient air
NMHC	non-methane hydrocarbons – a sub-category of VOC, defined by compounds containing H and C but excluding methane because of its relatively high background concentration in the atmosphere
NMMAPS	National Morbidity, Mortality and Air Pollution Study (US)
NOTLINE	University of Nottingham line source dispersion model
NRPB	National Radiological Protection Board
OAF	optical amplification factor – used to describe the response of UV to ozone changes
OCD	ozone column depth
ODP	ozone depletion potential
OECD	Organisation for Economic and Cultural Development
OTC	open-top chamber – field chamber for plant pollution exposure
PAH	polycyclic aromatic hydrocarbons – a family of carcinogenic chemicals, including benzpyrenes
PAMS	particle analysis by mass spectroscopy

PAN	peroxyacetyl nitrate – an irritant gas formed by the same photochemical processes as ozone
PAR	photosynthetically active radiation – in the waveband 400 –700 nm
PBL	planetary boundary layer – the vertical region of the Earth's atmosphere from ground level up to about 1500 m within which the physical and chemical interactions with the surface mainly occur
PCB	polychlorinated biphenyls – carcinogenic pollutants released from PCB handling and poor PCB incineration
PCDF	polychlorinated dibenzofurans (known as furans for short) – a toxic pollutant produced in small quantities by incinerators
PCDD	polychlorinated dibenzodioxins (known as dioxins for short) – a toxic pollutant produced in small quantities by incinerators
PCR	polymerase chain reaction
PEC	particulate elemental carbon
PIB	polyisobutylene – a 2-stroke petrol additive to reduce smoke production
PIXE	proton-induced X-ray emission
PM	particulate matter
PM_{10}	particulate matter having an aerodynamic diameter less than 10 μm
$PM_{2.5}$	particulate matter having an aerodynamic diameter less than 2.5 μm
POP	persistent organic pollutant
ppb	parts per billion, or parts per 10^9, by volume
ppm	parts per million, or parts per 10^6, by volume
ppt	parts per trillion, or parts per 10^{12}, by volume
PSC	polar stratospheric cloud–ozone depletion reactions occur on the surfaces of cloud particles
PSI	Pollution Standards Index (US)
PTFE	polytetrafluoroethylene – an inert plastic used for sample pipes when reactive gases such as ozone are present
PVC	polyvinyl chloride
QALY	quality-adjusted life years – method for assessing benefits of air quality improvements
QA/QC	quality assurance/quality control
RAF	reactivity adjustment factor – a measure of the ozone-forming potential of different fuel mixtures
RAG	radiatively active gas
RCEP	Royal Commission on Environmental Pollution (UK)
Re	Reynolds number
RH	relative humidity
RPK	revenue passenger kilometres
RVP	Reid vapour pressure

SAGE	Stratospheric Aerosol and Gas Experiment
SALR	saturated adiabatic lapse rate – the rate of decrease of temperature with height in the atmosphere applicable to a parcel of air that contains liquid water. Typical range 4–9.8 °C km^{-1}
SAR	Second Assessment Report (by IPCC on climate change)
SBLINE	University of Nottingham (Sutton Bonington campus) vehicle emission and dispersion model
SBS	sick building syndrome
SCA	specific collection area
SCC	squamous cell carcinoma
SED	standard erythemal dose (of UV radiation)
SEM	scanning electron microscopy
SI	Système International – the internationally recognised system of physical units based on the metre, kilogram, second and Coulomb
SIP	State Implementation Plan (US)
SMPS	scanning mobility particle sizer
SOF	soluble organic fraction
SOI	southern oscillation index
SOS	Southern Oxidants Study (US)
SST	supersonic transport
Stk	Stokes' number
STP	standard temperature and pressure – 0 °C and 1 atmosphere
STS	sidestream tobacco smoke – released from the cigarette between puffing
SUM60	sum of hourly-mean ozone concentrations >60 ppb
SUV	sport utility vehicle
SVP	saturated vapour pressure
TAR	Third Assessment Report (by IPCC on climate change)
TEA	triethanolamine – a strong absorbent for NO_2
TEAM	Total Exposure Assessment Methodology
TEM	transmission electron microscopy
TEOM	tapered element oscillating microbalance
TEQ	toxic equivalent – a standardisation of the toxicity of TOMPS
TOMPS	toxic organic micropollutants – generic term that includes PCDD, PCDF and other minority chemicals with recognised toxicity at low (ppt) concentrations
TOE	tonnes oil equivalent
TRL	Transport Research Laboratory (UK)
TSP	total suspended particulate – all the particles in the air, regardless of diameter
TWC	three-way catalyst – converts the three harmful gases in petrol-engined vehicle exhaust to carbon dioxide, nitrogen and water
UARS	upper atmosphere research satellite
ULPA	ultra low penetration air

ULSP   ultra low sulphur petrol
UNECE   United Nations Economic Commission for Europe – a group of
           countries, larger than the EU and including the US, that has a
           wide ranging remit to organise joint ventures in European affairs
UNFCCC   United Nations Framework Convention on Climate Change
USEPA   *see* EPA
UV   ultraviolet radiation, conventionally defined as occurring in the
           wavelength range below 400 nm. Subdivided into UVA, UVB and
           UVC
VEI   volcanic explosivity index
VOC   volatile organic compound – molecules, mostly containing hydro-
           gen and carbon, that are released from sources such as motor fuels
           and solvents. They are toxic in their own right and serve as precur-
           sors for ozone formation
WHO   World Health Organisation
WTP   willingness to pay – method for assessing the benefits of air quality
           improvements
XRF   X-ray fluorescence
ZEV   zero emission vehicles – presumed to be electric, and required by
           law to make up a certain proportion of the fleet in California

Chapter 1

Gaseous air pollutants: sources and control

The World Health Organisation (WHO) estimates that 500 000 people die prematurely each year because of exposure to ambient concentrations of airborne particulate matter. In the UK alone, this figure is around 10 000 people. World Health Organisation also estimated the annual health cost of air pollution in Austria, France and Switzerland as £30 billion, corresponding to 6% of the total mortality; about half this figure was due to vehicle pollution. In the US, the annual health cost of high particle concentrations has been estimated at £23 billion. Clearly, we are paying a high price, both in lives and money, for polluting the atmosphere.

Pollution (in the general sense) was defined in the Tenth Report of the Royal Commission on Environmental Pollution as:

> The introduction by man into the environment of substances or energy liable to cause hazard to human health, harm to living resources and ecological systems, damage to structure or amenity or interference with legitimate use of the environment.

This is a very broad definition, and includes many types of pollution that we shall not cover in this book, yet it contains some important ideas. Note that by this definition, chemicals such as sulphur dioxide from volcanoes or methane from the decay of natural vegetation are not counted as pollution, but sulphur dioxide from coal-burning or methane from rice-growing *are* pollution. Radon, a radioactive gas that is a significant natural hazard in some granitic areas, is not regarded as pollution since it does not arise from people's activities. The boundaries become more fuzzy when we are dealing with natural emissions that are influenced by our actions – for example, there are completely natural biogenic emissions of terpenes from forests, and our activities in changing the local patterns of land use have an indirect effect on these emissions. The pollution discussed in this book is the solid, liquid or gaseous material emitted into the air from stationary or mobile sources, moving subsequently through an aerial path and perhaps being involved in chemical or physical transformations before eventually being returned to the surface. The material has to interact with something before it can have any environmental impacts. This interaction may be, for example, with other molecules in the atmosphere (photochemical formation of ozone from hydrocarbons), with electromagnetic radiation (by greenhouse gas molecules), with liquid water (the formation of acid rain from sulphur dioxide), with vegetation (the direct effect of ozone), with mineral surfaces (soiling of buildings by particles) or with animals (respiratory damage by acidified aerosol). Pollution from our activities is called anthropogenic, while that from animals or plants is said to be biogenic. Originally, air pollution was taken to include only substances from which environmental damage was anticipated because of their toxicity or their specific capacity to damage organisms or structures; in the last decade, the topic has been broadened to include substances such as chlorofluorocarbons, ammonia or carbon dioxide that have more general environmental impacts.

1.1   UNITS FOR EXPRESSING POLLUTANT CONCENTRATION

Before we go any further, we must make a short detour to explain the units in which pollutant concentrations are going to be discussed throughout this book. Two sets of concentration units are in common use – volumetric and gravimetric.

If all the molecules of any one pollutant gas could be extracted from a given volume of the air and held at their original temperature and pressure, a certain volume of the pure pollutant would result. *Volumetric* units specify the mixing ratio between this pollutant volume and the original air volume – this is equivalent to the ratio of the number of pollutant gas molecules to the total number of air molecules. Owing to historical changes in the systems used for scientific units, there are at least three notations in common use for expressing this simple concept. Originally, the concentration would be expressed, for example, as parts of gas per million parts of air. This can be abbreviated to ppm, or ppmv if it is necessary to spell out that it is a volume ratio and not a mass ratio. Later, to make the volumetric aspect more explicit and to fit in with standard scientific notation for submultiples, the ratio was expressed as $\mu l\ l^{-1}$. Unfortunately, the litre is not a recognised unit within the Système International (SI). The SI unit for amount of substance (meaning number of molecules, not weight) is the mol, so that $\mu mol\ mol^{-1}$ becomes the equivalent SI unit of volumetric concentration. This is correct but clumsy, so ppm (together with ppb (parts per billion, 10^{-9}) and ppt (parts per trillion, 10^{-12})) have been retained by many authors for convenience's sake and will be used throughout this book. *Gravimetric* units specify the mass of material per unit volume of air. The units are more straightforward – $\mu g\ m^{-3}$, for example. Unlike volumetric units, gravimetric units are appropriate for particles as well as for gases.

These relationships are summarised in Table 1.1 for the typical concentration ranges of ambient gaseous pollutants.

Table 1.1 Abbreviations for volumetric and gravimetric units

	Volumetric				Gravimetric
Parts per million (micro)	10^{-6}	ppm	$\mu l\ l^{-1}$	$\mu mol\ mol^{-1}$	$mg\ m^{-3}$
Parts per billion (nano)	10^{-9}	ppb	$nl\ l^{-1}$	$nmol\ mol^{-1}$	$\mu g\ m^{-3}$
Parts per trillion (pico)	10^{-12}	ppt	$pl\ l^{-1}$	$pmol\ mol^{-1}$	$ng\ m^{-3}$

Both volumetric and gravimetric systems have their uses and their advocates. The volumetric concentration is invariant with temperature and pressure, and therefore remains the same, for example, while warm flue gas is cooling in transit through exhaust ductwork. When gas enters a leaf, the effects may depend primarily on the number of molecular sites occupied by the pollutant gas molecules – this is better indicated by the volumetric than by the gravimetric concentration. However, if concentration is being determined by extracting the gas onto a treated filter for subsequent chemical analysis, or health effects are being related to the mass of pollutant inhaled, the result would normally be calculated as a gravimetric concentration.

1.1.1 Conversion between gravimetric and volumetric units

Since both gravimetric and volumetric systems are in use and useful, we need to be able to convert between the two.

The basic facts to remember are that 1 mol of a pure gas (an Avogadro number of molecules, 6.02×10^{23}) weighs M kg, where M is the relative molar mass, and takes up a volume of 0.0224 m^3 at standard temperature and pressure (STP – 0 °C, 1 atmosphere).

For example, sulphur dioxide (SO_2) has $M = 32 \times 10^{-3} + (2 \times 16 \times 10^{-3})$ $= 64 \times 10^{-3}$ kg, so that *pure* SO_2 has a density (= mass/volume) of $64 \times 10^{-3}/0.0224 = 2.86$ kg m^{-3} at STP. But pure SO_2 is 10^6 ppm, by definition. Therefore

$$10^6\ ppm \equiv 2.86\ kg\ m^{-3}$$

$$1\ ppm \equiv 2.86 \times 10^{-6}\ kg\ m^{-3}$$

$$= 2.86\ mg\ m^{-3}$$

and

$$1\ ppb \equiv 2.86\ \mu g\ m^{-3}$$

Hence we can convert a volumetric concentration to its gravimetric equivalent at STP.

1.1.2   Correction for non-standard temperature and pressure

The temperature and pressure are unlikely to be standard, so we also need to be able to convert gravimetric units at STP to other temperatures and pressures. At STP, we have 1 m³ containing a certain mass of material. When the temperature and pressure change, the volume of the gas changes but it still contains the same mass of material. Hence we need only to find the new volume from the Ideal Gas Equation

$$\frac{P_1V_1}{T_1} = \frac{P_2V_2}{T_2} \tag{1.1}$$

where P_1, V_1 and T_1 are the initial pressure, volume and absolute temperature and P_2, V_2 and T_2 are the final pressure, volume and absolute temperature.

In our case

$P_1 = 1$ atmosphere

$V_1 = 1$ m³

$T_1 = 273$ K

and we need to find V_2.

Therefore, rearranging equation (1.1),

$$V_2 = \frac{T_2}{P_2} \times \frac{P_1V_1}{T_1} = \frac{T_2}{273\,P_2}$$

For example, the highest ambient temperature that we might find in practice is 50 °C, and the lowest ambient pressure at ground level might be 950 mbar.

Hence

$T_2 = 273 + 50 = 323$ K,

and

$P_2 = 950/1013 = 0.938$ atmosphere (because standard atmospheric pressure $= 1013$ mbar)

Therefore

$$V_2 = \frac{323}{273 \times 0.938} = 1.26 \text{ m}^3$$

Table 1.2 Conversion factors between volumetric and gravimetric units

Pollutant	Molecular weight M/g	To convert			
		ppb to μg m^{-3}		μg m^{-3} to ppb	
		0 °C	20 °C	0 °C	20 °C
SO_2	64	2.86	2.66	0.35	0.38
NO_2	46	2.05	1.91	0.49	0.52
NO	30	1.34	1.25	0.75	0.80
O_3	48	2.14	2.00	0.47	0.50
NH_3	17	0.76	0.71	1.32	1.43
CO	28	1.25	1.16	0.80	0.86

The original volume of 1 m^3 has expanded to 1.26 m^3. This is physically reasonable because we have raised the temperature and reduced the pressure – both
changes will increase the volume. The increased volume will still contain the
same number of molecules, which will have the same mass. Hence the concentration must decrease by the same factor, and 1 ppm of SO_2, for example, would
now be equal to 2.86/1.26 mg m^{-3} or 2.27 mg m^{-3}. The volumetric concentration, of course, would remain at 1 ppm.

For the pollutant gases discussed most frequently in this book, Table 1.2 gives
the conversion factors from ppb to μg m^{-3}, and *vice versa*, at 0 °C and 20 °C. For
example, to convert 34 ppb of SO_2 at 20 °C to μg m^{-3}, multiply by 2.66 to get
90 μg m^{-3}.

1.2   THE BASIC ATMOSPHERE

1.2.1   The origins of our atmosphere

What we experience today as our atmosphere is a transient snapshot of its evolutionary history. Much of that history is scientific speculation rather than established fact. The planet Earth was formed around 4600 million years ago by the
gravitational accretion of relatively small rocks and dust, called planetesimals,
within the solar nebula. There was probably an initial primordial atmosphere
consisting of nebula remnants, but this was lost to space because the molecular
speeds exceeded the Earth's escape velocity of 11.2 km s^{-1}. A combination of
impact energy and the radioactive decay of elements with short half-lives raised
the temperature of the new body sufficiently to separate heavier elements such as
iron, which moved to the centre. The same heating caused dissociation of hydrated
and carbonate minerals with consequent outgassing of H_2O and CO_2. As the Earth
cooled, most of the H_2O condensed to form the oceans, and most of the CO_2 dissolved and precipitated to form carbonate rocks. About one hundred times more

gas has been evolved into the atmosphere during its lifetime than remains in it today. The majority of the remaining gases was nitrogen. Some free oxygen formed (without photosynthesis) by the photolysis of water molecules. Recombination of these dissociated molecules was inhibited by the subsequent loss of the hydrogen atoms to space (hydrogen is the only abundant atom to have high enough mean speed to escape the gravitational attraction of the Earth). The effect of atomic mass makes a huge difference to the likelihood of molecules escaping from the Earth. The Maxwell distribution means that there is a most likely velocity which is relatively low, and a long tail of reducing probabilities of finding higher speeds. For example, a hydrogen atom at 600 K (typical temperature at the top of the atmosphere) has a 10^{-6} chance of exceeding escape speed, while the corresponding figure for an oxygen atom is only 10^{-84}. This process will result in a steady attrition of lighter atoms. The first evidence of single-celled life, for which this tiny oxygen concentration was an essential prerequisite, is shown in the fossil record from around 3000 million years ago. Subsequently, the process of respiration led to a gradual increase in the atmospheric oxygen concentration. This in turn allowed the development of O_3 which is thought to have been a necessary shield against solar UV. Subsequent evolution of the atmosphere has been dominated by the balance between production and consumption of both CO_2 and O_2.

1.2.2   Natural constituents of air

People tend to refer to air as though it consists of 'air' molecules, which is evidence of the spatial and temporal constancy of its properties that we take for granted. Consider first the molecular components that make up unpolluted air. Air consists of a number of gases that have fairly constant average proportions, both at different horizontal and vertical positions and at different times. Table 1.3 gives the proportions of the gases that are present at concentrations of around and above 1 ppm.

The average molar mass of dry air can be found by summing the products of the proportions by volume and molar masses of its major components.

Table 1.3 Proportions of molecules in clean dry air

Molecule	Symbol	Proportion by volume
Nitrogen	N_2	78.1%
Oxygen	O_2	20.9%
Argon	Ar	0.93%
Carbon dioxide	CO_2	370 ppm
Neon	Ne	18 ppm
Helium	He	5 ppm
Methane	CH_4	1.7 ppm
Hydrogen	H_2	0.53 ppm
Nitrous oxide	N_2O	0.31 ppm

i.e.

$$M_a = (0.781 \times 28.01) + (0.209 \times 32.00) + (0.0093 \times 39.95)$$
$$+ (0.00037 \times 44.01)$$
$$= 28.95 \text{ g mol}^{-1}$$

Mixed into the quite uniform population of atmospheric molecules is a large range of additional materials that vary greatly in concentration both in space and time:

- *sulphur dioxide* may be released directly into the atmosphere by volcanoes, or formed as the oxidation product of the dimethyl sulphide released by oceanic phytoplankton
- *oxides of nitrogen* are created when anything burns
- *nitrous oxide* is emitted from the soil surface by bacterial denitrification
- *hydrogen sulphide* is produced by anaerobic decay
- *ammonia* is released from animal waste products
- *ozone* is formed in the stratosphere, by the action of UV radiation on oxygen
- *ozone* is also found in the troposphere, via both diffusion down from the stratosphere and local natural photochemistry
- *volatile organic compounds* (VOCs) are emitted from many different species of vegetation, especially coniferous and eucalyptus forests
- *non-biogenic particles* are generated by volcanoes or entrainment from the soil
- *biogenic particles* include pollen, spores, and sea salt.

We shall come back to these additional materials, and others, throughout this book when we consider them as pollutants rather than as naturally occurring substances.

1.2.3   Water in the atmosphere

The proportions given in Table 1.3 are for dry air – without water vapour molecules. The gases listed have long residence times in the atmosphere, are well mixed and their concentrations are broadly the same everywhere in the atmosphere. Water is very different, due to its unusual properties at normal Earth temperatures and pressures. It is the only material which is present in all three phases – solid (ice), liquid and gas (water vapour). There is continuous transfer between the three phases depending on the conditions. We take this situation very much for granted, but it is nevertheless remarkable. Certainly, if we found pools of liquid nitrogen or oxygen on the surface of the Earth, or drops of these materials were to fall out of the sky, it would get more attention.

The proportion of water vapour in the atmosphere at any one place and time depends both on the local conditions and on the history of the air. First, the temperature of the air determines the maximum amount of water vapour that can be present. The water vapour pressure at this point is called the saturated vapour

Table 1.4 Variation of saturated water vapour pressure with temperature

Units	Temperature/°C				
	−10	0	10	20	30
Pa	289	611	1223	2336	4275
mbar	3	6	12	23	43

pressure (SVP), and varies roughly exponentially with temperature (Table 1.4). Various mathematical expressions have been used to describe the relationship; an adequate one for our purposes is

$$e_s(T) = 611 \exp\left(\frac{19.65T}{273 + T}\right) \text{Pa}$$

where $e_s(T)$ is the saturated vapour pressure in Pa and T is the air temperature in degrees Celsius.

Second, the ambient (meaning local actual) vapour pressure e_a may be any value between zero and e_s. The ratio of actual to saturated vapour pressure is called the relative humidity h_r, often expressed as a percentage. If water vapour is evaporated into dry air (for example, as the air blows over the sea surface or above a grassy plain), then the vapour pressure will increase towards e_s, but cannot exceed it. If air is cooled, for example by being lifted in the atmosphere, then a temperature will be reached at which the air is saturated due to its original water content. Any further cooling results in the 'excess' water being condensed out as cloud droplets. If the cooling occurs because the air is close to a cold ground surface, then dew results. The complexity of this sequence for any air mass is responsible for the variability of water vapour concentration in space and time. For comparison with the proportions given in Table 1.3 for the well-mixed gases, we can say that the highest vapour concentrations occur in the humid tropics, with temperatures of 30 °C and relative humidities of near 100%. The vapour pressure will then be 4300 Pa, corresponding to a mixing ratio of 4.3/101 = 4.3%. At the low end, an h_r of 50% at a temperature of −20 °C would correspond to a mixing ratio of around 0.1%. The global average mixing ratio is around 1%, so the abundance of water vapour is similar to that of argon.

1.3   THE VERTICAL STRUCTURE OF THE ATMOSPHERE

1.3.1   Pressure

The pressure of a gas is due to the momentum exchange of individual molecules when they collide with the walls of their container. In the atmosphere, only the molecules at the surface have got a container – the remainder simply collide with other gas molecules. At any height in the atmosphere, the upward force due to this momentum exchange must equal the downward force due to gravity acting on all the molecules above that height. Since this force decreases with height, pressure decreases with height.

Considering the force dF acting on a small vertical column of height dz, area A and density ρ, with the acceleration due to gravity of g, we have $dF = -g\rho A \, dz$ (the minus sign is because the force acts downwards, but z is usually taken to be positive upwards).

Hence

$$dp = df/A = -g\rho \, dz$$

But for an ideal gas at temperature T

$$\rho = mp/kT$$

where m is the molecular mass and k is the Boltzmann constant.

Hence

$$\frac{1}{p} dp = -\frac{mg}{kT} dz$$

which integrates to give the hydrostatic equation

$$p = p_0 \exp\left\{ -\int_0^z \frac{mg}{kT} dz \right\}$$

where p_0 is the surface pressure. The average atmospheric pressure at sea level over the surface of the Earth is 101.325 kPa = 1013.25 mbar (or hPa).

The equation in this form allows for the variation of m, g and T with height z. In practice, the atmosphere is remarkably shallow, having a thickness of only 0.2% of the Earth's radius up to the tropopause and 1.4% even at the mesopause. Hence for practical purposes g can be taken as constant and equal to its value at the surface, g_0. Also, the major components – nitrogen, oxygen and argon – are well mixed by turbulent diffusion, so m is nearly constant at 28.95. If T were constant as well, the integration would give

$$p = p_0 \exp\left(-\frac{mg_0}{kT} z \right)$$

The exponent in brackets in this equation must be dimensionless, so that kT/mg must have the same dimensions as z, which is length. Hence we can write

$$P(z) = P_0 e^{-\frac{z}{H}}$$

where $H = kT/mg$ is the scale height of the atmosphere. H corresponds to the height over which the pressure decreases to $1/e = 0.37$. At the surface, where the temperatures are around 20 °C or 290 K, the scale height is 8.5 km. At the tropopause, where T ~220 K, the scale height has fallen to 6 km.

The total atmospheric pressure is the sum of the partial pressures from each of the gases present according to its proportion by volume or mixing ratio (e.g. as % or ppm). Hence the great majority of the pressure is due to nitrogen, oxygen and argon. Superimposed on this long-term average situation are short-term variations due to weather. These may cause the sea level pressure to decrease to 95 kPa in the centre of a severe depression or increase to 106 kPa in an anticyclone.

Looked at the other way round, pressure = force per unit area, so that the surface atmospheric pressure is just the total force (= mass of the atmosphere $\times$ acceleration due to gravity) divided by the total surface area of the Earth ($4\pi R^2$).

$$P_{\text{sea level}} = \frac{M_{\text{atm}}\, g}{4\pi R^2}$$

By rearranging and substituting the known value of average sea level pressure (101.325 kPa), $g = 9.81$ m s^{-2} and $R = 6.37 \times 10^6$ m we can calculate the total mass of the atmosphere to be around 5.3×10^{18} kg. This is useful for understanding the significance of differing rates of input of atmospheric pollutants.

1.3.2   Temperature

The temperature structure of the atmosphere, which is shown in Figure 1.1, is more complex than the pressure structure, because it is the result of several competing processes. First, the Earth's surface is emitting longwave thermal radiation, some of which is absorbed and re-radiated by the atmosphere. Because the atmospheric pressure and density decrease exponentially with height, the absorption and emission decrease as well, which establishes a non-linear decrease of the equilibrium radiative temperature with height. Second, convective forces come into play. Below an altitude of 10–15 km, the lapse rate of radiative temperature exceeds the adiabatic lapse rate. This promotes overturning and defines the well mixed region known as the troposphere. The mixing process establishes the adiabatic lapse rate (see Chapter 6) of around 6.5 °C km^{-1} within this region. Above that altitude, the lapse rate of radiative temperature has dropped to a value well below the adiabatic lapse rate, resulting in the stable poorly-mixed conditions characteristic of the stratosphere. Third, warming due to solar energy absorption by the layer of ozone between 20 and 50 km reverses the temperature decline, so that air temperature increases up to the stratopause at 50 km, further increasing the atmospheric stability. The stratosphere is dry throughout because the main source of moisture is the troposphere, and any air moving into the stratosphere from the troposphere must have passed through the tropopause, where the very low temperatures act as a cold trap. Some additional water molecules are created

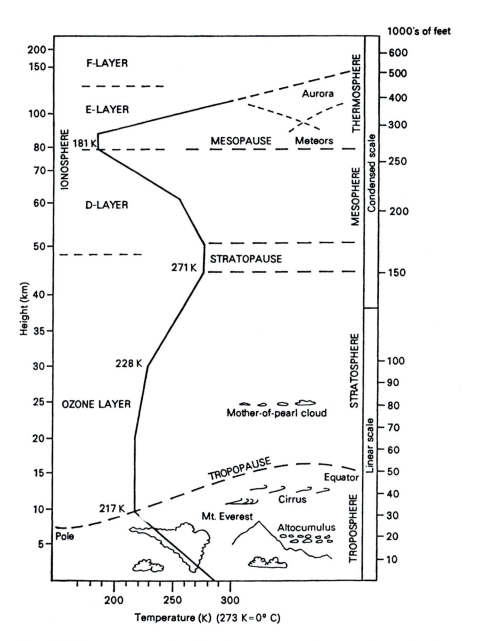

Figure 1.1 The average vertical profile of temperature in the atmosphere, related to other atmospheric features. Note that the vertical scale is linear up to 40 km, and logarithmic above.

Source: Lawes, H. D. (1993) 'Back to basics', *Weather* 48(10): 339–344.

by the oxidation of CH_4. Above the stratopause, in the mesosphere, the warming effect is offset by decreasing air density, and the temperature falls again. Although the temperature is falling, the rate of decrease is small at around 4 K km^{-1} and the mesosphere is also stable, but only just. Finally, above the mesopause at 90 km the air becomes so thin, and collisions so infrequent, that unequal energy partition between vibrational and translational modes result in very high temperatures. It is these high translational energies that allow light particles such as hydrogen atoms to escape from the Earth altogether.

1.4   ANTHROPOGENIC EMISSIONS

The three major groups of gaseous air pollutants by historical importance, concentration, and overall effects on plants and animals (including people), are sulphur dioxide (SO_2), oxides of nitrogen ($NO_x = NO + NO_2$) and ozone (O_3). Sulphur dioxide and nitric oxide (NO) are primary pollutants – they are emitted directly from sources. We shall start by looking at the main sources of these and other primary gases, and also consider some of the methods of control that can be used to reduce emissions and concentrations when required. Then we will move on to ozone, which is referred to as a secondary pollutant because it is mainly formed in the atmosphere from primary precursors, and emissions of the gas itself are negligible. Nitrogen dioxide (NO_2) is both primary and secondary – some is emitted by combustion processes, while some is formed in the atmosphere during chemical reactions. Production of SO_2 is commonly associated with that of black smoke, because it was the co-production of these two materials during fossil fuel combustion that was responsible for severe pollution episodes such as the London smogs of the 1950s and 1960s.

1.4.1   Energy consumption

During most of recorded history, the population of the world has grown slowly (Figure 1.2), reaching 200 million in AD 1, 250 million in AD 1000 and 450 million in 1600. In the seventeenth century, we started to learn how to keep people alive before we realised the consequences, and the rate of growth increased explosively. The population reached 6 billion in 2000, and is now forecast to reach 7.5 billion by 2020 and to stabilise at 9 billion in 2050. Ninety eight per cent of the future growth will be in developing countries, and most of that will be in urban areas. This rapidly increasing population has also been increasing its standard of living, underpinned by energy obtained from fossil fuels – initially from coal burning and later by oil and gas. Although increased energy efficiency in the developed nations stabilised the use per person after 1970, the continuing increase in total population is still driving up the total energy use. We each use about 2 kW on average – equivalent to one fan heater – although this single average conceals a wide range between

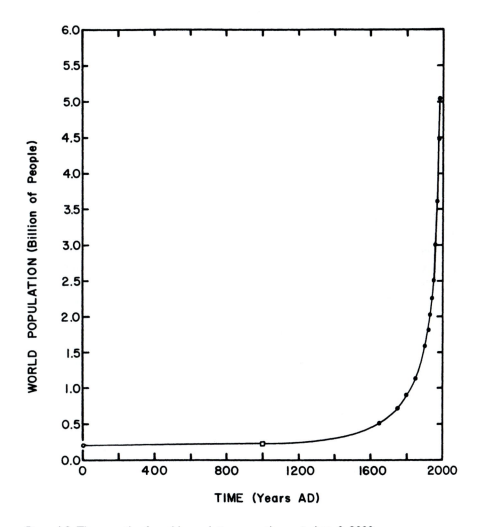

Figure 1.2 The growth of world population over the period AD 0–2000.
Source: Idso, S.B. (1989) *Carbon Dioxide and Global Change: Earth in Transition*, IBR Press, Arizona, USA.

intensive users such as the US (10 kW) or Western Europe (5 kW), and less developed countries having very small energy inputs. The combustion of fossil fuels to generate this energy converts carbon into carbon dioxide and releases it into the atmosphere. Guideline energy contents of fossil fuels, and of alternatives, are shown in Table 1.5.

Figure 1.3 shows the parallel growth in carbon release (as CO_2) in the nineteenth and twentieth centuries. Coal was the original fossil fuel used, then oil from 1900 and gas from 1930. Also visible are major disturbances to the drive for

Table 1.5 Typical energy contents of widely-used fuels

Energy source	Energy density/MJ kg^{-1}
Natural gas	51
Petroleum	37
Coal	
Anthracite	30
Bituminous	30
Sub-bituminous (brown)	20
Lignite	10–15
Animal dung (dry weight)	17
Wood (dry weight)	15

growth – the recession in the 1930s, the Second World War and the oil price crisis of 1974.

Coal is the fuel that underpinned successive industrial revolutions from the Bronze Age through to the eighteenth century. It is the most abundant fossil fuel, with huge reserves of some 1000 billion tonnes that are expected to last another 300 years at present rates of use. What were originally peat deposits became buried and compressed under accumulating sediments. The increased pressure and temperature caused the peat to pass through a series of stages called the coal series, characterised by decreasing moisture content and volatiles and a higher carbon content. The members of this series are called lignite, sub-bituminous brown coal, bituminous coal and anthracite. The earlier members of the coal series (from younger deposits,

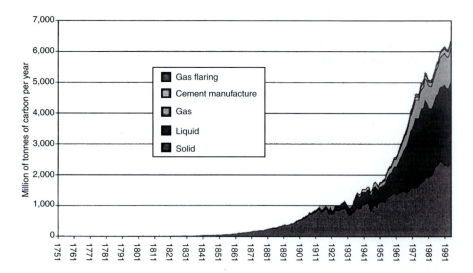

Figure 1.3 Global carbon release 1750–1995.
Source: Wuebbles, D. J., Jain, A., Edmonds, J., Harvey, D. and Hayhoe, K. (1999) 'Global change: state of the science', *Environmental Pollution* 100: 57–86.

Cretaceous rather than Carboniferous) are poor quality low-calorific value fuels which need to be burnt in large tonnages. However, coal use as a proportion of total is declining, because of the handling advantages of fluid fuels such as oil and gas.

The origin of the world's oil deposits is far less well understood than is coal's. Oil itself consists of thousands of different organic molecules, mainly hydrocarbons. Whereas coal formed from the action of sediments on accumulating vegetable matter in a terrestrial environment, it is thought that oil formed in a parallel process acting on accumulating microscopic animal remains in a marine environment. Although fluid, the oil could sometimes be trapped below a domed cap of impermeable rock. As well as these deposits of pure liquid oil, there are extensive deposits of oil shales and sands, in which the oil soaks through a permeable rock much like water through a sponge. These types of deposit will be much more expensive to extract. Crude oil as extracted is not useable as a fuel, but has to be refined by the process of fractional distillation. This process yields not only fuels such as heavy fuel oil, diesel, petrol and paraffin, but a wide range of chemicals which can be used to make plastics and other materials.

Natural gas, which is mainly the simplest hydrocarbon methane, is associated with both the formation of coal seams and oil deposits. With coal it is usually just a safety issue, being present in sufficient concentrations to foul the air and cause risk of explosions. With oil it is usually present in large volumes, overlying the oil deposit, that have been exploited as a fuel in their own right.

Table 1.6 does not include renewables such as wind turbines or biofuels. The per capita energy use varies widely between countries according to their stage of development, from 7–8 toe per year in North America (where 5% of the world's population uses 25% of the energy), and 4 toe in Europe, to 0.2 toe in India.

Developing countries also offer very different fuel combustion profiles, especially in the domestic sector. Whereas domestic energy supply in developed countries is likely to be from burning gas in a centralised boiler for a central heating system, developing countries are much more likely to be using fuels such as kerosene, wood, roots, crop residues or animal dung. Furthermore, these are likely to be burnt in an unflued indoor stove with poor combustion efficiency. Hence

Table 1.6 Global consumption of energy in 1998

Energy source	Consumption/Mtoe[*]	Consumption/%
Oil	3500	41.1
Natural gas	1900	22.4
Coal	2100	24.7
Nuclear	800	9.4
Hydro-electric	200	2.4
Total	8500	100

Note
* Mtoe = Million tonnes oil equivalent. All the different fuel consumptions are translated into a common currency by expressing them as the weight of oil that would contain the same energy. 1 Mtoe is 42 GJ.

the potential for pollutant emissions and enhanced indoor pollutant concentrations is much greater. Ironically, if the renewable fuels could be combusted without pollutant emissions, they would be superior to fossil fuels because they have no net impact on CO_2 concentrations. In some countries, for example China, national programmes have been used to increase the availability of higher efficiency flued stoves. Although, these still generate significant emissions to atmosphere, they certainly reduce the indoor concentrations and consequent health impacts.

1.4.2   Sulphur emissions

All fossil fuels contain sulphur, most of which is released as sulphur dioxide during combustion. Almost all the anthropogenic sulphur contribution is due to fossil fuel combustion. Different fuels offer a wide range of sulphur contents:

- Oil and its by-products contain between 0.1% sulphur (paraffin) and 3% (heavy fuel oil) in the form of sulphides and thiols. Petrol contains negligible sulphur in the context of overall mass emissions, although there can be an odour problem from conversion to hydrogen sulphide (H_2S) on catalytic converters.
- Coal contains 0.1–4% sulphur, mainly as flakes of iron pyrites (FeS_2). The average sulphur content of UK coal is 1.7%.
- Natural gas (mainly methane, CH_4) can be up to 40% H_2S when it is extracted from the well. The sulphur is taken out very efficiently at a chemical processing plant before distribution, so natural gas is effectively sulphur-free – one of the reasons for the 'dash for gas'.

Global sulphur dioxide emissions are estimated to have increased from 4 Mt (containing 2 Mt of sulphur) in 1860 to 150 Mt in 1990 (Figure 1.4). The emissions from the US and Europe increased steadily until the 1970s before coming under control. Sulphur emissions from the faster-growing Asian region have continued to increase, due largely to coal combustion. Emissions from China are now comparable to those from the US, and in 1990 emissions from China, US and Russia accounted for over half the global total.

The major natural sulphur emissions are in the reduced forms of H_2S (hydrogen sulphide), CS_2 (carbon disulphide) or COS (carbonyl sulphide), and the organic forms CH_3SH (methyl mercaptan), CH_3SCH_3 (dimethyl sulphide, or DMS) and CH_3SSCH_3 (dimethyl disulphide, or DMDS). Dimethyl sulphide is produced by marine phytoplankton and oxidised to SO_2 in the atmosphere; H_2S from decay processes in soil and vegetation; and SO_2 from volcanoes. Whatever their original form, much of these sulphur compounds eventually get oxidised to gaseous SO_2 or to sulphate aerosol. The natural sources are now heavily outweighed by human ones, principally fossil fuel combustion, as shown in Table 1.7. Since 90% of the biogenic emission is as DMS, and an even higher proportion of the human emission is as SO_2, we have a clear demarcation between the source types. Since most of the DMS comes from oceans in the southern hemi-

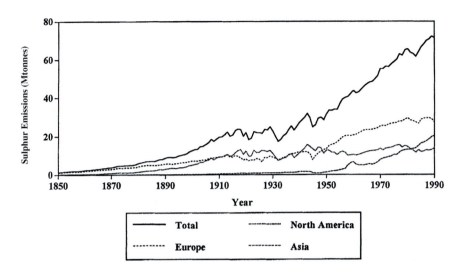

Figure 1.4 World and Regional sulphur production in the period 1850–1990.
Source: Lefohn, A. S., Husar, J. D. and Husar, R. B. (1999) 'Estimating histor-
ical anthropogenic global sulphur emission patterns for the period 1850–1990',
Atmospheric Environment 33: 3435–3444.

sphere, and most of the human SO_2 from fossil fuel emissions in the northern
hemisphere, we also have a geographical split. Note that the emission strengths
are given as mass of sulphur, not of sulphur dioxide. If these values are compared
with others expressed as SO_2, then they must be multiplied by the ratio of the
molecular weights, which is 2.0 (64/32) in this case.

Information about the budgets of reduced sulphur components is less soundly
based than that for oxidised S. Approximate total emissions are given in Table 1.8.

Table 1.7 Natural and anthropogenic sources of sulphur dioxide

Source	Global emission/Mt S a^{-1}
Natural	
DMS	16
Soils and vegetation	2
Volcanoes	8
Total natural	26
Anthropogenic	
Biomass burning	3
Fossil fuel combustion	70
Total anthropogenic	73
Total	99

Table 1.8 Global emissions of reduced S

Compound	Emission/Mt a^{-1}
Carbonyl sulphide (OCS)	1.3
Carbon disulphide (CS$_2$)	1.0
Hydrogen sulphide (H$_2$S)	8.0
Dimethyl sulphide (DMS)	16

During the last three decades, the types of fuel burned for energy in the UK have been changing quite rapidly, although the total has remained fairly constant. Figure 1.5 gives the breakdown by fuel type over the period 1970–1998. Coal consumption for power generation declined rapidly in the 1990s, while domestic coal use continued the decline that had started in the 1960s. Natural gas consumption rose steadily, not only for domestic space heating but in the industrial and power sectors as well. Consumption by vehicles (both petrol and diesel), and by aircraft, increased steadily.

About 66% of UK emissions are currently from power stations, and are emitted from chimneys several hundred metres above ground level. For example, a typical 2000 MW power station as operated by National Power or PowerGen burns up to 10 000 tonnes of coal per day, producing between 500 and 1000 tonnes of SO$_2$ per day. The total UK generating capacity of about 60 000 MW is responsible for the emission of about 1 Mt of sulphur dioxide per year.

For the UK, detailed statistics in Table 1.9 from the Department for Environment, Food and Rural Affairs give the annual amounts and percentages from various source categories. The total SO$_2$ emission has fallen dramatically since 1970, due to the combined influence of lower sulphur fuels, the switch from coal to gas and increased energy efficiency. Two-thirds of the emissions are due to

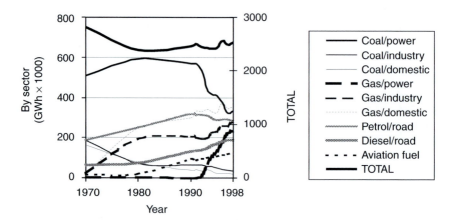

Figure 1.5 UK energy production by type 1970–1998.

Table 1.9 UK emissions of SO_2 in 1970 and 1998, by source category

Source	Emissions, 1970 kt S	Emissions, 1998 kt S	Percentage of 1998 total
Combustion for energy production			
Electricity	1457	536	66
Petroleum refining	127	86	12
Other combustion	151	6	1
Combustion in commercial and residential			
Residential	261	26	3
Commercial and agricultural	221	17	2
Combustion by industry			
Iron and steel production	217	23	3
Other industrial	705	74	9
Production processes	46	8	1
Extraction and distribution of fossil fuels	3	3	0
Road transport	22	12	1
Other transport and machinery	46	19	2
By fuel type			
Solid	1830	585	72
Petroleum	1282	185	23
Gas	96	30	2
Non-fuel	48	24	3
Total	3256	810	100

electricity generation, and transport contributes a negligible proportion (contrast this with NO_x in Section 1.4.3).

Emissions are not usually spread uniformly across the country. In the UK the National Atmospheric Emissions Inventory (NAEI) compiles very detailed maps of estimated emissions on a 10×10 km grid. These estimates are updated annually. Two factors drive SO_2 emissions – population and power stations. There are clusters of emissions around the large urban centres, because these areas are industrial as well as population centres. There are also clusters around the group of coal-fired power stations in the English East Midlands. Areas of low population density, such as the Highlands of Scotland and central Wales, have correspondingly low emission densities.

We will see in Chapter 5 how such a concentration of emissions also results in a concentration of deposition and effects. In industrialised countries, emissions have been falling in recent decades as industry moves away from coal and towards natural gas. Figure 1.4 showed that, even while global emissions continued to rise, European emissions peaked in about 1980 and have been falling since then. This trend was anticipated by UK emissions (Figure 1.6), which rose fairly consistently until the late 1960s before starting to decline.

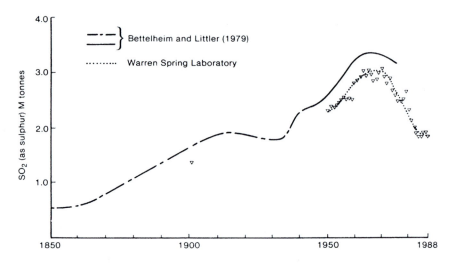

Figure 1.6 UK SO$_2$ emissions 1850–1988.
Source: Eggleston, S., Hackman, M. P., Heyes, C. A., Irwin, J. G., Timmis, R. J. and Williams, M. L. (1992) 'Trends in urban air pollution in the United Kingdom during recent decades', *Atmospheric Environment* 26B: 227–239.

Total UK emissions of SO$_2$ were 6.3 Mt in 1970; they declined to 3.7 Mt in 1980 and have continued to fall since. Figure 1.7 shows how these changes have been distributed between different source categories since 1970, together with forecast emissions until 2020. Ninety five per cent of UK SO$_2$ emissions in 1998 were due

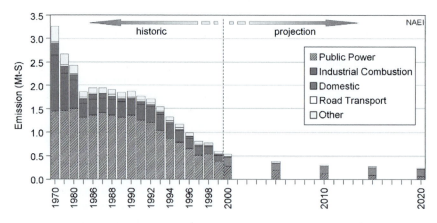

Figure 1.7 UK SO$_2$ emissions by source category 1970–2020.
Source: National Expert Group on Transboundary Air Pollution (2001) *Transboundary Air Pollution: Acidification, Eutrophication and Ground-level Ozone in the UK*, Centre for Ecology and Hydrology, Edinburgh.

to combustion of solid fuel or petroleum products, with emissions from these two sources having declined by 68 and 86% respectively between 1970 and 1998.

Emissions from the power station sector were almost constant until the early 1990s, and then fell steadily under the combined influence of desulphurisation and the switch to gas from coal. Industrial and domestic emissions fell throughout the period. This decline conceals a major redistribution of source types – the power stations have been moved out of the urban areas into greenfield rural sites near the coalfields, while domestic coal combustion for space heating has been almost completely replaced by gas central heating. Again, political influences can also be seen on the total emission curve. The sharp reduction in 1974 was caused by the oil price increase imposed by the Gulf States, and that in 1980 by the economic recession following the election of the Conservative Party. In more recent years the economic recovery, followed by protracted recession, has added to changes due to intended sulphur emission controls. Future emissions targets (for 2010) have been set by the 1999 Gothenburg Protocol, and declines already underway due to improved energy efficiency, reduced S in fuels and other factors are expected to achieve these.

1.4.2.1   Reduction of sulphur dioxide emissions

Burn less fuel! It is self-evident that, other things being equal, we can always reduce pollutant emissions by burning less fuel. However, for several hundred years, as we have already seen, the rising standards of living of the developed countries have been based fundamentally on production and consumption of energy that has mostly been derived from fossil fuel. We live in an increasingly energy-intensive society, and reductions to our quality of life in order to save energy are not yet politically acceptable. Many measures, such as improved thermal insulation and draught proofing of buildings, offer enormous potential for improved quality of life and reduction of emissions. Wonderful opportunities for change have been missed. For example, the price of oil was arbitrarily quadrupled by the Gulf States in the 1970s; this incentive could have been used to redirect society quite painlessly, raising the price of energy over a twenty-year period and encouraging people to live closer to their work and relatives, travel on public transport, and make their homes less energy-wasteful. Other attractive options, such as combined heat and power that can raise the overall energy efficiency of fossil fuel combustion from below 40% to above 80%, have also been largely ignored. Some countries have made more progress than others – in Copenhagen, for example, new houses are automatically supplied with hot water from the district heating scheme.

Fuel substitution. This involves the use of a lower-S fuel to reduce emissions, and is very logical, but may have other implications. For example, we have seen from the above data that power stations must be the first target for sulphur control. In the UK, a large investment in coal-fired stations was made in the 1960s, with

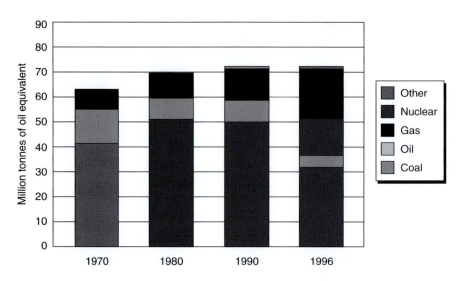

Figure 1.8 Contributions to UK power generation, 1970–1996.
Source: Environment Agency (2000) *Report into an Air Pollution Episode*, Environment Agency, Bristol, UK.

coal sourced from the nationalised British coal pits. Most British coal, particularly in the Trent valley where the power stations are concentrated, is not low-sulphur. When acid deposition and the sulphur content of coal became an issue in the 1970s, there was little flexibility for these stations to import low-sulphur coal. In the 1980s and 1990s, the British coal industry collapsed under both economic and political pressures, increasing the freedom of power stations to import low-sulphur coal. In addition, the privatised electricity generators are now free to construct gas-fired power stations that emit much less sulphur dioxide. The effect of these changes on the contribution to UK power generation can be seen in Figure 1.8.

Coal was dominant in 1970, and remained so until 1990. By 1996, however, gas generation had taken 21% of the market and coal had declined to around 45%. Although these changes do reduce SO_2 emissions, more is needed to meet the Government targets. Her Majesty's Inspectorate of Pollution (now part of the Environment Agency) set the goal of reducing the SO_2 emissions from power generation to 365 kt year^{-1} by 2005, whereas current (2001) emissions are around 800 kt. The effects of changes in sulphur content can be dramatic. In Hong Kong, the use of fuel oil containing more than 0.5% S by weight was prohibited in July 1990. Ambient SO_2 concentrations in the most polluted areas dropped from around 120 μg m^{-3} to around 30 μg m^{-3} within weeks.

Fuel cleaning. The coal used in large-scale generating plant is ground in a ball mill to the texture of a fine powder so that it can be blown down pipes and mixed with air before combustion. Since the sulphur-containing pyrites occur as

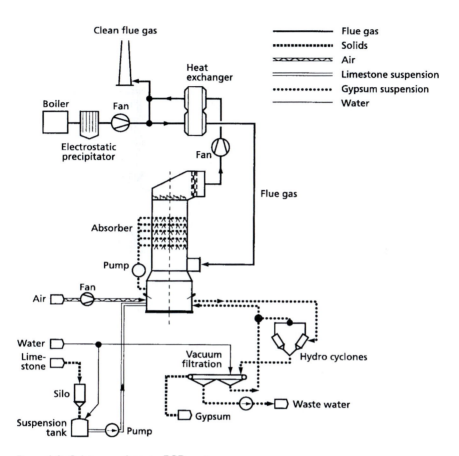

Figure 1.9 Calcium carbonate FGD system.
   Source: Johnsson, J. E. (1998) 'Urban sources of air pollution', In: J. Fenger,
   O. Hertel and F. Palmgren (eds) *Urban Air Pollution: European Aspects*, Kluwer
   Academic Publishers, Dordrecht, The Netherlands.

physically-distinct particles having a different density to the coal, a process such
as froth flotation can be used to separate the relatively dense flakes of pyrites
from the powdered coal. Reductions of 80% in S content can be achieved, but
40% is more typical and this process is not widely used.

   Flue Gas Desulphurisation (FGD). Chemical engineering plant can be built
into or added onto power stations to remove most of the sulphur dioxide from the
combustion (flue) gases before they are exhausted to atmosphere. Several different methods are available. In the most popular, a finely divided limestone slurry
is sprayed into the flue gas (Figure 1.9). The calcium carbonate from the limestone reacts with the SO_2 to produce hydrated calcium sulphate (gypsum), which
can be used to make plasterboard for the building industry.

$$CaCO_3 + SO_2 \rightarrow CaSO_4 \cdot 2H_2O$$

In a modification of the above, a lime slurry is atomised by a spinning disc. Although, the reaction between the absorbent and the acid gas takes place mainly in the aqueous phase, the water content and rate of evaporation are controlled to give dry calcium sulphite which is captured in a conventional dust collector and can be used for landfill or cement manufacture.

$$SO_2 + CaO + \frac{1}{2} H_2O \rightarrow CaSO_3 \cdot \frac{1}{2} H_2O$$

Finally, for plants located near the sea, the natural alkalinity of sea water can be utilised to form sulphuric acid in a packed bed absorber.

$$SO_2 + H_2O + \frac{1}{2} O_2 \rightarrow H_2SO_4$$

The acid effluent is normally disposed of at sea, which may or may not be environmentally acceptable. This process is a good example of the need to consider air, water and land pollution as a whole when evaluating different control strategies.

Currently 10% of UK generating capacity (6000 MW) has been fitted with FGD systems. One typical FGD plant, operating on a 2000 MW station at 90% efficiency:

- reduces SO_2 emissions by 130 000 tonnes a^{-1}
- uses 300 000 tonnes limestone a^{-1}
- produces 470 000 tonnes gypsum a^{-1} (10–15% of the UK market)
- costs £300M as a retrofit.

Although sulphur emissions are greatly reduced, other issues are raised. Limestone must be quarried in, and transported from, rural areas, and the gypsum may have to be disposed of to landfill if a market cannot be found for it. There is then a consequent risk of leachate creating a water pollution problem. The UK electricity providers have installed FGD systems on the minimum number of power stations necessary in order to meet European Union targets on sulphur emission reduction. West Germany, in contrast, installed FGD on 40 000 MW of capacity between 1982 and 1991, which reduced annual SO_2 emissions from 1.6 Mt to 0.2 Mt. In 1990, with the unification of East and West Germany, emissions rose sharply because the East Germans had been burning 300 Mt a^{-1} of high-sulphur soft brown coal. There is currently a second wave of FGD retrofits being made to existing power stations which is expected to bring total German emissions down from 6 Mt to 0.6 Mt by 2005.

Although the old industrialised countries have the wealth and level of technology to lower SO_2 emissions to whatever value they choose, we cannot be complacent on a global basis. For example, China currently burns around 1000 Mt coal and emits 15 Mt SO_2. By 2020, around 2000 Mt coal will be burned, emitting up to 55 Mt SO_2 depending on the sulphur content of the coal and the level of sulphur emission control installed on new plants. The 1999 Gothenburg Protocol of the Convention on Long Range Transport of Air Pollution has set a combined EU15 annual emissions ceiling of 4.044 Mt SO_2 to be achieved by 2010. Remarkably, this is similar to the SO_2 emission from the UK *alone* during the 1980s.

1.4.2.2   Sulphur trioxide

A small proportion of the fuel sulphur (typically up to a few percent) may be further oxidised to sulphur trioxide (SO_3). This has three main consequences: first, a blue mist of sulphuric acid droplets can be emitted from boiler chimneys; second, water vapour and SO_3 combine to condense out sulphuric acid on cool parts of the ductwork, accelerating the corrosion of metal parts; third, smoke particles accumulate on these sticky areas of the ductwork. Small lumps of this acidified deposit, known as 'acid smuts', are entrained into the gas stream, ejected from the chimney, and may then fall onto and damage local sensitive surfaces such as car paintwork or clothes. The conversion to SO_3 is catalysed by vanadium, so that some fuel oils that are high in vanadium can convert 20–30% of the initial sulphur into SO_3. Proprietary chemical oil additives are available that can help to control this problem.

1.4.3   Nitrogen oxide production

The two principal oxides of nitrogen are nitric oxide (NO) and nitrogen dioxide (NO_2). The sum of these two is known as NO_x (pronounced either as en-oh-ex or as 'knocks'). Despite their quite different physical properties, chemical affinities and environmental impacts, they are often lumped together. Combustion always produces a mixture of NO_2 and NO, although typically more than 90% of combustion NO_x production is in the form of NO.

Nitric oxide is formed in two distinct ways:

Thermal NO is formed from reactions between the nitrogen and oxygen in the air. The reactions can be summarised by the Zeldovitch mechanism:

$$N_2 + O \rightarrow NO + N$$

$$N + O_2 \rightarrow NO + O$$

This is highly endothermic, so that thermal NO production is at a maximum in the highest-temperature regions of a combustion chamber.

Fuel NO is formed from nitrogen in the fuel. Typically, fuel nitrogen contents are 0.5–1.5% in oil and coal, and rather less in gas.

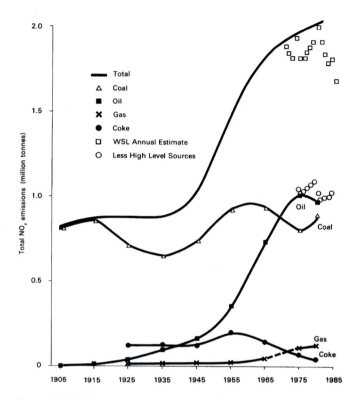

Figure 1.10 NO$_x$ emissions in the UK 1905–1985.
Source: UK Building Effects Review Group (1989) *The Effects of Acid Deposition on Buildings and Building Materials*, Department of the Environment, London, UK.

Figure 1.10 shows that total UK NO$_x$ emissions were quite steady until the 1940s, when they started to increase sharply. Emissions from coal combustion fell during the recession in the 1930s before reviving, while those from oil products increased steadily until curtailed in 1974. More recent data are shown in Figure 1.11 for the period since 1970, together with predicted emissions up to 2020. Note that the data in this figure are expressed as Mt nitrogen, whereas those in Figure 1.10 were in terms of Mt NO$_2$. Between 1970 and 1984 the total was almost constant, although there was a redistribution between sources, with motor vehicle emissions increasing and industrial emissions falling. Between 1984 and 1990 total emissions increased steadily, due entirely to the increase in vehicular emissions. After peaking in 1990, a decline started which is expected to continue to at least 2005 due to the fitting of low-NO$_x$ burners in power stations and to catalytic control of vehicle exhaust gases.

Nitric oxide is also emitted from soils, but it has proved harder to quantify the source strength. Nitrifying bacteria produce both nitric and nitrous oxide – anaer-

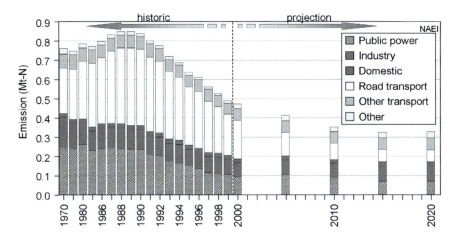

Figure 1.11 UK NO$_x$ emissions by type 1970–2020.
Source: National Expert Group on Transboundary Air Pollution (2001)
*Transboundary Air Pollution: Acidification, Eutrophication and Ground-level
Ozone in the UK*, Centre for Ecology and Hydrology, Edinburgh.

obic production of NO is thought to dominate overall. Guesstimates based on
rather sparse data suggest that UK soil production of NO may be equivalent to
2–5% of the production from fuel combustion.

Table 1.10 gives the detailed Department of Environment (DoE) statistics for
the UK emission of nitrogen by fuel combustion in 1970 and 1998. The annual
total is less than for sulphur dioxide, with a different balance between source
types. Power stations make up only a fifth; while road transport, which contributed only a tiny proportion of the SO$_2$ emissions, contributes nearly half
of the N emissions. The similarity of the road transport emissions in 1970 and
1998 conceals a large rise to a peak of 400 ktonnes in 1990 (Figure 1.11)
before they were controlled by catalytic converters. The emissions are in fact
quite evenly distributed amongst coal, petrol and diesel fuels. This balance
naturally affects the priority areas for control measures. It is important to note
that, despite the precision with which the emissions are quoted, the error on
these figures (and on those of corresponding emissions of other gases) has
been estimated to be 15–20%. The total weight of nitrogen emitted is some
500 kt, equivalent to more than half the total nitrogen used in UK fertiliser.
Hence it should be anticipated that the nitrogen, once returned to the surface,
will have a significant impact on both managed and natural ecosystems.

Table 1.11 shows the global distribution of emissions. Although most of the
direct emission will be as NO, the source strengths are given as NO$_2$ equivalent
since all the NO is potentially available for oxidation to NO$_2$. Within this global
total of about 150 Mt, European emissions account for about 20 Mt. The figures

Table 1.10 UK emissions of NO_x in 1970 and 1998, by source category

Source	Emissions/ 1970 kt N	Emissions/ 1998 kt N	Percentage of 1998 total
Combustion for energy production			
Electricity	247	111	21
Petroleum refining	13	16	3
Other combustion	19	16	3
Combustion in commercial and residential			
Residential	19	22	4
Commercial and agricultural	22	10	2
Combustion by industry			
Iron and steel production	23	7	1
Other industrial	98	47	9
Production processes	4	2	0
Extraction and distribution of fossil fuels	0	0	0
Road transport	234	243	46
Other transport and machinery	76	60	11
Total	760	534	100

Table 1.11 Global emissions of NO_x by source type

Source category	Source strength/ kt NO_2 a^{-1}	Percentage of total
Surface sources		
Fuel combustion		
Coal	21 000	13.7
Oil	10 200	6.6
Gas	7 600	4.9
Transport	26 300	17.1
Industrial	4 000	2.6
Soil release	18 100	11.8
Biomass burning		
Savannah	10 200	6.6
Deforestation	6 900	4.5
Fuel wood	6 600	4.3
Agricultural refuse	13 100	8.5
Atmospheric sources		
NH_3 oxidation	10 200	6.6
Lightning	16 400	10.7
High-flying aircraft	1 000	0.7
NO_y from the stratosphere	2 000	1.4
Total emission	153 600	100

show that some 70% of the total release is due to human activities – or to put it another way, we are emitting more than twice as much as the whole Planet. An indication of the uncertainty in compiling these estimates is that the global total is predicted to lie in the range 60–300 Mt. As with SO_2 and ammonia, there are large variations in emission density with location – by a factor of at least 30 between the heavily populated regions of the Netherlands, Germany and the UK on the one hand, and sparsely inhabited regions such as Scandinavia and northern Scotland on the other.

In Asia, just three countries (China, India and Japan) accounted for over 75% of total energy consumption in 1990, with hard coal predominating. The total NO_x emission was estimated as 19.2 Tg, with 15.8 Tg from area sources and 3.3 Tg from large point sources. By 2020, total energy use is expected to increase by a factor of over four, with oil and gas taking an increasing share of the market. Emissions from some individual sub-regions and cities will increase tenfold over this period. Low emissions on a per capita basis can indicate either an immature economy which is not energy-intensive, or a mature economy which is energy-intensive but has extensive pollution controls in place. Many Asian economies will be more interested in robust growth than in pollution control, and it may take decades for emissions to be reduced to Western levels.

1.4.3.1   Nitrogen dioxide formation

As has been described in Section 1.4, only a small proportion of the NO_2 found in the atmosphere was released in that form from sources. The remainder has been created in the atmosphere as part of the same photochemical activity that is responsible for ozone formation. The nitric oxide from fossil fuel combustion reacts with ozone

$$NO + O_3 \rightarrow NO_2 + O_2$$

During daylight, the NO_2 absorbs blue and UV radiation <420 nm and decomposes back to NO and O_3, resulting in a photochemical equilibrium between the four gases. In rural areas, away from the NO sources, the NO_2 concentration is usually considerably higher than that of NO. In urban areas, the O_3 becomes depleted and the balance moves in favour of NO. The production of ozone, which requires more complex cycles involving organic compounds and radicals, is discussed in Section 1.8.1.

The principal sink for NO_x is oxidation to nitric acid HNO_3.
In the daytime

$$NO_2 + OH + M \rightarrow HNO_3 + M,$$

and at night

$$NO_2 + O_3 \rightarrow NO_3 + O_2$$

$$NO_3 + NO_2 + M \rightarrow N_2O_5 + M$$

$$N_2O_5 + H_2O \rightarrow 2HNO_3$$

The resulting lifetime of NO_x is only about a day. Furthermore, HNO_3 is highly soluble in water and easily removed by precipitation, so it has not been clear what acts as the effective NO_x reservoir which is required to explain its occurrence in regions remote from anthropogenic sources. The likely route is thought to be via peroxyacetyl nitrate ($CH_3C(O)OONO_2$, usually referred to as PAN). Peroxyacetyl nitrate is produced in the troposphere by photochemical oxidation of carbonyl compounds in the presence of NO_x. For example, from acetaldehyde CH_3CHO

$$CH_3CHO + OH \rightarrow CH_3CO + H_2O$$

$$CH_3CO + O_2 + M \rightarrow CH_3C(O)OO + M$$

$$CH_3C(O)OO + NO_2 + M \rightarrow PAN + M$$

Since PAN is only slightly soluble in water, it is not removed effectively by precipitation processes. The main loss route is by thermal decomposition which regenerates NO_x:

$$PAN \xrightarrow{heat} CH_3C(O)OO + NO_2$$

This decomposition process has a time constant of 1 h at 295 K but several months at 250 K. Hence it is thought that PAN generated close to NO_x sources moves into the upper troposphere and lower stratosphere, where the low temperatures allow it to be transported long distances before releasing NO_x. This process maintains NO_x concentrations at 50–100 ppt throughout the remote troposphere.

1.4.3.2   Reduction of nitrogen oxide emissions

Burn less fuel! Fewer motor cars, smaller engines, more public transport, more home insulation and similar measures . Since the same combustion processes are involved, all the arguments and methods that applied to SO_2 will reduce NO_x production correspondingly.

Low-nitrogen fuel. There is not much to be gained by using low-N coal or oil, since they tend to have similar N content. Also, a perversity of thermodynamics means that low-N fuels convert their N more efficiently to NO than do high-N fuels, offsetting the expected benefit.

Peak temperature reduction. The endothermic nature of thermal NO generation means that production can be lessened by reducing peak temperatures everywhere within the combustion zone; it is better to have a uniform distribution at the average temperature than to have high-temperature peaks and low-temperature troughs. A technique called Flue Gas Recirculation can also be used, in which a small proportion of the flue gases (which are inert because their oxygen has been used up) are fed back to dilute the air–fuel mixture before combustion and reduce the intensity of the flame.

Air–fuel ratio. Fuel NO can be reduced by making the combustion fuel-rich (i.e. by reducing the air–fuel ratio). However, this also reduces the thermal efficiency, which is of paramount importance to large-scale users such as electricity generators, and means that more fuel has to be burnt. Hence it is not widely used.

'Low-NO$_x$' burners. Use burner aerodynamics and combustion chamber design to slow the rate at which fuel and air are mixed and burnt, giving long lazy flames and the minimum number of hot-spots. N is still released from the fuel, but under reducing conditions so that it forms N_2 rather than NO. These burners can be retrofitted to existing plant, and have the potential to reduce fuel-NO by 40%. The EC Large Combustion Plant Directive specifies emission limits for new plant. The national total emission from existing plant that has a capacity greater than 50 MW (thermal) also has to be reduced by 15% (1993) and 30% (1998) from 1980 values. In order to meet these targets in the UK, twelve of the coal-fired power stations in England and Wales, representing more than 70% of total coal-fired capacity, have been retrofitted with low-NO$_x$ burners.

Flue Gas Denitrification. Conventional coal and oil-fired power stations may be used in areas with tight NO$_x$ emission limits. Also, nitrogen-rich fuels such as sewage, refuse, some heating oils and waste wood can generate high NO$_x$ emissions. For such emissions, two types of flue gas denitrification (deNox) system are in use. The first, used at lower nitrogen loads, operates without a catalyst at temperatures of 850–1000 °C. Ammonia or urea is injected into the flue gas at the combustion chamber outlet. The second again involves the injection of ammonia or urea, but this time at lower temperatures (250–350 °C) in the presence of a catalyst such as vanadium pentoxide or titanium dioxide.

$$4NH_3 + 4NO + O_2 \rightarrow 4N_2 + 6H_2O \text{ (typically 80–90\% efficient)}$$

This process is expensive; it has been fitted to about 150 plants in Germany, the US and Japan, but is not used in the UK. There may be up to 5 ppm of ammonia slip, which can in turn be controlled with transition metal ions in the catalyst washcoat. The selective catalytic converter is upstream of any dust removal equipment, so the operating conditions for such a catalyst are poor, with very high dust loadings and acid gas concentrations, and the pore structure can get blocked with sulphates and fly ash. Partial regeneration may be possible by soot blowing.

The 1999 CLRTAP Gothenburg Protocol caps EU15 NO$_x$ emissions in 2010 at 6.65 Mt (2.02 Mt as N). EU15 countries have not been as successful in reducing their N emissions as they have been for S. This reflects the fewer opportunities for reducing N emissions via low N fuel, fuel substitution and flue gas deNox.

1.4.4 Ammonia

A further class of emissions is often grouped as NH_y, meaning the sum of ammonia (NH_3) and ammonium (NH_4). The three main sources of atmospheric ammonia are livestock farming and animal wastes, with emissions primarily due to the decomposition of urea from large animal wastes and uric acid from poultry wastes. The overall total emission from these sources has increased with the intensification of agriculture, which has also changed the nature of the emissions from area to point sources. Emissions from housed and field animals are relatively steady in nature, while operations such as slurry and manure spreading result in more intense short-term emissions. Ammonia emissions can be changed by feed N content, the conversion of feed N to meat N (and hence the N content of animal wastes), and the management practices applied to the animals.

Globally, animal emissions are roughly equal to the combined emissions from other sources (Table 1.12). The largest single source of emissions in the UK is release to atmosphere for the few days immediately following slurry application to fields. Table 1.13 below gives a more detailed estimate of the principal UK sources of ammonia. As indicated by the total range, there is a high degree of uncertainty attached to these estimates, with different authors quoting ranges for each term that can vary by an order of magnitude. The total figure represents around 380 kt N, over half of the 500 kt N due to NO_x. The pattern is reflected in emission estimates for other countries – the percentages due to cattle, sheep and pigs are around 88 in Germany, 94 in the Netherlands and 83 in Japan. The total European emission is about 7 Mt NH_3 year^{-1}, of which some 80% is thought to be due to animal manure from intensive livestock production. This agricultural connection has a sharp influence on the geographical distribution of sources. Unlike emissions of NO or SO_2, those of ammonia are at their greatest in rural areas associated with intensive livestock production – the Netherlands, northern parts of Germany and France, South-west and North-west England, Wales and Ireland. The nature of the emission is also different – whereas combustion emissions are produced at very high concentrations from point

Table 1.12 Global ammonia emissions

Source	Global emission/Mt N a^{-1}
Vegetation	5
Oceans	7
Biomass burning	2
Fertiliser application	6
Livestock	25
Total	45

Table 1.13 UK emissions of ammonia by source type

UK source category	Source strength/ kT NH_3 a^{-1}	Percentage of total
Agricultural		
Cattle	200	48
Sheep	70	17
Pigs	30	7.3
Poultry	24	5.8
Horses	2	0.5
Fertiliser application	40	9.7
Others		
NH_3 and fertiliser production	8	
Human sweat and breath	10	
Sewage treatment plants	5	
Sewage sludge disposal	5	
Domestic coal combustion	5	
Motor vehicles	1	
Domestic pets	12	
Total	412 (range 180–780)	

sources, ammonia emissions are released at very low initial concentrations from large area sources such as pastures. The potential for ammonia release is indicated by the increase in fertiliser use in the second half of this century – from 200 kt N in 1950 to 1500 kt N in 1990. Much of this fertiliser has been used to increase pasture and feed grain production for cattle and sheep. Clover in pasture is used to fix around 200 kg N ha^{-1}, and this again increases the amount of N from animal wastes.

Ammonia is also volatilised directly from fertilised crops and grassland. When urea is used, this loss can amount to 5–20% of the applied N; losses from ammonium nitrate (which makes up around 90% of UK applications) are probably less than 5%. The picture is further complicated by the fact that ammonium is present in plant tissues; when the atmospheric concentration is less than a critical value (known as the compensation point), ammonia can diffuse to atmosphere from the stomata. Conversely, if the concentration is above the compensation point, then there is deposition. Hence the emission rate is influenced by plant N status, stomatal aperture and microclimate. There may be both diurnal and annual cycles, with emission being favoured after fertiliser application, during daylight and during leaf senescence, and deposition at night or when the leaf canopy is wet.

As can be seen from Table 1.13, there are sundry other sources of atmospheric ammonia. The most significant are losses from the industrial plants where ammonia and fertilisers are produced, human sweat and breath (around 0.2 kg person^{-1} a^{-1}), and wastes from domestic pets such as cats and dogs. These emissions are insignificant compared to those from agriculture.

Table 1.14 Summary of nitrogen emissions from the UK by species

Nitrogen species	Emission/kt N a^{-1}
NO_x	500
NH_3	380
N_2O (see Chapter 11)	110
Total	990

Table 1.14 gives an approximate summary for the total N emissions in the UK from NO_x, NH_3 and N_2O.

1.4.5   Volatile organic compounds (VOCs)

The general category of VOCs includes non-methane hydrocarbons (NMHC, such as alkanes, alkenes and aromatics), halocarbons (e.g. trichloroethylene) and oxygenates (alcohols, aldehydes and ketones). Historically, measurements of atmospheric HC concentration have been expressed in terms of the non-methane component (NMHC), because the methane concentration was regarded as a stable natural background. However, it is now recognised that methane is also a man-made pollutant from intensive animal and rice production, that the concentration is increasing globally, and that it plays an important role in ozone photochemistry. Emissions of NMHCs are larger than those of methane; nevertheless, since they are more reactive, the typical atmospheric concentrations are considerably lower. Within the VOC, hydrocarbons have received the most attention for their role in photochemistry. More recently, the range of molecules of concern has expanded to include other groups such as chlorinated and oxygenated hydrocarbons. Non-methane hydrocarbons can undergo many transformations in the atmosphere, most of which involve reactions with NO_3 or OH radicals, or with O_3 – during these reactions, the NMHC may form more stable or soluble species, or may become converted from gas to particle or *vice versa*.

Data for NMVOC emissions in 1998, taken from the UK Atmospheric Emissions Inventory, are given in Table 1.15. The major emission categories are solvent use (which includes paints, adhesives, aerosols, metal cleaning and printing) and road transport. Substantial VOC emissions occur during processes such as painting (evaporation of solvents), oil production (flaring and venting of gas), oil refining (flaring and fugitive emissions), distribution of oil or refinery products (evaporation from storage, displacement losses when venting tanks), dry cleaning (final drying of clothes), use of aerosol sprays (both in the product and from the propellant), inefficient combustion of bituminous coal in domestic grates, production of alcoholic drinks (breweries and distilleries) and arable farming (crop growing, silage manufacture, sludge spreading).

The total emission, at about two million tonnes per year, is considerably greater than that of NO_x or SO_2. The emissions are dominated by industrial

Table 1.15 UK emissions of NMVOC

Source category	Source strength in 1970/kt a^{-1}	Source strength in 1998/kt a^{-1}	Percentage of total in 1998
Combustion for energy production			
Public power	7	6	0
Commercial and residential	296	41	2
Production processes	372	289	15
Extraction and distribution of fossil fuels			
Gas leakage	3	19	1
Offshore oil and gas	5	167	9
Petrol distribution	81	113	6
Solvent use	576	532	27
Road transport			
Combustion	523	394	20
Evaporation	112	134	7
Other transport and machinery	62	48	2
Waste	10	29	1
Nature	178	178	9
By type of fuel			
Solid	309	31	2
Petroleum	592	444	23
Gas	6	18	1
Non-fuel	1365	1466	74
Total	2272	1958	100

processes, solvents and road transport. They increased steadily between 1970 and 1990, then started to decline, albeit slowly, due largely to control of vehicle exhaust emissions (Figure 1.12).

More detailed surveys have been used to speciate the NMVOC emissions between sectors. The UNECE VOC protocol, for example, identifies three groups of VOC according to their potential for photochemical ozone production. Group I (highest potential) includes alkenes, aromatics, low alkanes and isoprene, Group II ethanol and high alkanes, and Group III methanol, acetone, benzene and acetylene.

Although a notional figure of 80 kt has been included for forests in Table 1.15, we know far less about biogenic VOC emissions than we do about anthropogenic ones. Plants synthesise many organic molecules as an integral part of their bio-chemistry – hydrocarbons include hemiterpenes such as isoprene (C_5H_8), the monoterpenes ($C_{10}H_{16}$) such as α-pinene and β-pinene, the sesquiterpenes ($C_{15}H_{24}$) such as humulene. Oxygenated compounds include C_1 such as CH_3OH, $HCHO$; C_2 such as C_2H_5OH, and C_3 such as acetone (CH_3COCH_3). Some of these VOCs are released to atmosphere – the turpentine smell characteristic of pine forests, the monoterpenes from mint and the aldehydes from mown grass are well-known examples. Some information is available about emissions of ethylene (ethene) which regulates physiological processes such as fruit ripening and leaf

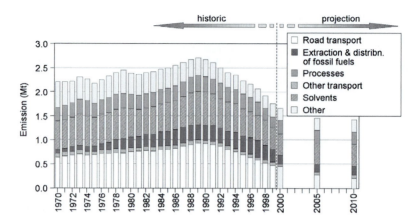

Figure 1.12 UK NMVOC emissions by type 1970–2010.
Source: National Expert Group on Transboundary Air Pollution (2001) *Transboundary Air Pollution: Acidification, Eutrophication and Ground-level Ozone in the UK,* Centre for Ecology and Hydrology, Edinburgh.

abscission, and isoprene, but less is known about the rest. Ethylene has been heavily studied, and is known to be associated with plant growth and develop-ment, seed germination, flowering, fruit ripening and senescence. There is a nice little interaction with air pollution – damage due to chemicals such as ozone and sulphur dioxide can stimulate the production of stress-ethylene, which is then in-volved in the production of new atmospheric compounds. Deciduous trees tend to be isoprene emitters, while conifers favour monoterpenes, although the exact role of these substances in plant biochemistry is not clearly understood. The global fluxes are huge – some 400 Mt C in both isoprene and monoterpene. Hydrocarbons make up the majority, although smaller quantities of partially oxi-dised compounds such as alcohols, aldehydes, ketones and acids are also emitted. Altogether around 1000 different compounds (some of which are themselves families with thousands of members) are known to be emitted (Table 1.16).

Table 1.16 Global biogenic VOC emissions

Compound	Emission rate/Mt C a^{-1}
Methane	160
Dimethyl sulphide	15–30
Ethylene (ethene)	8–25
Isoprene	175–503
Terpenes	127–480
Other reactive VOCs	~260
Other less reactive VOCs	~260

In the US, for example, detailed inventories have shown that isoprene emissions exceed anthropogenic hydrocarbon emissions. Hence O_3 production will not be controlled by reductions in anthropogenic HC emissions alone, but will need greater control of NO_x emissions. Once released, the VOCs are involved in a great multiplicity of oxidation reaction pathways, although there is much uncertainty about the details of these at present.

1.4.5.1   Reduction of VOC emissions

The potential for reduction of VOC emissions falls into two categories – mobile and stationary sources. Mobile sources are discussed in Chapter 3. In the UK, many stationary processes are prescribed under the Environmental Protection Act 1990, and emissions from these will fall steadily as the legislation comes into effect. It was predicted that stationary source emissions would fall by an overall 36% from a 1988 base by 1999. Many of the emissions come from small-scale operations that are hard to control – for example, surface cleaning is performed by dipping parts into open tanks of solvent. There are estimated to be 150 000 to 200 000 sites in the EU at which solvent degreasing is carried out in the metals and electrotechnical industrial sectors.

There are many methods for industrial VOC control:

Condensation – The gas stream is cooled to a temperature at which the partial pressure of the required VOC component exceeds its dew point, so that it condenses out as a liquid. The procedure is most effective for compounds with high boiling points. If the gas stream also contains compounds that solidify at the reduced temperature, they will block the condenser.

Adsorption – The VOC is adsorbed onto a material such as activated carbon which has a large surface area per unit volume. When the adsorber is saturated, the VOCs are stripped off by steam, which is in turn condensed.

Absorption – In absorbers, or scrubbers, the gas stream is put in intimate contact with an absorbing solution by means of spraying the liquid through the gas or bubbling the gas through the liquid. The solution absorbs gas until the concentration is in equilibrium under Henry's law.

Thermal and Catalytic Incineration are discussed in detail below.

Flaring – The organic stream is burned in an open flame. With proper design the flame stability is insensitive to gas flow rate, so that wide variations (1000:1) can be handled. This makes the method useful in plant start-up or shut-down situations, or in an emergency when a process stream has to be dumped.

Biological treatment – Microorganisms are used to break down the large organic molecules, using some of them as their energy supply. The method can have high efficiency for dilute gas streams, and is most effective on alcohols, ethers, aldehydes and ketones.

1.4.5.2   Air pollution control by catalysts

A catalyst is a material that increases the rate of a chemical reaction by lowering the activation energy for a particular step in the reaction sequence; the catalyst does not itself get changed in the reaction. The reaction products can be controlled by choice of catalyst – for example, platinum will catalyse the full oxidation of ethylene to carbon dioxide and water, whereas vanadium catalyses partial oxidation to aldehyde. Catalytic control of air pollutants has been given a huge role in the reduction of motor vehicle emissions; there are also many industrial process emissions that can be reduced by catalytic conversion. The effectiveness of catalytic reactions partially depends on access by the reactants to the catalytic sites. Hence catalysts are usually presented as a coating on a porous support matrix having large surface area. Common matrix materials, such as Al_2O_3, SiO_2 and TiO_2, are applied as a washcoat to an inert mechanical substrate. This washcoat is precipitated from solution and then heat treated to form a network of 2–5 μm particles interlaced with pores 1–10 μm across. Such a washcoat can have a surface area as high as 200 $m^2\ g^{-1}$. Choice of washcoat may be critical for particular gases – for example, alumina reacts with sulphur dioxide and is unsuitable for catalysing sulphurous exhaust gases. The washcoat is then impregnated with the catalyst by soaking it in a solution containing a salt such as $Pt(NH_3)^{+2}$. The exhaust gases to be treated could be passed directly through the catalysed washcoat. However, at high flow rates this generates high pressure drops and hence energy costs (or equivalently, a loss of power). To reduce the pressure drop, the washcoat is applied to an inert ceramic monolith having a honeycomb structure of 50–60 channels cm^{-2}, each channel being 1–2 mm across and giving a surface area density of 0.3 $m^2\ g^{-1}$. Temperature is another important factor. Catalysis of relatively cold gases will be limited by the chemical reaction rates at the catalyst sites, while for hot gases the rate-limiting process will be diffusion of reactant and product molecules to and from the sites.

Process VOC emissions can be controlled by commercially-available thermal or catalytic oxidisers. With thermal oxidisers, the incoming contaminated air is preheated in a heat exchanger before passage through a burner section. In the burner section, the air is heated to around 800 °C in the presence of excess oxygen, oxidising the VOCs to CO_2 and water. The hot gases leave through a heat exchanger which preheats the inlet gases and increases energy efficiency. The heat exchanger may be either continuous (recuperative) or cyclical (regenerative). With catalytic oxidisers, a thin-walled honeycomb lattice is coated with a washcoat to provide high surface area; and then impregnated with a fine dispersion of platinum-group metals. A range of catalysts is available to cover methane, ethane, propane and other VOCs. Inlet concentrations are typically a few hundred ppm. The catalytic converters operate at lower temperatures of 150–350 °C, so they have lower energy requirements, although again much of the energy expenditure can be recovered with an efficient heat exchanger. Choice of system is influenced by factors such as expected working life, inlet gas composition and temperature, and process intermittency. With either type of system, VOC destruction efficiencies up to 99% can be achieved.

An interesting application of catalytic control is for ozone in the cabins of high-altitude jets. The WHO guideline for personal exposure to ozone is 0.1 ppm. During very high pollution episodes in Los Angeles the concentration has risen to 0.6 ppm. Yet long haul passenger aircraft routinely fly at altitudes at which the ozone concentration is several ppm. Although this air is very thin, if it is compressed for use as the source air for cabin air conditioning the ozone content could be unacceptably high. Ozone abaters based on a palladium catalyst are carried aboard such aircraft.

1.4.5.3   Incineration

Volatile organic compounds and many other substances such as halogenated compounds, dioxins and furans, and N- or S-containing materials can also be treated by incineration, which effectively uses the waste gas as the air supply to a high temperature burner. Temperatures of 700–1200 °C are reached, with lower temperatures resulting in higher CO emissions and higher temperatures resulting in higher NO_x emissions. In a typical example from the chemical industry, a waste gas stream is treated for the removal of acetone, isopropanol, toluene, cresole, ethanol, acetic acid, dichlorobenzol, methyl bromide, epoxy propane, methanol, acrylonitrile and dimethyl ether. Quite a soup!

For either incinerators or oxidisers, where the process gas stream involves a large flow at a relatively low concentration, a concentrator can be used as a pre-treatment. A concentrator consists of a large wheel coated in an absorbent such as zeolite (e.g. a sodium or potassium aluminosilicate). The wheel rotates slowly. A part of the wheel is immersed in the process gas and absorbs the pollutant. Another part is immersed in a low volume flow of hot gas which desorbs the pollutant. Then this low volume flow is treated in the usual way.

1.4.6   Carbon monoxide

Complete combustion of any fuel containing carbon would lead to the production of carbon dioxide (Table 1.17). There is always an associated production of carbon monoxide, which is a toxic gas that affects the transport of oxygen in the blood stream. In the atmosphere at large, the concentrations are negligible. Under the restricted ventilation conditions sometimes found in towns, concentrations can be a significant health hazard.

In the UK in 1998, the total anthropogenic emission was 4.7 Mt, with about 73% coming from road vehicles. As with NO_x and NMVOC, this source has fallen rapidly since 1990 with the increasing use of catalytic converters on petrol-engined cars.

Table 1.17 Global CO budget

Sources	Range of estimates/Tg CO a^{-1}
Fossil fuel combustion/industry	300–550
Biomass burning	300–700
Vegetation	60–160
Oceans	20–200
Methane oxidation	400–1000
Oxidation of other hydrocarbons	200–600
Total sources	1800–2700
Sinks	
Tropospheric oxidation by OH	1400–2600
Stratosphere	~100
Soil uptake	250–640
Total sinks	2100–3000

1.4.7 Hydrogen chloride

Most gaseous hydrogen chloride enters the atmosphere through the dechlorination of airborne sea salt particles by pollutant gaseous nitric acid or sulphuric acid (Table 1.18).

$$HCl_{(aq)} + HNO_{3\ (gas)} \rightarrow HCl_{(gas)} + HNO_{3(aq)}$$

$$HCl_{(aq)} + H_2SO_{4\ (gas)} \rightarrow HCl_{(gas)} + H_2SO_{4(aq)}$$

Since both pollutants and sea salt are needed, the highest HCl fluxes are found around the UK, and in Northeast Europe. The other major sources are various types of combustion, except for a small contribution by diffusion down from the stratosphere of HCl that has been formed there from CFCs. HCl dissolves in water to form the chloride ion Cl$^-$.

Table 1.18 Sources of atmospheric HCl

	Flux/Tg Cl a^{-1}
Natural	
Volcanoes	2.0
Anthropogenic	
Seasalt dechlorination	50.0
Fossil fuel combustion	4.6
Biomass burning	2.5
Incineration	2.0
Transport from stratosphere	2.0
Total	63.1

There is no upward trend to HCl concentrations, so the deposition must be assumed equal to the emissions. In the UK, HCl is a minority air pollutant, with total emissions of around 90 kt year^{-1}, 75% of which is due to coal-burning power stations. There may be localised acidity production in rainfall downwind of such sources.

1.5   PRIMARY EMISSION SUMMARY

Approximate overall global totals for emissions of the different gases are presented in Table 1.19, with corresponding values for Europe in Table 1.20.

In Table 1.21, we summarise also the sources of the different gases shown in Table 1.20. The range of different sources, and of the balances between them, for the different pollutants means that the spatial variation of emissions can also be very high.

Figures 1.13a–d show the contrasting spatial distributions of the emission density of four pollutants in the US in 1998. Sulphur dioxide (a) is largely confined to the east, where the population is dense and the power is generated from coal. Nitrogen oxides (b) are similar, except that stronger emissions come from California where the vehicle density is highest. Ammonia (c) is strongly

Table 1.19 Global natural and anthropogenic emissions of different species

Species	Natural/Mt a^{-1}	Anthropogenic/Mt a^{-1}	Total/Mt a^{-1}
CO_2	7×10^5	0.23×10^5	7.2×10^5
CH_4	160	375	535
CO	430	1815	2245
SO_2	15	146	161
N_2O	26	16	42
NO_x	24	85	109
NH_3	15	30	45
H_2S	1–2	4	5

Table 1.20 EU* Total emissions of different species

Gas	Emission/Mt a^{-1}
CO_2	3250
CH_4	27.2
CO	46.1
SO_2	14.4
N_2O	1.4
NO_x	12.9
NH_3	3.7
NMVOC	17.4

Note
* The EU 15 plus Croatia, Iceland, Malta, Norway and Switzerland.

Table 1.21 Percentage contributions of different processes to the totals in Table 1.20

Source	CO	NMVOC	CO₂	SO₂	NOₓ	NH₃	CH₄	N₂O
Public power	1.23	0.86	32.32	50.88	18.93	0.09	0.18	4.57
Combustion plants	12.9	3.27	19.40	7.18	4.44	0.04	1.53	2.12
Industrial combustion	5.76	0.36	18.29	16.41	9.45	0.05	0.19	2.29
Production processes	5.39	5.87	4.71	3.99	1.75	2.83	0.17	22.12
Fuel extraction and distribution	0.23	6.00	0.08	0.22	0.88	–	13.82	–
Solvent use	–	23.56	0.13	–	–	0.06	–	0.68
Traffic	61.4	27.97	20.74	3.34	47.37	0.96	0.62	3.96
Other mobile sources	7.55	5.03	3.95	1.89	15.87	–	0.07	1.09
Waste treatment and disposal	4.66	1.28	2.09	0.58	0.84	1.09	30.44	1.65
Agriculture	0.74	17.17	−1.80	–	0.29	94.65	39.52	34.45
Nature	0.21	8.64	0.09	15.50	0.17	0.23	13.47	27.06

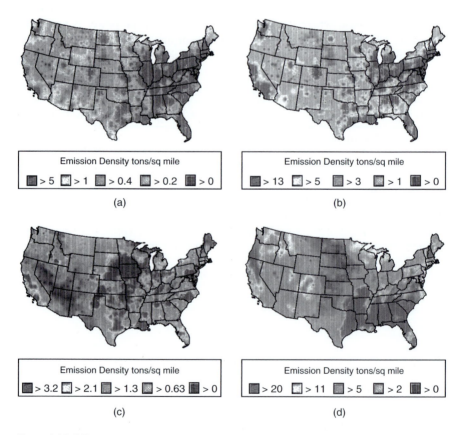

Emission Density tons/sq mile
■ > 5 ▨ > 1 ▨ > 0.4 ▨ > 0.2 ■ > 0

(a)

Emission Density tons/sq mile
■ > 13 ▨ > 5 ▨ > 3 ▨ > 1 ■ > 0

(b)

Emission Density tons/sq mile
■ > 3.2 ▨ > 2.1 ▨ > 1.3 ▨ > 0.63 ■ > 0

(c)

Emission Density tons/sq mile
■ > 20 ▨ > 11 ▨ > 5 ▨ > 2 ■ > 0

(d)

Figure 1.13 US emission maps.
a) SO₂, b) NOₓ, c) NH₃, d) VOC.

associated with the main livestock and grain agricultural areas, while biogenic VOC emissions (d) are mainly generated by the forests in the warm south-east.

1.6   ADSORPTION AND ABSORPTION OF GASES

We have seen above that SO_2 and NO_x can be removed from gas streams by specific chemical plant. There is a much wider range of equipment available that is used to reduce gas concentrations before release to atmosphere. In some cases, these create a useable by-product – for example, nitric acid can be recovered when NO_x is removed.

1.6.1   Adsorption

Gaseous air pollutants such as organics are captured on the surface of a bed of porous material (the adsorbent) through which the gas flows. The adsorbents used are materials such as activated carbon (charcoal made from wood or coconut shells), silica gel (sodium silicate), alumina, synthetic zeolites and clays such as fullers' earth (magnesium aluminium silicates). Activation refers to the thermal pretreatment of the adsorbent so as to maximise its surface area per unit volume. This large specific surface area is an essential requirement for good adsorption systems. Commonly-used adsorbers such as silica gel and activated charcoal have surface areas of up to 2 km^2 per kg, and adsorption capacities of up to 0.5 kg kg^{-1}. The adsorption process may be physical, (due primarily to Van der Waals short-range forces), or it may be chemical. When it has become saturated, the adsorbent is regenerated by heating (often with steam) with or without a reduction in pressure. Adsorption is most effective at high concentration and low temperature. The relationship between the concentrations in the gas and solid phases at a particular temperature is known as the adsorption isotherm. In general, the concentration in the solid phase increases with molecular weight and pollutant concentration and decreases with increasing temperature. The adsorption is most effective for non-polar (hydrophobic) compounds such as hydrocarbons, and less so for polar (hydrophylic) compounds such as organic acids and alcohols.

1.6.2   Absorption

The gas stream is given intimate contact with an absorbing solution by means of bubbling or spraying. The absorption may either be chemical (by reaction) or physical (by dissolution). Solvents in common use include water, mineral oils and aqueous solutions. Gases that are commonly controlled by absorption include HCl, H_2SO_4, HCN, H_2S, NH_3, Cl_2 and organic vapours such as formaldehyde, ethylene and benzene. The absorption process depends on the equilibrium curve between the gas and liquid phases and on the mass transfer between the two, which in turn is a function of the driving force divided by the resistance.

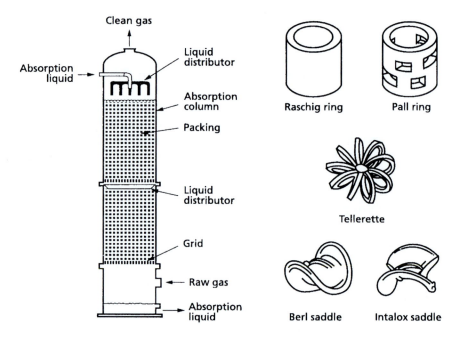

Figure 1.14 A packed absorption tower.
Source: Johnsson, J. E. (1998) 'Urban sources of air pollution', In: J. Fenger, O. Hertel and F. Palmgren (eds) *Urban Air Pollution: European Aspects*, Kluwer Academic Publishers, Dordrecht, The Netherlands.

In commercial absorbers, the gas and solvent are brought into intimate contact, with equilibrium being achieved if the results are to be predictable, before the cleaned gas and solvent are separated. In stagewise absorption, such as tray scrubbers, the gas is bubbled through the liquid to achieve one absorption step under one set of equilibrium conditions before going on to the next step. A vertical casing contains several trays one above the other.

Solvent cascades from the top downwards, while gas moves upwards (countercurrent absorption). Each tray has many small vents across its surface through which the gas bubbles, into and through the layer of solvent. These bubbles provide the high surface area and thin gas–solvent interface necessary for efficient movement of the gas into the solvent. In continuous differential contact systems, such as packed towers, the interfacial surface is maintained on a subdivided solid packing (Figure 1.14). Very high efficiencies, greater than 99.9%, can be achieved.

A common application of gas scrubbers is for odour control. Some gases have extremely low odour thresholds, which are unrelated to their toxicity, and scrubbers must operate at consistently high efficiencies to avoid odour complaints. Often, the odorous contaminant is absorbed into a solvent, then reacted with an oxidising agent such as potassium permanganate to fix it as a less harmful compound.

1.7   INCINERATION AS A SOURCE OF AIR POLLUTANTS

There is a serious waste disposal problem in developed industrial countries. Much of our solid wastes are currently buried in landfill sites, but the supply of suitable holes in the ground is drying up. Hence there is pressure to use alternative methods of disposal, of which incineration is one. Potentially incineration is attractive because it can recover thermal energy for electricity generation or district heating schemes. Sewage sludge can be burnt alongside coal in power stations, or in purpose-built incinerators. In one example of the latter, sewage sludge is pumped into a filter press dewatering plant which converts it into a dry cake. The cake is then incinerated in two fluidised bed furnaces, the waste heat from which is used to generate steam and hence electricity to run the process. Incineration can generate about three times the energy per kg waste compared to the energy recovered from the decomposition of landfill.

In the UK, households generate around 30 million tonnes of municipal solid waste (MSW) annually, and the quantity is growing by about 3% per year. Eighty three per cent goes to landfill, and less than 10% is incinerated or recycled. There are 13 MSW incinerators, burning an annual total of 2 Mt. Local Authorities will be required to double their recycling rates within the next 3 years, almost triple in five years, and achieve 30% by 2010. There is also a target to recover or recycle 45% of MSW by 2010. The EU Landfill Directive will require the UK to reduce by 2020 the amount of biodegradable municipal waste sent to landfill to 35% of the 1995 value. Currently, only around 8% of municipal and 5% of commercial and industrial waste is incinerated in the UK, compared with 32% in France and 54% in Denmark. Ideally, these incinerators would serve as municipal combined heat and power centres, burning the local waste and returning the hot water and electricity to the local community.

There are four competing technologies

> The simplest form is *mass burn*, in which the mixed wastes are combusted on a moving grate and energy is recovered from the hot combustion gases. Around 7 MW can be generated for an annual throughput of 100 000 tonnes.
> *In fluidised bed* combustion, the fuel is cleaned of non-combustible material and shredded before being combusted in an updraft of air through a bed of sand and dolomite.
> *Pyrolysis* involves heating the waste to 400–800 °C in the absence of oxygen; this breaks down complex molecules which can then be burned. *Gasification* is similar, but the low-grade gaseous fuel is burned in a turbine.

The EU intention is to divert MSW from landfill to recycling and incineration, and this process will require the commissioning of up to 50 new incinerators in the UK by 2015. But incineration produces emissions other than CO_2 and H_2O, such as acid gases (HCl, HF, SO_2), NO_x, heavy metals (cadmium, mercury, arsenic, vanadium, chromium, cobalt, copper, lead, manganese, nickel, thallium and tin), dioxins, PCBs, PAH and other particles. Also there is residual solids' ash, (typically 10–20% of mixed urban waste) which is usually very enriched in aluminium, calcium, silicates, iron, sodium, magnesium and potassium, and which may contain 1–10% lead and zinc. Some of the volatile metals such as cadmium and lead are vaporised and emitted. Polyvinylchloride (PVC) combines with lead to make lead chloride, which is volatile and emitted. If the combustion conditions for MSW are wrong, emissions of dioxins and furans can be increased significantly. The tight control required is shown by the EU emission limit for dioxins, which is 0.1 ng TEQ m^{-3}. Municipal solid waste incinerators are currently responsible for only 3% of UK dioxin emissions, but since the total fell from 1078 to 325 kg year^{-1} between 1990 and 1998, any increase due to incineration will be unwelcome. Production of these toxics can be minimised by keeping the combustion temperature above 1100 °C, with a gas residence time in the combustion chamber of at least 1 s and a surplus of oxygen. The UK Hazardous Waste Incineration Directive, which came into force in 2000, brings operating conditions and emissions within a stricter regime. Best available technology must be used for controlling stack emissions and the incineration temperature must exceed 850 °C.

Pyrolysis and gasification are competing with incineration for market share. These processes rely on heating the wastes in the absence of oxygen (like producing charcoal from wood or coke and coke-oven gas from coal), so that they are decomposed into energy-rich gas, oil and char. The gas 'syngas' can be used as a fuel or as a petrochemical feedstock.

1.7.1   Persistent organic pollutants (POPs)

Persistent organic pollutants is a class of pollutants that has only been defined quite recently. It covers materials such as dioxins, furans, PCBs and organochlorine pesticides such as DDT. The materials involved are persistent in the environment, with half-lives of years in the soil or sediment and days in the atmosphere. Hence concentrations may still be increasing decades after emissions have started. They favour soil organic matter and fats rather than aqueous media for storage, and hence enter the food chain and can be magnified by bioaccumulation. They readily volatilise into the atmosphere and condense onto particles, so they can be involved in long range transport. The most common effects reported are of reproductive impairment or carcinogenicity; because there are mixtures of POPs and metabolites present, it is often not possible to specify directly the causative agent.

The term dioxins covers 210 organic compounds, comprising 75 polychlorinated dibenzo-*p*-dioxins (PCDDs) and 135 polychlorinated dibenzofurans (PCDFs). The most common source is the incomplete combustion of plastics such as PVC. The dioxins contain 1–8 chlorine atoms substituted round one of two base structures. The 17 compounds that contain chlorine in all of the 2, 3, 7 and 8 positions are the ones that have toxic effects on humans. There are also 12 of the 209 polychlorinated biphenyls (PCBs) which have recognised toxic effects similar to those of dioxins.

The different compounds have toxicities that vary by 1000:1, and there is an internationally-agreed weighting scheme so that the overall toxicity of a mixture can be quantified. The most toxic is 2,3,7,8 tetrachloro dibenzo-p-dioxin (2,3,7,8 TCDD). The quantity of 2,3,7,8 TCDD which would have the same toxicity as the mixture is known as the International Toxic Equivalent (I-TEQ). Dioxins have very limited solubility in water, and low vapour pressures. In the atmosphere they are normally associated with particles, and are extremely persistent. Typical ambient concentrations are in the region of 10–100 fg I-TEQ m^{-3}, where 1 fg (femtogram) = 10^{-15} g. A very wide range of processes contribute to dioxin emissions, with no single category being dominant. In 1998, the four largest categories in the UK were sinter plants (13%), open fires and field burning (20%), clinical waste incineration (7%) and non-ferrous metal production (7%).

1.8   SECONDARY GASEOUS POLLUTANTS

1.8.1   Tropospheric ozone

Natural ozone mainly occurs in the stratosphere, between heights of 15 and 50 km. It is formed from the action of UV photons (with wavelengths of <242 nm) on oxygen molecules. This 'ozone layer', which is sometimes known as 'good ozone' and which accounts for about 90% of all atmospheric ozone, is discussed in more detail in Chapter 12. Natural ozone is also present in the troposphere, where it has a 'background' concentration of 10–20 ppb. Some of this ozone has diffused down from the stratosphere. Current estimates suggest that this stratosphere–troposphere exchange amounts to between 400 and 500 × 10^9 kg a^{-1}, and can constitute 30–40% of the total tropospheric ozone burden, although it is uncertain how much actually penetrates down to the surface. The natural background concentration can be raised by at least an order of magnitude when additional ozone is formed in the troposphere from other pollutant gases. It cannot be formed directly in the same way as stratospheric ozone because the lowest wavelength UV to reach the surface is 280 nm and has insufficient energy to dissociate the oxygen molecule.

The remainder forms photochemically from the action of UV photons on natural NO$_x$ in the following manner:

$$NO_2 + h\nu(\lambda < 400 \text{ nm}) \rightarrow NO + O$$

This is referred to as NO$_2$ photolysis. Photons with a wavelength greater than 400 nm do not have enough energy to break up the NO$_2$ molecule, while those having a wavelength shorter than 280 nm have been absorbed so effectively in the stratosphere that the tropospheric flux is negligible.

$$O + O_2 + M \rightarrow O_3 + M$$

(M is any 'third body' molecule, also known as a chaperone, such as N_2 or O_2, which is required for energy absorption but which does not take part in the reaction as such). Left to its own devices, the generated O_3 reacts readily with NO:

$$NO + O_3 \rightarrow NO_2 + O_2 \qquad (1.2)$$

(NO_2 production)

This cycle of reactions generates a small contribution (a few ppb) to the background ozone concentration.

In the unpolluted atmosphere, equilibrium between the above three reactions gives low and stable O_3 concentrations. The O_3 concentration never gets very high because it is consumed immediately by the NO. Additional ozone can be formed in the troposphere as a secondary pollutant when there are increased concentrations of NO_2 for any reason. In the 1950s, an air pollution problem was identified in the Los Angeles area of California – high concentrations of ozone were being formed in the air and causing breathing and eye irritation and visible damage to vegetation. The exact formation mechanisms turned out to be very complex, involving reactions between hundreds of different hydrocarbons, radicals, NO and NO_2.

Two main routes for the production of additional NO_2 are the reactions of NO with either the hydroperoxy radical HO_2:

$$HO_2 + NO \rightarrow NO_2 + OH$$

or with an organic peroxy radical RO_2.

$$RO_2 + NO \rightarrow NO_2 + RO$$

where R is an organic radical – any atomic group consisting only of C and H (e.g. C_2H_5)

For example, in the case of methane (CH_4), R = CH_3.
Hence

$$OH + CH_4 \rightarrow CH_3 + H_2O$$

and

$$CH_3 + O_2 + M \rightarrow CH_3O_2 + M$$

Then photolysis releases oxygen atoms which combine with oxygen molecules to generate elevated ozone concentrations. The sequence of reactions for methane would be

$$OH + CH_4 \rightarrow CH_3 + H_2O$$

$$CH_3 + O_2 + M \rightarrow CH_3O_2 + M$$

$$CH_3O_2 + NO \rightarrow CH_3O + NO_2$$

$$NO_2 + h\nu \; (280\text{--}430 \; nm) \rightarrow NO + O$$

$$O + O_2 + M \rightarrow O_3 + M$$

$$CH_3O + O_2 \rightarrow HO_2 + HCHO$$

$$HO_2 + NO \rightarrow OH + NO_2$$

$$NO_2 + h\nu \; (280\text{--}430 \; nm) \rightarrow NO + O$$

$$O + O_2 + M \rightarrow O_3 + M$$

$$CH_4 + O_2 + O_2 \rightarrow HCHO + O_3 + O_3,$$

with a corresponding sequence for other organic compounds.

Hence the summer daytime polluted troposphere contains a mixture of NO, NO_2 and O_3 in concentrations and proportions that depend not only on the source strengths of the precursors, but on the availability of the reactive components, mixing conditions, life histories and weather conditions. Also note that the OH radical is regenerated during the sequence; since it effectively acts as a catalyst, tiny concentrations (ppt) can play a major role in atmospheric chemistry. Essentially, any hydrocarbons present can be broken down by the hydroxyl radical (OH) to form organic peroxy radicals (RO_2) which react easily with NO to form NO_2, depleting NO from equation (1.2) above and shifting the equilibrium in favour of O_3 production.

First, the OH hydroxyl radical reacts with an organic molecule – that then reacts with molecular oxygen to form an organic peroxy radical RO_2.

$$OH + RH \rightarrow R + H_2O$$

Then

$$RO_2 + NO \rightarrow RO + NO_2$$

(RO is an organic alkoxy radical)

$$NO_2 + h\nu \rightarrow NO + O$$

$$O + O_2 + M \rightarrow O_3 + M$$

During the daytime, this reaction scheme is the source of NO_2 for NO_2 photolysis, giving O_3 as the by-product. Figure 1.15 shows the results of a laboratory experiment in which initial concentrations of 500 ppb of propene, 400 ppb of NO and 100 ppb of NO_2 were irradiated in a chamber for 7 h. The propene and NO concentrations declined steadily as they were consumed by the photochemical reactions. The NO_2 concentration climbed to a peak after 90 min before declining again. Ozone formation increased steadily during the period of the experiment, reaching nearly 600 ppb by the end. There were also significant concentrations of two further secondary pollutants, formaldehyde (HCHO) and peroxyacyl nitrate (PAN), which are both toxic and irritant. Analysis in terms of reaction rates and sources of OH has shown that around two days of strong sunshine would be sufficient to generate the elevations of around 60 ppb O_3 that are experienced in regional photochemical pollution episodes.

At night, and in the winter in temperate latitudes, there are fewer UV photons, so that NO can no longer be formed by photolysis, and continues to be used up by reaction with O_3. The concentrations of NO and O_3 therefore decline. The excess NO_2 reacts with organic molecules, and through various intermediates, generates nitric acid (HNO_3). These nitric acid molecules can be transferred to water droplets and contribute significantly to the acidification of rain.

The requirement for UV photons and the short lifetime of O_3 molecules in the atmosphere combine to create a characteristic diurnal cycle of ozone concentrations in the lower troposphere. In Figure 1.16, a diurnal cycle of ozone production is shown in Los Angeles air. The morning rush-hour is marked by an increase in the primary emissions of NO and CO (and hydrocarbons, which are not shown). With increasing solar radiation, the photochemical reactions accelerate. The NO_2 concentration peaks in mid-morning, followed by ozone in the early afternoon. Hundreds of organic species are present in the polluted atmosphere; each has a different rate coefficient in the scheme above, and elaborate computer modelling is necessary

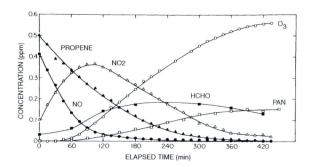

Figure 1.15 Laboratory reaction of NO_x and propene to produce O_3.
Source: Winer, A. M. (1986) 'Air pollution chemistry', in: R. M. Harrison and R. Perry (eds) *Handbook of Air Pollution Analysis*, Kluwer Academic Publishers, Dordrecht, The Netherlands.

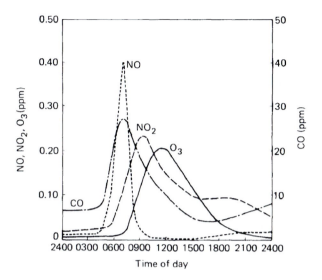

Figure 1.16 Diurnal cycles of ozone production and precursor concentrations in Los Angeles.
Source: Strauss, W. and Mainwaring, S. J. (1984) *Air Pollution*, Edward Arnold, London, UK. Report, European Communities, Luxembourg.

to predict photooxidant formation. Detailed modelling of these photochemical reactions, combined with single-site measurements of both precursors and products, is now enabling predictions to be made of the proportionate contribution of individual hydrocarbons to overall ozone generation. In one case study covering reactions of 34 different hydrocarbons, it was calculated that methane, butene, carbon monoxide, ethylene and isoprene together accounted for over 50% of the ozone production, and that 19 of the 34 modelled species had to be included to account for 90%. The potential for ozone production by the different hydrocarbons can be ranked in comparison to ethylene (taken as 100). For example, propene is 103, propane 42, octanes 53, benzene 19 and methanol 12. We are also able to model the ozone production rate due to specific VOCs. In one example, a total rate of ozone formation of 1.323 ppb h^{-1} was calculated, with contributions of 0.249 from isobutene, 0.147 from ethylene and 0.116 from isoprene.

The ozone production sequence described in the first part of this section is dominant in the highly polluted, strongly irradiated atmospheres that were characteristic of the US in the 1950s and 1960s. Ozone can also be generated by more complex cycles involving the oxidation of methane and/or carbon monoxide, in the presence of nitrogen oxide catalysts.

For example

$$CH_4 + 8O_2 + h\nu \rightarrow CO + 4O_3 + 2OH + H_2O$$

The OH goes on to attack methane

$$CH_4 + OH \rightarrow CH_3 + H_2O$$

which starts the cycle again. The CO may go on to *produce* O_3 if NO_x concentrations are high, or to *deplete* O_3 if they are low.

For example:

$$HO_2 + O_3 \rightarrow OH + 2O_2$$

$$OH + CO \rightarrow H + CO_2$$

$$H + O_2 + M \rightarrow HO_2 + M$$

$$\text{Net } CO + O_3 \rightarrow CO_2 + O_2$$

An example of a cyclical catalytic process for ozone production is

$$NO_2 + h\nu \ (\lambda < 400 \text{ nm}) \rightarrow NO + O$$

$$O + O_2 + M \rightarrow O_3 + M$$

$$OH + CO \rightarrow H + CO_2$$

$$H + O_2 + M \rightarrow HO_2 + M$$

$$HO_2 + NO \rightarrow OH + NO_2$$

$$\text{Net } CO + 2O_2 + h\nu \rightarrow CO_2 + O_3$$

Methane, which is present at average global background concentrations of 1700 ppb as against 20 ppb for O_3, is involved in many complex oxidation reactions which affect the overall balance of OH radicals and O_3 molecules.

The first reaction step for methane is with the hydroxyl radical:

$$OH + CH_4 \rightarrow CH_3 + H_2O$$

The resulting methyl radicals are then involved in further oxidation reactions.

One of the major advances in atmospheric chemistry in the last decade has been realisation of the importance of the OH hydroxyl radical in many atmospheric reactions that remove pollutants, despite its low concentration of a few ppt. Hydroxyl radicals oxidise many of the trace constituents in a low-temperature burning of atmospheric rubbish which dominates the daytime chemistry of the troposphere.

Production of OH is by reaction of water vapour with $O(^1D)$, which is the oxygen atom in the excited singlet state in which it is most reactive.

$$O_3 + h\nu \rightarrow O_2 + O(^1D)$$

$$O(^1D) + M \rightarrow O + M$$

$$O(^1D) + H_2O \rightarrow 2OH$$

Because of the dependence of OH production on solar UV photons, it was originally thought that OH reactions would only be significant in the stratosphere where the UV flux is strong. The creation of $O(^1D)$ needs a photon of less than 320 nm wavelength, but the flux of photons of less than 300 nm at the bottom of the atmosphere is very small, so the window is tiny. In fact, it turned out that there is sufficient OH generated in the troposphere (partly due to the larger mixing ratio of water found there than in the stratosphere) for it to play a significant role in oxidation of species such as CO and CH_4. The lifetime of OH in the presence of gases such as CO is only a few seconds, so the atmospheric concentration is highly variable and dependent on local sources and sinks. The global mean OH concentration has been estimated, from its reaction with the industrial solvent methyl chloroform (CH_3CCl_3), as $\sim 1 \times 10^6$ molecules cm^{-3}. The solvent is one of the gases responsible for attrition of the ozone layer by delivering chlorine into the stratosphere, and its main oxidation route is via OH. OH production rate was perceived as a problem in the 1970s, because it was realised that insufficient O_3 was diffusing from the stratosphere into the troposphere to supply OH production, and that hence all OH would be immediately reacted with an excess of CO

Table 1.22 Tropospheric ozone budgets

	Tg O_3 a^{-1}
Sources	
Chemical production (total)	3000–4600
Of which HO_2 + NO	70%
CH_3O_2 + NO	20%
RO_2 + NO	10%
Transport from the stratosphere	400–1100
Total sources	3400–5700
Sinks	
Chemical loss (total)	3000–4200
Of which $O(^1D)$ + H_2O	40%
HO_2 + O_3	40%
OH + O_3	10%
Other reactions	10%
Dry deposition	500–1500
Total sinks	3400–5700

and CH_4 molecules, leaving these and other molecules such as HCFCs to accumulate. In fact this is prevented by NO_x, which not only allows the regeneration of OH but also produces new O_3 from which additional OH can be produced.

Overall, the global O_3 budget is currently estimated to be as in Table 1.22. It should be noted that many of these values have been arrived at by complex 3-D models of atmospheric photochemistry, and the resulting uncertainty is indicated by the ranges of the values quoted.

1.8.2   Wintertime NO_2 episodes

In London, in December 1991 there was a serious air pollution episode involving elevated concentrations of NO_2. Although the prevailing meteorological conditions restricted dispersion (as they had in the coal smog episodes in 1952 and 1962), the air pollution community was surprised by the concentration increase at the time, because the winter conditions did not favour photochemical generation of NO_2. It was subsequently realised that the NO_2 was due to the direct combination of NO with molecular oxygen:

$$NO + NO + O_2 \rightarrow 2NO_2$$

This is a second-order reaction for which the rate increases as the square of the NO concentration. Hence, although the NO_2 production rate is negligible at low NO concentrations, it dominates at high NO concentrations. Figure 1.17 shows the relationship between hourly mean NO_x and NO_2 concentrations at a roadside site in London for 12 months that included the December episode. The NO_2 concentration can be seen ramping up when the NO_x concentration rises above 200 ppb.

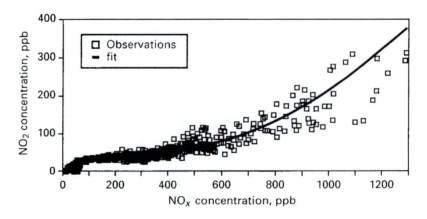

Figure 1.17 Interdependence of NO_x and NO_2 concentrations.
Source: Williams, M. L. (1999) 'Patterns of Air pollution in developed countries', In: S. T. Holgate, J. M. Samet, H. S. Koren and R. L. Maynard (eds) *Air Pollution and Health*, Academic Press, London, UK.

1.8.3   Reduction of ozone concentrations

In principle, ozone concentrations can be reduced by controlling the concentrations of the precursor hydrocarbons and NO_x:

- Reduce NO/NO_2 emissions from power stations (with low NO_x burners) and vehicle exhausts (with catalysts).
- Reduce VOC emissions from vehicle exhausts (with catalysts), fuel systems, solvent evaporation.

However, the detailed effects of any control measure depend on the balance between the precursor gases. This topic has taken a high profile in the US, where ozone concentrations have been an issue since the 1950s. Laboratory and modelled experiments on irradiated gas mixtures confirm the general form given in Figure 1.18 for the relationship between the VOC and NO_x concentrations that prevail in the morning rush hour and the peak ozone concentration that results in the early afternoon.

In the top-left part of the diagram, known as the NO_x-inhibition region, reduction of VOCs at constant NO_x reduces peak O_3 effectively. However, reduction of NO_x at constant VOC *increases* O_3. This interaction has been reported in

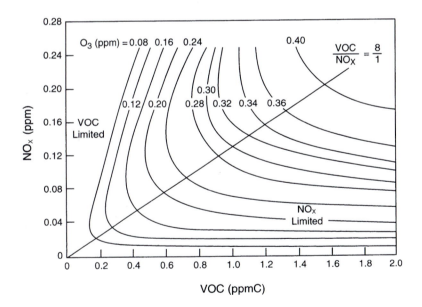

Figure 1.18 Dependence of peak O_3 concentration on the initial concentrations of VOC and NO_x.
Source: Finlayson-Pitts, B. J. and Pitts Jnr, J. N. (1993) 'Atmospheric chemistry of tropospheric ozone formation: scientific and regulatory implications', *J Air and Waste Management Assoc.* 43: 1091–1100.

measurements from Belgium, in which it was shown that weekend O_3 concentrations were higher than weekday ones because vehicle NO_x emissions were reduced. In the centre part of the diagram, known as the knee region, reductions in either VOCs or NO_x reduce peak O_3. In the lower right part, known as the VOC-saturation region, O_3 formation is insensitive to VOC concentration. The critical parameter for evaluation of the effectiveness of precursor reductions is therefore the VOC:NO_x ratio, which is about 10:1 (i.e. 10 ppb carbon per 1 ppb NO_x) in the middle of the knee region.

As an airmass moves away from an urban area, photochemical reactions and meteorological processes decrease the NO_x concentration faster than that of VOC, so that the VOC:NO_x ratio increases and the chemistry tends towards the NO_x-limited case. NO_x reduction is probably the most effective precursor control in this situation. Application of these concepts to the European atmosphere gave predictions that UK O_3 concentrations would *increase* by 6 ppb if the NO_x concentration was *reduced* by 60%, while the same level of NO_x reduction in Norway would reduce O_3 by 1 ppb. Although these ideas look quite straightforward on the ozone isopleth diagram, it is a quite different matter to derive sensible control procedures, because of the complexity of the photochemical processes in an urban airmass. There is a further complication due to biogenic VOCs, which can make up a significant proportion of the total. If, for example, the VOC:NO_x ratio falls in the knee region but the VOCs are mainly biogenic, then VOC reduction alone may be insufficient to meet target air quality standards. In California, current knowledge about the reactivity of different VOCs with OH to generate ozone has been incorporated into new legislation on vehicle exhausts (Chapter 13). Another factor which must be taken into account is the accumulation of pollutants in special meteorological circumstances. Whereas the high NO concentrations generated by urban vehicles should reduce O_3 concentrations, some Mediterranean cities such as Barcelona, Marseilles, Rome and Athens experience their highest O_3 concentrations right in the city centre. This is due to the trapping of coastal emissions in a recirculating land–sea breeze system. For example, in Athens during the summer, a special synoptic situation favours northerly winds (from the land to the sea). The weakening of these winds (Etesians) allows the development of the land–sea-breeze circulation. The sea breeze stratifies the air above the city, trapping pollutants near the ground. Furthermore, the pollutants are advected out to sea and then back over the city in a daily cycle, with the concentration building up each time.

1.9   OTHER AIR POLLUTANTS

The pollutants described above are the gaseous toxic compounds that have received the greatest attention and legislation this century. As with the composition of air discussed in Section 1.2, there are many other air pollutants that may be important in particular situations.

Fluorine is produced as gaseous HF by chemical processes such as fertiliser production and steel manufacture. It accumulates in grass and causes the bone disease fluorosis in herbivores.

Heavy metals such as arsenic, cadmium, chromium, manganese, mercury, nickel and vanadium.

Formaldehyde is an irritant chemical released from, for example, urea formaldehyde cavity wall insulation.

Polychlorinated dibenzo furans (PCDF) and *Polychlorinated dibenzo dioxins* (PCDD) are complex molecules produced by high-temperature combustion processes such as municipal waste incineration. They are carcinogenic, though their true health effects are hard to evaluate owing to low concentration.

Polychlorinated biphenyls (PCB) are released during the production and disposal of PCBs used for transformer insulation in the electrical industry. These are also carcinogenic.

Asbestos covers a range of mineral fibres released from building thermal insulation and brake linings. Inhalation of the fibres can result in potentially fatal lung disorders.

Secondary cigarette smoke is probably responsible for several hundred deaths per year in the UK.

More information is given on these substances later in this book. At least 3000 different chemicals have been identified in air samples. However, it can safely be assumed that the number of chemicals actually in the Earth's atmosphere is at least equal to the number that has ever been produced on the Earth, plus additional ones formed by subsequent reactions.

1.10   EMISSION INVENTORIES

Before we can hope to understand the budgets of different materials in the atmosphere, we have to be able to quantify the emission of each pollutant into the atmosphere: this leads to a cascade of finer and finer spatial resolution, from global totals through regional and national to local, with the finest resolutions currently $1°$ latitude $\times$ $1°$ longitude for global maps, 50×50 km (EMEP), 20 km $\times$ 20 km for regional maps and 1 km $\times$ 1 km for local maps. The general principle of emission inventories is to form the triple sum

$$\Sigma\Sigma\Sigma \ N_{s,a} \ P_{s,t} \ \Phi_{i,s}$$

where $N_{s,a}$ is the number of sources of type s in area a, $P_{s,t}$ is the production from the sources of type s in time period t (e.g. tonnes steel/year), $\Phi_{i,s}$ is an emission factor for the species i from source s (e.g. kg SO_2/tonne of steel).

There is a wide range of values of Φ depending on the conditions. For example, the emission factor for black smoke production during coal combustion ranges from 0.25 kg/tonne coal for a power station to 40 kg/tonne coal for an open domestic fire. Furthermore, there is a great range of uncertainty in the values for specific processes. For example, some industrial processes run continuously at a nearly constant rate under tightly controlled conditions, and pollutant emissions are measured and logged continuously. In this way, a substantial and reliable data base is accumulated from which an emission factor can be calculated with some confidence. At the other extreme, we may be trying to quantify particle emissions from ploughing, where there will not only be great natural variability due to soil type, windspeeds, soil moisture etc., but few measured results are also available. Depending on the activity being considered, the emission factor may involve a wide range of statistics, such as process output or sales volume in place of simple consumption. Many processes are far from continuous, and become increasingly intermittent as the time resolution is increased. For example, a power station might operate at a similar average output from year to year. But from month to month, there will be sharp variations in output due to factors such as seasonal load changes and boiler servicing. Hence the concepts of emission factors and budgets are usually applied to time-scales of a year or more. Again, it must be emphasised that there is no bank of coal consumption meters that can be read at will. Instead, an exhaustive survey is done by a group of researchers who gather, collate and analyse all the available information on a particular topic, such as sulphur emissions. The published output from that research then becomes the 'state of the art' until it is updated in due course. Furthermore, the information about emissions of a particular substance is improved retrospectively, so that, for example, the 1990 estimate of 1970 sulphur emissions might be different to (and hopefully better than) the 1970 or 1980 estimates.

In the UK, the NAEI collates emissions of over 30 major pollutants. They apply hundreds of different emission factors to fuel consumption and other national statistics to calculate totals, disaggregated breakdowns, spatial distributions and temporal trends of pollutants. Estimates are made of emissions from seven categories of stationary combustion source; mobile sources including road vehicles, off-road vehicles, aircraft and others such as trains and shipping; seven categories of industrial process; offshore oil and gas; construction; solvent use.

The UK Department for Environment, Food and Rural Affairs (DEFRA) publishes an emission factor database, so that, for example, a local authority that has compiled a list of processes and activities within its area that might generate emissions can make a first estimate of the emissions and decide whether a more detailed assessment is required. Urban inventories have also been compiled for ten of the major urban areas in the UK. These give contributions from line, area and point sources within 1×1 km squares, separated according to type of source.

The major inventory for Europe is organised by the European Environment Agency under the project name CORINAIR. Eight pollutants (SO_2, NO_x, NMVOC, CH_4, N_2O, CO_2, CO and NH_3) are covered, emitted from 30 countries by 250 activities within 11 main source categories. The time-lag involved in collecting and collating this amount of statistical data is significant, and

CORINAIR inventories are published 4–5 years after the year to which they apply. A further series of global inventories, known as GEIA (Global Emissions Inventory Activity) is being prepared within the framework of the International Global Atmospheric Chemistry project. In the US, the Department of Energy and Environmental Protection Agency compile similar inventories.

FURTHER READING

Commission of the European Communities (1991) *Handbook for Urban Air Improvement*, CEC, Brussels, Belgium.

Department of the Environment (1995) *Air Quality: Meeting the Challenge*, DoE, London, UK.

Department of Environment, Transport and the Regions (2000) *Review and Assessment: Estimating Emissions*, Local Air Quality Management, Technical Guidance 2, DETR, London.

European Environment Agency (1999) *Tropospheric Ozone in the European Union: The Consolidated Report*, EEA, Copenhagen.

European Environment Agency (2000) *Emissions of Atmospheric Pollutants in Europe, 1980–1996*, EEA, Copenhagen.

Fenger, J., Hertel, O. and Palmgren, F. (eds) (1998) *Urban Air Pollution: European Aspects*, Kluwer Academic Publishers, Dordrecht, The Netherlands.

Heumann, W. L. (1997) *Industrial Air Pollution Control Systems*, McGraw-Hill, New York, USA.

Jacob, D. J. (1999) *Introduction to Atmospheric Chemistry*, Princeton University Press, Princeton, New Jersey.

Masters, G. M. (1998) *Environmental Engineering and Science*, Prentice Hall, New Jersey, USA.

Monteith, J. L. and Unsworth, M. H. (1990) *Principles of Environmental Physics*, Edward Arnold, London, UK.

National Society for Clean Air and Environmental Protection (annual) *NSCA Pollution Handbook*, NSCA, Brighton, UK.

Seinfeld, J. H. and Pandis, S. N. (2000) *Atmospheric Chemistry and Physics*, Wiley, New York.

The Watt Committee on Energy Report No. 18 (1988) *Air Pollution, Acid Rain and the Environment*, Elsevier Applied Science Publishers, London, UK.

van Aardenne, J. A., Carmichael, G. R., Levy II, H., Streets, D. and Hordijk, L. (1999) 'Anthropogenic NO_x emissions in Asia in the period 1990–2020', *Atmospheric Environment* 33: 633–646.

van Loon, G. W. and Duffy, S. J. (2000) *Environmental Chemistry – a Global Perspective*, Oxford University Press, Oxford.

Warneck, P. (1988) *Chemistry of the Natural Atmosphere*, Academic Press, San Diego, USA.

Warren Spring Laboratory (1993) *Emissions of Volatile Organic Compounds from Stationary Sources in the UK*, DoE, London, UK.

Wellburn, A. (1994) *Air Pollution and Climate Change*, Longman, Harlow, UK.

Wells, N. (1997) *The Atmosphere and Ocean*, Wiley, Chichester.

Chapter 2

Airborne particles

Particles in the atmosphere may be either primary or secondary, solid or liquid. They come into the atmosphere, and leave it again, by a wide variety of routes. After looking at general aspects of particles such as definition of size and the way they behave in the atmosphere, we will describe the sources of particles and methods for controlling emissions. Particles less than 10 μm in diameter have very low sedimentation speeds under gravity, and may remain in the air for days before eventually being washed out by rain or impacted out onto vegetation or buildings. They are an important environmental pollutant, being responsible for loss of visual range, soiling of surfaces and health effects on people. Knowledge of their concentrations, size distributions and chemical composition is therefore needed, but for various reasons these are quite hard to obtain correctly. There is an important historical difference in the way that gases and particles have been treated as pollutants. Measurement methods for gases rapidly became specific to the gas, so that concentrations of SO_2, NO_2 etc. have been quantified. In contrast, the predominant method for particle concentration measurement has been filtration, from which mass loading is identified regardless of chemical composition.

2.1 PARTICLE TERMINOLOGY

Many different terms are used to describe the origins and characteristics of particles. They tend to be used somewhat casually, and to have different meanings in popular and scientific contexts. Here are some of them:

- *Suspended particulate matter* (SPM), and *total suspended particles* (TSP). Both mean total airborne particles; often as measured by a high-volume sampler without a size-selective inlet.
- *Particulate matter* (PM). Sometimes used in this form as an abreviation, but more commonly as PM_{10} or $PM_{2.5}$. PM_{10} is the mass concentration of particulate matter (PM) due to particles that pass through a size-selective inlet that has 50% efficiency at an aerodynamic diameter of 10 μm. $PM_{2.5}$ is the corresponding concentration for a cut diameter of 2.5 μm.

- *Fine particles.* In general, those smaller than a few μm. Sometimes used synonymously with PM$_{2.5}$.
- *Ultrafine particles* or *Nanoparticles.* Those smaller than about 0.2 μm, for which the size is normally expressed in nm.
- *Aerosol.* Any solid or liquid particles suspended in the air.
- *Grit.* The coarsest fraction of material usually formed by a mechanical process such as crushing or grinding. Size range is sometimes defined.
- *Dust.* As for grit but smaller, typically greater than 1 μm.
- *Smoke.* Particles formed by incomplete combustion, as a mixture of carbon and condensed volatiles. Usually less than 1 μm.
- *Black smoke.* Suspended particulate as determined by the reflective stain method. Size range not specified, but later measurements have shown no particles sampled above 10 μm aerodynamic diameter, and a 50% cut at about 4 μm, so the measurement would correspond to the respirable fraction.
- *ACGIH and ISO conventions.* The human breathing system has evolved to filter out large particles at an early stage, and the proportion of particles reaching the lungs depends strongly on particle size. The American Conference of Government Industrial Hygienists (ACGIH) and the International Standards Organisation (ISO) have defined particle fractions on this basis. They are discussed in more detail in Chapter 10.

2.2   PARTICLE SIZE DISTRIBUTIONS

2.2.1   Relative sizes of particles and gas molecules

Figure 2.1 shows the relative scale of air molecules (typically 0.3 nm diameter), the average separation between them (typically 3 nm, or 0.003 μm), and the average distance they move between collisions (a statistical value called the mean free path, which is about 70 nm). Figure 2.2 shows the relative scale of air molecules (now represented by dots), the space between them, and the edges of spherical particles having diameters of 100 nm (0.1 μm, typical of the peak of the number concentration in vehicle exhaust), 500 nm (0.5 μm, typical of the accumulation mode), and 5 μm. Note that single molecules are always referred to as molecules, and that the term particles does not normally include molecules unless they are in clusters.

In between single molecules and 'particles' we have an interesting no-man's-land. Below sizes of about 20 nm we lose the continuum that we take for granted, the surface of the particle starts to become associated with the surface of individual atoms and molecules, and most of the molecules that make up the particle are on its surface. At 5 nm, the particle may contain only a few hundred molecules. Such quasi-particles have extremely high Brownian diffusion velocities, and aggregate extremely fast with other small particles. Hence their short lifetimes. The size ranges of some other families of atmospheric particles are shown in Figure 2.3. The finest metallurgical fumes are condensed volatiles that are so small that they

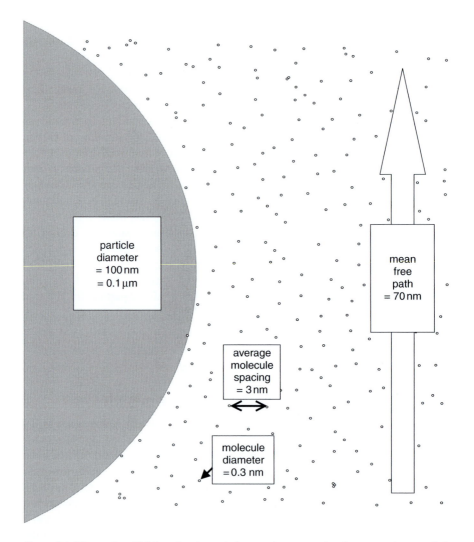

Figure 2.1 The scale of 'air' molecules relative to the separation between them and the mean free path.

overlap with the molecular size range, although these are often quickly aggregated into longer chains. Accumulation mode particles are mainly the secondary particles generated in the atmosphere as the result of photochemical reactions between primary gases. These primary gases may themselves be anthropogenic (sulphur dioxide and ammonia, for example) or natural (conversion of low-volatility hydrocarbons). Sea salt is the evaporated residue of innumerable droplets, injected into the atmosphere whenever a wave breaks at sea. The salt particles remain in the atmosphere until they eventually serve as the nuclei on which cloud droplets form.

If all the particles in a volume of gas are the same diameter, the size distribution is said to be monodisperse. Monodisperse aerosols are only found when

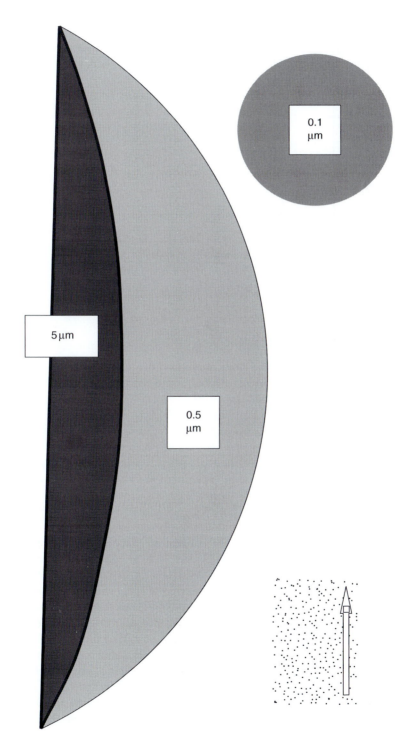

Figure 2.2 The scale of 'air' molecules and the mean free path relative to particles of different sizes.

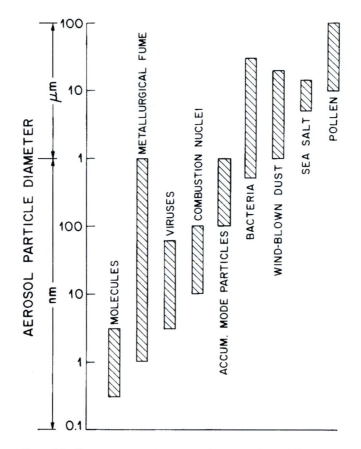

Figure 2.3 Characteristic size ranges of particles from different sources, with molecular sizes shown for comparison.
Source: Leygraf, C. and Graedel, T. (2000) *Atmospheric Corrosion*, J. Wiley and Sons, New York.

they have been made artificially for calibrating sampling equipment; even then they soon start to become polydisperse, due to collisions and agglomeration between the original particles. Atmospheric aerosol distributions are always polydisperse. Imagine that we have 'frozen' a cubic metre of air in space and time, and that we can move around within it to examine each particle in turn. At the smallest scale would be the molecules of nitrogen, oxygen, water vapour and argon, all rather less than 0.001 μm (1 nm) in diameter, that make up about 99.999% of the total 1.3 kg mass in that cubic metre. Then there would be a lower concentration of larger molecules such as organics, blending into a rapidly-increasing concentration of particles not much larger than the molecules – around 0.01 μm (10 nm) in diameter – formed by the condensation of volatilised material – for examples, iron vapour released during steel

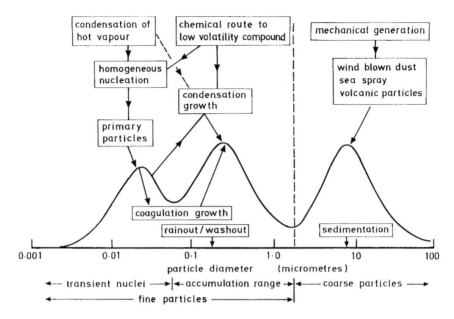

Figure 2.4 The trimodal size distribution, showing general relationships between the three common size ranges and the particle sources.
Source: Harrison, R. M. (1999) 'Measurements of concentrations of air pollutants', In: S. T. Holgate, J. M. Samet, H. S. Koren and R. L. Maynard (eds) *Air Pollution and Health*, Academic Press, London.

manufacture or volcanic eruptions, or VOCs from trees. These are clusters of molecules, and mark the start of the nucleation range, which extends from molecular diameters up to about 0.1 μm (Figure 2.4).

These very tiny particles have high thermal energies and coagulate rapidly with others, so that this size range has a high number density and short residence time. The main bulk of atmospheric aerosol mass lies between 0.1 and a few μm. There are many different types: aggregated chains of smaller particles, salt crystals (from the evaporation of ocean spray) that will go on to become condensation nuclei for cloud droplets, fly ash from power station chimneys, black carbon from inefficient fuel combustion, photochemical ammonium salts. This is the so-called accumulation diameter mode, where particle concentrations tend to increase because deposition is inefficient. Finally, there will be a smaller number of larger particles (bigger than a few μm) which will be lost soon after we unfreeze our sample because they will impact easily or sediment fast. They might include some giant sea-salt particles, mineral dust eroded from bare soil surfaces, volcanic emissions or biological material such as spores, pollen or bacteria. These three mechanisms – nucleation, accumulation and mechanical production – generate the classical trimodal size distribution.

2.2.2   Specification of size

Now imagine that we are going to measure each particle in the cubic metre, and build up a size distribution. In some cases, we will be able to catch the particles on a filter, put the filter under an optical or electron microscope, and compare the observed dimensions with a graticule or scale calibrated in μm or nm. Even if this is possible, the particle is unlikely to be spherical, and we will have to decide on a convention, such as the average of the smallest and largest dimensions. In the great majority of cases, direct measurement will not be possible, and we will have to use some other property of the particle to characterise its size in terms of an equivalent diameter. The two most common such properties are the equivalent aerodynamic diameter and the equivalent light scattering diameter. The aerodynamic diameter is the diameter of a spherical particle of density 1000 kg m^{-3} (the same as water) which is equivalent in following airflow streamlines. The light-scattering diameter is the diameter of a spherical particle of specified refractive index which scatters the equivalent amount of light. These definitions work reasonably well for some particles that are 'almost' spherical, but very poorly for others such as asbestos fibres that are very long and thin, or flat crystalline platelets. A further complication is that it is often not in fact the distribution of number with diameter that is needed. For light scattering effects, we need to know the distribution of surface area with diameter, while for inhalation studies we need the distribution of mass with diameter. This is where life really gets complicated! Particles may have very different densities and shapes (from a density of about 7000 kg m^{-3} for a long-chain iron aggregate, down to a density of 800 kg m^{-3} for spherical organic condensates) so it may not be meaningful to average any properties over the size distribution as a whole.

2.2.3   Presentation of size distributions

Figure 2.5 gives a stylised example of one common way of presenting size distributions in graphical form. The vertical axis is scaled in *dN/dlog D*, *dA/dlog D* or *dV/dlog D*, and the horizontal axis in *log D*. The range of sizes from a given sample is expressed as the normalised number distribution – the number of particles per unit size interval. Note that there has to be an interval – if we only give a single value of diameter, there will be a vanishingly small number of particles that have exactly that diameter. Hence the concentration is always expressed as μg m^{-3} μm^{-1} or equivalent. Both the number concentration and the size scales are logarithmic because of the wide ranges of the values involved – there are hundreds or thousands of particles per cm^3 in polluted tropospheric air, compared with around ten in the stratosphere. The number distribution – which we obtain simply by counting the number of particles within each diameter range as described above – is the left-most peak in Figure 2.5, showing that the number population peaks at 0.01 μm. Since a 10 μm particle has 10^6 times the surface area of a 0.01 μm particle, and 10^9 times the volume, the area and volume

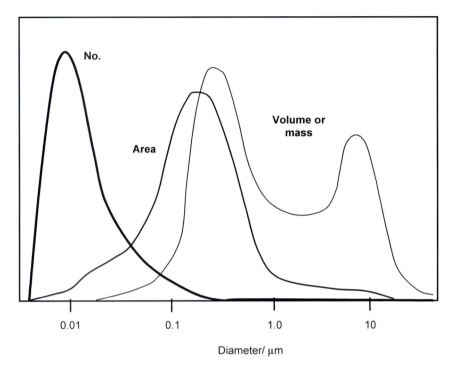

No.

Area

Volume or
mass

0.01   0.1   1.0   10

Diameter/ μm

Figure 2.5 A simple distribution of particle number with diameter, and its transformation
into surface and volume (or mass) distributions.

dependencies can give very different significance to different diameters. These
curves, for the same number distribution, are also shown on Figure 2.5. Note that
0.2 μm particles now dominate the area distribution, and for the volume distribution there is an additional peak at 8 μm that was not even visible on the number
distribution because it is due to a very small number of these larger particles. This
has implications for particle sampling, because if we really want to know the rate
of inhalation of particle mass (for health effects studies, for example), then we
must be sure to sample this tiny proportion with sufficient accuracy.

The size distributions discussed in Section 2.2.1 were in the differential form,
for which the concentration refers to the actual concentration at that size or within
that size range. An alternative widely-used presentation is the cumulative distribution. All the concentration values are accumulated, starting at the smallest
particles, so that the concentration refers to particles larger than a given diameter,
rather than at that diameter. In this formulation, it is easy to calculate, for example, the mass loading of particles smaller than 2.5 μm.

Even nominally monodisperse populations, such as are found for latex calibration aerosols, will always have a range of diameters smaller and larger than the
mean. These very tight distributions can often be represented by normal, or

Gaussian, distributions, characterised by an arithmetic mean diameter d_{mean} and a standard deviation σ, such that

$$d_{\text{mean}} = \frac{\Sigma\, n_i d_i}{\Sigma\, n_i} \quad \text{and} \quad \sigma = \left[\frac{\Sigma n_i (d_{\text{mean}} - d_i)^2}{(\Sigma n_i - 1)} \right]^{0.5}$$

where n_i is the number of particles of diameter d_i, usually approximated by the number in a size class with mean diameter d_i. This curve is symmetrical about the mean diameter. However, ambient particle distributions usually have a long tail out to larger sizes (Figure 2.6(a)), which can often be approximated by the log-normal distribution, in which it is the logs of the diameters that are normally distributed, rather than the diameters themselves (Figure 2.6(b)).

In that case we have

$$\log d_{\text{geo}} = \frac{\Sigma n_i \log d_i}{\Sigma n_i} \quad \text{and} \quad \log \sigma_{\text{geo}} = \left[\frac{\Sigma n_i (\log d_{\text{geo}} - \log d_i)^2}{(\Sigma n_i - 1)} \right]^{0.5}$$

where d_{geo} is the geometric mean diameter.

The actual distribution is then given by

$$n(\log d) = \frac{N}{(2\pi)^{1/2} \log \sigma_{\text{geo}}} \exp \left[-\frac{(\log d - \log d_{\text{geo}})^2}{2 \log^2 \sigma_{\text{geo}}} \right]$$

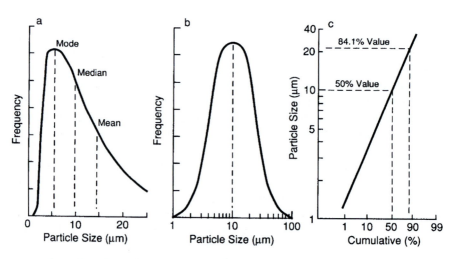

Figure 2.6 Differential and cumulative size distributions.
Source: Lippmann, M. (2000) 'Introduction and background', In: M. Lippmann (ed) *Environmental Toxicants*, Wiley, New York.

Log-normal distributions have the convenient feature that if they are plotted in the cumulative form on log-probability axes (Figure 2.6(c)), they are transformed into straight lines from which σ_{geo} can be estimated:

$$\sigma_{geo} = \frac{d_{84.13}}{d_{50}} = \frac{d_{50}}{d_{15.87}}$$

where d_n is the diameter exceeded by n% of the sample.

2.2.4   General features of real size distributions

Real atmospheric particle size distributions are the outcome of many competing processes – condensation, sedimentation, evaporation, agglomeration, impaction, gas-to-particle conversion. We can therefore expect the distributions to be very variable, and background aerosol has a wide range of characteristics, depending on the history of the airmass. There are often only two modes in the mass distribution. The nucleation mode will still be present in these distributions, but since they are given by mass not number, the peak is often not significant. The coarse mode corresponds to particles larger than about 2–3 μm in diameter, and the accumulation mode to those between 0.1 and 2.5 μm.

The coarse mode is composed mainly of primary particles generated by mechanical processes such as crushing and grinding, or by entrainment of soil particles from fields or roads. Sedimentation velocities are significant at these larger diameters, so settling occurs with a time constant of a few hours within the atmospheric boundary layer unless strong winds stir the air effectively. The accumulation mode particles, in contrast, are mainly from gas-to-particle conversion and agglomeration of ultrafine particles. Deposition velocities are low and residence times long. The ultrafine range ($d_p < 0.1\ \mu$m) may stem from local combustion sources or photochemical homogeneous nucleation. The flux through the population is rapid, and mainly due to Brownian diffusion towards, and attachment to, larger particles in the accumulation mode. The timescale for this process would range from a few minutes to a few hours, depending on the diameter of the ultrafine particles and the concentration of the accumulation particles.

In situations for which a single log-normal distribution cannot represent the real data, several log-normal distributions may be combined (Figure 2.7). The distributions shown here were used to model biomass aerosol in a calculation of radiative forcing. Three log-normal distributions, with mode radii of 0.08, 0.33 and 0.96 μm respectively, and standard deviations of around 1.4, were used. As discussed in Section 2.2.3, the number distribution (top graph) is dominated by the smallest mode, and the volume (or mass) distribution by the largest.

With increasing sophistication of imaging techniques combined with chemical analysis, we are now getting a better idea about what all these particles actually

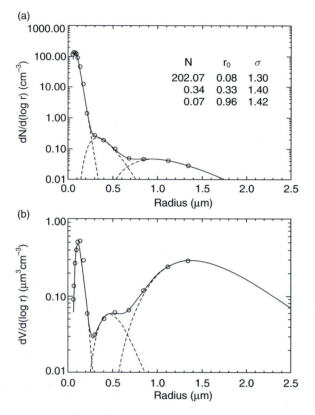

(a)

(b)

Figure 2.7 Combination of three log-normal distributions.
Source: Grant, K. E., Chuang, C. C., Grossman, A. S. and Penner, J. E. (1999) 'Modeling the spectral optical properties of ammonium sulfate and biomass burning aerosols: parameterization of relative humidity effects and model results', *Atmospheric Environment* 33: 2603–2620.

look like. Figure 2.8 shows images of various particle types collected in a sampling campaign on the North Sea coast of Germany.

2.3   AEROSOL MECHANICS

Particles are emitted into the air, or may form in the air. They spend some time in the air before being deposited from the air either naturally or artificially. In this section we shall look at some of the fundamental principles that control how particles behave under various conditions. Particles are deposited to surfaces by three main mechanisms: sedimentation, Brownian diffusion and impaction/interception. Each of the three processes operates most effectively in a different particle size range.

Figure 2.8 Scanning electron microscope images of particles: (A) Irregularly shaped sea-salt particle; (B) sea-salt particle with typical cubic morphology (1), irregularly-shaped sea-salt particle (2), and alumosilicate particle (3); (C) aged sea-salt particle; (D) biological particle (1), Si-rich fly ash particle (2), and various soot particle(diameter of primary particles below 100 nm); (E) soot agglomerate(1) with alumosilicate particles (2, 4), and carbonaceous material (3); (F) alumosilicate particle(clay material); (G) two alumosilicate fly ash particles; (H) agglomeration of iron oxide spheres; (I) calcium sulphate(presumably gypsum) particle (1), spherical iron oxide particle (2), and titanium oxide particle (3); (J) typical soot agglomerate; (K) biological particle(1) and sea-salt (2); (l) agglomerate of carbonaceous material(1, 2), alumosilicates (3), Si-rich fly ash (4), and calcium sulphate (5, 6).

Source: Ebert, M., Weinbruch, S., Hoffmann, P. and Ortner, H. M. (2000) 'Chemical characterization of North Sea aerosol particles', *J Aerosol Science* 31(5): 613–632.

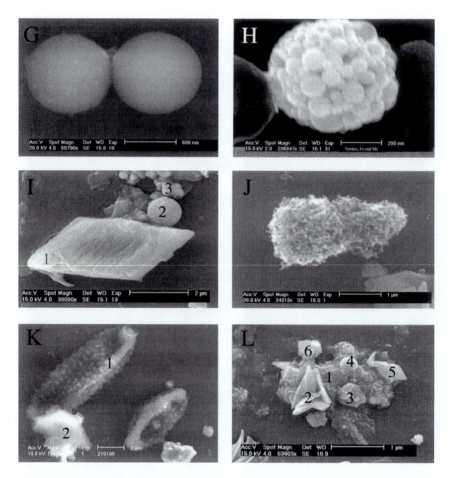

Figure 2.8 Continued.

2.3.1   Drag force

If a particle is moving through the air, or the air is moving past a particle, then a viscous drag force acts on the particle. Provided that the flow past the particle is smooth (laminar) and not turbulent, then this drag force F_D is given by

$$F_D = 3\pi\eta Vd$$

where η is the gas viscosity. For air at tropospheric temperatures and pressures, η is around 1.8×10^{-5} Pa s, V is the relative speed between the gas and the particle, d is the particle diameter. We can determine whether the flow is smooth by calculating the Reynolds number Re $= \rho_g Vd/\eta$, where ρ_g is the gas density. Flow starts to become turbulent for Re > 1.

Table 2.1 Properties of airborne particles*

Diameter/ μm	Slip correction C_C	Terminal speed V_{term}/m s^{-1}	Relaxation time τ/s	Mobility B/m N^{-1}s^{-1}	Diffusion coefficient D/m^2 s^{-1}	Coagulation coefficient K/m^3 s
0.001	224	6.75E-9	6.89E-10	1.32E+15	5.32E-6	3.11E-16
0.01	23	6.92E-8	7.05E-9	1.35E+13	5.45E-08	9.48E-16
0.1	3	8.82E-7	8.99E-8	1.72E+11	6.94E-10	7.17E-16
1.0	1.2	3.48E-5	3.54E-6	6.77E+09	2.74E-11	3.35E-16
10.0	1.02	3.06E-3	3.12E-4	5.95E+08	2.41E-12	3.00E-16
100.0	1.002	2.49E-1	3.07E-2	5.87E+07	2.37E-13	2.98E-16

Note
* At 20 °C and 1 atm, for particle density 1000 kg m^{-3} (e.g. water).

The drag equation assumes that the particle is large compared to the spaces between the molecules in the gas. This latter distance, known as the mean free path λ, is about 0.07 μm for standard atmospheric temperature and pressure. As particle diameter decreases, there becomes an increased probability that the particle will move a distance without experiencing a collision, effectively slipping through the gaps between the molecules. Then we must add the Cunningham slip correction factor C_C where

$$C_C = 1 + \frac{2.52\lambda}{d}$$

for

$$d > 0.1 \ \mu m$$

and

$$C_C = 1 + \left[2.34 + 1.05 \exp\left(-0.39\frac{d}{\lambda} \right) \right] \text{ otherwise.}$$

Table 2.1 shows that this slip correction is worth 2% for a 10 μm particle and 20% for a 1 μm particle, so it is an important factor in the design of cascade impactors that typically have several stages operating within that size range.

2.3.2 Sedimentation

Every particle in the Earth's gravitational field experiences a force due to gravity towards the centre of mass of the Earth. If, for example, a stationary particle is released into the atmosphere, the gravitational force will accelerate it downwards. As its speed increases, the drag force described in Section 2.3.1 will increase. The gravitational force acts downward, but the drag force acts upward. The net force, which is the difference between the two, therefore decreases as the particle accelerates. Since the gravitational force remains constant, a speed will eventually be reached at which the two forces are equal and opposite and no more acceleration can take place. This is called the terminal or sedimentation speed. Particles that fall out of the air under the action of gravity do so at their terminal speed.

$$F_{\text{grav}} = mg = \rho_{\text{part}} \frac{\pi d^3}{6} g \text{ (for a spherical particle).}$$

When $F_{\text{drag}} = F_{\text{grav}}$, we have

$$3C_C \pi \eta V_{\text{term}} D_{\text{part}} = \rho_{\text{part}} \frac{\pi d^3}{6} g.$$

Hence

$$V_{\text{term}} = \frac{\rho_{\text{part}} d^2 g C_C}{18 \eta}, \text{ which is Stokes' Law.}$$

Values of V_{term} are given in Table 2.1. The other columns in Table 2.1 refer to particle properties that are discussed in Section 2.3.

Figure 2.9 shows the dependence of this terminal speed on particle radius, not only for the particles discussed above ($r < 30 \ \mu\text{m}$), but also for larger particles such as cloud droplets and raindrops. The curve is linear up to a particle diameter of 50 μm and a Reynolds number (the ratio of inertial to viscous forces) of 1 – this is the region in which Stokes' law is valid. For larger particles, the curve starts to depart from linear as Stokes' law starts to fail. Large raindrops become non-spherical due to the gravitational and fluid forces acting on them, and depart a little from this curve (dashed line). If the particle is non-spherical and of density different to water, a Stokes diameter can be assigned to it which is the diameter of a sphere having the same density and sedimentation velocity.

2.3.3   Brownian diffusion

All the particles discussed here are very much larger than atmospheric molecules such as nitrogen and oxygen, and are continually bombarded all over their surfaces by such molecules. For very small particles, two effects become important. First, the rate of collisions with molecules becomes small enough that there is a reasonable probability of the collisions from one direction not being instantaneously balanced by collisions from the opposite direction. Second, the particle mass becomes small enough that the resultant momentum of these colliding molecules can exert a significant impulse in the unbalanced

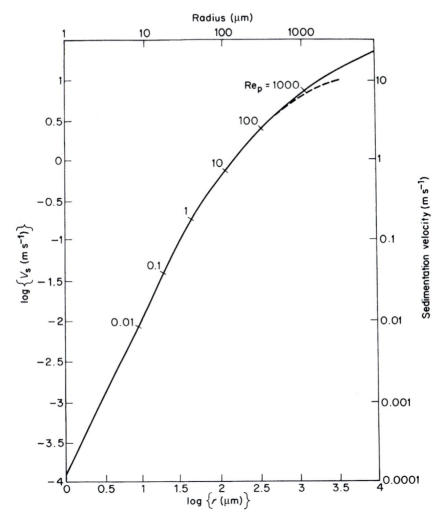

Figure 2.9 The variation of terminal speed with radius for spherical particles with density 1 g cm^{-1} falling in air.
Source: Monteith, J. L. and Unsworth, M. H. (1990) *Principles of Environmental Physics*, Edward Arnold, London.

Table 2.2 Ratio of Brownian and gravitational displacements in 1 s

Particle diameter/μm	Displacement ratio
0.01	4800
0.1	42
1.0	0.21
10	7×10^{-4}

direction. The resultant erratic motion of the particle gives it a mean square displacement

$$\overline{x^2} = 2Dt$$

where D is the Brownian diffusion coefficient for the particle ($m^2 \, s^{-1}$) given by

$$D = \frac{kT}{6\pi\mu r}$$

where k is Boltzmann's constant, T is the absolute temperature and μ is the dynamic viscosity of air. Furthermore, the effective value of μ itself falls when the particle diameter is significantly less than the mean free path of the air molecules (about 70 nm at sea-level temperature and pressure). As a result, $D \propto r^{-1}$ when $r \gg 70$ nm, and $D \propto r^{-2}$ for $r \ll 70$ nm.

Hence although the individual steps in Brownian displacement involve an erratic random walk, the statistical average is a well-behaved function of time and particle radius. We can see from Table 2.2 that Brownian motion is really important compared to gravity for particles of 0.01 μm or even 0.1 μm, but trivial for 10 μm particles.

This aspect of particle behaviour is also significant when particles have to be sampled through pipes before measurement. Even if the designer takes meticulous care to avoid impaction of large particles by bends and fittings, small particles will diffuse to the walls of the pipes and distort the size distribution.

2.3.4   Coagulation

When there is relative movement of particles within a cluster by Brownian or turbulent diffusion, there will be collisions between particles and this will sometimes result in their combination to form a single new larger particle. This process is called coagulation. The new particle may be spherical if the original particles were liquid, as with sulphuric acid aerosol, but it is more likely to be an untidy asymmetric clump of discrete parts, held together by electrostatic and molecular forces. As we would expect, coagulation is fastest for smaller particles, and the denser the particle cloud, the more collisions and coagulation there will be. The basic rate equation is

$$\frac{dN}{dt} = -K_0 N^2$$

where

$$K_0 = 4\pi d_{\text{part}} D = \frac{4kTC_C}{3\eta}$$

Table 2.3 Concentration halving times and diameter doubling times for various initial concentrations. $K_0 = 5 \times 10^{-16}$ m^3 s^{-1}

Initial concentration N_0/cm^{-3}	Time to reach 0.5 N_0	Time for particle diameter to double
10^{14}	20 µs	140 µs
10^{12}	2 ms	14 ms
10^{10}	0.2 s	1.4 s
10^{8}	20 s	140 s
10^{6}	33 min	4 h
10^{4}	55 h	16 days
10^{2}	231 days	4 years

and N is the number concentration of particles. For air at ambient temperatures, $K0 \approx 3 \times 10^{-16} C_c$ m^3 s^{-1}.

For a monodisperse particle cloud with initial concentration N_0, the number concentration decreases with time according to

$$N(t) = \frac{N_0}{1 + N_0 K_0 t}$$

The particle diameter increases correspondingly, although at a slower rate because it takes eight original particles to produce one new particle of double the diameter (Table 2.3).

$$d(t) = d_0(1 + N_0 K_0 t)^{1/3}$$

2.3.5   The influence of shape and density

All the values given in Section 2.3.2 refer to spherical particles having the same density as water (1000 kg m^{-3}). Although a significant proportion of particles in the atmosphere have these characteristics, many do not, and this will affect their behaviour in the air.

2.3.5.1   Shape

Any non-spherical particle can be given an equivalent diameter d_{equ}, which is simply the diameter of a sphere that has the same volume and density as the particle. It can then be assigned a dynamic shape factor F (Table 2.4), which corrects for the influence of shape on viscous drag.

$$F_D = F \cdot 3\pi\eta V d_{equ}$$

Table 2.4 Dynamic shape factors

Shape	Dynamic shape factor, F	Ratio of length to diameter		
		2	5	10
Sphere	1.00			
Cube	1.08			
Cylinder				
Axis vertical		1.01	1.06	1.20
Axis horizontal		1.14	1.34	1.58
Cluster of three spheres	1.15			
Dusts	1.05–1.88			

so that

$$V_{\text{term}} = \frac{\rho_{\text{part}} d^2 g C_C}{18 \eta F}$$

2.3.5.2   *Aerodynamic diameter*

The aerodynamic diameter d_a is an equivalent diameter which is used to compare the dynamic properties of particles of different shapes and/or densities. It is the diameter of the spherical particle of density 1000 kg m^{-3} that has the same terminal speed as the particle in question.
i.e.

$$V_{\text{term}} = \frac{\rho_0 d_a^2 g}{18 \eta}$$

If the particle is spherical, and neglecting slip, then

$$d_a = \sqrt{\frac{\rho_{\text{part}}}{1000}} \, d_{\text{part}}$$

For example, a spherical particle of density 4000 kg m^{-3} and diameter 2 μm would have an aerodynamic diameter of 4 μm.

For a non-spherical particle of actual diameter d_{part}, density ρ_{part}, dynamic shape factor F, Cunningham slip factor C_{part} and flow Reynolds number Re_{part}, the aerodynamic diameter is calculated as:

$$d_a = \left[\frac{\rho_{\text{part}} C_{\text{part}} \text{Re}_{\text{water}}}{\rho_{\text{water}} C_{\text{water}} \text{Re}_{\text{part}} F} \right] d_{\text{part}}$$

2.3.6   Relaxation time

When a particle experiences a change in the forces acting on it, it does not instantaneously switch to the new equilibrium condition, but approaches it exponentially, because the force causing the acceleration gradually decreases to zero as the equilibrium is approached. In general, the speed $V(t)$ at any time t is given by

$$V(t) = V_{\text{final}} - (V_{\text{final}} - V_{\text{initial}})e^{\frac{-t}{\tau}}$$

where V_{final} and V_{initial} are the final and initial speeds, respectively, and τ is the time constant of the process, also called the relaxation time. τ can be thought of as the characteristic timescale on which the particle adjusts to the new conditions, and is given by

$$\tau = \frac{\rho_{\text{part}} d_{\text{part}}^2 C_C}{18\eta}$$

Note on units and dimensions
It is worth checking that an expression like the one above really does give a time in units that everyone can understand. The symbols on the righthand side of the equation have the following units

ρ_{part} density – kg m^{-3}

d_{part} diameter – m

C_C dimensionless factor

η viscosity – $\text{N s m}^{-2} = \text{kg m}^{-1}\text{ s}^{-1}$.

If we combine these we get

$$\frac{(\text{kg m}^{-3})\text{m}^2}{\text{kg m}^{-1}\text{ s}^{-1}}$$ which yields *seconds* after cancellation.

For the situation described above, when a particle is accelerating from rest to its terminal speed under gravity, we have

$$V(t) = V_{\text{term}} - (V_{\text{term}} - 0)e^{\frac{-t}{\tau}} = V_{\text{term}}\left(1 - e^{\frac{-t}{\tau}}\right)$$

Table 2.5 Times for particles to reach 95% of their terminal velocity

Particle diameter/μm	Time to achieve 0.95 V_{term}/ms
0.01	0.00002
0.1	0.00027
1.0	0.011
10.0	0.94
100.0	92

At $t = 1\tau$, the particle will have achieved $1-(1/e) = 0.63$ of V_{term}, and at $t = 3\tau$, $V(t) = 0.95\ V_{term}$. Hence after a time of any more than 3τ has elapsed, the particle will effectively have reached its terminal speed. This happens extremely quickly for all the particles that we shall be dealing with in this book, as shown in Table 2.5. Even a 100 μm particle achieves its terminal speed in less than 100 ms.

2.3.7   Stopping distance

Whereas the relaxation time can be thought of as the characteristic timescale over which changes in particle motion occur, the stopping distance S corresponds to the characteristic length scale over which the particle decelerates to rest after it has been projected into still air at an initial speed V_0 (Table 2.6). It is related to relaxation time very simply:

$$S = V_0\tau$$

As with relaxation time, the concept can be useful to gain a qualitative feel for a particle's behaviour under different conditions. For example, the smallest scale of atmospheric turbulence is around 10 mm. Since most airborne particles are <10 μm, and the stopping distance for 10 μm particles is only a few mm, they will normally follow the flow fluctuations very closely.

Table 2.6 Stopping distances for spherical particles

Particle diameter/μm	Stopping distance (mm) from an initial speed $V_0 = 10$ m s^{-1}	Time to travel 0.95 of stopping distance/s
0.01	7.0×10^{-5}	2.0×10^{-8}
0.1	9.0×10^{-4}	2.7×10^{-7}
1.0	0.035	1.1×10^{-5}
10.0	2.3	8.5×10^{-4}
100.0	127	0.065

2.3.8 Impaction and interception

When air blows round an obstacle, the streamlines curve (Figure 2.10). The trajectory of each particle is then determined by the balance between inertial forces
(which tend to keep the particle on a straight line path), pressure gradient forces
(which tend to keep it following the stream lines) and viscous forces (which act
to slow down movements across the stream lines). The smallest particles have
negligible inertia, and will follow the streamlines as well as the gas molecules

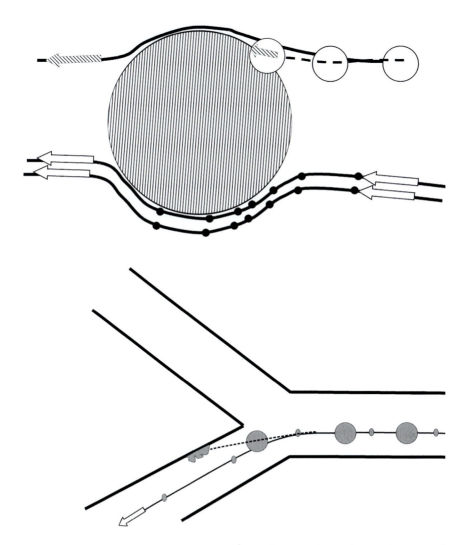

Figure 2.10 Streamlines of air flowing round a cylinder and a bend, showing how particles
are deposited from streamlines onto the surface.

themselves do. The larger a particle is, the more its inertia will tend to take it across the streamlines to be deposited on the object. This is called impaction, and is shown happening to the particle on the upper streamline of Figure 2.10. The same phenomenon applies when particles are deviated by bends in pipes – in the respiratory system, for example – as is shown in the lower diagram of Figure 2.10. For any given combination of obstruction, particle population and airspeed we can expect some of the particles to impact and some to flow round. Hence we can define the efficiency of impaction, c_p, by

$$c_p = \frac{\text{number of collisions on the obstacle}}{\begin{array}{c}\text{no. of particles that would have passed through}\\ \text{the space occupied by the obstacle if it hadn't been there}\end{array}}$$

Hence for a cylinder of length L and diameter D, oriented normal to an airflow of speed U and particle concentration C

$$C_p = \frac{\text{no. of collisions}}{\text{ULDC}}$$

The probability of a particle impacting onto an obstruction is indicated by the ratio of its stopping distance to the radius R of the obstruction. This special parameter is called the Stokes number. The curves marked A and B on Figure 2.11

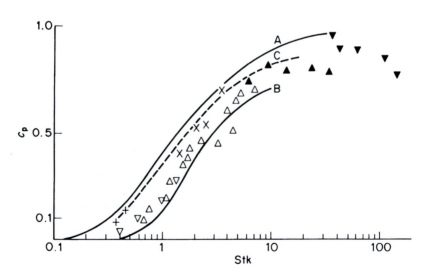

Figure 2.11 Variation of collection efficiency with Stokes' number.
Source: Monteith, J. L. and Unsworth, M. H. (1990) *Principles of Environmental Physics*, Edward Arnold, London.

show the theoretical variation of impaction efficiency with Stokes No. (Stk) for Re > 100 and Re = 10 respectively.

The dashed curve C is from experimental results with droplets, while the individual points are from measurements with spores. One visible example of these relationships is the collection of snow flakes on different parts of trees. The air flow round trunks and large branches has low curvature, so that snowflake accelerations are small and they remain on the streamlines – deposition to the tree is small. For small branches and twigs, however, the local curvature is high, accelerations are large and deposition to the tree much greater.

We can gain a semi-quantitative feel for the likelihood that impaction will occur from the Stokes number of the particle/obstacle combination. For example, a 1 μm particle in a 2 m s^{-1} wind has a stopping distance of 25 μm. Hence impaction will only occur on obstacles smaller than about 80 μm across. If the main obstacle is a leaf, the particle will avoid it. However, some leaves are covered in thin hairs, and the particle can be impacted on these. As the windspeed increases, particles may tend to bounce off even after they have been impacted.

Interception is a special case of impaction which applies when a particle remains on its streamline, but nevertheless collides with the obstruction because it passes so close that physical contact is unavoidable. Interception efficiency increases with r/R, and this relationship has been used very effectively to sample small fog droplets with very fine threads such as spiders' webs, for which R ≪ r.

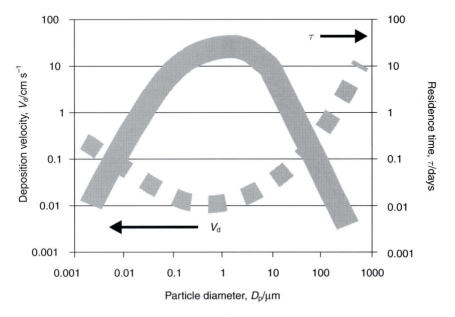

Figure 2.12 Variation of deposition velocity and residence time with particle diameter.

In natural situations, all of the deposition processes can operate simultaneously; the importance of each one is determined by the relationship of factors such as particle size, object size and windspeed. Figure 2.12 shows schematically the variation of particle deposition velocity and residence time with diameter. For particles less than about 0.01 μm, Brownian diffusion is dominant. For particles larger than about 1 μm, sedimentation becomes important. In the range 0.01–1 μm, neither of those processes is important, and experimental results confirm that deposition is slow. This is just the size range of pollutant aerosol such as ammonium sulphate and ammonium nitrate, and accounts for the long residence times that allow them to build up significant haze until washed out by rainfall. The concentration of particles in the atmosphere at any place and time is the net result of production and deposition; as discussed in Section 2.2, these processes often lead to a tri-modal size distribution.

2.3.9   Stokes number and impactors

The Stokes number Stk is also useful when considering the effect of a jet, for example, on particles passing through it. Stk is simply the ratio of the particle stopping distance S (at the jet throat speed, in this case) to the characteristic diameter of the jet.

$$Stk = S/d_c = \tau U_0/d_c$$

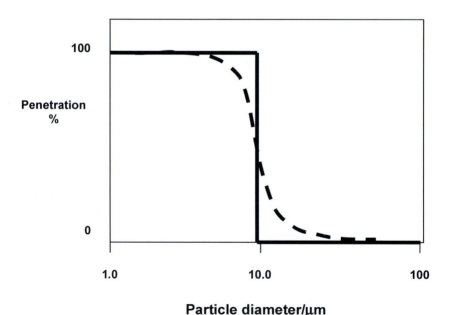

Figure 2.13 Ideal variation of collection efficiency with particle diameter.

All combinations of jet diameter, air speed and particle aerodynamic diameter can then be summarised by one graph of collection efficiency against Stk.

An ideal impactor would have the step-function variation of collection efficiency shown by the solid line in Figure 2.13. For values of Stk less than some critical value, all particles would follow the streamlines through the impactor without being collected. For greater Stk values, all particles would be collected. Life is never this neat, and in practice there is always a smoothing of the step function as indicated by the dashed line, so that some 'undersized' particles get collected and some 'oversize' ones get through. It turns out that for round jets, this critical value of Stk is 0.5. At this value of Stk, 50% of the particles are collected and 50% go through, so it is called Stk_{50}. The particle diameter corresponding to this value of Stk is called d_{50}, and is the 'cut diameter' of the jet under those conditions. Note that the cut diameter is only fixed if the density, pressure and flow rate of the air are fixed, and will have to be recalculated if there are changes in these parameters.

$$Stk = \frac{\tau U}{d_{jet}/2} = \frac{\rho_{part} d_{part}^2 U C_C}{9\eta d_{jet}}$$

At the cut diameter, $Stk = Stk_{50}$ and $d_{part} = d_{50}$
then

$$d_{50} C_C^{1/2} = \left(\frac{9\eta d_{jet} Stk_{50}}{\rho_{part} U} \right)^{1/2}$$

or, since the volume flow rate Q through the jet $= \pi d_{jet}^2 U/4$

$$d_{50} C_C^{1/2} = \left(\frac{9\pi \eta d_{jet}^3 Stk_{50}}{4\rho_{part} Q} \right)^{1/2}$$

2.3.10 Thermophoresis

When particles are in a temperature gradient in a gas, there will be more energetic collisions from gas molecules on the high temperature side of the particle than there are from molecules on the low temperature side. The result is a net force acting on the particle in the direction in which temperature is falling. For most natural particles in naturally-occurring temperature gradients, the effect is insignificant. In some situations, however, it can be very important. When particles are being formed in vehicle exhaust gases, high temperature gradients exist between the hot gas and the cool walls. The particles are also very small – typically a few nm. Thermophoretic deposition may then be a significant factor.

2.3.11   Erosion and resuspension

The blowing wind moves soil and other surface deposits around and can lift them into the atmosphere and transport them great distances. The process is known as deflation and can contribute to:

- sandstorms, desert and dune formation
- erosion of agricultural soils
- geomorphology of wind-blown deposits such as loess
- atmospheric particle burden
- loss of visibility
- ocean sediments
- 'red rain' events.

Three stages are involved – suspension from the surface into the air, transport and deposition.

If the particle happens to protude above the general ground level, then it will be subject to a drag force from the wind, and can be lifted off if the instantaneous shear stress is greater than the gravitational force. If the particle is too small (<100 μm) then it remains within the viscous sub-layer where the flow is aerodynamically smooth. Also, adhesive forces start to add significantly to the gravitational forces.

It may not be obvious how a wind flowing parallel to the surface can generate a lifting force. One common mechanism is saltation (from the Latin for jump or leap). Once particles such as sand grains have been set in motion, they strike other grains on the surface. Sometimes they will bounce up into the air flow after such a collision. The particles' relaxation time is comparable to the time they take to return to the surface at their sedimentation speed, so that they accelerate to the windspeed at that height (which might be several m s^{-1} only a few cm above the surface). This gives them momentum for the next collision, and so on. Theory and experiment show that the rate of saltation depends on the cube of the windspeed. Saltating grains will usually be in the 50–300 μm diameter range, moving amongst larger grains, which either remain stationary or roll along, and amongst smaller ones. When the smaller ones are ejected by the saltating grains, they become airborne and leave the area, possibly to form loess soils or red rain or ocean sediment thousands of km downwind. In addition, the sand blasting effect of the windborne grains on larger rocks can create a new supply of smaller grains for removal. Although the saltation process is most easily visualised for a dry sand region such as a desert or beach, it will contribute to soil erosion for all soils, depending on the fetch upwind, the size distribution of the soil particles, the moisture content and the extent of vegetation.

2.4 PARTICLE SOURCES

2.4.1 Wind erosion

Much of the primary natural aerosol is generated from the erosion of soil by the wind. The processes described in Section 2.3.11 can generate highly episodic erosion fluxes, concentrated into short periods when strong winds over bare soil follow extended dry periods. It will be seen from Figure 2.9 that typical sedimentation speeds are about 0.3 cm s^{-1} for a 10 μm particle and 30 cm s^{-1} for a 100 μm particle (the size of the smallest drizzle droplets). Typical updraft speeds in stormy conditions can be metres s^{-1}, so that particles can remain suspended for significant periods and travel long distances before deposition. The principal sources are in the northern hemisphere, where the Saudi Arabian peninsular, the US south west, the Gobi and the Sahara are the major contributors to a total entrainment of 1500 Mt each year. This process is going on all over the world all the time, but is not noticed until revealed by a particular event. About 100 Mt of dust is blown off the Sahara each year, and a significant proportion of the rain events in southern Europe involve 'red rain'. General wind-entrained soil particles tend to fall into two distinct size ranges. The coarser fraction is mainly quartz grains of 10–200 μm which tend to sediment out close to the source, while clay particles are in the range <10 μm and can get transported further. Airborne measurements in a plume from the Sahara have shown that the mass median diameter was over 50 μm close to the source, and that this fell steadily with distance until it was only 2–3 μm after 5000 km. These particles will tend to be the most common crustal oxides of iron, aluminium, silicon and titanium. The chemical composition can be significant – it has been estimated that the 1.4 Mt of calcium contained in the 24 Mt of dust eroded from European soils each year would be sufficient to neutralise 10% of the annual SO$_2$ emission. The flux can be huge – measurements have shown source strengths of up to 10^{-6} cm^3 of particles per cm^2 of soil per second, equivalent to around 1 mm day^{-1}. On a larger scale and longer term, a belt of loess (wind-blown deposits from adjacent areas) up to a hundred metres deep runs from France to China.

2.4.2 Sea salt

Another major natural source of airborne particles is the evaporation of droplets of sea water. Every time that a wave breaks or a bubble bursts in the world's oceans, many droplets are flung into the air. For those that remain suspended, the water will evaporate away and leave a salt crystal. Table 2.7 shows that around 3 billion tonnes of sea salt particles are generated in this way each year. The mass median diameter of these salt particles near the sea surface is around 8 μm. Their chemical composition closely tracks that of the dissolved species in the surface seawater, i.e. 55% Cl^-, 31% Na^+, 7.7% SO_4^{2-}, 3.7% Mg^{2+}, 1.2% Ca^{2+} and 1.1% K^+. These ions exist mainly as the salts NaCl, KCl, $CaSO_4$ and Na_2SO_4.

2.4.3   Other natural sources

Further primary natural sources include volcanoes, forest fires, spores and pollen. Natural processes also operate to convert gases into particles. The conversion may be physical – for example, the condensation of terpenes and pinenes in forested areas, or chemical – the production of sulphuric acid aerosol from sulphur dioxide, dimethylsulphide (CH_3SCH_3) and hydrogen sulphide. Even in the stratosphere, sulphuric acid aerosol is formed via the photodissociation of carbonyl sulphide (COS) into CO and S.

2.4.4   Anthropogenic sources

Human sources of particles are also legion. All activities such as combustion, melting, grinding, crushing, ploughing or spraying produce particles; a variable proportion of these will be fine enough to remain suspended in the air, depending on the initial size distribution. Human biomass burning has made up 95% of all biomass combustion over the most recent several decades, and generates a variable mixture of soot, sulphates, nitrates and hydrocarbons. Small diesel engines

Table 2.7 Global particle production

Source	Production rate/ 10^6 t a^{-1}
Natural	
Primary	
Sea salt	3000
Mineral dust	1500
Volcanic dust	300
Forest fires	100
Total natural primary	4900
Secondary	
Sulphates	40
Nitrates	30
Hydrocarbons	20
Total natural secondary	90
Total Natural	5000
Anthropogenic	
Primary	200
Secondary	300
Total man-made	500
Grand Total	5500

(cars and vans) emit around 5 g particles per litre of fuel, while this figure rises to around 12 g for heavier duty vehicles. In contrast, petrol-engined vehicles without catalysts emit only around 0.5 g. The effects of these emissions, of course, depend also on the size distribution. Secondary human sources include sulphate and nitrate particles converted from the primary gas emissions, and organics from VOC emissions.

Hence of the total 5.5×10^9 t produced, 90% is of natural origin. Nevertheless, anthropogenic particles have had a profound effect on the atmosphere, both because their production is concentrated in urban and heavily inhabited areas, and because they are mostly formed as small secondary particles with long atmospheric residence times.

In the UK, the balance between different source types is as given in Table 2.8. Road transport is the largest single contributor, closely followed by production processes. However, the total from stationary combustion (both for electricity generation and other applications such as space heating) made up 48% of the total. Although coal is thought of as a larger source of particle emissions than oil, the two have similar proportions overall due to the high total contribution from motor vehicles.

Details for other individual processes are available from the UK Environment Agency, which compile a Chemical Release Inventory (now relaunched as the Pollution Inventory) for major processes that come within Part A of the Environmental Protection Act (1990). The inventory includes estimates of total particulate matter (not PM_{10}), the ten largest of which are shown in Table 2.9 for 1998, when the total of such emissions was 19 800 tonnes. These figures do not include some thousands of much smaller Part B processes which are controlled by Local Authorities.

The particle form that has received the most attention from the public and legislators is black smoke – the 'soot' associated with incomplete combustion of fossil fuels. Figure 2.14 graphs the decline in black smoke emissions in the UK between 1970 and 1999. The dominant source had been domestic coal combustion, which decreased steadily as gas-fired central heating replaced the domestic hearth and smoke-control regions were established in urban areas. In 1989 the new smoke source – diesel road vehicles – equalled domestic emissions in magnitude for the first time, and is now the main source of black smoke, although that is in decline.

2.5   REDUCTION OF PRIMARY PARTICLE EMISSIONS

There are many competing techniques for control of particle emissions. They all have one common principle – the application of a force to cause the deposition of particles to a receiving surface. The techniques can be classified into sedimentation, filtration, centrifugation, washing and electrostatic precipitation. Each of these methods is worth a book in itself, and only the barest details will be given

Table 2.8 PM_{10} emissions from UK sources, 1970–1998 ht year^{-1}

BY UN/ECE category	1970	1980	1990	1998	1998 (%)
Combustion in energy production	88	84	76	31	19
Public power	67	76	70	23	14
Petroleum refining plants	5	5	3	4	2
Other comb. &Trans.	16	3	3	4	3
Combustion in Comm/Inst/Res	237	109	56	31	19
Residential Plant	216	98	48	26	16
Comm/Pub/Agri Comb.	21	11	8	5	3
Combustion in Industry	74	31	25	17	10
Iron and steel combustion	8	2	1	1	0
Other Ind. Comb.	66	29	24	16	10
Production processes	42	34	44	38	23
Processes in industry	12	10	11	11	6
Quarrying	22	21	29	23	14
Construction	8	3	4	4	2
Road transport	46	56	67	40	25
Combustion	44	53	63	35	22
Brake and tyre wear	2	3	4	5	3
Other transport and machinery	6	5	5	4	3
Waste	1	3	2	1	1
Land use change	1	1	1	0	0
Fuel type					
Solid	350	189	128	52	32
Petroleum	86	79	79	47	29
Gas	4	6	9	12	7
Non-fuel	54	48	60	52	32
Total	495	322	276	163	100

Table 2.9 UK particle emissions from different processes

Process	Particle emission/tonnes a^{-1}
Acids	8473
Cement and lime	4456
Incineration	2031
Petroleum processing	1614
Ceramics	723
Carbonisation	626
Mineral fibres	536
Organic chemicals	511
Inorganic chemicals	347
Fertiliser	132

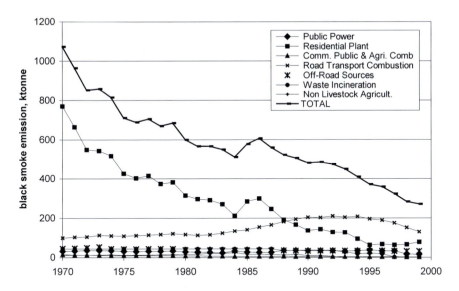

Figure 2.14 UK black smoke emissions by type 1970–1999.

here. All particle collectors have a collection efficiency $E(D_p)$, for particles of diameter D_p, given by

$$E(D_p) = 1 - \left[\frac{\text{concentration of particles of diameter } D_p \text{ in the exhaust gas}}{\text{concentration of particles of diameter } D_p \text{ in the inlet gas}} \right].$$

The term in square brackets is also known as the particle penetration. For most collectors, the efficiency decreases as the diameter decreases. Hence the particle diameter for which the efficiency is 0.5, called the cut diameter, is often used to parameterise and compare collector efficiencies. Any collector will also have an overall efficiency E for a given inlet particle diameter distribution, but if the process that is producing the particles changes then the diameter distribution and overall efficiency may change too. Schematic diagrams of the different types of collector, except for settling chambers, are given in Figure 2.15. These collecting devices are discussed in more detail in the following sections.

2.5.1 Sedimentation

The gas is passed horizontally through a large-volume chamber, so that the average speed is low and the residence time long. Particles which are large enough (typically >20 μm) to sediment under gravity fall directly into hoppers.

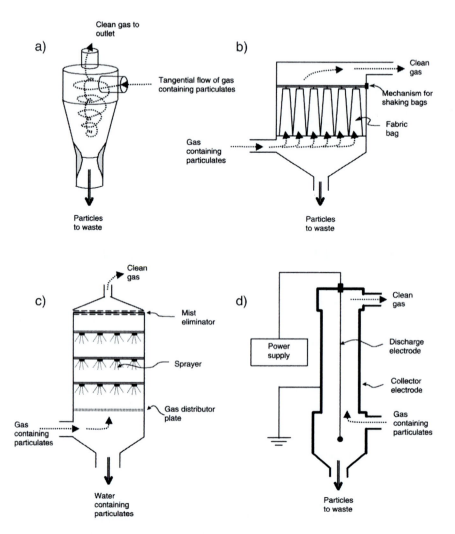

Figure 2.15 Schematic diagrams of different particle collectors. a) Cyclone, b) filter, c) spray tower, d) precipitator.
Source: van Loon, G. W. and Duffy, S. J. (2000) *Environmental Chemistry – A Global Perspective*, Oxford University Press, Oxford.

$$E(D_p) = 1 - \exp\left[-\frac{u_t(D_p)L}{uH} \right]$$

where $u_t(D_p)$ is the terminal speed of particles of diameter D_p in still air, u is the horizontal gas speed through the settling chamber, L, H are the length and depth of the chamber respectively.

For laminar flow, the terminal speed is given by

$$u_t = \frac{\rho_p g D_p^2 C_C}{18\mu}$$

where ρ_p is the particle density, g the acceleration due to gravity, and μ the dynamic viscosity of the gas.
Hence

$$E_c(D_p) = 1 - \exp\left[-\frac{g\rho_p L D_p^2}{18\mu u H}\right]$$

2.5.2 Centrifugation

For particles less than 20 μm diameter, gravitational force is not strong enough to deposit them within a practical time. The force can be multiplied by passing the air through a cyclone; a spiral path within a cone generates centripetal force and moves the particles to the wall of the cone, where they slide down to the vertex. The cleaned air leaves the cyclone from the central core to the cone (Figure 2.15(a)). Cyclones are cheap, reliable and straightforward gas-cleaning devices with a wide range of applications. They are not only used for air pollution control, but for process control (e.g. capture and return of entrained fluidised-bed materials), as pre-cleaners for other air pollution control equipment, and as separators for liquid droplets after a wet scrubber or similar. High-efficiency cyclones offer an efficiency of over 90% for aerodynamic diameters of larger than 2 μm. The simplest theoretical model, which assumes that the flow within the cyclone is everywhere laminar, gives

$$D_{50} = \left[\frac{9\mu W}{2\pi N_e \rho_p V_g}\right]^{\frac{1}{2}}$$

where
W = width of entry duct, N_e = effective number of turns of gas on spiral path, V_g = gas velocity at entry.

More sophisticated models allow for turbulent flow within the cyclone and the finite depth of the boundary layer at the walls. Although cyclones could in principle be used for removing particles less than 2 μm, the high velocities required to generate the centripetal forces increase the pressure drop across the cyclone, which in turn increases the energy used and operating cost.

2.5.3 Filtration

Fabric filters – known as baghouses in the USA – work just like domestic vacuum cleaners. The particle-laden air is drawn through a permeable cloth, and the tortuous path of the particles through the fibres (Figure 2.16) results in their deposition. The collection mechanisms for filters include sieving, impaction, interception and diffusion.

As a layer of particles builds up, the collected particles themselves act as the deposition sites for new particles. The accumulating layer also increases the pressure drop required to draw the air through, which reduces the gas flow rate. In the most basic systems, such as those used in heating ventilation and air-conditioning (HVAC), the used filter is disposed of and a new one fitted. On industrial systems (Figure 2.15(b)), the flow is stopped intermittently and the layer of particles shaken or blown off into hoppers. The three main media types that are used for industrial gas cleaning are needle-punched felt, woven synthetic or cotton, and filter paper. Fabric filters offer the highest efficiency for collection of small particles, and the efficiency increases as diameter decreases. Efficiencies of 99.9% at 0.1 μm are commonly reported.

2.5.4   Washing

Water droplets can be used to collect particles. The droplets are sprayed into the gas stream so that there are high relative speeds between the particles and the droplets and the impaction efficiency is high. The droplets themselves are

Figure 2.16 The surface of a fibrous filter.
Source: JIT Stenhouse (1998) 'Fibrous filtration', In: I. Colbeck (ed) *Physical and Chemical Properties of Aerosols*, Blackie, London.

relatively large (e.g. number distribution mode at 50 μm), and therefore easy to collect. The smaller the particles, the smaller the droplets and the higher the relative speeds required. In the most common design, the dirty gas, containing water droplets and particles, flows through a contraction or venturi (Figure 2.15(c)). As the gas flow accelerates, the particles go with it due to their small inertia, but the larger droplets lag behind. Hence the particles collide with the droplets. As with other processes based on impaction, the particle collection efficiency decreases with particle diameter. Downstream of the venturi, the gas and particles slow down faster than the droplets, so the collision process continues. A cyclone separator will usually follow immediately after the venturi. Wet scrubbing is typically used to collect sticky emissions not suitable for dry systems, to recover water-soluble dusts and powders, or to collect both particles and gases simultaneously. The pressure drop through the system, which is an important parameter in the calculation of operating costs, can be derived from formulas. No such simple relationships exist for collection efficiency owing to the wide variety of droplet dispersion designs in use. A collection efficiency of 0.99 can be achieved for 1 μm particles through a pressure drop of 20 kPa.

2.5.5   Electrostatic precipitation

The electrostatic precipitator (ESP) is widely used for collecting emissions from certain process industries – notably coal-fired power plants, cement kilns, pulp and paper mills, steel mills and incinerators. The ESP casing (Figure 2.15(d)) has large cross-sectional area to slow the gas down (0.9–1.7 m s^{-1}), establish laminar flow and give a long residence time. Within the casing, the gas flows between vertical parallel plates which are the positive electrodes. Sandwiched in alternate layers between the plates are vertical frames carrying wires, the negative electrodes, which are maintained at 20–100 kV. Uniform distribution of velocities over the cross section is important, since collection efficiency depends on residence time. During its passage between the plates, the corona discharge between the electrodes imparts net negative charges to the particles, which then migrate to the positive plates under the influence of the electrostatic field. When a layer of particles has accumulated on the plates, it is knocked off (rapped) into collecting hoppers. The detailed theory of operation is very complex, involving the corona current, attachment of ions to particles of different diameters and electrical properties, and movement of the charged particles at their terminal speeds to the collecting plate. These processes can be summarised in the Deutsch equation, by which

$$E = 1 - \exp(-w_e\,SCA)$$

where w_e = effective migration speed for the particle population (i.e. the mean terminal speed). This varies from 0.01 to 0.2 m s^{-1} depending on the material and the conditions. w_e is not a measurable speed, but a factor used to para-meterise an ensemble of particle properties, such as diameter distribution and resistivity, appropriate to a process such as steel-making. SCA = specific collection area (plate area per unit volume flow rate). This is typically in the range 20–200 s m^{-1}.

There are several persistent problems with achieving high collection efficien-cies. Dusts with high electrical resistivity ($>2 \times 10^{11}$ ohm-cm) are slow to take on a charge as they pass through the ESP. The ESP therefore has to be longer. Such particles are also slow to discharge when they reach the positive electrode plates, so they adhere under electrostatic forces, and build up an insulating layer on the plates which generates back corona, and reduces current flow. Dusts with resistivity $<10^8$ ohm-cm, on the other hand, are not held onto the plates and can be easily reentrained into the gas stream. Resistivity can vary greatly with oper-ating conditions such as temperature so stable process control is a major factor in attaining consistent collection efficiency. Physical design constraints mean that some particles can pass through the precipitator without being charged. The rap-ping process itself ejects particles back into the gas flow, and if this occurs near the discharge then there is little chance to recapture the particles. Efficiencies of commercial large-scale precipitators tend to be a minimum of about 0.95 at around 0.5 μm, and to increase for both larger and smaller particles.

2.5.6   Comparison of efficiencies

To some extent, the overall particle collection efficiencies of any of these control devices can be increased to whatever level is required. Nevertheless, the different methods have certain common characteristics – bag houses offer high collection efficiencies but need high pressure drops, electrostatic precipitators have large volumes, cyclones have reasonably low pressure drops but cannot deliver high efficiencies for submicrometre particles. These characteristic collection efficien-cies are illustrated in Figure 2.17.

Figure 2.18 emphasises in a different way that each of the particle collection devices has a 'natural' size range within which it works with the best combination of installation and operational costs.

2.6   SECONDARY PARTICLES

The same summertime photochemical conditions that can generate ozone in the troposphere from pollutant precursors (Chapter 1) can also generate particles. Chemical analysis of collected material has shown that the principle species found are:

- ammonium sulphate
- ammonium bisulphate
- ammonium nitrate
- ammonium chloride

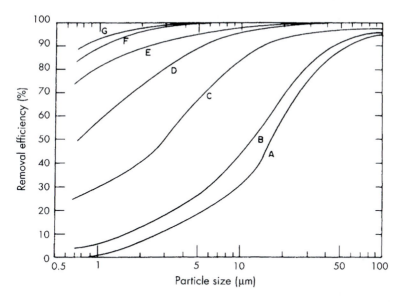

Figure 2.17 Collection efficiencies of different control devices. a) Baffled settling chamber, b) simple cyclone, c) high efficiency cyclone, d) electrostatic precipitator, e) spray tower, f) venturi scrubber, g) fabric filter.
Source: Peirce, J. J., Weiner, R. F. and Vesilind, P. A. (1998) *Environmental Pollution and Control*, Butterworth-Heinemann, Boston, USA.

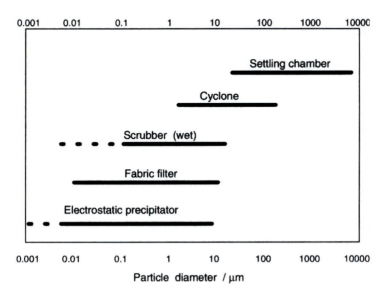

Figure 2.18 Optimum particle size ranges of different collecting devices.
Source: van Loon, G. W. and Duffy, S. J. (2000) *Environmental Chemistry – A Global Perspective*, Oxford University Press, Oxford.

- sulphuric acid
- sodium nitrate
- elemental carbon.

The main source of secondary particles is the atmospheric oxidation of SO_2 by the hydroxy radical to form sulphur trioxide (SO_3). The latter reacts with water vapour to form sulphuric acid (H_2SO_4) vapour, which in turn nucleates to form a sulphuric acid mist. As with gas-phase photochemistry, the OH radical is recycled and small concentrations can exert a powerful influence.

The other major player in secondary particle formation is ammonia (NH_3). This alkaline gas is released at low concentrations over large land areas from animal wastes that have been spread on fields, or from animal urine which contains urea, or from nitrogenous fertilisers such as ammonium nitrate (NH_4NO_3) or urea itself ($CO(NH_2)_2$). Urea hydrolyses to generate carbon dioxide and ammonia.

$$CO(NH_2)_2 + H_2O \rightarrow CO_2 + 2NH_3$$

The ammonia is readily taken up on the surfaces of both sulphuric acid droplets and of acidified particles, where it forms ammonium sulphate or ammonium bisulphate. The formation of particles from vapour naturally generates extremely small particles (well below 1 μm), and does so rather fast because of their high surface area to volume ratio. These small particles coagulate rapidly to move into the accumulation size mode, with sizes that give them long residence times in the atmosphere and high efficiency at scattering light and reducing visibility.

The ammonium nitrate and ammonium chloride particles are generated by corresponding reactions between gaseous ammonia and nitric acid vapour or hydrogen chloride, respectively. Long-term records of suspended particulate show that European sulphate concentrations rose slowly but steadily from the 1950s, peaked in the late 1970s and have since declined. This pattern reflects the improved control of SO_2 emissions over the same period. Particulate nitrate, on the other hand, has continued to increase, although concentrations are expected to fall again as vehicle NO_x emissions decline. On a larger scale, measurements from the EMEP network have shown a gradient of particulate sulphate concentration from peak values in south-east Europe of around 3 μg S m^{-3}, through about 1.5 μg S m^{-3} down to around 0.5 μg S m^{-3} in Scotland. This sulphate probably contributes about 6 μg m^{-3} to the background PM_{10} concentration in both rural and urban areas of southern Britain, and about one-third of that in the north. The corresponding values for nitrate PM_{10} would be about two-thirds of these figures. Hence, in industrial countries that also have intensive animal husbandry and/or crop production, a characteristic haze of fine particles is produced. The long range 'import' of secondary particles can also make it hard to achieve standards for PM_{10} concentration.

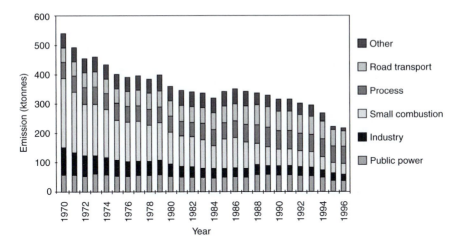

Figure 2.19 UK PM$_{10}$ emissions 1970–1996.
Source: Air Pollution Expert Group (1999) *Source Apportionment of Airborne Particulate Matter in the United Kingdom*, Department of the Environment, London.

2.7   TRENDS IN PARTICLE EMISSIONS

Size-specific emission inventories have been constructed for particles by using information on measured size distributions from different sources such as vehicles, industrial processes and electricity production. Figures 2.19 and 2.20 show

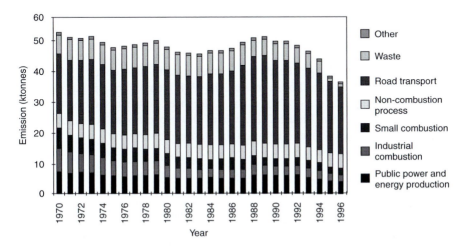

Figure 2.20 UK PM$_{0.1}$ emissions, 1970–1996.
Source: Air Pollution Expert Group (1999) *Source Apportionment of Airborne Particulate Matter in the United Kingdom*, Department of the Environment, London.

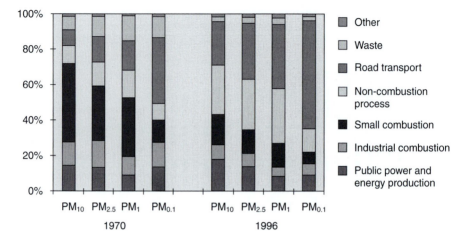

Figure 2.21 Contributions of different sectors to UK particle emissions.
Source: Air Pollution Expert Group (1999) *Source Apportionment of Airborne Particulate Matter in the United Kingdom*, Department of the Environment, London.

two examples of estimated trends in UK particle emissions over the period 1970–1996.

The PM_{10} graph confirms the fairly steady reduction in total emissions over that period, due largely to control of small combustion sources. On the other hand, the finest fraction represented by $PM_{0.1}$ emissions has been much more resistant to control, only falling significantly with the introduction of mandatory catalytic converters and lead-free petrol in 1993. There are considerable differences in the per capita emissions from different countries, with the EU figure typically 5–10 kg per head, and former Eastern bloc countries typically 10–30 kg per head. Note, however, that the $PM_{0.1}$ emissions are an order of magnitude less than those of PM_{10}. The differences between sectors are more clearly demonstrated in Figure 2.21, where the proportional contributions of different sources are compared between the start and end years.

FURTHER READING

Airborne Particles Expert Group (1999) *Source Apportionment of Airborne Particulate Matter in the United Kingdom*, Department of the Environment, London.

Colbeck, I. (ed.) (1998) *Physical and Chemical Properties of Aerosols*, Blackie, London.

Friedlander, S. K. (2000) *Smoke Dust and Haze*, Oxford University Press, New York.

Harrison, R. M. and van Grieken, R. (eds) (1998) *Atmospheric Particles*, Wiley, Chichester.

Hinds, W. C. (1999) *Aerosol Technology: Properties, Behaviour and Measurement of Airborne Particles*, Wiley, New York.

McMurry, P. H. (2000) 'A review of atmospheric aerosol measurements', *Atmospheric Environment* 34: 1959–1999.

Peirce, J. J., Weiner, R. F. and Vesilind, P. A. (1998) *Environmental Pollution and Control*, Butterworth–Heinemann, Boston, USA.

Quality of Urban Air Review Group (1996) *Airborne Particulate Matter in the United Kingdom*, Department of the Environment, London.

Raes, F. *et al.* (2000) 'Formation and cycling of aerosols in the global troposphere', *Atmospheric Environment* 34: 4215–4240.

Willeke, K. and Baron, P. A. (eds) (1993) *Aerosol Measurement: Principles, Techniques and Applications*, van Nostrand Reinhold, New York, USA.

Chapter 3

Mobile sources

Pollution from motor vehicles has become an issue simply because of the steady increase both in the number of vehicles in use and the distance travelled by each vehicle each year. Since the 1960s, the number of motor vehicles in the world has been growing faster than its population. In 1950 there were 50 million cars for 3.5 billion people. There are now 600 m cars for 6 billion people, with a global production of 45 million cars per year. By 2020 there will be a billion cars. The net growth rate for all motor vehicles is now around 5%, compared to a population growth rate of 1–2% (Figure 3.1).

The population is set to stabilise at 9–10 billion between 2030 and 2050, but the vehicle fleet will continue to grow. Vehicle production in India is increasing at 15–25% per year (depending on category) and the vehicle population of Beijing is

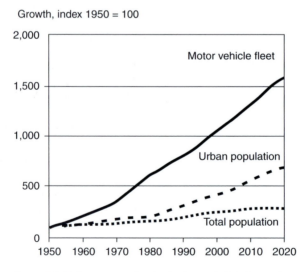

Figure 3.1 Relative growth of vehicles and global population, 1950–2020.
Source: Fenger, J. (1999) 'Urban air quality', *Atmospheric Environment* **33**: 4877–4900.

predicted to increase from 2.3 million in 2000 to 4.3 million in 2010. During the past 50 years, mobile sources have taken over from static combustion of fossil fuel as *the* major threat to air quality. There are four main polluting emissions from motor vehicles: carbon monoxide (CO), NO_x, hydrocarbons (HC) and particles. In this Chapter we will discuss the sources and control of motor vehicle emissions as a group, because the pollutants are emitted together. It must also be borne in mind that the emissions from the vehicles themselves are not the end of the story. Any full estimation of emission potential should be made on a life-cycle basis – including the emissions that are created before and after use by processes such as mining and processing the steel, making the plastics, spraying the paint, making and distributing the fuel, and disposing of the vehicle at the end of its useful life. For petrol vehicles, for example, fuel production might account for up to 70% of the total lifetime HC emissions, whereas fuel use will make up a similar proportion of lifetime CO_2 emissions.

Although most attention has been paid to vehicular emissions because of their impact on the urban environment, there are substantial emissions from other types of transport. In general they have less impact on people, and people are less aware of them, than for vehicle emissions simply because they occur in more remote locations. Nevertheless they make a significant contribution to some aspects of chemical budgets in the atmosphere. Emissions from planes, boats and trains will be discussed after those from motor vehicles.

3.1   MOTOR VEHICLE EMISSIONS

In the UK we now drive 540 billion km per year in 26 million motor vehicles, and this figure is expected to increase to 653 billion km by 2010. As a consequence, the use of motor vehicles now generates more air pollution than any other single human activity. It is the fastest-growing source of CO_2, and in urban areas accounts for the bulk of emissions of CO, HC and NO_x. Both people and vehicles are concentrated in urban areas, where the emission densities per unit area are the highest and the dispersion conditions the worst. Vehicles are now recognised as having the major environmental impact on our towns and cities. Table 3.1 shows the contribution of road transport to urban pollution in selected cities of the world.

The contrast between urban and national pollutant emissions is highlighted in Table 3.2, in which the proportional emissions from different source categories are detailed for London and for the UK as a whole. Thus road transport contributed 33% of CO_2 in London, but only 19% in the country as a whole. The differences are largely due to the absence from London of power stations and coal burning, and the presence of intense concentrations of motor vehicles.

3.1.1   Combustion of fuel

Most of the world's vehicles use internal combustion engines, burning petrol or diesel or gas hydrocarbon fuels for their energy supply. If a pure hydrocarbon fuel

Table 3.1 Relative contribution of road transport to pollution in different cities

City	Population (1990)	No of vehicles	Vehicles/100 population	Relative contribution of road transport to urban pollution in the year given/%					
	millions	millions	population	SO_2	CO	NO_2	TSP	HC	Year
Bangkok	7.16	1.760	25	–	99	30	–	–	1990
Beijing	9.74	0.308	3	–	–	–	–		
Bombay	11.13	0.588	5	35	89	52	–	–	1990
Buenos Aires	11.58	1.000	9	38	99	48	69	98	1989
Cairo	9.08	0.939	10	8	99	29	–	–	1990
Calcutta	11.83	0.500	4	5	48	60	2	–	1990
Delhi	8.62	1.660	19	5	88	83	5	–	1990
Jakarta	9.42	1.380	15	26	99	73	–	–	1989
Karachi	7.67	0.650	9	4	31	36	7	–	1989
London	10.57	2.700	26	1	95	75	76	–	1988
Los Angeles	10.47	8.000	76	60	90	70	89	–	1987
Manila	8.40	0.510	6	23	97	91	82	–	1990
Mexico City	19.37	2.500	13	22	97	76	2	52	1989
Moscow	9.39	0.665	7	1	80	20	1	–	1990
New York	15.65	1.780	11	–	–	–	–	–	
Rio de Janeiro	11.12	–	–	–	98	92	31	–	1978
Sao Paulo	18.42	4.000	22	73	94	82	31	70	1988
Seoul	11.33	2.660	23	12	56	78	30	17	1990
Shanghai	13.30	0.147	1	–	–	–	–	–	
Tokyo	20.52	4.400	21	–	–	–	–	67	1985

Table 3.2 Pollutant emissions in London compared to national averages

Category	Percentage of total emissions in category (London (1991)/National (1990))					
	CO_2	SO_2	Black smoke	CO	NO_x	VOC
Road transport	33/19	22/2	96/46	99/90	76/51	97/41
Other transport	3/1	1/2	2/1	1/0	4/7	1/1
Electricity generation	2/34	0/72	0/6	0/1	1/28	0/1
Other industry	13/26	43/19	1/13	0/4	5/9	1/52
Domestic	30/14	1/3	0/33	0/4	6/2	1/2
Other	6/19	32/2	2/1	0/1	8/3	0/3

was burned under ideal conditions, then the simple reaction taking place would be similar to:

$$C_xH_y + \left(x + \frac{y}{4}\right)O_2 \rightarrow xCO_2 + \frac{y}{2}H_2O$$

so that carbon dioxide and water would be the only combustion products. For example, pure octane (C_8H_{18}) would react as follows

$$C_8H_{18} + 12.5O_2 \rightarrow 8CO_2 + 9H_2O$$

Also, the masses of the reactants and products are related exactly by their molar masses. Hence the masses of CO_2 and H_2O are specified by the mass of fuel burned. For example, the mass of carbon in the fuel is given by

$$[C] = [C_xH_y] \times \frac{12}{(12x + 1y)}$$

because the molar masses of carbon and hydrogen are 12 and 1 respectively. This mass of carbon would combine with the corresponding mass of oxygen (molar mass = 32) as follows:

$$[C] + \left([C] \times \frac{32}{12}\right)O_2 = [CO_2]$$

Taking octane as an example again, a molecule with a mass of 114 combines with 12.5 molecules of oxygen having a mass of 400, a ratio of 3.5:1 by mass. Since oxygen makes up about 23% of air by mass, this represents around 15:1 of air to fuel. This proportion is known as the stoichiometric ratio; although it varies depending on the proportions of different hydrocarbons, a global value of 14.7:1 is accepted for the general mixtures used in petrol and diesel engines.

   In practice, this perfect combustion never occurs. First, the oxygen does not come alone, but mixed with nitrogen and a variety of other components of the atmosphere. Second, the fuel is not a single pure hydrocarbon, nor even a blend of pure hydrocarbons, but a rather impure mixture of hydrocarbons and other compounds that contain sulphur, nitrogen, lead and other minor constituents. Third, perfect combustion implies perfect mixing – i.e. the availability of the correct ratios of reacting molecules in the same place at the same time – and internal combustion engines cannot achieve this. As a result of these factors, some atmospheric and fuel nitrogen is oxidised to NO_x, some fuel is imperfectly combusted to form new HC or CO or carbon particles instead of CO_2, some fuel is not combusted at all and is emitted as VOC, and the minor contaminants are emitted in various forms.

A carbon balance can still be made, providing that all the carbon-containing components are included:

$$[\text{Fuel}] = (12 + r_1) \times \left\{ \frac{[\text{CO}_2]}{44} + \frac{[\text{CO}]}{28} + \frac{[\text{HC}]}{(12 + r_2)} + \frac{\alpha[\text{PM}]}{12} \right\}$$

where square brackets indicate masses.

r_1 and r_2 are the hydrogen to carbon ratios of the fuel and HC emissions, respectively. These may be taken as equal to a first approximation, though they vary between fuel types, being typically 1.8 for petrol and 2.0 for diesel.

α is the proportion of carbon in the PM emissions. It will usually be acceptable to assume that $\alpha = 1$, because PM makes up such a small proportion of total carbon emissions. The typical PM emission for a diesel is only a small fraction of 1% of its CO_2 emission factor.

No such mass balance can be made with certainty for the other components. A variable proportion of the nitrogen in the air will be oxidised, depending on exact combustion conditions such as peak temperature reached during the combustion cycle. It is usually assumed that all the sulphur in the fuel will be emitted, although some may be as SO_2 and some adsorbed to carbon particles. It is also presumed that 75% of the lead in the fuel will be emitted, with the remainder being retained in the exhaust system, engine parts and lubricating oil.

3.1.1.1   Petrol engines

The majority of petrol or spark-ignition engines operate on a four-stroke cycle of *suck* (draw in petrol and air mixture), *squeeze* (compress the mixture), *bang* (ignite the mixture with a spark) and *blow* (exhaust the waste gases). Petrol is a complex and variable mixture of many different hydrocarbons ranging from C_7 to C_{35}. A typical composition is shown in Table 3.3.

Table 3.3 Typical petrol composition

Component	Percent by volume
n-hexane	31.5
n-octane	21.2
2-methylpentane	10.5
methylcyclopentane	7.2
toluene	6.0
m-xylene	5.2
benzene	4.4
styrene	3.9
ethylbenzene	3.5
3-ethylpentane	3.4
o-xylene	2.2
n-nonane	1.0

In addition to four-stroke petrol engines, there are two-stroke engines, which are widely used for engines up to 0.5 l capacity and are more efficient because they fire every cycle rather than on alternate cycles. However, there are two disadvantages for air quality. First, the air–fuel mixture is introduced to the cylinder on the same piston movement that exhaust gases are discharged, so that unburned hydrocarbons can be lost. Second, the engine is often lubricated by means of oil mixed with the petrol, which leads to a serious emissions problem, often characterised by visible blue smoke from the exhaust. Two-stroke engines on motorcycles and small three-wheeler vehicles such as Tuk-Tuks have become very popular in developing countries, especially India, Indonesia and Malaysia where 60–70% of the vehicles on the roads are two or three wheelers. Measurements in Bangkok have shown that total suspended particle emissions from two-stroke motor-cycles are similar to those from heavy goods vehicles and buses. Emissions from motorcycles, lawnmowers, chainsaws, strimmers, jetskis, outboards and other small petrol-engined devices are becoming proportionally more significant as their number increases and emissions from large vehicles decrease.

3.1.1.2   Diesel engines

Diesel engines also operate on a four-stroke cycle in which the fuel is injected into the compressed air. The air–fuel mixture is leaner (more air per unit fuel), the compression ratio is higher (typically 15:1 instead of 9:1) and the ignition is spontaneous due to the high temperature achieved by the compression. Although diesel engines are more efficient than petrol ones, they are both still enormously wasteful. Of the initial energy content of the fuel, only about 30% is converted into useful mechanical energy, with about 35% each going into heating the exhaust gases and the cooling water. A further 5% is spent on flexing the tyres and on friction in the drive train, leaving just 25% for propelling the vehicle.

3.1.2   Combustion emissions

For both petrol and diesel engines, complete and incomplete combustion of the fuel generates a complex mixture of gaseous and particulate pollutants, many of which are harmful to human health.

3.1.2.1   Emission concentrations

The order-of-magnitude emission concentrations generated by the above processes in the exhaust gases of a non-catalysed petrol engine and a diesel engine are indicated in Table 3.4. The concentrations are very sensitive to the engine operating mode. With the emission concentrations being some 10^4–10^5 times typical ambient backgrounds, and released only a few tens of centimetres above ground level, excellent dispersion is essential.

Table 3.4 Exhaust gas emission concentrations

Fuel and pollutant	Emissions (ppm by volume in the exhaust)			
	Idling	*Accelerating*	*Cruising*	*Decelerating*
Petrol				
CO	69 000	29 000	27 000	39 000
HC	5 300	1 600	1 000	10 000
NO_x	30	1 020	650	20
Diesel				
CO	trace	1 000	trace	trace
HC	400	200	100	300
NO_x	60	350	240	30

3.1.3   Specific pollutants

3.1.3.1 NO_x

Petrol is a low-N fuel, so there is little fuel NO. Thermal NO is created in high temperature regions via the Zeldovitch mechanism discussed in Chapter 1. The ratio of NO to NO_2 is typically >0.95 for petrol engines, >0.8 for diesel.

3.1.3.2   CO

Imperfect combustion in fuel-rich regions of the combustion chamber means that some carbon is only partially oxidised to CO. The emitted concentration may be several percent. As shown in Table 3.2, motor vehicles emit about 90% of total CO in the UK.

3.1.3.3   Combustion HC

Hydrocarbon emissions arise both as unaltered components of the original fuel and through partial oxidation of the fuel molecules. Benzene, for example, may arise via both routes. The main families of molecules are alkynes, alkenes, alkanes and aromatics. Alkanes contain single hydrogen–carbon bonds and are referred to as saturated. Alkenes contain both single and double bonds, and alkynes, single, double and triple bonds. Molecules containing double or triple bonds are referred to as unsaturated. Several hundred different hydrocarbons have been distinguished, with detailed proportions depending on combustion conditions. Diesel engines emit HCs of higher molecular weight (typically C_8 to C_{28}) than do petrol engines (typically C_5 to C_{12}). There are also non-hydrocarbon VOC emissions, which are also referred to as oxygenates. The two largest HC groups are ketones and aldehydes (including formaldehyde HCHO, acrolein CH_2CHCHO and acetaldehyde CH_3CHO). The oxygenates are a small proportion of total VOC for conventionally-fuelled engines, but the proportion is larger if the fuel is methanol or natural gas. All these VOCs are ozone precursors to varying degrees, and a small number are toxic in their own right (benzene, toluene, xylene and 1,3-butadiene in particular).

Reasons for the HC emissions are complex, but have been grouped into the following categories:

Imperfect combustion

Wall cooling – the thin boundary 'quench' layer in close proximity to the water-cooled combustion chamber wall fails to reach optimum combustion temperature.
Crevice trapping – dead spaces in parts of the combustion chamber where the flame front does not penetrate, such as between the sides of the piston and the cylinder wall, or the threads of the spark plugs. Any porous deposits on the combustion chamber walls can have a similar effect by protecting the absorbed mixture from getting burned.
Poor mixing – any region of the mixture which is oxygen-deficient. This situation is most likely to occur under transient engine operating conditions.

Leakage

Piston blow-by – leakage of air–fuel mixture past the piston rings and into the sump.
Oil seep – leakage of lubricating oil in the reverse direction.

3.1.3.4 Benzene

Although there are some industrial sources of benzene, two-thirds of atmospheric benzene in the UK is derived from road vehicles, either in the exhausts or by evaporation (see Section 3.1.3.5). In large urban areas, this may rise to over 90%. Furthermore, 90% of the exhaust emissions, and almost all the evaporative emissions, are from petrol engines.

The rural benzene concentration is around 0.5 ppb, while the normal range of urban concentrations is 2–10 ppb. This means that the range of daily benzene intake varies by at least a factor of ten, from a few tens to a few hundreds of μg. At the low end of this range the atmospheric intake would be dominated by that from food, while at the high end it would exceed that due to food, and be similar to that due to smoking 20 cigarettes per day.

3.1.3.5 Evaporation HC

In addition to the combustion HC which are emitted from the exhaust pipe, HC may evaporate directly into the air from various sources on the vehicle. The latter include:

Evaporation of fuel directly from the carburettor.

Fuel tank displacement – every time an empty tank is filled, the corresponding volume of air saturated with petrol vapour is displaced into the atmosphere. This only occurs when the vehicle is refuelled. In addition, every time that a fuel tank warms up during the day after cooling down at night, a further, though smaller, volume of saturated vapour is expelled. These are known as diurnal, or breathing, losses.

Hot soak losses – evaporation that continues from the engine or its fuel supply system after the engine is switched off

Running losses – evaporation of fuel while the engine is running

The composition of the vapour is determined by the vapour pressure of the components as well as by the composition of the liquid fuel. For the petrol composition given in Table 3.3, for example, the vapour composition at 20 °C would be as given in Table 3.5.

Clearly, the processes in which the hydrocarbons are exposed to the high temperatures of the combustion zone may lead to the generation of new molecules, whereas evaporation will simply release original fuel molecules. Evaporative emissions can be estimated from the following equation:

$$E_{evap,j} = 365 \times a_j \times (e_d + S_c + S_{fi}) + R$$

where $E_{evap,j}$ are the annual emissions due to vehicles in category j, a_j is the number of vehicles in category j, e_d is the mean emission factor for diurnal losses, S_c and S_{fi} are hot soak emission factors for vehicles fitted with carburetors and fuel injection, respectively, R is an emission factor for running losses.

The functions given in Table 3.6 have been used to calculate evaporative losses, where RVP is the Reid Vapour Pressure of the fuel in kPa, T_a is the average monthly ambient temperature, in °C, $T_{a,min}$ is the monthly average of the daily minimum temperature, ΔT is the monthly average of the daily increase in temperature.

Table 3.5 Typical petrol vapour composition

Component	Vapour composition (% by volume)
n-hexane	49.6
2-methylpentane	19.0
Methylcyclopentane	10.9
n-octane	6.0
Benzene	5.4
3-ethylpentane	3.4
Toluene	3.2
m-xylene	0.9
Ethylbenzene	0.7
Styrene	0.5
o-xylene	0.3
n-nonane	0.1

Table 3.6 Functions for calculating evaporative losses

Emission factor and units	Function
Diurnal (g day^{-1})	9.1 exp (0.0158 (RVP-61.2) + 0.0574 ($T_{a,min}$ − 22.5) + 0.0614(ΔT − 11.7)
Warm soak (g per soak)	exp(−1.644 + 0.01993 RVP + 0.07521 T_a)
Hot soak (g per soak)	3.0041 exp(0.02 RVP)
Warm running losses (g km^{-1})	0.1 exp(−5.967 + 0.04259 RVP + 0.1773T_a)
Hot running losses (g km^{-1})	0.136 exp(−5.967 + 0.04259 RVP + 0.1773T_a)

Table 3.7 Example of annual evaporative emissions

Source	Emission/g a^{-1}
Diurnal breathing	3100
Warm soak	2168
Hot soak	2436
Warm running	693
Hot running	628
Total	9025

To get a rough idea of the significance of these evaporative emissions compared to exhaust emissions, suppose a car uses fuel with an RVP of 70 kPa. The monthly average ambient, minimum and daily increases in temperature are 15, 10, and 10 °C respectively. The car is driven to work and back every weekday, each of which ends in a warm soak (engine not fully warmed up). On the weekends there are four further short trips which end in warm soaks and two longer ones that end in hot soaks. The total mileage during the year is 16 000 km, of which 60% is warm and 40% hot. This driving profile would result in the evaporative emissions given in Table 3.7.

The total is a little less than half the corresponding rough estimate of exhaust hydrocarbon emissions (Table 3.8). Although this type of calculation is very crude, it does show that evaporative hydrocarbon emissions must be addressed just as seriously as exhaust emissions if the total atmospheric concentration is to be controlled. The functions also show the sensitivity of the emissions to RVP. Indeed, in some countries limits are being set on petrol RVP to control emissions.

3.1.3.6   The importance of air–fuel ratio

All the gaseous emissions vary systematically with air–fuel ratio, as shown in Figure 3.2. This variation is central to understanding recent developments in engine management and emission reduction.

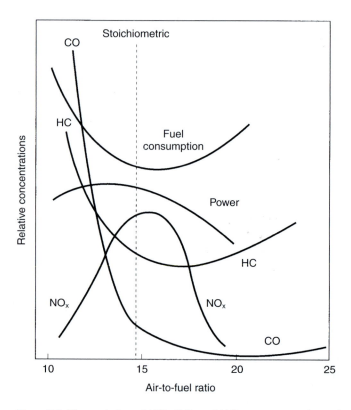

Figure 3.2 The variation of NO, CO, and HC emissions with air–fuel ratio.
Source: Masters, G.M. (1998) *Environmental Engineering and Science,*
Prentice Hall, New Jersey.

The chemically optimum air–fuel ratio, regardless of air pollution production, is called *stoichiometric* (just the right weight of oxygen to exactly combine with the available fuel on a molecule-for-molecule basis), and is indicated by the vertical dashed line. As mentioned in Section 3.1.1, this ratio is typically about 14.7 kg air per kg fuel, although it varies with fuel composition. At this ratio, NO production is close to a peak, while CO and HC are both low. With richer mixtures (lower air–fuel ratio), there is not enough oxygen to fully combust the available fuel, so CO and HC increase. Less energy is released, so less NO is created in the cooler conditions. When the air–fuel ratio is increased above 16 (leaner combustion), there is excess oxygen so that CO and HC stay low. However, excess air is also being imported into the combustion chamber, warmed up and exhausted. This reduces the temperature and again lowers NO production. Running the engine at non-stoichiometric conditions also reduces thermal efficiency, so that more fuel is burned to offset any reductions in emission concentration.

The variation of stoichiometric ratio with fuel composition has led to the development of the lambda scale. Lambda (λ) is just the air–fuel ratio divided by the stoichiometric value for that particular fuel. Hence λ is unity for a stoichiometric mixture of air with any fuel (regardless of composition), less than one for a rich mixture and greater than one for a lean mixture.

3.1.3.7 Particulate pollutants

Particles

Smoke (mainly particles of condensed carbonaceous material, typically 1 μm or less in diameter) is now the major toxic particle emission from vehicles. Well-maintained petrol engines have a low emission rate, although the predominance of petrol engines means that their total contribution is still important. Diesel vehicles are the largest source, accounting for about 40% of total UK smoke emissions.

Diesel aerosol consists mainly of highly agglomerated solid carbonaceous material and ash, together with volatile organic and sulphur compounds. Solid carbon is formed during combustion in locally fuel-rich regions because there is not enough oxygen to form CO_2. Metal compounds in the fuel generate a small amount of inorganic ash. Figure 3.3 shows a typical composition for diesel exhaust particulate.

For many years, it was assumed that the relevant metric for particles was the mass concentration. Recently attention has focussed on the number concentration of very tiny particles, called nanoparticles, typically having aerodynamic sizes below 0.1 μm. These nanoparticles constitute an insignificant proportion of the mass in the ambient aerosol, although they may dominate the particle population before it leaves the exhaust pipe. They are mostly volatile, except those formed from metallic fuel additives such as lead. They are formed very rapidly as the exhaust is diluted into the ambient air, and the number formed is strongly dependent on details of the dilution process. They may be important for health effects if, for example, an alveolar cell is neutralised by any particle, whatever its size. Measurements of particles made before this was realised generally paid no attention to particle number, and most of the measurement techniques were insensitive to particles as small as nanoparticles.

There is a complex set of processes that go on within the exhaust gases as they leave the exhaust pipe. Almost all the original particles are present as solid carbonaceous agglomerates. There are also volatile organics and sulphur compounds in the gas phase. As the exhaust gases dilute and cool, these may be transformed into particles by nucleation, adsorption and condensation. New particles are formed either by homogeneous nucleation (direct gas-to-particle conversion at high supersaturations to form new particles), or by heterogeneous

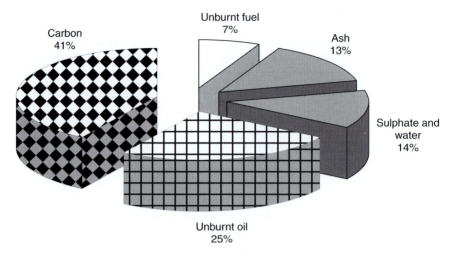

Figure 3.3 Composition of diesel exhaust emissions.
        Source: Kittelson, D. B. (1998) 'Engines and nanoparticles: a review', *J Aerosol Science* 29(5/6): 575–588.

nucleation (deposition of vapour molecules onto a starter particle at lower saturation ratios).

The volatiles are also adsorbed onto existing carbonaceous particles. Initially, a single layer of HC molecules will become attached to the surface of a carbon particle. Then additional layers may attach if the saturation ratio (the ratio of actual vapour pressure to saturated vapour pressure) for that component is high enough. These are physical processes, not chemical binding, so the organic molecules may desorb later. The particles may be thought of as sponges that soak up gaseous volatile organic particle precursors. Hence, if the concentration of these substrate particles is reduced by, for example, a filter in the exhaust pipe, then the concentration of homogeneous nanoparticles may be increased. Laboratory experiments have shown that nanoparticle concentrations may be changed by three or four orders of magnitude (1000–10 000 times) by manipulating the dilution conditions. It is important to note that natural dilution into the ambient air is only one aspect. When measurements of exhaust gas particle concentrations are taken, the exhaust is usually diluted before measurement to bring the concentrations within the operating range of the analyser. Typically, the exhaust gases are diluted by 30–50 times within five seconds. For vehicles on the road, the exhaust gases are diluted by around 1000 times in 1 s. Hence laboratory measurements may generate size distributions that are an artifact of the measurement process. Furthermore, petrol (spark-ignition) engines have been shown to emit as many nanoparticles, around 10^{14} km^{-1}, as diesel engines under high speed cruising conditions. This understanding may change the common perception that it is diesel engines that are mainly responsible for particle pollution from vehicles.

The analytical techniques used for particulate determination have contributed to the nomenclature for these particle emissions. The insoluble fraction is in the solid phase and consists mainly of carbon, plus some incombustible ash from the engine oil and fuel additives. Fuel sulphur compounds are emitted as sulphur dioxide or sulphur trioxide gases, which then form sulphate particles. Most of the sulphur in the fuel is oxidised to SO_2, but a small proportion is oxidised to SO_3 and leads to the formation of sulphate and sulphuric acid aerosol. The sulphuric acid/sulphate fraction depends on the sulphur content of the fuel – hence the importance of clean diesel. A small fraction of the fuel, together with atomised and evaporated engine oil, escape oxidation and appear as volatile or soluble organic compounds. This soluble organic fraction (SOF) is a complicated mixture of high molecular weight organic compounds, HCs, oxygenates, polycyclic aromatic compounds containing oxygen, nitrogen and sulphur, and can vary widely, between 10 and 90%, both between engines and from one engine to another. It is called SOF because it is the material that will be dissolved in methylene chloride after all particles have been collected on a filter. The SOF depends a lot on maintenance and load, and is often identified as a 'smokey' exhaust. Both the sulphates and SOF are collectively referred to as the volatile component, because the proportion of them that resides in the particulate form depends strongly on the thermodynamic history of the emissions. Conversely, the SOF/sulphate fraction can be identified by strong heating of the sample.

The ambient concentration of particles from diesel exhausts is a few µg m^{-3}. In confined environments, such as underground mines, this can rise by three orders of magnitude. An idealised particle size distribution from a diesel exhaust is shown in Figure 3.4. Both number and mass weightings are shown. There are three modes. The nuclei mode, between 5 and 50 nm, consists mainly of the volatile organic and sulphur compounds that form during exhaust dilution and cooling, together with solid carbon and metallic compounds from the combustion process. This mode dominates the number distribution, but contributes only a few percent to the mass. The accumulation mode, between 50 nm and 1 µm, contains carbonaceous agglomerates and their adsorbed materials and makes up most of the mass. The coarse mode contains around 10% of the mass, made up of a small number of larger particles which have been re-entrained from deposits on the walls of the exhaust system.

The compounds responsible for the mutagenicity, and possibly carcinogenicity, of diesel soot extract once the particles have been deposited in the lungs have been the subject of intensive research in Europe and the US. Polycyclic aromatic hydrocarbons, especially benzo[a]pyrene, have been prime suspects. More recently many nitrated polycyclics, such as the nitropyrenes, have also been invoked.

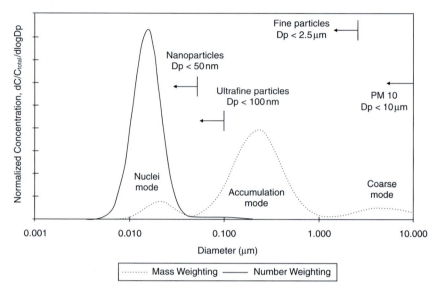

Figure 3.4 Particle size distribution in diesel exhaust.
Source: Kittelson, D. B. (1998) 'Engines and nanoparticles: a review', *J Aerosol Science* 29(5/6): 575–588.

Lead

For a long time, lead was added to petrol to improve its combustion properties. In a petrol engine, the ratio of the maximum cylinder volume when the piston is at the bottom of its stroke to the minimum volume at the top is called the compression ratio. Thermodynamically, there is an efficiency advantage in raising the compression ratio. The efficiency of the cycle is given by:

$$\frac{\text{work out}}{\text{energy in}} = 1 - \left(\frac{1}{\text{compression ratio}}\right)^{\frac{R}{c_V}}$$

where R is the Gas Constant and C_v the specific heat at constant volume. However, if the compression ratio is raised too far, the engine acts like a diesel, heating the petrol/air mixture enough to detonate it before the spark does. This condition is called knocking or pinking and damages the engine. It has been found that different blends of petrol, namely those with a higher proportion of certain hydrocarbons such as branched-chain alkanes, are much less susceptible to knocking. One particular fuel, called 2,2,4-trimethylpentane or isooctane, was found to be the most knock-resistant and hence other fuels have been given an octane number to compare with it. A high octane number means that the engine can be designed for a high compression ratio and hence extract more power from the fuel. It has also been found that the octane rating of otherwise low-octane petrol can be increased by adding certain organic lead compounds, especially tetraalkyl leads (PbR_4, where R is either CH_3 or CH_2CH_3). These compounds have been added to petrol since the 1920s at the rate of around 1 g l^{-1}. After combustion, lead particles are emitted in the exhaust gases. Some of them are emitted directly in the 0.1–1.0 μm diameter size range and a proportion is deposited in the exhaust pipe and emitted later as coarser flakes >10 μm.

Unfortunately, lead is a toxic heavy metal to which the nervous systems of babies and children are particularly sensitive. Adults absorb 0–15% of ingested lead from the gastrointestinal tract, children up to 40%. The unexcreted fraction is distributed between blood, soft tissue and mineralised tissue (bones and teeth). About 95% of the body burden of lead in adults, and about 70% in children, is located in the bones. Tetraalkyl lead is absorbed through the respiratory tract and through the skin, then metabolised in the liver to trialkyl lead, the most toxic metabolite. It has effects on neurological development, particularly in children. Lead is also transported across the placental barrier into the foetus, and there may be release of maternal lead stored in bones during pregnancy. Lead uptake may also be affected by nutritional status – deficiencies in Ca, Fe or Zn have been associated with high blood lead concentrations. Although there have been many routes by which significant lead intakes to the body could be made, such as lead-based paint and lead water pipes, airborne particles became significant due to the growth of motor vehicle use. By the mid 1970s, 160 000 tons of lead were being used each year in American petrol alone. Although fatalities and illness due to lead poisoning were common in the plants at which the tetraethyl lead was manufactured, the benefits were sufficient that it took 50 years of scientific debate, research and campaigning to get lead banned in America, and after that in Europe. By the 1970s the use of motor cars had grown to the extent that there was serious concern about the lead exposure of people living near main roads, busy intersections or other areas with high traffic densities. As a consequence, legislation was introduced both to cut the lead content of petrol, and to limit the concentration of lead in the air. In the UK, the maximum allowable lead content of leaded petrol was reduced from 0.45 to 0.40 g l^{-1} in 1980, and to 0.15 g l^{-1} in 1986. The current average lead content of leaded petrol is 0.11 g l^{-1}. An economic incentive to reduce the consumption was provided by the Government in the form of a slightly lower excise duty on unleaded petrol, and further pressure by the increased use of catalysts, which cannot be used with leaded petrol because the lead poisons the precious metals. The combined effect of all these measures caused a rapid increase in the use of unleaded petrol between 1989 and 1999, to take over 90% of the petrol market (Figure 3.5). Total road vehicle fuel use remained fairly constant over the same period, although diesel fuel use increased by 50%.

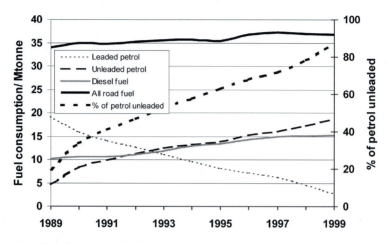

Figure 3.5 Changes in fuel use and proportion of unleaded petrol used in the UK, 1989–1999.

Consequently, although total petrol consumption was flat, emissions of lead and airborne lead concentrations in urban areas both fell by over 95% (Figure 3.6).

In complementary controls on the airborne lead concentration, EU Directive 82/884/EEC imposed a limit value of 2.0 μg Pb m^{-3}, and a guideline value of 0.5–1.0 μg m^{-3} (both as annual means). The concentrations are monitored at

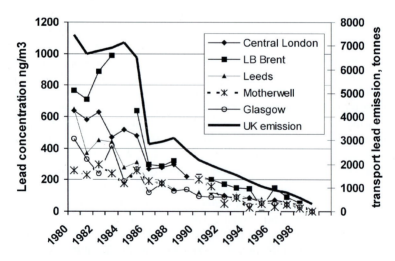

Figure 3.6 Changes in the amounts of lead emitted by vehicles in the UK, and in the airborne lead concentrations at urban sites, 1980–1999.

nine sites in the UK to check compliance with the Directive. The limit value has not been exceeded at any of the sites for the last 12 years, while the lower guideline value has been exceeded at one site. Typical urban lead concentrations are 0.1–0.2 $\mu g \; m^{-3}$, while typical rural ones are 0.01–0.05 $\mu g \; m^{-3}$. Total inhaled mass loadings of lead are trivial compared with the amounts ingested via food and drinking water, although there may be some concern that nanoparticulate lead can penetrate very effectively into the blood directly through the alveoles. Although blood lead levels in developed industrial countries have declined steadily in recent years, this reflects the combination of many independent controls on sources such as lead water pipes, leaded paint and leaded solder on tin cans, as well as leaded petrol. Effects of elevated blood lead have only been shown conclusively at concentrations above around 10 $\mu g \; dl^{-1}$, and since the transfer rate by inhalation is believed to be around 4 $\mu g \; dl^{-1}$ per $\mu g \; m^{-3}$ of atmospheric concentration, the range of urban concentrations given above is unlikely to result in detectable effects. An increase in blood lead from 10 to 20 $\mu g \; dl^{-1}$ has possibly been shown in epidemiological studies to result in a decrease of IQ by two points, although this remains contentious. The issue is now more urgent in developing countries, where low-lead petrol has taken longer to be accepted. Average blood lead levels up to 20–30 $\mu g \; dl^{-1}$ have been measured in urban/industrial populations in Peru, India, China and Mexico.

Polycyclic aromatic hydrocarbons

Another class of materials emitted from vehicle exhausts that has caused concern due to their carcinogenicity is the polycyclic aromatic hydrocarbons (PAH) such as benzo-*a*-pyrene (BaP) and benzo-*e*-pyrene (BeP). Measurements in Copenhagen have shown peak concentrations of around 4 ng m^{-3}, and that the winter-time contribution of traffic to PAH is around 90% for street air on a weekday, falling to 60% on the weekend. Diesel buses in Denmark now have to use a fuel having a low distillation end-point, which gives reduced particle and PAH emissions. The USEPA evaluation is that 0.3 ng BaP-equivalent m^{-3} will cause one additional lung cancer per 10^6 individuals, while the corresponding WHO guideline is that a lifetime exposure to 1 ng m^{-3} will cause 90 additional respiratory cancers per 10^6 individuals. High exposures to PAH can be tested by sampling the urine for PAH metabolites such as 1-hydroxypyrene and β-napthylamine.

Road dust emissions

Casual observation tells us that unpaved roads can be major sources of dust and fine particles. More surprisingly, perhaps, some measurements have shown that, even on paved surfaces, roads themselves can be significant sources of PM_{10}. The source is loose material, called silt, lying in the irregularities in the road surface. The passage of a vehicle, with the associated turbulent air movement due to the car itself and its tyres, entrains some of this silt into the air. It has been estimated that around 30% of the PM_{10} emissions in California are due to this process.

A simple model has been derived by the USEPA to predict road surface emissions. The model states that

$$E = k(sL)^{0.65}W^{1.5}$$

where E is the emission factor (g per vehicle kilometre), k is an empirical constant, s is the fraction of the total surface silt loading that passes through a 75 μm sieve, L is the total surface silt load (g m^{-2}), W is the vehicle weight.

The model is intuitively and empirically rather than mechanistically based. The empirical constants were derived by fitting the model to limited specialised data sets, and the model makes poor predictions if used in new situations without optimisation.

3.1.4   Reduction of motor vehicle emissions

3.1.4.1   Burn less fuel!

Fewer vehicles, smaller lighter cars with smaller more efficient engines, more public transport, shorter journeys, park and ride, bicycle lanes, pedestrianisation schemes – these are all common-sense ways to reduce emissions, but we have been slow to adopt them because we are still hooked both on the freedom of movement and flexibility of access that the motor vehicle provides, and on the power and speed of unnecessarily large cars.

3.1.4.2   Combustion optimisation

Detail design improvements to the cylinder head can give better mixing, less dead space and more optimum air–fuel ratio throughout the volume of the chamber. 'Lean-burn' engines have been designed that operate at a high air–fuel ratio of 18 or more (Figure 3.2) to give the best compromise of low NO, CO and HC. These conditions lead to instability of combustion (e.g. stalling) and need advanced fuel and air management. Although large research programmes have been carried out by individual manufacturers, lean-burn has really lost out to catalytic converters for reliable emission reduction, and it is clear from Figure 3.2 that it cannot compete with the potential reductions of 90+% achieved by catalytic converters. Lean-burn engines still generate unacceptable NO emissions, and there has been considerable research into a catalyst which will clean up this NO efficiently even under the oxygen-rich conditions. Some manufacturers are experimenting with direct injection, in which petrol is injected into the cylinders instead of being mixed into the inlet air.

3.1.4.3   Emissions recycle

The sump (or crankcase) is sealed and combustion gases that have leaked past the pistons (blow-by) are piped back to the inlet manifold so that the fuel and oil vapours are burnt instead of being vented to atmosphere.

3.1.4.4   Vent controls

Several measures have been adopted or are being considered in different countries:

- Small canisters (typical volume 1 l) of a vapour adsorbent such as activated charcoal are fitted to fuel tank inlets to control breathing emissions. These are standard in the US. The canister is purged to recover the fuel when the vehicle is running. The canister may overload if the vehicle is standing for many days.
- The fuel filling system is redesigned so that the pump nozzle seals the tank inlet to prevent direct emission of vapour. The displaced gases are either routed through an absorbent or pumped back into the underground storage tank.
- The petrol composition is changed to reduce fuel volatility (often reported as Reid vapour pressure (RVP), which is the vapour pressure of the fuel at a standard temperature of 34.4 °C). Butane is usually the first component to be reduced.

3.1.4.5   Exhaust gas recirculation (EGR)

A small fraction of the exhaust gases is directed back into the air inlet. This dilutes the air–fuel mixture and reduces combustion temperature, thereby reducing thermal NO_x production.

3.1.4.6   Catalytic converters

When the mixture of HCs and air in the combustion zone is burned, reactions proceed at different rates for the different components. All the reactions are ended (quenched) at the same moment when the combustion stroke ends and the exhaust stroke starts. Hence the exhaust gas contains a partly combusted mixture, at a temperature too low for normal combustion to continue. It is the purpose of catalytic converters to complete the combustion at a lower temperature. Catalytic conversion of the main pollutant gases in the exhaust has been used in America and Japan for many years and since 1 January 1993 has been mandatory in the countries of the European Union (see Chapter 13 for more details). Around 75% of the UK petrol-engined car fleet is currently equipped with catalysts; this figure will increase to 98% by 2008 as older vehicles are replaced. Other countries are gradually following suit – for example, India legislated for compulsory use of catalytic converters in its four largest cities from April 1995.

Construction

The catalytic converter typically consists of a ceramic support matrix (cordierite $Mg_2Al_4Si_5O_{18}$ is commonly used) and alumina washcoat to provide a very large surface area, on which the precious metal catalysts are deposited. The catalyst is presented to the exhaust gases either as a packed bed of beads of catalysed wash-coat, or as coated honeycomb. There is a power reduction of up to 10% (and hence associated increase in fuel consumption) from engines fitted with convert-ers, due to the higher pressure drop across the exhaust system and consequent reduction in gas throughput. The catalysts are usually platinum, palladium and rhodium; endless combinations and physical arrangements of these have been tried in order to increase the conversion efficiency and reduce the cost. The first catalyst contacted as the exhaust gas flows through is rhodium, which promotes the reduction of NO, using one of the incompletely oxidised components such as CO, H_2 or HCs.

e.g.

$$NO + CO \rightarrow \frac{1}{2}N_2 + CO_2$$

or

$$NO + H_2 \rightarrow \frac{1}{2}N_2 + H_2O$$

or

$$\left(2 + \frac{n}{2}\right)NO + C_yH_n \rightarrow \left(1 + \frac{n}{4}\right)N_2 + yCO_2 + \left(\frac{n}{2}\right)H_2O$$

Reduced components such as HCs and CO are then oxidised by passing them over a second catalyst such as palladium or platinum. There must be excess oxygen in the exhaust gases for this to occur.

e.g.

$$C_yH_n + \left(1 + \frac{n}{4}\right)O_2 \rightarrow yCO_2 + \left(\frac{n}{2}\right)H_2O$$

or

$$CO + \frac{1}{2}O_2 \rightarrow CO_2$$

or

$$CO + H_2O \rightarrow CO_2 + H_2$$

Hence all the oxidising agents in the exhaust gas are caused to react with all the reducing agents, which is a clever trick. Other materials, such as ceria (CeO_2) or zirconia are added to increase oxygen storage capacity or to promote HC conversion in rich exhaust conditions.

The lambda window

As this type of catalyst converts all three gases at once, it is known as a three-way catalyst (TWC). To make this work, the O_2 content of the exhaust gases must be maintained within narrow limits; this is achieved by measuring the equilibrium O_2 partial pressure with a zirconia/platinum electrolytic sensor (the lambda sensor) and feeding a control signal back to the fuel injector. The air–fuel mixture must be kept very close to, and slightly on the rich side of, stoichiometric at all times. Figure 3.7 shows that a catalytic converter can only achieve a conversion efficiency above 90% for *all three* gases if the air–fuel ratio is always within a very narrow range close to stoichiometric.

On the one hand, this means that precise mixture control is essential; on the other, the potential emission reductions associated with lean burn are not available. Overall, provided that the exact combustion conditions are maintained, the benefits of 90% reduction of toxic emissions will be combined with those of minimum fuel consumption due to the engine operating at its maximum thermal efficiency. If the combustion conditions are not so tightly controlled, the emission reduction can easily fall away to between 50 and 70%, and European standards cannot then be met.

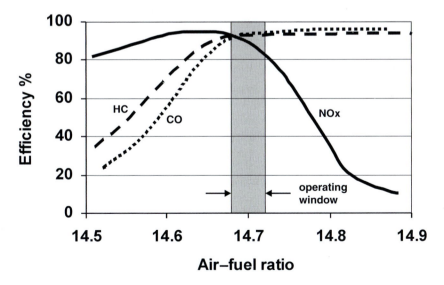

Figure 3.7 The variation of catalytic conversion efficiency with air–fuel ratio.

The nature of a feedback like this is to develop a small oscillation in the output, such that even when the engine is operating under constant conditions (cruising) the exhaust gas composition will move from slightly excess oxygen to slightly deficient. To handle this, an oxygen-labile material such as cerium oxide (CeO_2) is added to the washcoat mixture. Excess oxygen is stored in the CeO_2 by conversion to CeO_3 when there is some spare, and released when it is deficient. Oxygen storage materials can make up 10–20% of the washcoat, as compared to only 0.1–0.2% precious metals.

Other catalyst issues

The chemical reactions occurring on the catalyst surfaces are combustion, and the temperature has to be high enough for these to occur. Conversion efficiency is zero if the engine exhaust system temperature has not reached its 'light-off' value of around 300 °C. Since this may take several minutes, and around 20% of journeys are of less than 1 km, there are important consequences for emissions. Likely solutions will be either to electrically preheat the TWC or to adsorb the HCs in a charcoal trap before desorbing them into the TWC when it has warmed up. This poor conversion by the cold catalyst is compounded by poor fuel vaporisation in the cold engine, so that some fuel remains in droplet form and HC emissions are increased.

Hydrogen sulphide formation in catalytic converters has been an issue since the 1980s. Although the concentrations are small and probably no threat to health, the low odour threshold and unpleasant smell of H_2S has made the emissions noticeable. Emissions often occur as a spike when rapid deceleration to idle follows a period of cruise at high speed. Sulphur is stored in the TWC matrix during lean or stoichiometric operation, and released during rich transients. The exact chemistry is not yet clear, but minor washcoat components such as ceria are also believed to be involved.

Sulphur is higher in diesel fuel than it is in petrol, and this again increases particle emissions. Hence particle traps will be more practicable where lower-sulphur fuel is available. The average sulphur content of diesel fuel in the UK, which was 0.2% in 1994, was reduced to 0.05% (50 ppm) in 1996; this was expected to reduce particle emissions overall by about 15%. Although the sulphur content of petrol is only 10–20% of that of diesel, the greater proportion of petrol cars means that total sulphur emissions (as particles or as SO_2) are similar. A refined and desulphurised diesel fuel, sometimes known as City Diesel, which had been used in Scandinavia for some years, was introduced to the UK in 1996. City diesel is claimed to reduce sulphur and particle emissions by 99% and 40% respectively. In the US, ultra-low sulphur diesel fuels are being required as a 'technology enabler' to pave the way for the use of sulphur-intolerant particle traps.

Low-sulphur petrol is also necessary to reduce NO_x emissions down to the pending EURO IV limit of 0.25 g km^{-1}. Achieving this limit requires the use of storage catalytic converters, which employ a base coat of barium or potassium oxides. Nitrogen oxides are converted to nitrates and stored temporarily in the base coat. Intermittently the engine fuel mixture is turned from slightly lean to slightly rich, which converts the nitrates to N_2. The storage capacity is greatly reduced by the presence of S, which forms sulphates in the base coat irreversibly.

The current trends with catalyst development are to fit the catalyst closer to the manifold so that it warms up faster and runs hotter, tighten up the feedback so that the engine spends more of its time close to the optimum, and replace rhodium and/or platinum with palladium because it is cheaper.

3.1.5   Diesel exhausts

In California in 2000, diesel engines made up around 3% of the vehicle population and travelled 5% of the miles, yet 35% of the vehicle NO_x and 56% of the vehicle exhaust PM were attributed to these vehicles. Diesel engines operate under different combustion conditions to those of petrol engines. The leaner mixture and higher thermal efficiency lead to lower exhaust gas temperatures (200–500 °C), with highly oxidising conditions which make for low HC/CO emissions. These emissions can be further reduced with a diesel oxidising catalyst (DOC, see Section 3.1.5.3). However, the same oxidising conditions limit the potential for reducing the NO to N_2. DOC can also reduce particle emissions by removing SOF, although creation of sulphate particles from SO_2 can more than offset this improvement. Turbocharging reduces both NO_x and PM emissions by around 30%, more if combined with aftercooling. The big issue with diesel emission control is how to reduce the concentration of NO without increasing that of particles, and *vice versa*. For example, retarding the fuel injection reduces the peak flame temperature and hence NO_x formation. However, it also lowers fuel efficiency, giving higher overall PM emissions. Emission controls can be acceptably cost-effective. For example, the successive legislation enacted in the US to control mobile source emissions, including diesel, cost typically US$ 1500 per tonne of ozone precursors.

3.1.5.1   Diesel particulate filters

The simplest approach to reducing diesel particulate emissions is the diesel particulate filter (DPF), or trap. These devices take many forms, such as ceramic or metal foams, wire mesh or ceramic wool; they all trap particles by the normal range of filtration processes. As the particles accumulate, they create a pressure drop which reduces engine efficiency, so they must be burned off before the pressure drop becomes excessive. The burning off may occur spontaneously if the exhaust gases ever attain a high enough temperature, but usually some form of assistance is needed. This may take the form of electrical or combustion heating, or the engine may be run temporarily in an inefficient mode to artificially raise the exhaust gas temperature, or a catalyst may be included in the filter to promote combustion at lower temperatures. The holy grail of the catalyst method is the continuously regenerating trap (CRT), which filters and burns particles continuously. Ideally, NO is oxidised to NO_2, which then supplies the oxygen to burn the carbon, thus solving both of the most serious diesel emission problems at a stroke. These devices cannot work with high sulphur fuels, which promote the formation of sulphate particulate and block the filter. Even at 50 ppm S in the fuel, the catalyst can increase the exhaust gas sulphate loading by a factor of ten.

3.1.5.2   DeNO$_x$

There are emerging technologies to deal with the NO_x emissions from lean combustion systems independently of the particles. These are known as $DeNO_x$ methods. With storage $DeNO_x$, NO_x is scavenged from the exhaust gas by an adsorber and stored as nitrates. The adsorber is regenerated by operating the engine rich for a short time. In passive lean $DeNO_x$, existing HCs in the exhaust gas are used to reduce the NO_x, while in active lean $DeNO_x$ HCs or ammonia are injected to achieve the same end.

3.1.5.3   Diesel oxidation catalyst (DOC)

The lowering of the sulphur content of diesel fuel has opened up the possibility of treating at least some of the particulate emissions catalytically. The DOC reduces HC, CO and PM, but needs a low-sulphur fuel (below 50 ppm S by weight). The more active the catalyst, the greater the reduction in gas-phase emission, but the greater the sulphation of sulphur to produce particulate. Diesel oxidation catalysts has been used on off-road vehicles in the US since the 1960s, and fitted to over 1.5 million trucks and buses since 1990. Efficiency can be further increased by 10–15% by the use of a fuel-borne catalyst. Various combinations of washcoat and catalyst have been tried. Such catalysts are not prone to thermal deactivation because they operate at lower temperatures than TWCs – typically 650 °C rather than 950 °C. They can be poisoned more easily by P and Zn from the engine oil, more of which is burned in diesel than in petrol engines.

Test programmes in the US have shown that various combinations of these technologies can be used to reduce the emissions from modern heavy goods vehicle (HGV) engines below proposed standards, and that doing so will often improve fuel economy and reduce noise as well. It is not cheap, typically costing a few thousand dollars per tonne of particulate matter reduction.

3.1.6   Vehicle maintenance

Vehicle maintenance is another major factor in emission control. Roadside measurements have been made in several countries with Fuel Efficiency Automobile Test (FEAT), which is a non-dispersive infrared transmissometer-based instrument that can be used to determine the concentrations of CO, HC, NO_x and particles in the air above the road surface. Since it has a response time of only 1 s, the measurements can be related to the passage of individual vehicles. By making certain assumptions about dilution, the emission concentration can also be derived. Such measurements consistently show that 1% of vehicles is responsible for 5–10% of the total CO emissions, and 10% of vehicles for 50% of the emissions. Contrary to expectation, this 'dirty 10%' is not just older or high-mileage vehicles, but is evenly distributed across model years. Although some countries such as the US have a long history of vehicle emission legislation (catalysts were first introduced there in 1971), and strict enforcement policies (biannual smog checks), the dirty 10% continues to evade control.

3.1.7   Vehicle emission calculations

There are many practical issues that need to be considered when calculating potential emissions from a given length of road. The overall rate of emission will vary not only with obvious factors such as traffic density (flow rate in vehicles per hour), vehicle speed, driving mode (accelerating, decelerating, cruising or idling), mix of vehicle types (cars, buses, goods etc.), and engine types (petrol or diesel or other), but also with less obvious ones such as engine maintenance, meteorology (air temperature and humidity), and engine operating temperature.

We can make order-of-magnitude calculations of quantities such as vehicle emissions, to get a general idea of how significant they are compared to other factors. In broad terms, combustion of 1 l of fuel generates 100 g CO, 20 g VOC, 30 g NO_x and 2.5 kg CO_2, together with smaller quantities of other materials such as sulphur oxides, lead and fine particles. Consider the example of a single car which is driven an average 16 000 km per year with a fuel economy of 16 km l^{-1}, hence using 1000 l of fuel. This single car will generate the emissions shown in Table 3.8 (assuming no emission controls). Also shown in Table 3.8 are the corresponding emissions from the UK fleet of 25 million vehicles and the global fleet of 600 million vehicles. Although this back-of-the-envelope calculation ignores all factors such as different vehicle types and different driving patterns in different countries, it does show how serious vehicle emissions are on a global scale, not merely as a local pollution problem in urban environments.

Table 3.8 Vehicle exhaust emissions

Pollutant	Per litre of fuel/g	Single car/kg per year	UK fleet/Mt per year	Global fleet/Mt per year
CO	100	100	2.5	60
VOC	20	20	0.5	12
NO_x	30	30	0.8	18
CO_2	2500	2500	50	1500

Share of the total daily traffic load

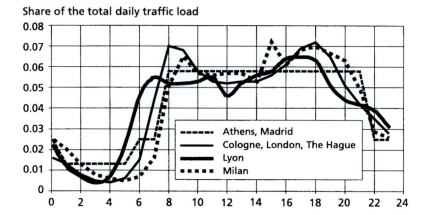

Figure 3.8 Typical patterns of traffic movements in European cities.
Source: Samaras, Z. and Sorensen, S. C. (1998) 'Mobile sources', In:
J. Fenger, O. Hertel and F. Palmgren (eds) *Urban Air Pollution: European
Aspects*, Kluwer Academic Publishers, Dordrecht, The Netherlands.

3.1.7.1   Diurnal variations

The primary factor controlling vehicle emissions from a length of road is the number of vehicles passing along the road per unit time. This variation follows a characteristic cycle related to human social patterns – most people are at home in the early hours of the day, go to work by car or public transport sometime between 6 and 9 am, go about their business during the day, and go home again in the evening. Some of those people will go out again in the evening. Of course, there are many exceptions to these patterns, such as for shift workers. Nevertheless, the resulting pattern of vehicle movements is quite reliable (Figure 3.8).

3.1.7.2   Variations with speed

Cars and light duty vehicles

There is a considerable database of direct measurements of emission rates from cars and their variation with speed. The European research programme CORINAIR has published equations relating emissions to vehicle type and speed. In 1999 a major programme of work resulted in the Methodology for Calculating Transport Emissions and Energy Consumption (MEET). Much of the work has been performed on chassis dynamometers, in which the vehicle is run on a rolling road. Two kinds of analysis are made. In the first, the straightforward variation of emissions with speed are measured under constant conditions. In the second, average emissions are determined while the vehicle performs a standard cycle of accelerating, cruising and decelerating in a pattern that is intended to simulate urban, motorway or cross-country driving or a combination of these. Although the cycles look quite complicated, and some have been derived from direct measurements of driving patterns in the appropriate environment, they usually remain less transient and complex than real driving patterns. Clearly, a wide range of speed cycles may all generate the same average speed, so that comparison between vehicles depends on standardisation of the test cycle. Naturally, different organisations and countries have tended to develop their own standards, and it has taken some years for any uniformity to be achieved. This averaging process facilitates comparisons between different vehicles. There may also be variations due to the condition of the specific vehicle tested (whether it is representative of those on the road), and the analysis itself. For example, gas analysers with response times of one minute would be quite adequate to measure steady-state emissions, but would not track the transient changes during rapid acceleration/deceleration cycles. The analysis becomes even more difficult when dealing with particles, where it has been shown that the details of exhaust gas dilution may affect the production of the smallest particles by orders of magnitude.

The outcome of many such tests is a set of algebraic relationships that enables the emissions to be predicted if the speed is known. Quadratic equations of the general form emission (g km^{-1}) = a + bV + cV^2 are fitted to the data. For CO_2 the equations are more complex, with combinations of logarithmic and inverse functions being used. The fit is often not good because of the large amount of variability from test to test on the same vehicle, and from vehicle to vehicle in the same category.

Note that the emissions are expressed in g km^{-1}. In real life, an idling engine will emit an infinite mass per km when the vehicle is stationary, and this situation is not handled by the equations. Indeed, some of them give zero, some of them give a constant, and some of them give infinity when $V = 0$ is substituted into them. This is why all the speed ranges start at a non-zero speed, either 5 or 10 km h^{-1}. Very low speeds (traffic creep) or idling are best handled by assuming that the mass emission per unit time at the lowest speed considered in the equation is also correct for a stationary vehicle. A stationary queue of traffic is then treated as a line source at the appropriate total emission rate per unit length and time (g m^{-1} s^{-1}). For example, a single petrol car with the smallest size of engine, fitted with emission control to EURO I requirements, has a mass emission rate for CO given by

$$9.846 - 0.2867 \ V + 0.0022 \ V^2$$

This equation is valid down to 10 km h^{-1}, at which speed the mass emission rate is

$$9.846 - (0.2867 \times 10) + (0.0022 \times 10^2) = 7.2 \text{ g km}^{-1}$$

At 10 km h^{-1}, this corresponds to 7.2 × 10 = 72 g h^{-1}. Hence a stationary car of this type will emit 72 g h^{-1} = 0.2 g s^{-1}, and can be treated as a point source of this strength. A queue of 100 such cars over a 1 km length of road would be treated as a uniform line source of strength $100 \times 0.2 = 20$ g km^{-1} s^{-1}.

At the next level of sophistication, the instantaneous emissions are related both to the speed and to the product of speed and acceleration (with units of m^2 s^{-3}), since the latter gives a better indication of the power supplied. A matrix is created which specifies an emission factor for each combination of these two factors. This type of model is useful not only for estimating normal road emissions, but for estimating the impact of measures such as traffic calming, when repeated acceleration/deceleration cycles may increase emissions compared to the uncalmed case.

Heavy duty vehicles

The term heavy duty vehicles (HDVs) covers HGVs, buses and coaches. They are almost entirely diesel-fuelled. Far fewer emission data are available for heavy duty vehicles, and they are mostly from engine test beds, in which the engine alone, rather than the complete vehicle, is studied. The MEET Report recommended that emissions from HGVs be described by polynomial functions of the general form

$$E = K + aV + bV^2 + cV^3 + \frac{d}{V} + \frac{e}{V^2} + \frac{f}{V^3}$$

where E is the emission rate in g km^{-1}, K is a constant, a–f are coefficients, V is the mean speed of the vehicle in km h^{-1}.

Figure 3.9 shows examples of the relationships discussed above, taken from the UK National Atmospheric Emissions Inventory (NAEI) database of emission factor estimates for 1999. As speed increases, emissions per km generally decrease, except for NO$_x$ from petrol engines which is low anyway. The decrease in emissions of PM$_{10}$ and VOC is very marked. Conversely, there is a sharp increase in these emissions when traffic slows down, in urban rush-hours for example. A car might emit ten times the weight of CO, and seven times the weight of VOC, in crawling 1 km at 10 km h^{-1} as it would covering the same distance at a cruising speed of 60 km h^{-1}. For vehicle speeds above 60–80 km h^{-1}, emissions start to rise again. Note that in the graphs described here, the motorbike emissions are constant. This is only because there is no data available from NAEI on the variation of motorbike emissions with speed.

3.1.7.3   Effect of engine temperature

Figure 3.9 ignores one major source of error in estimating vehicle emissions, which is lack of knowledge of the engine operating temperature. Many urban engines operate at reduced efficiency (i.e., higher fuel consumption and less complete combustion) because they are not yet warmed up, and catalytic converters do not achieve their potential 90% efficiency until they reach an

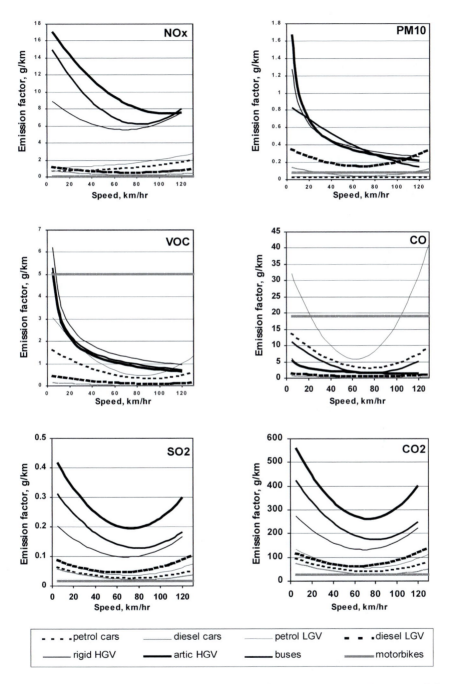

Figure 3.9 Variations of exhaust emission factors with vehicle speed for NOₓ, PM₁₀, VOC, CO, SO₂ and CO₂.

operating temperature of about 300 °C. The range of average journey lengths in different European countries is only 6–15 km; of the average UK trip length of 10 km, 3.4 km is 'cold' and 6.4 km 'hot'. Typically, 80% of car journeys are under 8 km, 33% are under 2 km, and it has been estimated that 75% of total CO is emitted from the first 1 km of a 10 km journey. The excess emissions are highest when the engine is started, and gradually decrease as the operating temperature approaches its stable value at the fully warmed-up condition. Hence the vehicle does a 'cold' distance, the length of which also depends on the average speed, because the faster it is driven, the greater distance it will go before it is warmed up. Thus if a car travels slowly over the cold distance, it will be warmed up in 5–10 km. Alternatively, if it drives at the UK motorway speed limit of 70 mph ($= 112$ km h^{-1}), then it can take over 30 km to warm up.

In covering this cold distance, the vehicle will emit 'excess' of each of the pollutants compared to the amounts that it would have emitted covering the same distance in the warmed-up condition. If the vehicle trip length exceeds the cold distance, then the excess emissions due to the cold start are maximum and are fixed. If the trip length is less than the cold distance, which is quite likely given the short average trip lengths described above, then the excess emissions are expressed as a function of the ratio of the trip length to the cold distance

$$\text{distance correction} = \frac{1 - e^{-a\delta}}{1 - e^{-a}}$$

where δ is the ratio of the trip distance to the cold distance, and a is a constant that varies with pollutant and vehicle type. Surveys have shown that on average, 60–70% of total annual mileage is driven without the engine having achieved its warmed-up operating conditions.

Air temperature, whether high or low, also affects emissions. This is allowed for by a correction factor that multiplies with the emission factor. The correction factor takes the form $AT + B$, where A and B are constants that depend on the operational state of the engine and the pollutant concerned, and T is the ambient temperature in °C.

As TWC efficiencies increase and legislated emissions fall, cold emissions make up an increasing proportion of total trip emissions and increasing effort is being spent to control them. In the US FTP-75 test cycle, for example, which lasts 1900 s, around 80% of the total HC has been emitted after typically 300 s. Various improvements to engine management are being made or discussed. These involve pre-heating the engine block to improve initial efficiency, improving the evaporation of fuel droplets in the inlet manifold, or boosting the oxygen content of the mixture. Possible improvements to TWCs include adjusting the catalyst mix to reduce light-off temperature, placing the catalyst closer to the engine to increase warm-up rate, insulating it to retain heat while not in use, and storing the cold-start emissions for later treatment.

3.1.7.4   Effect of operating mode

A further source of error, as we have seen in Table 3.4, is the operating mode of the engine – whether it is idling, accelerating, cruising or decelerating. Idling and cruising are steady-state operations that can be reproduced fairly precisely and for which the air/fuel supply and exhaust controls can be properly engineered. In urban areas, driving is likely to involve repeated acceleration/braking cycles associated with intersections, roundabouts, traffic lights, queues or car parks. Under acceleration, the engine enters a transient regime during which current management practices are unlikely to be able to maintain stoichiometric combustion, or to deliver exhaust gases to the catalyst within the narrow tolerances needed for efficient operation. A summary of modal effects is seen in Table 3.9, in which the emissions during each mode are expressed as a fraction of those at idle. During acceleration, and to a lesser extent during cruising, the combustion

Table 3.9 Relative emissions under different operating modes

Mode	Gas		
	CO	HC	NO$_x$
Idle	1	1	1
Acceleration	0.6	0.4	100
Cruising	0.6	0.3	60
Deceleration	0.6	10	1

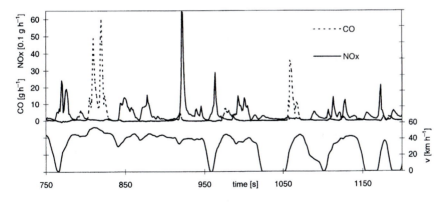

Figure 3.10 Short-term emission changes from a petrol-engined car fitted with a catalytic converter.
Source: de Haan, P. and Keller, M. (2002) 'Emission factors for passenger cars: application of instantaneous emission modeling', *Atmospheric Environment* 34: 4629–4638.

is efficient and there is excess air, so HC and CO are low. However, combustion chamber temperatures are high so NO_x is also high. During deceleration, there is some excess fuel so HC emissions are increased.

Even when driving conditions are relatively steady, emissions are likely to vary very abruptly, as shown in Figure 3.10. The upper graph shows CO and NO_x emissions from a vehicle driving the speed profile shown in the lower graph. Although some emission peaks are associated with obvious events such as changing gear, others have no such clear relationship.

3.1.7.5   Catalyst degradation

In extreme cases, contamination of the catalyst by lead can render the converter useless very quickly. For all catalyst-equipped cars, thermal ageing and contamination will result in a gradual decrease in conversion performance. This is internal and uncorrectable by maintainance. European and American legislation allows for this deterioration by specifying that the emission limits are not exceeded within a distance of 80 000 or 100 000 km. The test for this performance is undertaken on a continuously operating engine, whereas an engine in actual use will undergo many stop/start, hot/cold cycles that normally act to reduce lifetime. The MEET Report gives equations to describe the deterioration, which occurs rather uniformly up to around 100 000 km before levelling off.

3.1.7.6   National vehicle fleets

It is clear from the above discussion that accurate prediction of the emissions from a fleet of vehicles in an urban area requires a very high level of detailed information. Not only must we know the flow rate and mix of vehicles (the engine type and size, and years of legislation in which they were manufactured), but also their engine operating temperatures, speeds and acceleration modes. If these parameters are not known for a particular situation, then we can use default values from national average statistics.

Figure 3.11 shows the estimated average composition of road traffic in the 15 member states of the EU in 1995, in units of billion vehicle·kilometres by each emission-related vehicle category. As vehicles age and are replaced, the new ones are made to a higher emission standard.

It can clearly be seen from Figure 3.12 how successive waves of legislation impact the car fleet, rising to a peak usage and then declining as the stock is replaced, until eventually the whole fleet conforms with EURO IV standards of emission control. Assuming, that is, that a new set of standards has not been introduced in the mean time. Also note that the total number of cars continues to rise, so that emissions per car have to reduce just to stand still.

As these successive improvements are made to control emissions, emissions per vehicle are expected to come down as shown in Table 3.10 and Table 3.11.

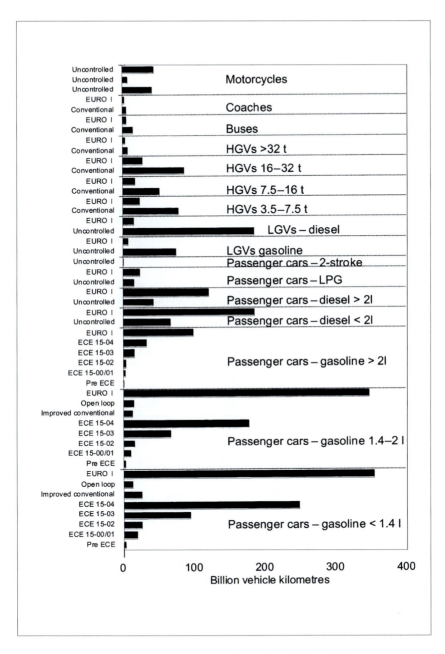

Figure 3.11 Average composition of road traffic in the EU15 in 1995.
Source: European Communities, (1999) *MEET – Methodology for Calculating Transport Emissions and Energy Consumption*, European Communities, Brussels.

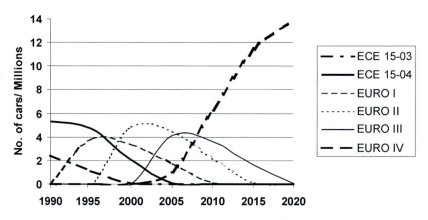

Figure 3.12 Composition of petrol cars (<1.4 L) in the UK by class of emission legislation, 1990–2020.

Table 3.10 Reductions in petrol engine emissions due to EURO II–IV, relative to EURO I/%

Control level	CO		VOC		NO$_x$	
	Excess cold	Hot	Excess cold	Hot	Excess cold	Hot
EURO II	30	5	40	55	55	55
EURO III	51	24	67	73	73	73
EURO IV	80	62	84	88	88	87

Table 3.11 Reductions in diesel engine emissions due to EURO II–IV, relative to EURO I/%

Control level	CO		VOC		NO$_x$		PM	
	Excess cold	Hot	Excess cold	Hot	Excess cold	Hot	Excess cold	Hot
EURO II	0	0	30	30	40	30	30	40
EURO III	35	45	51	51	58	51	51	64
EURO IV	55	56	76	76	79	76	76	84

In the US, there are around 230 million road vehicles, of which 68% are passenger cars and 26% LGVs. Thirteen percent of the cars, and 19% of the LGVs, are more than 15 years old. Since legislated and actual emissions have decreased significantly over that period, overall emissions will again decrease as these older vehicles are replaced.

3.1.8   Fuel composition

3.1.8.1   Basic specification

The phasing out of lead additives has created another conflict. Engines have been designed around the fuel octane rating, which was lowered by the reduction in lead content (Table 3.12). The octane ratings of unleaded petrol were maintained by an increase in the proportion of aromatics – as lead content fell from 0.7 to 0.15 g l^{-1}, average aromatic content rose from 28 to 42%, but is currently back down to 35%. Under the EU Fuel Quality Directive 98/70/EC, tight and very detailed environmental specifications are being introduced for both petrol and diesel which cover various aspects of combustion and evaporation emissions (Table 3.13).

In terms of overall use, there were 10 kt of lead and 6 Mt of aromatics in UK petrol in 1985; by 1993 the lead content had dropped to 1600 tonnes while the

Table 3.12 EU environmental specifications for petrol

Parameter	Limit by the year 2000 (maximum value unless otherwise indicated)
Research octane number	95 minutes
Reid vapour pressure/kPa	60.0
Hydrocarbon analysis	
Olefins/% v/v	18.0
Aromatics/% v/v	42 (35 by 2005)
Benzene/% v/v	1
Oxygen content/% m/m	2.7
Oxygenates	
Methanol/% v/v	3.0
Ethanol/% v/v	5.0
Isopropyl alcohol/% v/v	10.0
Tertiary-butyl alcohol/% v/v	7.0
Iso-butyl alcohol/% v/v	10.0
Ethers with five or more carbon atoms/% v/v	15.0
Other oxygenates/% v/v	10.0
Sulphur/mg kg^{-1}	150 (50 by 2005)
Lead/g l^{-1}	0.005

Table 3.13 EU environmental specifications for diesel

Parameter	Limit by the year 2000 (maximum value unless otherwise indicated)
Cetane number	51 min
Density at 15 °C/kg m^{-3}	845
Polycyclic aromatic hydrocarbons/% m/m	11
Sulphur/mg kg^{-1}	350 (50 by 2005)

aromatics had risen to 9 Mt. Emissions of benzene, which is carcinogenic, have been shown to rise from 60 mg km^{-1} for leaded fuel (35% aromatics, 1.5% benzene content) to 70 mg km^{-1} for 95 octane 'premium' unleaded (40% aromatics, 2% benzene) and 85 mg km^{-1} for 98 octane 'super plus' unleaded (47% aromatics, up to 5% benzene). The recommended ambient limit for benzene in the UK is 5 ppb as an annual running mean, and the Government has the objective of reducing this to 1 ppb as soon as practicable. There has been considerable public awareness of this issue in England, with one supermarket chain recently offering a 'low-benzene' (0.5%) unleaded petrol. Unfortunately, we have inadequate methods at present for comparing the low-level risks to the population of toxic materials such as lead and benzene – it is only safe to say that the less we have of either, the better.

In North and South America the oil industry has taken a different route to maintain the combustion properties of petrol. The organometallic manganese compound MMT (methylcyclopentadienyl manganese tricarbonyl) has been added at much lower concentrations than for lead – typically 18 mg l^{-1}. This material is emitted as Mn_3O_4 particles, and about 300 tonnes per year are currently being used in Canadian petrol. Manganese is both an essential trace element and a toxic metal which is known to impair neuromotor performance. A prolonged campaign was fought to ban its use in the USA, which was lost in 1995 because of lack of evidence of its effects at environmental concentrations. Although there is some public concern, it is not thought to be a health hazard, since airborne particle monitoring has shown that a guide value of 0.1 μg m^{-3} is not exceeded, and only 2% of Mn intake is from the air.

3.1.8.2   Fuel sulphur content

The maximum permitted sulphur content of diesel (for road vehicles) in the EU fell rapidly from 3000 ppm in 1989 to 350 ppm in 2000, and will fall to 50 ppm by 2005. Petrol sulphur was limited to 500 ppm in 1993, is 150 ppm in 2001, and will also fall to 50 ppm by 2005. In the US, diesel S will be reduced from 500 to 15 ppm by June 2006. Actual levels in fuel are already below these limits, and there are emission-control incentives for decreasing the sulphur content of both petrol and diesel still further – to below 10 ppm. First, NO_x emissions from catalyst-equipped vehicles may be reduced by around 40% when the petrol sulphur content is reduced from 130 ppm to 30 ppm; second, low-sulphur fuels enable the use of new generation catalysts which in turn enables more fuel-efficient engine technology; third, low sulphur emissions reduce the formation of sulphate nanoparticles (<50 nm) which may have the most important health effects. The reduction of S from 50 to 10 ppm will typically reduce particle emission factors by 35 and 20% for petrol and diesel respectively. While all this is going on, of course, it has to be remembered that the extra refining involves the release of additional CO_2 which will offset the efficiency gains at the point of use.

Ultra Low Sulphur Petrol (ULSP) is currently available at UK petrol stations. Although it will reduce tailpipe NO_x emissions by 5%, this will have negligible impact on UK air quality. The long-term benefit is that ULSP will enable the introduction of more fuel-efficient petrol direct-injection engines.

The sulphur content of fuel used in other applications can be far higher than the figures given above. For diesel used elsewhere (known as gas oil), including marine diesel used in EU waters, the maximum sulphur content in the EU is being reduced from 0.2% by mass (2000 ppm) to 0.1% (1000 ppm) in 2008. For fuel oil used in combustion plants the limit is 1% (10 000 ppm) from 2003. The operator can also apply for a permit which requires compliance with a sulphur dioxide emission limit of 1700 mg/Nm^3 from the same date.

3.1.8.3   Fuel reformulation

For both petrol and diesel fuel, there has been considerable interest in whether toxic emissions can be reduced by processing the fuel at the refinery. In some cases the effect is obvious – reducing the benzene content of petrol, for example, decreases benzene emissions. Addition of an oxygenate decreases CO emissions. Particle emissions from diesel decrease with sulphur content. Understanding of both these relationships is already being used to reduce emissions. California has required the use of a low sulphur (31 ppm), low Reid vapour pressure (51 kPa), narrow distillation range fuel with reduced aromatics, olefins and benzene and enhanced MTBE. In other cases the effects are less clear-cut. For example, increasing the cetane number (an equivalent ignition value) of diesel fuel can reduce CO and VOC emissions, but increase particle emissions. Overall, it seems unlikely that fuel reformulation as currently practised will significantly reduce emissions.

3.1.9   Diesel vs petrol

In recent years, public awareness of environmental issues, the introduction of lead-free petrol and the good fuel economy of diesel engines have combined to give diesel a rather green image. Partly as a consequence of these factors, UK market penetration by diesel cars increased from 3% in 1984 to 20% in 1993, and is expected to reach 30% by 2005. Currently around 8% of the total UK car fleet is diesel, and this figure is expected to increase to 20% during the next decade. The popularity of diesel cars varies greatly from country to country, being influenced by public perception, the price of fuel and the price differential between diesel and petrol (both fuel and cars). In the US, fuel is very cheap and the fuel-economy advantage of diesel is not sufficient to out-weigh the higher first cost and reduced smoothness of diesel engines. In France, the tax regime results in generally higher fuel costs and in diesel fuel costing around 65% that of petrol. Potential purchasers of new cars regularly use their annual mileage to calculate the pay-back period for recovering the

Table 3.14 Emission factors for petrol and diesel cars

Vehicle type	Average on-road emissions/g km^{-1}			
	CO	HC	NO$_x$	Particles
Petrol, unleaded	27.0	2.8	1.7	0.02
Petrol, TWC	2.0	0.2	0.4	0.01
Diesel	0.9	0.3	0.8	0.15

higher initial cost of a diesel car. As a consequence of these financial incentives, diesel sales in Europe are currently 35% of new car registrations, and expected to rise to 40% by 2005.

Table 3.14 compares emissions from cars with warmed-up petrol engines (with and without catalysts) and diesels. One disadvantage of diesel engines is that they tend to have high emissions from a cold start. The cold/hot ratios for the usual pollutants are 1–2, similar to those for non-TWC petrol engines and much lower than those for TWC engines. On the other hand, diesel engines achieve operating temperature in half the distance of petrol engines.

3.1.9.1   Gaseous emissions

Owing to the excess oxygen in the air–fuel mixture, carbon monoxide emissions from diesel vehicles are lower even than those from petrol vehicles fitted with TWC. Hydrocarbon emissions are similar, although there is evidence that those from diesel engines have higher proportions of the light hydrocarbons such as ethylene and propylene that are important precursors of ozone (Table 3.15). Diesel engines operate on very lean mixtures (up to 20:1), and with present technology diesel emissions can only be treated with an oxidation catalyst (which will convert HC to H_2O and CO_2, and CO to CO_2) but not with the full TWC (which requires the air/fuel ratio to be very close to stoichiometric). Hence NO$_x$ emissions are higher than from a TWC, although there is an improvement by a factor of two over the non-TWC exhaust. Nevertheless, the increasing proportion of diesel cars will eventually offset the reduction in NO$_x$ concentrations being made by mandatory fitting of TWCs to petrol cars. The

Table 3.15 Concentrations of untreated emissions from diesel and petrol engines

Gas	Petrol/ppm	Diesel/ppm
CO	20 000–50 000	200–4000
HC	10 000	300
NO$_x$	600–4000	200–2000
Aldehydes	40	20

required breakthrough in this area is the development of a deNox catalyst that will convert NO efficiently at the high oxygen concentrations found in diesel exhaust. Prototype versions of such catalysts are being trialled currently, but do not have acceptable lifetimes or conversion efficiencies (especially when igh-sulphur fuel is used). Diesel cars are very low emitters of carcinogenic aromatics such as benzene and toluene, but high emitters of high molecular weight hydrocarbons such as the PAH (e.g. benzo(*a*)pyrene). Evaporative emissions are inherently very low from diesel vehicles owing to the low vapour pressure of the fuel.

3.1.9.2 Particulate emissions

Petrol-engined cars are usually thought of as negligible particle emitters (except in the case of leaded petrol), although this is not necessarily the case. The carbon particles from leaded petrol are mainly organic rather than elemental, whereas those in diesel exhaust gases are two-thirds elemental. In the UK, a 'blackness factor' has traditionally been used to quantify the different effectiveness of particles from different sources in reducing the reflectivity when sampled on to a filter paper. This factor cannot be related to health effects, but bears some relationship to soiling of surfaces. The scale is based on a factor of one for coal smoke; particles from petrol combustion have a factor of 0.43, and those from diesel have a factor of 3 – a ratio of 7. Experimental determinations from an

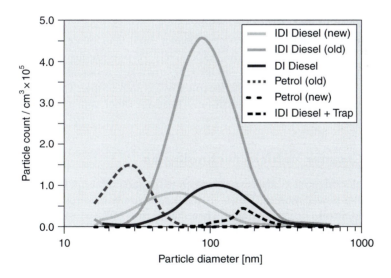

Figure 3.13 Particle emissions from petrol and diesel engines.
Source: Airborne Particles Expert Group (1999) *Source Apportionment of Airborne Particulate Matter in the United Kingdom*, Department of the Environment, London.

urban mix of vehicles in Leeds have given an overall factor of 0.92. In terms of blackness, typical emission factors of 1.5 and 6 mg (kg fuel)$^{-1}$ for petrol and diesel respectively become transformed to 0.7 and 18 mg (kg fuel)$^{-1}$ – a ratio of 26.

Figure 3.13 shows typical particle size distributions (by number) for different engine types. The old petrol engine peaked at around 30 nm, while the line for the new petrol engine is hardly visible along the diameter axis. The indirect injection (IDI) diesel engine gets significantly worse with age, but can be improved greatly by the use of a particle trap.

3.1.10   Impact of control measures

At the present rate of renewal of the UK vehicle fleet, the petrol-engined vehicles will have become almost completely catalyst-equipped by 2015. This change will significantly reduce emissions of CO, NO_x and HC. On the other hand, the proportion of the fleet that uses diesel engines will continue to increase. First, these vehicles cannot at present be fitted with three-way catalysts, so their emissions will be uncontrolled. Second, they will have much higher particle emissions than their petrol counterparts. During the past quarter-century, the impact of emission controls (amongst other changes) has gradually been bringing down the concentrations of SO_2, CO and NO_2 in urban areas, as seen in Figure 3.14 for Geneva. However, the same figure also shows the increasing trend in O_3 concentrations, which have been lower in urban areas but are now matching rural levels.

Figure 3.15 and Figure 3.16 show predictions of the overdue but nevertheless dramatic effects of catalytic converters, cleaner fuels and other measures on vehicular emissions of NO_x and PM_{10}, respectively. At present, just over halfway through the time period shown, emissions of these two key pollutants are less than they were in 1970. By 2025 they will be typically 25–35% of their values in 1970. The pattern and scale of reduction will be echoed for other vehicular pollutants such as CO and HCs.

3.1.11   Cleaner vehicle technologies

There is a major ongoing global research effort to try and replace petrol and diesel fuels with other power sources. This is seen as necessary both to reduce pollutant emissions and also to keep cars going when the oil runs out.

Electric – electric vehicles were serious contenders with the internal combustion engine in the early development of the motor car. Their clear attractions of quiet operation and no emissions at the point of use have been offset by one main drawback – the low energy density compared to petroleum-based fuels has meant that vehicles have had to be rather heavy, and this has made them uncompetitive on range and performance. There has been a long period of research into better batteries – sodium/sulphur and nickel/metal hydride, for

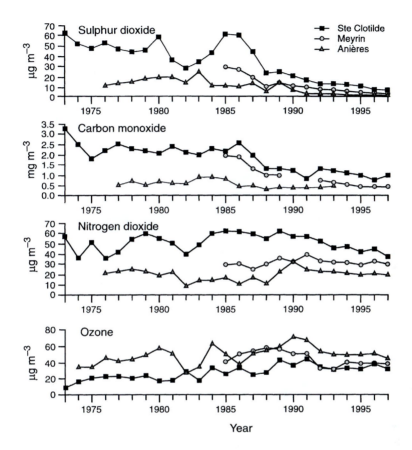

Figure 3.14 Gaseous pollutant concentrations in and around Geneva.
Source: Cupelin, F. and Zali, O. (1999) 'Motor vehicle pollution control in Geneva', In: D. Schwela and O. Zali (eds) *Urban Traffic Pollution*, Spon, London.

example, instead of lead/acid – but progress has been slow. In California, where there is still a serious problem with photochemical pollutants despite over twenty years of catalyst use, an increasing proportion of vehicles sold will have to be 'zero-emission vehicles' (ZEVs, assumed to be electric). There are still emissions associated with electric vehicles, but they occur at the power station rather than at the vehicle itself. Emission factors, which are now expressed as gram of pollutant emitted per GigaJoule of electricity generated, therefore vary greatly from country to country, depending on the way the electricity is produced. Examples of the extreme cases, together with the European average emission factor, are given in Table 3.16.

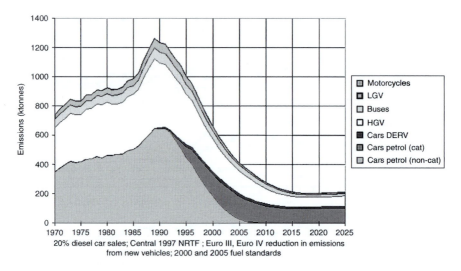

Figure 3.15 UK urban road transport NO$_x$ emissions, 1970–2025.

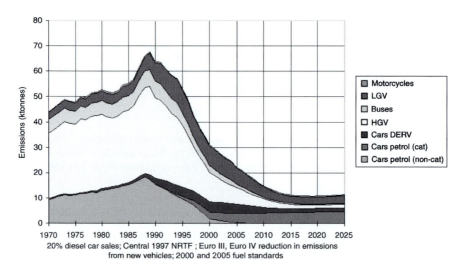

Figure 3.16 UK urban road transport PM$_{10}$ emissions, 1970–2025.

These generation emission factors are divided up amongst the vehicle fleet, with a proportion being allocated to each vehicle according to its use. The effective emission factor F for each vehicle type can then be written in the form

Table 3.16 Emission factors from electricity generation

Country	Emission factor/g GJ^{-1}						
	CO_2	CO	NO_x	NMHC	SO_2	CH_4	PM
Norway (mainly hydro)	1700	0.6	2.8	0.2	3.7	6	0.2
France (mainly nuclear)	17600	3.2	61	3.2	184	36.1	7.9
UK (mainly coal and gas)	167800	27.4	632	20.2	1446	459	70
European average	127400	21.3	326	22.6	745	283	39

$$F = aV^2 + bV + c$$

where a, b, c are coefficients and V is the average vehicle speed in km h^{-1}.

Table 3.17 gives the coefficients for the European average emission factors from electricity generation. For example, a passenger car travelling at an average speed of 50 km h^{-1} would have the emission factors given in Table 3.18.

In urban areas there would clearly be a huge reduction in local pollutant burdens if a substantial proportion of the vehicles was electric, because the power stations are in the country. The arguments are less clear-cut in other ways. Electricity generation by fossil fuel combustion is only 40% efficient, at best.

Table 3.17 Electric vehicle emission coefficients

Pollutant	Passenger cars			Light duty vehicles		
	a	b	c	a	b	c
CO_2	0.0233	−3.249	198	0.419	−5.848	356
CO	0.0000	−0.001	0.03	0.0000	−0.001	0.06
NO_x	0.0001	−0.008	0.51	0.0001	−0.015	0.91
VOC	0.0000	−0.001	0.04	0.0000	−0.001	0.06
SO_2	0.0001	−0.019	1.16	0.0002	−0.034	2.09
CH_4	0.0001	−0.007	0.44	0.0001	−0.013	0.79
PM	0.0000	−0.001	0.06	0.0000	−0.002	0.11

Table 3.18 Example of emission factor calculation

Pollutant	Electric passenger car emission factor/g km^{-1}	Petrol passenger car (EURO I) emission factor/g km^{-1}
CO_2	87.2	141
NO_x	0.51	0.39
SO_2	0.28	−

While this may be slightly more efficient than individual internal combustion engines, it will not result in significant reduction of CO_2 emissions. Unless DeNox and FGD equipment is fitted to all fossil fuel stations, there may well be an increase in overall NO_x and SO_2 emissions, although these will be released from tall stacks in rural areas rather than from exhaust pipes.

Hybrid – the need for a combination of low emissions in congested urban areas and extended range out of town has prompted the development of hybrid vehicles, in which electric and conventional power work together. The electric motor can also act as a generator, so that when the vehicle is braking some energy is recovered and stored in the battery.

Hydrogen – there is increasingly vocal lobby to make hydrogen the fuel of the future. Burning hydrogen creates water as the major waste product. There is no HC, CO, CO_2 or particle production, although there is some thermal NO_x if air is used to supply the oxygen. If the hydrogen is made from petrol, gas or biomass, then there are substantial emission possibilities at the point of manufacture. If it is made by the electrolysis of water, then energy of some type will have to be used. Although the use of hydrogen in the internal combustion engine would reduce emissions, the engine itself is only 20–25% efficient and even greater emission reductions would be obtained by the use of fuel cells and electric motors.

Fuel cell – electricity is generated directly from the low-temperature chemical reaction of hydrogen and oxygen. In the front-runner proton exchange membrane (PEM) cell, a thin polytetrafluorethylene (PTFE) membrane is sandwiched between an anode, where the fuel enters, and a cathode to which oxygen (from air) is supplied. The membrane has been treated to make it permeable to protons, but not electrons. The hydrogen atoms split into their constituent protons and electrons, and because the former are separated by the membrane, the latter are driven round the external circuit to power a motor or other device. Current fuel cells obtain power densities of 1.5 kW l^{-1}, which is marginal if enough power output is to be fitted within a car engine compartment. However, this power density is increasing, and the fuel cells already operate at twice the efficiency of the internal combustion engine. A variety of fuels can be used to supply the hydrogen, with the leading contender being liquid organic fuels such as methanol and liquified natural gas (LNG), from which the hydrogen is recovered *in situ* using a reformer. Ideally, a hydrocarbon should be used that produces maximum hydrogen for minimum carbon dioxide, and methane has the highest H/C ratio of all the hydrocarbons. Potentially, pure hydrogen could be stored and used, but this would demand an even bigger change in society infrastructure for the generation, delivery and storage of the gas. Hydrogen does not liquify until the temperature is below -253 °C, so could only be stored on board as a highly compressed gas. The state-of-the-art vehicle at present stores about 80 l of hydrogen on board at pressures up to 355 bar, giving the car acceptable but not startling performance and a range of up to 250 km. If the hydrogen is obtained from hydropower this would sever the link with fossil fuels and greatly reduce emissions of both CO_2

and air pollutants. Estimates have shown that gasoline CO_2 emissions could be reduced by 70% overall by the use of hydrogen produced from the large-scale reforming of natural gas. Larger reductions would require the energy for the hydrogen production to be generated by greener means such as solar, wind or hydropower. Eventually, we can imagine a hydrogen-based economy with the fuel being generated from the water and electricity at hydroelectric power stations. Dream on!

3.1.12   Alternative fuels

Whereas the cleaner technologies described in the previous section all require a radically different engine, there is a parallel development of alternative (non-petroleum) fuels that are intended to be used in a conventional engine but to have particular advantages over conventional fuels.

Natural gas – predominantly methane (75–95%) plus ethane (5–20%), propane (3–20%) and butane (1–7%). Used either in single or dual-fuel spark-ignition engines, when they can be swapped with diesel. CO, NO_x and particle emissions are reduced, HC are increased, and fuel consumption is about the same. The λ window is smaller than for petrol engines, so that it is harder to maintain TWC efficiency.

Methanol – methanol (CH_3OH) is a good fuel for lean-burn 'petrol' engines. It has a high octane number and low vapour pressure (hence low evaporative emissions). It can be manufactured from a range of feedstocks – natural gas, crude oil, biomass and urban refuse. However, it has so far failed to live up to its promise. The high heat of vaporisation makes it difficult to vaporise fully in the combustion chamber, so that 10–15% petrol has to be added to ensure reliable cold starting. This offsets the emission advantages of pure methanol, which emits less CO, HC, NO_x and CO_2 than petrol. However, the ability to make methanol from non-fossil resources is a major advantage. Emission problems with formaldehyde and unburned methanol have been found.

Ethanol – major programmes have developed ethanol (CH_2OH) as a vehicle fuel in some countries – notably in Brazil, where it is made from sugarcane wastes. It can be used alone or blended with petrol. In fact, it was the fuel recommended by Henry Ford for the Model T. Today, around 4 million cars (one-third of the fleet) run on ethanol in Brazil. In 1996–1997, 273 Mt of sugarcane were used to produce 13.7 Mm^3 of ethanol (as well as 13.5 Mt sugar). It has similar operational advantages and disadvantages to methanol, and emissions are not significantly reduced.

Biodiesel – biodiesels are fuels derived from renewable lipid feedstocks, such as vegetable oils and animal fats, or from the fermentation of vegetable matter. In a process called esterification, the large branched triglyceride lipid molecules in oils and fats are broken down into smaller chain molecules (esters and glycerol). The glycerol is removed, and the remaining esters are similar to those in petroleum diesel. They are not a new idea – Rudolf Diesel demonstrated in

Table 3.19 Emissions due to alternative fuels

If you replace	CO	HC	NOx	PM
Petrol by natural gas in an LDV with catalyst	Decreases (0.4–0.5)	Increases (1.5–2.0)	Decreases (0.4–0.6)	n/a
Petrol by methanol in an LDV with catalyst	No change (0.7–1.1)	Decreases (0.5–0.8)	Decreases (0.8–0.9)	n/a
Diesel by methanol in HDV	Decreases (0.5–0.8)	Decreases (0.4–0.6)	Decreases (0.4–0.75)	Decreases (0.1–0.2)
Diesel by biodiesel in HDV	Decreases (0.75–0.8)	Decreases (0.2–0.8)	Increases (1.1–1.2)	No change (0.6–1.2)

1900 that his engine would run on groundnut oil. They have received much press and research attention because of the perceived advantages of growing fuel sustainably rather than using up fossil supplies, especially if this could be combined with lower emissions (Table 3.19). Again they have not yet fulfilled their promise. It is more difficult to make a biofuel diesel engine run smoothly and start reliably, and emission performance is not startling.

In general, emissions of regulated air pollutants are reduced. Whether these alternative fuels become widely adopted depends on the economics of production and distribution, as well as on air quality. There are major side issues, such as the fertiliser use which is necessary to achieve high plant productivity.

Dimethyl ether (DME) is a liquified gas with handling characteristics similar to propane and butane, but with a much lower autoignition temperature. Hence it is suitable as an alternative fuel for diesel engines. The DME molecule consists of 2 methyl groups separated by an oxygen atom (H_3C-O-CH_3). On a large scale, DME can be produced in a similar manner to methanol via synthesis gas (a mixture of CO, CO_2 and H_2) from a variety of feedstock including (remote) natural gas, coal, heavy residual oil, waste and biomass. Particulate emission is very low, because of the absence of carbon–carbon bonds and the significant oxygen content. Formation of components such as PAH and Benzene, Toluene, Xylene (BTX) is reduced. Low NO_x emission is possible through fuel injection 'rate shaping' and exhaust gas recirculation.

3.2   TRAINS

Even within a rather compact and uniform region such as the European Union (EU), there is a great range of train types and operating regimes. The motive power may be either electric (so that the emissions are from power station point sources) or diesel electric (emissions from the train itself). The powers, weights and duty cycles of the trains and engines all vary greatly. As a consequence, the emission factors vary by an order of magnitude, and the calculation of total emissions needs detailed knowledge of train operating parameters, including factors such as mean distance between stops. Such calculations usually start with an estimate of energy consumption in kJ per tonne·km, which is independent of the motive power. This is then multiplied by the appropriate emission factor. The diesel engines used on diesel-electric trains are quite similar to those used on HGVs, and emissions will be reduced by similar proportions by the application of EURO II, III and IV technology.

The MEET Report calculated the following emission factors for all EU trains in 1998 and 2020 (Table 3.20). It was expected that the share of traffic powered by electricity would increase from 65 to 80% over this period, although a much lower proportion of the rail network would be electrified. It can be seen that electrical power generation leads to higher SO_2 emissions than diesel because of the higher sulphur content of the fuel; that other emissions are generally lower from power stations, and that both sources will be reduced by new fuel standards and emission controls.

In the UK, the Department for Transport, Local Government and the Regions (DTLR) specifies locomotive emission factors so that local authorities can

Table 3.20 EU train emission factors

Pollutant	Electrical power generation g $(kW\,h)^{-1}$		Diesel locomotive emissions g $(kW\,h)^{-1}$	
	1998	2020	1998	2020
SO_2	2.7	0.8	1.0	0.03
NO_x	1.2	0.35	12	3.5
HC	1.1	0.55	1.0	0.50
CO	0.08	0.04	4.0	0.50
PM	0.14	0.07	0.25	0.08

Table 3.21 DTLR locomotive emission factors*

Train type	Pollutant				
	CO	NO_x	PM_{10}	SO_2	HC
Pacer (diesel)	10.1	12.8	0.1	1.8	0.5
Sprinter (diesel)	1.4	19.2	0.1	1.3	0.7
Inter-city 125 high speed train	28.1	97.4	8.5	16.3	15.1
Class 47 locomotive plus seven passenger coaches	39.9	127.6	5.1	13.1	11.6
Class 37 freight (per locomotive, typical load)	24.5	51.8	5.1	15.1	12.6

Note
* Units in g km^{-1} per powered vehicle unless otherwise stated.

compute emissions in their regions. The currently-recommended values are given in Table 3.21.

In the US, where locomotive diesel engines are unregulated and have very high NO_x emissions, they account for about 10% of mobile source emissions of this gas. Emission factors are in the range 80–110 g (kg fuel)$^{-1}$, which is about three and ten times the corresponding values for the HGV and LGV fleets respectively. Future regulation will reduce these emissions by 55% after 2005.

3.3   SHIPPING EMISSIONS

There are around 86 000 ships in the world fleet, of which about half are cargo vessels, a quarter fishing vessels and 10% tugs. Shipping generally uses rather poor-quality diesel-type fuels variously termed bunker fuel oil (BFO, a heavy residue from petroleum refinery processing), marine gas oil and marine diesel oil (MDO). Large ships such as container ships, passenger ships and ferries typically use 60–80 tonnes of fuel each day, while at the other end of the size spectrum fishing vessels would use under a tonne per day. The calculation of emission factors involves a comprehensive accounting for shipping fleets in terms of fuel use and duty cycle (Table 3.22). There are good data available on tonnage and fuel use from organisations such as Lloyd's Register. The most sophisticated models allow for the different phases of a vessel's trip, such as berthing, cruising and manoeuvering, with specific emission factors for each.

Hence a medium-sized ferry, powered by medium speed diesel engines, burning 20 tonnes of 1% S fuel each day, and having a 50% duty cycle, will emit annually around 200 tonnes NO_x, 25 tonnes CO, 11 000 tonnes CO_2, 8 tonnes VOCs, 4 tonnes PM and 70 tonnes SO_2.

Although BFO may be up to 3% S, the sulphur contents of marine fuels are being reduced by legislation. Directive 93/12/EEC set a limit of 0.2% for the sulphur content of gas oils and 1% for fuel oils. The International Maritime

Table 3.22 Shipping emission factors (kg (tonne fuel)$^{-1}$) for different engine types

Engine type	NO_x	CO	CO_2	VOC	PM	SO_x
Steam turbines (BFO)	6.98	0.431	3200	0.085	2.50	20S*
Steam turbines (MDO)	6.25	0.6	3200	0.5	2.08	20S
High speed diesel engines	70	9	3200	3	1.5	20S
Medium speed diesel engines	57	7.4	3200	2.4	1.2	20S
Low speed diesel engines	87	7.4	3200	2.4	1.2	20S
Gas turbines	16	0.5	3200	0.2	1.1	20S

Note

* In the table above, S is the sulphur content of the fuel expressed as per cent by mass. Hence there would be S kg sulphur per 100 kg fuel, and 10S kg per tonne fuel. If all the S were emitted as SO_2, there would be 20S kg SO_2 per tonne fuel, because the molar mass of SO_2 is twice that of S.

Organisation is also introducing global limits for sulphur content, sulphur emissions, and NO_x emissions. This reduction is overdue, since shipping emissions fall outside the remit of, for example, the Gothenburg Protocol, and have not been effectively controlled. With the ongoing reduction in land-based emissions, the maritime S emissions have now taken on a more significant aspect. For example, North Sea shipping emissions alone in 1990 were similar in magnitude to the entire UK emission ceiling for 2010 set in the Gothenburg Protocol. Cost-effective measures to reduce acid deposition in continental Europe will include reductions to the S-content of bunkering fuels.

3.4   AIRCRAFT EMISSIONS

There has been sustained growth in civil aviation for the last 50 years, and it now plays a major role in global freight and personal transport. Today there are around 15 000 commercial aircraft operating over routes of around 15 million km, serving 10 000 airports and carrying 1.4 billion passengers each year. In 1995 a total 2500 billion revenue passenger km (RPK) were flown. Around 40% by value of the world's exports are transported by air. Demand has grown at around 5% each year during the last 20 years, and is expected to grow at a similar rate over the next 15, so that by 2015 there will be 7000 billion RPK flown. Predictions even further ahead are less certain, but estimates of around 25 000 billion RPK by 2050 have been made. This growth is expected to require a further 2–4 000 new airports. Because passenger and freight transport is concentrated over a relatively small number of high-density routes in and between Europe and North America, the fuel burned and resulting emissions are unevenly distributed as well (Figure 3.17).

The propeller-based fleet of the 1950s and 1960s was superseded by jets and then by turbofans. Cruise speeds have increased from 100 knots to 500 knots, and cruise altitudes for long-haul flights from 3 km for propeller-powered aircraft to 10–13 km today. The height increase has been designed mainly to reduce drag forces (and hence fuel consumption) and is environmentally significant because it means that a substantial proportion of the emissions is released into the stratosphere, where the increased atmospheric stability and reduced role of washout and rainout give the emissions a much longer residence time. Figure 3.18 shows the improvement in efficiency that has occurred over the last 50 years. During the period 1950–1997, average fuel efficiency (as measured by aircraft seat km per kg fuel) increased by 70%, and this is expected to be followed by a further 45% improvement by 2050. Aircraft fuel burn per seat has also decreased by about 70% during the period of jet passenger flight.

In addition to the commercial flights discussed above, flights are made by some 55 000 military aircraft (combat, trainer, tanker, transport), although the proportion of military flights has been falling and will continue to do so.

Aviation fuel currently makes up 2–3% of total fossil fuel consumption, which corresponds to 20% of all transportation fuel, with 80% of this used by civil

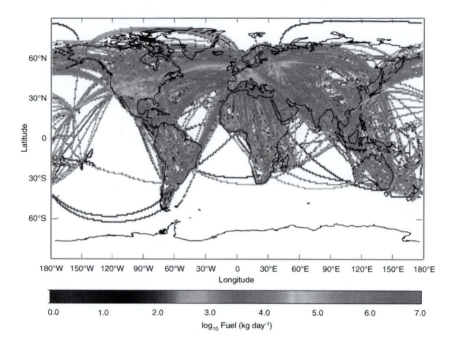

Figure 3.17 Global distribution of aviation fuel consumption.
Source: Intergovernmental Panel on Climate Change (1999) *Aviation and the Global Atmosphere,* Cambridge University Press, Cambridge.

aircraft. All these aircraft emit the usual range of combustion gases and particles into the atmosphere either at the airports, in the troposphere or stratosphere, and the pollutant burden has therefore been increasing too.

3.4.1   Emission calculations

Calculation of overall emissions, as with other sources, requires a detailed inventory of aircraft in use, flight patterns and emission factors. Operations such as taxing should be included. Taking off (and, to a lesser extent, landing) may be especially important because the engines are operated at high power and the emissions are released close to the ground.

Aircraft emission factors for different operational modes are given in Table 3.23. The emission indices for CO_2 and H_2O are simply those for stoichiometric combustion of kerosene (paraffin) which is the universal aviation fuel and of fairly uniform composition. All jet fuels are composed primarily of hydrocarbons as a blend of saturates, with no more than 25% aromatics. A fuel may contain up to 0.3% S by weight, although the general level is less than 0.1% and falling, with

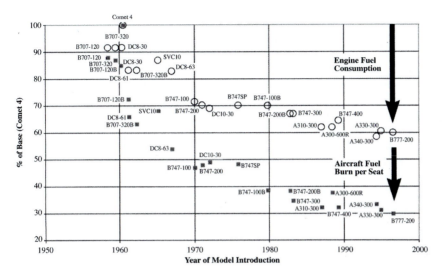

Figure 3.18 Reductions in engine fuel consumption and aircraft fuel burn per seat, 1950–1997.
Source: Intergovernmental Panel on Climate Change (1999), *Aviation and the Global Atmosphere,* Cambridge University Press, Cambridge.

Table 3.23 Typical emission indices (g kg^{-1}) in different operational modes

Species	Operating mode		
	Idle	Take-off	Cruise
CO_2	3160	3160	3160
H_2O	1230	1230	1230
CO	25 (10–65)	<1	1–3.5
HC (as CH_4)	4 (0–12)	<0.5	0.2–1.3
NO_x (as NO_2)			
Short haul	4.5 (3–6)	32 (20–65)	7.9–11.9
Long haul	4.5 (3–6)	27 (10–53)	11.1–15.4
SO_x	1.0	1.0	1.0

average reported values in the range 0.04–0.06%. Emissions of NO_x, CO, HC and soot, as with other fuel-burning devices, are strongly influenced by operational factors such as power setting and the inlet air-conditions. At low power, the engine is running cool, air–fuel mixing is least effective and CO/HC emissions highest. At full power, combustion temperatures are higher and the endothermic NO_x production greater. In principle there is scope for reducing emissions by

Table 3.24 Annual total fuel consumption and pollutant emissions in 2050

Approx traffic/10^9 RPK	Fuel burned/Tg	CO_2/Tg C	NO_x/Tg NO_2
20 000	800	800	11
40 000	2000	1700	14

varying fuel composition. For example, a fuel with higher H:C ratio will produce less CO_2. In practice, aviation kerosene is such a uniform composition around the world that there is little room for manoeuvre.

Emission data for each engine type are available from the International Civil Aviation Organisation (ICAO) Engine Exhaust Emission Databank. European emissions have been computed in a standard format from

$$TE_p = \sum_r \sum_p SE_{j,p,r} \times N_{j,r}$$

where

$$SE_{j,p,r} = \int_{D_1(A)}^{D_2(A)} FC_j(D_r) \times EI_{j,p}(D_r)dD_r$$

TE_p is the total emission of pollutant p (kg per period), $SE_{j,p,r}$ is the specific emission of pollutant p from aircraft/engine combination j on route r (kg per aircraft), $N_{j,r}$ is the number of aircraft of category j on route r during the period, $FC_j(D_r)$ is the fuel consumption of aircraft in category j (kg km^{-1}), $EI_{j,p}(D_r)$ is the emission index for pollutant p from category j (kg of pollutant per kg fuel), D_r is the distance between airports on route r, $D_2(A) - D_1(A)$ is the distance flown within area A.

Table 3.24 shows the total emissions of CO_2 and NO_x from aircraft (depending on the growth scenario) predicted by 2050.

3.4.2   Aerosol precursors

The exhaust plume of a jet aircraft contains around 10^{17} particles per kg of fuel burned, plus gaseous combustion products. These together constitute the precursors from which the particulate aerosol is constructed. Three distinct families of particle are formed: volatiles, non-volatiles (soot/metal) and ice/water (contrails). The volatile particles, which are only 1–10 nm in diameter, form initially from sulphuric acid, ions such as NO_3, and water vapour. Due to their small diameters they grow rapidly by coagulation and uptake of water vapour. Figure 3.19 shows particle size distributions in jet aircraft exhausts.

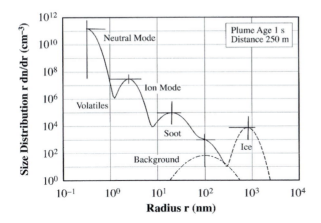

Figure 3.19 Particle size distributions in jet aircraft exhaust.
Source: Intergovernmental Panel on Climate Change (1999) *Aviation and the Global Atmosphere,* Cambridge University Press, Cambridge.

The smallest diameters (neutral mode) are formed by homogeneous nucleation of hydrated sulphuric acid molecules, and the next larger mode (ion mode) by rapid scavenging of charged molecular clusters by chemi-ions. These particles are not constant once they have been formed, but interact with others from the three families by coagulation, freezing, condensation and evaporation. Although the total aerosol emission is small in comparison to ground-level sources, the effects may be disproportionate due to the high release altitude.

Soot includes all primary, carbon-containing particles generated from incomplete combustion in the engine. Soot formation in gas turbines is complex and not yet fully understood. A simplified scheme is believed to involve the formation of nucleation liquid droplets with high carbon:hydrogen ratios, which then undergo rapid surface growth to form small soot particles. Some of the latter are charged and grow by aggregation to form soot. Soot emissions are specified on a mass basis (typically 0.04 g/kg fuel), and measured as smoke number, but neither of these measures involves the size distribution which is critical for environmental impact. Recent measurements have shown that the size distribution is log-normal and peaks at 20–30 nm, with concentrations of order 10^6–10^7 cm^{-3}, representing 10^{14}–10^{15} particles per kg fuel burned. The particles are nearly spherical and have diameters of 20–60 nm. Although initially separate, they soon aggregate into chains.

The sulphur content of aviation fuel is typically 400 ppm (i.e. 0.4 g S/kg fuel). Most fuel sulphur is expected to be emitted as SO_2, with 2–10% in the higher oxidised forms of H_2SO_4 and SO_3. The turbofan uses the hot exhaust gas flow to drive a fan which draws a 'bypass' flow through the engine and contributes to the thrust. The bypass ratio has been raised to increase engine efficiency, and along with this trend the combustion temperatures and pressures have also been increased. These high-bypass, more fuel efficient engines reduce CO_2 and H_2O emissions (due to greater fuel efficiency), and also HC and CO emissions. However, NO_x emissions are increased, and there is an inevitable trade-off.

3.4.3   Contrails

If the thermodynamic conditions are right, some of the water vapour forms condensation trails (contrails) of liquid/ice condensate. The relative humidity is elevated in the plume anyway because of the high water vapour content of the exhaust gases. As the plume cools it mixes with the ambient air. If cooling predominates, the temperature falls below the local dewpoint and condensation occurs. If mixing predominates, the dewpoint falls faster than the temperature and no condensation occurs. The activation temperature for contrail formation from exhausts of current engines is about 220 K (-53 °C). As engines are made more efficient, more of the fuel energy is used to propel the aircraft, and less to heat the exhaust gases. Hence the exhaust gas temperatures are closer to their dewpoint, and contrails are formed at higher ambient temperatures and over a larger range of altitudes. As efficiency has been increasing and specific fuel consumption decreasing, the critical altitude at which contrails will form in saturated air has decreased from around 8.5 km to 7.7 km.

The ice particles nucleate on exhaust soot and volatile particles. These will usually evaporate rapidly (seconds to minutes) into air of low relative humidity; while they do so they change the size and composition of the remaining liquid aerosol particles. If the relative humidity is above the ice saturation, the contrails can persist and grow by further deposition of ambient water vapour onto the crystals. Eventually the ice water content of the cirrus crystals is around 100 times the original exhaust water mass. In this sense the exhaust particles are 'seeding' the process. The cirrus can then have a significant impact on atmospheric transparency, which affects visibility, amenity and global heat budget.

Contrails are not visible until the exhaust is about 50 m behind the aircraft. This takes around 250 ms, during which the ice crystals grow large enough to give the plume a visible opacity. Persistent contrails have similar properties to cirrus clouds, which are characterised by large ice crystals that have significant terminal velocities and fall out of the cloud to give it a feathery edged appearance. Since aged contrails are indistinguishable from cirrus, it is hard to quantify their occurrence after they have lost their original linear structure. Correlations with fuel use have indicated that cirrus cover is increased by around 4% on average for the highest rates of fuel use above an altitude of 8 km. Although the modelled global mean cover is only 0.1%, the occurrence is naturally concentrated on the high density airline routes in and between North America and Europe (93% of aviation fuel is used in the Northern Hemisphere), where the local cirrus coverage can rise to 10%.

The increase in high-altitude cloud cover may contribute to the recorded decrease in diurnal surface temperature range that has been seen on all continents, although changes in other parameters such as aerosol loading and cloud height would also be implicated. High-flying aircraft can also disturb existing natural cirrus cloud, to which the exhaust adds additional freezing nuclei to enhance crystal production. Contrails in general will cause both a negative shortwave flux change (due to reflection of solar radiation) and a positive longwave flux change (due to absorption of upgoing terrestrial radiation). The net change due to 100%

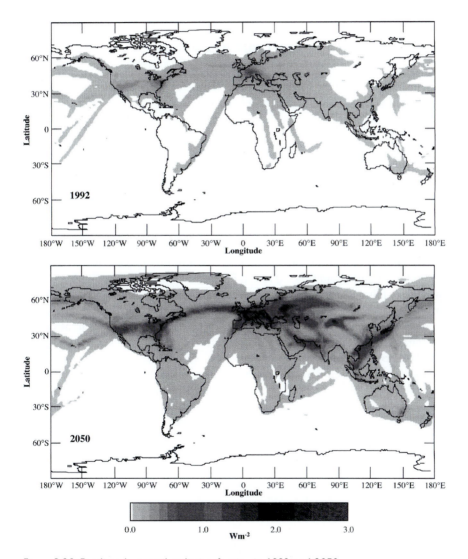

Figure 3.20 Predicted contrail radiative forcing in 1992 and 2050.
Source: Intergovernmental Panel on Climate Change (1999) *Aviation and the Global Atmosphere,* Cambridge University Press, Cambridge.

cirrus cover is estimated to be $+50$ W m^{-2} at the top of the atmosphere. When allowance is made for the small proportion of the Earth's surface which is directly affected by contrails, the maximum calculated forcing is around 0.7 W m^{-2}, which is similar in magnitude to other terms due to greenhouse gases and aerosol (see Chapter 11). The IPCC has calculated the likely changes in both the contrail cover and the resulting radiative forcing by 2050; the significant increases are shown in Figure 3.20

3.4.4   Dispersion of aircraft emissions

The normal cruise heights of medium and long-haul aircraft are between just-below and just-above the tropopause, and the height of the latter changes with latitude and season. Since the troposphere and stratosphere have very different mixing environments, predictions about plume dispersion vary from case to case. After release from the engine, the jet aircraft exhaust wake passes through three distinct regions in which mixing affects the potential impact of the constituents on the atmosphere. In the jet and wake vortex regions, which take up to 10 s and 10–100 s respectively, mixing is suppressed and only a small amount of ambient air is entrained. Hence high concentrations of emitted species react with small proportions of ambient air at rather elevated temperatures. When the wake vortex disintegrates, the plume region extends over a timescale between minutes and tens of hours and a distance scale up to thousands of km. Here the plume mixes into the ambient air in a Gaussian manner. During this process it may grow to 50–100 km in width and 0.3–1.0 km in height, with an eventual dilution of 10^8. Throughout this dilution process photochemical reactions and particle condensation and coagulation will further change the pollutants that had been emitted initially. NO$_x$, sulphur oxides, soot and water all have the potential to affect the concentration of atmospheric ozone. The effects do not necessarily act in the same direction. For example, increases in NO$_x$ in the upper troposphere and lower stratosphere are expected to increase ozone concentrations, whereas sulphur and water increases are expected to decrease ozone. Higher in the stratosphere, NO$_x$ increases will *decrease* ozone. So far, however, there is no direct observational evidence that aircraft movements have affected ozone. Aircraft emissions are calculated to have increased NO$_x$ at cruise altitudes by some 20%, although this change is substantially smaller than the observed natural variability. The effect is mainly a contribution to global warming; it is estimated that the small amount of NO$_x$ emitted into the free troposphere by aircraft (only about 3% of total emissions) generates global warming equal to that of the surface emissions.

3.4.5   Airport emission standards

Aircraft movements are concentrated at airports, where the emissions are closest to the surface and most likely to be noticed and to have effects on people and the

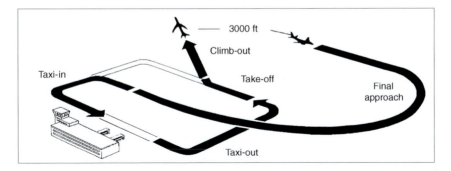

Figure 3.21 ICAO landing and take-off cycle.
    Source: Intergovernmental Panel on Climate Change (1999) *Aviation and the Global Atmosphere,* Cambridge University Press, Cambridge.

local environment. The ICAO has drawn up measurement procedures and compliance standards to control pollutant emissions from aircraft in the vicinity of airports. These cover soot (measured as smoke number), HC, CO and NO_x. The basis of measurement is the landing and take-off (LTO) cycle, which is illustrated in Figure 3.21.

Emissions are measured for appropriate power settings (approach 30%, taxing 7%, take-off 100% and climb-out 85%) then summed over the cycle, weighted by the time spent in each mode. Finally, the mass of pollutant released is normalised by dividing by the engine thrust in kN for comparison with the ICAO standards. As with motor vehicle emissions, the standards are being gradually tightened to improve emission performance. In the UK, DTLR has calculated emissions from specific airports based on their workload, fleet make-up and other factors (Table 3.25).

3.5   DIFFERENT MODES OF TRANSPORT

Comparisons between the air pollution implications of different transport modes are also of interest. Which comparisons are relevant depends partly on the route length and terrain. Long-distance passenger travel is dominated by aircraft, but most freight is carried on the surface. Trains compete with aircraft on land-based trips of up to several hundred kilometres. Roads and railways compete over shorter distances. Ships are involved at all distances, but need water. Ships also have specialist niches such as ferries and long distance movement of freight and raw materials. The air pollution impact can be quantified in various ways. We have seen how the overall emission or emission factor can be calculated per car, per train etc. In order to compare emissions per passenger·km (or per tonne·km for freight) we also need to know the capacity and load factor (proportion occupied).

Table 3.25 UK airport emissions (tonne a^{-1})

	SO_x	NO_x	CO	CO_2	NMVOC	Benzene	Movements per year	Annual arrivals	Annual departures	Fuel use
City 1994–1995	1	19	23	24 914	3	0	16 945	8505	8440	2107
Gatwick 1994–1995	1417	2626	1608	543 569	613	12	186 005	93 296	92 709	97 277
Heathrow 1994–1995	6666	5060	4275	1 280 312	926	18	438 113	219 147	218 965	322 988
Liverpool 1995	3	36	482	138 283	33	1	94 049	47 454	46 595	4532
Bristol 1994–1995	3	71	173	79 846	14	0	54 305	26 885	27 420	5129
Manchester 1994–1995	37	899	847	47 523	223	4	162 620	80 900	81 721	63 859
Birmingham 1994	6	165	329	8381	46	1	69 104	NA	NA	NA

Table 3.26 Emissions due to different passenger transport modes

Mode	CO_2/g (passenger km)$^{-1}$	NO_x/g (passenger km)$^{-1}$
Air	125	0.4
Rail	20	0.05
Road	70	0.23

Table 3.27 Emissions due to different cargo transport modes

Mode	CO_2/g (tonne km)$^{-1}$	NO_x/g (tonne km)$^{-1}$
Water	65	0.83
Train	42	0.08
Road	92	0.81

Typical rail journeys work out at about 40% of the emissions of the various regulated pollutants per passenger km, compared to those from a petrol-engined car conforming with EURO I. The MEET Report considered the CO_2 and NO_x emissions created by two typical trips in Europe. First, a passenger travelled from Vienna to Rome by air, rail or road (Table 3.26). Trains were the clear winner. Second, a freight cargo was moved from Rotterdam to Zurich by water (mainly), train or road. Again the train is the winner (Table 3.27), especially for NO_x emissions. The inland water transport carries far more than a goods vehicle, and is more fuel efficient. However, the emission per tonne of fuel from the ship engine is also much higher. In practice the decisions on what transport mode to employ are rarely made on pollution grounds, but on speed, cost or convenience. For example, the freight cargo had to be transhipped by road for the last part of its journey by either water or rail.

In terms of the CO_2 emitted per passenger km, long-haul aircraft are similar to passenger cars at 30–40 g C per passenger km. However, one long-haul flight can cover as many km in a day as the average car covers in a year, so the total emissions must be considered. For freight, aircraft cannot compete. Air freight transport generates 100–1000 g C per tonne km, as against 1–10 for ships, 1–20 for trains and 10–100 for HGVs.

FURTHER READING

Colvile, R. N., Hutchinson, E. J., Mindell, J. S. and Warren, R. F. (2001) 'The transport sector as a source of air pollution', *Atmospheric Environment* 35: 1537–1565.

Eastwood, P. (2000) *Critical Topics in Exhaust Gas Aftertreatment*, Research Studies Press Ltd, Baldock, UK.

Eggleston, H. S., Gaudioso, D., Gorisen, N., Joumard, R., Rijkeboer, R. C., Samaras, Z. and Zierock, K.-H. (1992) *CORINAIR Working Group on Emission Factors for Calculating 1990 Emissions from Road Traffic. Volume I: Methodology and Emission Factors.* Final Report on Contract B4-3045 (91) 10PH, CEC, Brussels, Belgium.

European Commission (1999) *MEET – Methodology for Calculating Transport Emissions and Energy Consumption*, EC, Luxembourg.

Heck, R. M. and Farrauto, R. J. (1995) *Catalytic Air Pollution Control: Commercial Technology*, Wiley, New York.

House of Commons Transport Committee (1994) *Transport-Related Air Pollution in London*, HMSO, London, UK.

Intergovernmental Panel on Climate Change (1999) *Aviation and the Global Atmosphere*, Cambridge University Press, Cambridge.

Lloyd, A. C. and Cackette, T. A. (2001) 'Diesel engines: environmental impact and control', *Journal of Air and Waste Management Association* 51: 809–847.

Schwela, D. and Zali, O. (eds) (1999) *Urban Traffic Pollution*, Spon, London.

United Kingdom Quality of Urban Air Review Group (1993a) *Urban Air Quality in the United Kingdom*, DoE/HMSO, London, UK.

United Kingdom Quality of Urban Air Review Group (1993b) *Diesel Vehicle Emissions and Urban Air Quality*, DoE/HMSO, London, UK.

Watkins, L. H. (1991) *Air Pollution from Road Vehicles*, HMSO, London, UK.

Chapter 4

Measurement of gases and particles

So far we have discussed the emission and control of trace gases and particles. In this chapter we will look more closely both at how to sample them correctly and how to measure their concentrations. Trace gases, by definition, normally have concentrations in the ppb and ppt ranges, and therefore require more specialist techniques and sensitive methods than those in general use for gases such as water vapour or carbon dioxide that occur at percentage or ppm concentrations. Although filtration provides the simplest method of measuring particles, other techniques enable real-time measurements, or size distributions, or chemical analyses to be made. For both gases and particles, correct sampling is important to avoid corruption of the ambient concentration.

4.1 METHODS OF DESCRIBING POLLUTANT CONCENTRATION

As with many variable quantities, a wide variety of methods is in use for describing the amount of material present. Suppose that we have a continuous record of concentration measurements of one pollutant at one location, taken every minute for a year. This would give us 525 600 individual values. Although this is a comprehensive data set, it is not a lot of use in this form. We must process the data and present it in such a way that the information we want is made clear without having to wade through the half-million values each time. We might wish to decide, for example, on the effects that might result from the pollutant, or compare the concentration record with previous records from the same location or with records from other locations. In the remainder of this section, a brief description is given of the common methods in use for describing pollutant occurrence. In Chapter 7, these methods are used on a real data set.

The most straightforward measure of such a signal is the time-weighted average – the arithmetic mean of all the original 1-min values. This gives us a single value with which to represent the concentration record. It does not tell us anything about changes during the course of the year, or about the magnitude or duration of either high or low episodes.

The next stage of processing might be to average the 60 values from each hour, and to plot these out as a complete sequence for the year. A 1-h average is usually regarded as being free from atypically short-term spikes. Such a record would also help to clarify any annual cycles due to changes in source strength or dispersion meteorology, although it is likely that many such annual cycles would need to be averaged to smooth out random variations.

If a systematic daily variation is suspected, as with ambient ozone concentrations or pollutants due to vehicle movements, then this can be clarified by calculating the average concentration during each of the 24 h of the day from all 365 available values. Naturally this process can be refined to give weekly, monthly or seasonal means.

Many air pollutant concentrations are highly variable, with a low average and a large range between near-zero minimum values and maxima during episodes. Furthermore, effects on organisms such as people and plants often depend on short episodes of high concentration. How can we summarise this breadth? The standard method is to examine the frequency distribution of the data, by making a tally of the numbers of hourly or daily means that fall in the ranges 0–4, 5–9, 10–14…ppb, for example. The number in each range is then converted to a proportion or percentage of the total number of readings – this is the frequency. A common result of such an analysis of air pollutant concentrations is that the data approximate to a log-normal frequency distribution – the log of the frequency is normally distributed with concentration. This fact has certain implications for the way in which the data are summarised, and a new set of statistical values must be used. The geometric mean replaces the arithmetic mean, and the geometric standard deviation replaces the standard deviation.

If all the proportions within each concentration range are summed and plotted against concentration, we have the cumulative frequency distribution. This is sig-moidal on linear scales, but can be transformed to a straight line on log-probability axes. The gradient of this line then gives the value of σ. From the line itself we can derive the proportions of the time for which any particular concentration is exceeded, or the concentration exceeded for certain proportions of the time (e.g. 50%, 10% and 2%). The latter values are often used in pollution control legislation.

The various parameters that can be identified from a given distribution are strongly correlated. For example, the 98 percentile hourly-mean NO_2 concentrations measured by UK urban monitors have been consistently around 2.3 times the corresponding annual means.

If the record is averaged over different periods, it is often found that $c_{max,t}$, the maximum average over any period t, is related to the 1-h maximum by an equation of the form

$$c_{max,t} = c_{max,h}\, t^q$$

where q is an exponent with a typical value of -0.3.

For example, if $c_{\text{max, 1 h}}$ (the maximum 1-h average) = 100 ppb, then $c_{\text{max, 1 day}}$ (the maximum 24-h average) = $100 \times (24)^{-0.3} = 39$ ppb and $c_{\text{max, 1 month}} = 100 \times (720)^{-0.3} = 9$ ppb.

4.2   SAMPLING REQUIREMENTS

There are many different reasons why air samples might be needed:

- Determination of community air quality as related to local health, social and environmental effects – for example, NO_x measurements in an urban area.
- Investigation of the influence of specific emission sources or groups of sources on local air quality – for example, CO near a motorway interchange.
- Research into dispersion, deposition or atmospheric reaction processes – for example, measurements to validate dispersion software using the Gaussian dispersion equation.

Before any sampling and measurement programme, the objectives must be defined clearly. These will in turn lead to specifications of sample point locations, frequency of measurement, and other factors. There are no fixed rules on any of these aspects – it is down to experience and common sense. Sample inlets will normally be away from surfaces such as walls and hedges that might affect the concentration. For human relevance, they will usually be at about 2 m above the ground; however, lower heights will be appropriate for babies or children, and for growing vegetation the inlet may need to be adjusted periodically to remain close to, but not within, the growing canopy. There should be no major obstructions, or interfering local sources, close upwind. The site must be both accessible and secure from large animals or vandalism.

4.3   GAS SAMPLING

Most measurement methods require the pollutant, or the air containing the pollutant, to be sampled. The purpose of sampling is to get some air to the analyser without changing the concentration of the gas to be measured, and in a condition that the analyser can accept. We can divide the available methods into three categories: pumped systems, preconcentration and grab samplers. There are also combinations of these.

4.3.1   Pumped systems

A sample is drawn continuously from the ambient air and delivered directly to the analyser (Figure 4.1). Sampling the gas will involve the inlet itself

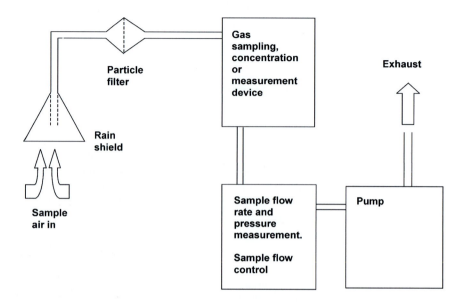

Figure 4.1 Component parts of a gas sampling system.

(normally shrouded to prevent the ingress of rain), tubing, the analyser, flow measurement and a pump. There are many potential sources of error in this combination:

- There are always particles in the air. If we draw these particles in, they may be deposited to the sample tube walls and to the internal parts of the analyser; they may produce a spurious signal from the analyser equivalent to a gas concentration (for example, a flame photometric sulphur analyser will respond equally whether the sulphur is in sulphate particles or in sulphur dioxide); or they may decompose to release a measurable gas concentration. The particles are therefore usually trapped on a filter close to the sample inlet, but this may introduce errors. For example, if ammonium nitrate particles are trapped on a filter, they can decompose and increase the measured concentration of ammonia gas. When sampling ozone, the build-up of particles on the filter can absorb ozone before it even enters the sample tube.
- As the air sample is drawn through tubing to the analyser, it can be adsorbed on the walls or on connecting fittings in transit. The severity of this effect varies with the gas and the tubing material – ozone must be sampled through glass, stainless steel or polytetrafluorethylene (PTFE, an inert plastic), SO_2 is less reactive and CO_2 is quite inert. The contact time is important here – samples taken down long lengths of tube must have high flow rates to minimise the residence time, which should be kept below 10 s.

- There may be reactions between the gas of interest and other species while the gas is flowing along the tube. For example, O_3 and NO and NO_2 are in a dynamic photochemical equilibrium in the atmosphere. As soon as they are sampled, the conditions are changed and the concentrations will start to change in consequence. The magnitude of this effect varies from system to system. When performing multi-point samples with one analyser, the analyser must be time-shared between the different locations. It is then usual to draw air rapidly and continuously through main sample lines, past a point that is physically close to the analyser. At this point, solenoid valves are used to extract samples cyclically to the analyser from each of the tubes in turn. In this way the dead-time spent by air in tubes is minimised, as are the opportunities for adsorption and chemical reactions.
- The pump will usually come last in the flow path to avoid interfering with the air sample before measurement. The gas sampling system will then be below atmospheric pressure. Hence there is a risk of leaks occurring, in which air from other locations, having a different concentration of the gas to that of the original sample, is drawn into the sample line and delivered to the analyser. Such leaks can occur at any of the connection fittings required for items such as particle filters, solenoid valves or flow meters, and can be very hard to detect in a complex sampling system which may involve dozens of such connections. Air can only flow down a pressure differential, and the pressure in such a system therefore becomes steadily more negative from the sample inlet towards the pump. Fittings near the pump are thus the most at risk from leaks.
- As well as particles, ambient air always contains water vapour. Provided that the water remains in the vapour state, no problems should result. If the sampling system allows the air to cool below the dew-point temperature, then the vapour will eventually become saturated and the excess water will condense out in the sample lines. This can cause problems. First, there is clearly a risk of liquid water being drawn right through the sample system – this is not good news for a £10 000 gas analyser. Second, any drops of liquid water lining the sample tubes will be sites for rapid absorption of some gases such as ozone and sulphur dioxide, possibly nullifying a major investment in PTFE tubing. One possible solution is to remove some of the water vapour by passing the air sample through an adsorbent such as silica gel. Although this works well for relatively unreactive gases present at ppm concentrations, such as carbon dioxide, it is likely to cause unacceptable loss or corruption of ambient trace gases. The preferred solution is to heat the sample lines so that the air is about 10 °C above ambient temperature. Proprietary sample tube is available that has concentric heating elements and insulation built into the tube itself, although it is expensive. An effective alternative can be constructed by wrapping heating tape around the tube before placing it inside plastic drainpipe for physical protection and insulation. The temperature should be regulated thermostatically to maintain the constant elevation above ambient.

4.3.2   Preconcentration

Gas analysers have sufficient sensitivity and speed of response to give a continuous measurement of the pollutant concentration as the sample is delivered to them. Many other techniques simply capture the pollutant, or a chemical derived from it, from the sample for later quantitative analysis in the laboratory by standard methods such as titration and spectrophotometry.

4.3.2.1   Absorption

The common theme is that a specific chemical reaction between the gas and captive absorbing molecules retains all the gas from the sample air. There are many different practical arrangements:

Bubblers

The sample is bubbled through an absorbing chemical solution; the gas is absorbed (a process which is controlled by the equilibrium partial pressure of the dissolved gas over the liquid surface), and preferably then also undergoes an irreversible chemical reaction which prevents it from being subsequently desorbed. If the collection efficiency is too low, then a fritted glass diffuser can be used which reduces the bubble size and hence increases the interfacial contact area and effective contact time, at the expense of higher pressure drop and lower flow rate.

Ideally, 100% of the pollutant molecules will be absorbed while passing through the solution. Absorbers such as bubblers can still be used at lower efficiencies provided that the efficiency is known and constant. Alternatively, the concentration can sometimes be measured by using two bubblers in series. For example, a volume of air V containing a concentration C of the pollutant is sampled. The total mass of pollutant in the sample is then $M = VC$. As the sample passes through the first absorber, a mass M_1 of pollutant is retained, at an efficiency $E = M_1/VC$. The mass of pollutant passed through the second absorber is reduced to $VC - M_1$, of which M_2 is retained.

$$E = \frac{M_1}{VC} = \frac{M_2}{VC - M_1} \quad \text{and hence} \quad C = \frac{M_1^2}{V(M_1 - M_2)}$$

The method assumes that the efficiency is the same in both absorbers. In practice, the efficiency is itself often a function of the concentration, and is lower at lower concentrations.

Impregnated filters

The sample is drawn through a filter paper that has been impregnated with a chemical that reacts with the gas. Often several different filters will be used. The first is a high-efficiency untreated particle prefilter, which prevents particles from interfering with the gas measurements on subsequent filters. It is important that this prefilter does not absorb a significant proportion of the gases of interest – for example, glass fibre filters collect low concentrations of both SO_2 and HNO_3 rather efficiently, while nylon filters absorb HNO_3 and HCl. The second and third filters might be impregnated with two different chemicals to absorb two different gases – for example, KOH for SO_2, NaCl for HNO_3, or H_2SO_4 for NH_3. The collection efficiency of impregnated filters is usually less than 100%, and can vary with atmospheric humidity.

Passive samplers

This category covers any preconcentration technique which relies on the gas diffusing to an absorber without pumping. Passive samplers rely on diffusion of the sample rather than pumped flow. They are cheap to purchase and do not require electrical power, so they are useful for large-area surveys. On the down side, they are labour and laboratory intensive and provide average concentrations over days to weeks, so the results are not directly comparable with standards. They were originally developed so that workers could carry out a full day's work with an unobtrusive sampling device clipped to their clothing. The sample would then be analysed to give the average concentration during the period. The principle has been adapted to give measurements of several ambient gases – NO_2, VOCs, SO_2 and NH_3.

In the simplest type of diffusion tube, a linear molecular diffusion gradient is set up, between the atmospheric concentration at the lower end of an inert plastic tube (7 cm long and 1 cm diameter) and zero concentration at an absorbent-coated substrate at the upper end (Figure 4.2). The gas molecules diffuse through the air molecules down the gradient and are absorbed on the substrate. The rate of diffusion depends only on the constant molecular diffusivity of the gas, since the long thin tube inhibits turbulent transport on to the substrate. However, high windspeeds can cause turbulent transport up the tube to be a significant source of error. The tube is left in position for a period of days to weeks, capped to prevent further absorption, and returned to the laboratory for analysis.

A quantity Q mol of the gas is transferred onto the substrate in time t, where Q is given by

$$Q = D_{12}At(C_a-C_0)/z$$

D_{12} is the diffusion coefficient of gas 1 (the pollutant) in gas 2 (air), A is the cross-sectional area of the plastic tube, C_a and C_0 are the ambient (at the open end of the tube) and substrate concentrations of the pollutant, respectively. C_0 is usually assumed to be zero, z is the length of the tube, or the diffusion pathlength.

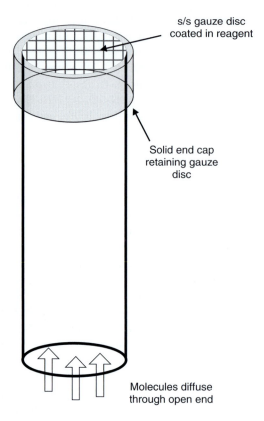

s/s gauze disc
coated in reagent

Solid end cap
retaining gauze
disc

Molecules diffuse
through open end

Figure 4.2 A diffusion tube.

Hence Q is proportional to the average concentration of the pollutant at the mouth of the tube over the exposure period. The same issues arise as for other sampling methods – particularly linearity, specificity, and the effects of temperature, windspeed and humidity. Some of the absorbers used, such as triethanolamine (TEA), will absorb more than one gas. This does not matter provided that the detection method is specific to the gas of interest. More complex designs involving diffusion screens have been used in order to compress the tube into a badge geometry, and tubes packed with solid adsorbent have been used for sampling organics.

Denuder tubes

Denuder tubes are a cunning method of avoiding the potential interference between the particle and gas phases of a pollutant – for example, with nitric

Table 4.1   Examples of passive sampler combinations

Gas	Absorbent	Analytical method
Ammonia	Oxalic acid	Colorimeter or Spectrophotometer
Hydrogen sulphide	Silver nitrate	Fluorimeter
Nitrogen dioxide	Triethanolamine (TEA)	Colorimeter or Spectrophotometer
Ozone	Sodium nitrite	Colorimeter, Spectrophotometer or Ion chromatograph
Sulphur dioxide	Sodium hydroxide	Colorimeter
VOCs	Activated charcoal or Tenax	Gas chromatograph or GC-MS

acid vapour and particulate nitrates. The air sample is drawn through a hollow tube, the inside of which is coated with an absorbent for the gas. The flow rate and tube diameter are selected so that the flow is laminar. Under these conditions, the molecular diffusivity of the gas is thousands of times greater than the Brownian diffusivity of the particles, so that the gas moves to the walls much more quickly. The length of the tube is selected so that by the time the air leaves the tube, most of the gas has been absorbed and most of the particles remain suspended. The particles can then be filtered separately. In a more sophisticated arrangement called an annular denuder, the air sample flows between two concentric tubes. The outer surface of the inner tube and the inner surface of the outer tube are both coated with the absorbent, which gives higher sampling efficiencies, shorter tubes, higher flow rates or shorter sampling times.

4.3.2.2   Adsorption

The gas molecules are bound to the surfaces of a solid collecting substrate by intermolecular forces. Subsequently they are stripped off by heating the substrate to 300 °C, or by solvent extraction, and delivered into a gas chromatograph or equivalent analysis system. The most common adsorbent substrates are activated charcoal, organic polymers and silica gel. The most widely used commercial product is Tenax, which has been used to sample a large range of organic compounds such as PAH, vinyl chloride, and hydrocarbons including benzene.

4.3.2.3   Condensation trapping

The air sample is passed through a chamber cooled almost to the temperature of liquid oxygen (-183 °C) so that volatiles condense out but the air itself passes through unchanged. Cooling will normally be arranged in a sequence of stages of reducing temperature, so that the majority of the water vapour is collected first.

4.3.3   Grab sampling

For various reasons, it may be impracticable to either measure the concentration continuously or to extract the pollutant gas from an air sample for subsequent analysis. In these cases, the fall-back technique is to capture a sample of the air itself, and take it back to the laboratory. There are several different ways in which this is put into practice.

Evacuated bottles – a stainless steel, glass or inert plastic bottle is evacuated in the laboratory and sealed securely. The bottle is opened in the ambient air, and fills with the sample.

Syringes – the plunger on an empty syringe is pulled out, drawing in a sample of ambient air.

Bellows – a collapsed cylindrical bellows is expanded, creating a volume at low pressure and drawing in the ambient air sample.

Bags – a bag made of an inert material is filled by pumping the air sample into it. The major disadvantage of this method is that the air must pass through the pump before entering the bag, so that reactive gases may be corrupted. In order to avoid this limitation, the bag can be held inside a bottle. The space *outside* the collapsed bag can then be evacuated by pumping out the air. This draws an air sample into the bag. The sample cycle can be repeated many times.

All of these techniques need very careful assessment and management to be successful. The bags and bottles need to be made of inert material and scrupulously clean and leak-free. For some applications, preconditioning of the sample holder with high concentrations of the pollutant gas may be necessary to reduce uptake of the gas from subsequent samples. The samples must be returned to the laboratory as rapidly as possible for analysis, before they degrade. By their nature, grab samples tend to be used at remote sites to which a sampling expedition is a major undertaking. If any of these quality assurance aspects is insecure, it may only be discovered that the samples are worthless after returning home. The risks can be minimised by a careful validation programme – comparison of concurrent real-time and grab samples, and sequential analysis of stored samples over a period of time.

4.4   GAS CONCENTRATION MEASUREMENT

There are seven factors that need to be balanced when choosing how a particular trace gas is to be measured:

Specificity – does the method measure only the gas of interest, or does it have a response to some other gas or gases. If so, is this interference or cross-sensitivity likely to be significant at the likely concentrations of all the gases involved? When the peroxide bubbler technique first became used for SO_2 determination, it was known that it would also respond to alkaline gases such as ammonia, and to other acid gases such as nitrogen dioxide, but there was no problem because the SO_2 concentrations far outweighed the others. When SO_2 concentrations decreased, this imbalance became less certain and a more SO_2-specific analysis was introduced.

Sensitivity – will the method measure the highest and lowest concentrations expected? If it measures the highest, then the lowest may have a large error due to instrument noise. If it measures the lowest with sufficient sensitivity for the noise not to be a problem, then there might be a risk of it going off-scale during a pollution episode. This is a very frustrating event, since it is always the episodes that create the greatest interest.

Reliability – is the instrument to be used continuously or intermittently? If continuously, will it be visited daily, weekly or less frequently? Is it important to obtain a complete record, or will it be acceptable to have the analyser out of action for maintenance periods?

Stability – this is an important issue for continuous analysers that are unattended for long periods. They will be calibrated (zero and span values confirmed by measuring the output from standard gas sources) and then left to operate unattended. During the following measurement period, the zero and span outputs will both drift away from their settings, so that an error is introduced into the measurement. The speed of this drift, and the relative magnitude of the resulting errors, will determine the frequency of the calibrations.

Response time – for grab sampling or preconcentration, the response time is the time over which the sample is taken. An almost instantaneous sample can be obtained by opening a valve on an evacuated container, or sample entry can be extended over any longer period. No information is available on concentration variations that are shorter than the sample period. In the same way, preconcentration methods such as bubblers give an average value over the length of the sample period. In principle, continuous analysers are capable of tracking the actual concentration of the gas. How closely this ideal is met in practice depends on how rapidly the concentration changes and the response time of the analyser. Response times are quoted by manufacturers in terms of the time to reach 90% or 95% of a step change in the input – times for modern analysers are typically one or two minutes.

Precision and Accuracy – these words are sometimes used interchangeably, but they have fundamentally different meanings, as shown on the three archery targets of Figure 4.3.

In Figure 4.3(a), the arrows have high precision (they have all landed within a small area of the board) but low accuracy (they are all well away from the centre). In Figure 4.3(b), the arrows have high accuracy (the average is close to the centre), but low precision (they are scattered all over the board). Figure 4.3(c)

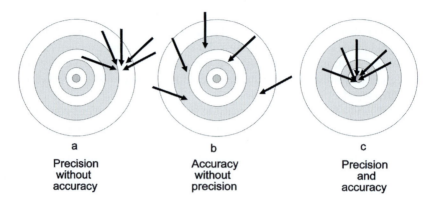

a	b	c
Precision without accuracy	Accuracy without precision	Precision and accuracy

Figure 4.3 The difference between precision and accuracy.

shows the ideal, with both precision and accuracy being high. Analyser manufacturers quote overall bands within which the errors due to all sources are expected to fall.

Cost – although last on this list, cost is bound to weigh heavily in the selection process. The range of possible costs is enormous. Preconcentration methods such as bubblers and diffusion tubes cost only a few pounds to obtain the sample itself, but have substantial operational costs due to labour and laboratory facilities. Continuous analysers cost c. £10 000, and substantial additional investment is also needed – in data logging and computing equipment, servicing and spares, air-conditioned cabin and calibration facilities.

It is not possible to give here a comprehensive review of all the different measurement methods available for specific gases. We will simply give a few examples of the key techniques in common use.

4.4.1   Wet chemical methods

The air sample is bubbled at a low rate (typically a few litres min^{-1}) through a solution that will absorb the required gas. At the end of the sample period, which is typically hours or days, the solution is returned to the laboratory for analysis by one of a wide range of standard techniques.

Example – SO_2

Air is bubbled through dilute hydrogen peroxide (H_2O_2). Any SO_2 is converted to sulphuric acid (H_2SO_4). The amount of acid created can be measured by titration or by pH change. In this form the method is not specific to SO_2, since other acid-forming gases give an increase in apparent concentration, while alkaline gases such as ammonia give a decrease. Specificity can be improved by analysing colorimetrically for the sulphate ion. The method was adopted for the National Survey monitoring sites – although it was adequate at the high concentrations of the 1960s, the errors are too great at the lower concentrations now prevailing. The standard wet chemical method for low concentrations is known as West-Gaeke, and involves more complex chemistry to form red-purple pararosaniline methyl sulphonic acid, which is determined colorimetrically.

4.4.2   Real-time pumped systems

Any physical or chemical effect of which the magnitude can be related reliably and repeatably to the gas concentration can in principle be used. In this section we describe techniques in which an air sample is pumped continuously through the instrument.

Example 1 – pulsed fluorescence for sulphur dioxide

The sample air is irradiated with intense UV (wavelength 214 nm) to stimulate fluorescence in a waveband centred on 340 nm (Figure 4.4). The flux density of the emission is measured by a photomultiplier. The sample air must normally be conditioned before the measurement to prevent spurious responses from obscuring the true signal. First, water vapour (which quenches the fluorescence) must be reduced by a diffusion drier; second, hydrocarbons (which augment the fluorescence) must be reduced by passing the air through a scrubber. The main initial drawback of this method was the slow response time, but engineered improvements have made pulsed fluorescence the current method of choice for SO_2.

Example 2 – UV absorption for O_3

Ultraviolet radiation at 254 nm from a low-pressure mercury vapour lamp traverses the length of a tube through which the sample air is pumped continuously (Figure 4.5). The attenuation of UV flux density at the far end of the tube depends on the O_3 concentration. The attenuation is compared with that due to air from which the ozone has been scrubbed by activated charcoal. The Beer–Lambert law ($I = I_0 e^{-ax}$) is then used to relate the attenuation to ozone concentration using the known extinction coefficient for ozone. This is a simple and reliable method for ozone determination.

Example 3 – chemiluminescence for NO

A supply of ozone is generated within the analyser and reacted with the NO in a sample cell (Figure 4.6). Nitrogen dioxide is generated in an excited state, and

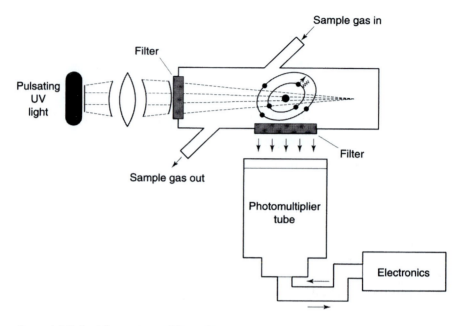

Figure 4.4 Pulsed fluorescence SO₂ analyser.

Source: Harrison, R. M. (1999) 'Measurements of the concentrations of air pollutants', In: S. T. Holgate, J. M. Samet, H. S. Koren and R. L. Maynard (eds) *Air Pollution and Health*, Academic Press, London.

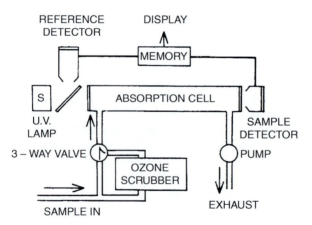

Figure 4.5 UV absorption O₃ analyser.

Source: Harrison, R. M. (1999) 'Measurements of the concentrations of air pollutants', In: S. T. Holgate, J. M. Samet, H. S. Koren and R. L. Maynard (eds) *Air Pollution and Health*, Academic Press, London.

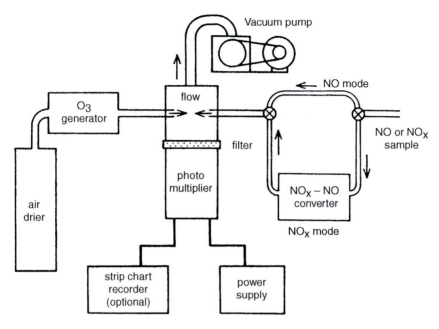

Figure 4.6 Chemiluminescent NO_x analyser.
Source: Harrison, R. M. (1999) 'Measurements of the concentrations of air pollutants', In: S. T. Holgate, J. M. Samet, H. S. Koren and R. L. Maynard (eds) *Air Pollution and Health*, Academic Press, London.

decays to the ground state with photon emission. The flux density is measured with a photomultiplier. In most commercial analysers, the NO measurement described above is alternated with a measurement on air that has been passed through a molybdenum catalyst to convert any NO_2 into NO. Hence the uncatalysed measurement gives NO directly, while the difference between that and the catalysed measurement gives NO_2. It is not only NO_2 that is reduced to NO, and a variable proportion of other N-containing gases such as HNO_3 will be detected as well. By using a higher temperature catalyst, ammonia can be oxidised to NO to give the NH_3 concentration by difference. Since typical ambient NH_3 concentrations are only a few percent of NO_x, the errors on the NH_3 value are unavoidably large.

Example 4 – infrared (IR) absorption for CO_2 (and many other gases)

Infrared radiation is attenuated while passing along a tube filled with the sample. By selecting specific wavebands to coincide with the absorption bands of different gases, several hundred different gases can be measured – but the species of gas must be known first. Portable infrared gas analysers are available that can determine the concentration of many gases in air. They employ a continuous IR source, a variable frequency interference filter to select a narrow frequency range at which the absorptivity of a particular gas is known, and a long optical path folded by means of mirrors into a compact case. The lower detection limit is typically around 1 ppm, so these analysers are mainly of use in the workplace environment.

4.4.3   Real-time remote systems

Some detectors measure the optical properties of the gas, and have been designed so that the reflected or transmitted signal is received after an extended path-length through the air. This arrangement offers the advantages of eliminating the possibility of sample degradation during passage to and through an instrument, and of integrating the concentration over a region of space rather than sampling at one point.

Light detection and ranging (LIDAR) describes a family of active remote sensing methods. The most basic technique is long-path absorption, in which a beam of laser light is reflected from a distant retroreflector and returned to a detector which is co-located with the source. A retroreflector is a corner cube which works in the same way as a 'cat's eye' road marker as used in the UK, and returns the incident beam in exactly the opposite direction that it came from, no matter what that direction. The wavelength of the radiation is chosen so that it coincides with an absorption line of the gas of interest. The concentration of that gas is found by applying the Beer–Lambert law to the reduction in beam flux density over the path length. No information is obtained on the variation in density of the gas along the path length; if the gas is present at twice the average concentration along half the path, and zero along the other half, then the same signal will be received as for uniform distribution at the average concentration.

Example 1 – differential absorption lidar (DIAL)

Pulses from a tuneable laser are directed into the air at two wavelengths, and the backscattered signals from the air molecules, gas molecules and particles are measured by a cooled detector. One laser wavelength (λ_{min}) is just to one side of an absorption band for the gas of interest, and calibrates the backscatter of the lidar system for molecular (Rayleigh) and aerosol (Mie) scattering that occurs whether the gas of interest is present or not. The second laser wavelength (λ_{max}) is tuned to an absorption band, so that the difference between the two can be used to derive absorption due to the gas alone.

The ratio of the scattered flux density at the two wavelengths is given by

$$\frac{I(\lambda_{max})}{I(\lambda_{min})} = \exp(-2RN\sigma)$$

Table 4.2 Absorption wavelengths and cross sections for dye lasers

Molecule	Peak absorption wavelength/nm	Absorption cross section/ 10^{-22} m^2
Nitric oxide	226.8, 253.6, 289.4	4.6, 11.3, 1.5
Benzene	250.0	1.3
Mercury	253.7	56000
Sulphur dioxide	300.0	1.3
Chlorine	330.0	0.26
Nitrogen dioxide	448.1	0.69

where σ is the absorption cross section of the target species at wavelength λ_{max}, R is the range and N is the number concentration of the gas.

By measuring the time for the back-scattered signal to return, the range can also be determined to within a few metres over a distance of 2000 m. The technique has been used for studying plume dispersion and vertical concentration profiles, as well as spatial distribution in the horizontal plane. The most widely used sources are CO_2 lasers emitting in the 9.2–12.6 μm band, within which they can be tuned to emit about 80 spectral lines. For example, O_3 absorbs strongly at 9.505 μm, and NH_3 at 10.333 μm. In the UV-visible, dye lasers pumped by flash lamps or by laser diodes are used. By using different dyes, they are tunable over the whole visible band. Some examples of the wavelengths used are given in Table 4.2.

Example 2 – differential optical absorption spectroscopy (DOAS)

A detector at one end of an atmospheric path (typically 200–10 000 m in length) scans across the waveband of a UV/visible source, such as a high-pressure xenon arc lamp that has a known broad spectrum, at the other end. The physical arrangement can be bi-static (with the receiver at one end and the transmitter at the other) or monostatic (a retroreflector returns the beam to the receiver, which is co-located with the transmitter). Gases that are present on the optical path absorb radiation according to the Beer–Lambert Law, and the absorption varies differently with wavelength for each gas. Variations across the spectrum are compared to stored reference spectra for different gases, and the equivalent amounts and proportions of the gases adjusted in software until the best match is achieved. The main wavelength range used is 250–290 nm, in the UV, with typical spectral resolution of 0.04 nm. Several different gases can be measured simultaneously. Sensitivity is high, and 0.01% absorption can be detected, equivalent to sub-ppb concentrations of many gases over a pathlength of 1 km. Detectable gases include SO_2, NO, NO_2, O_3, CO_2, HCl, HF, NH_3, CS_2, Cl_2, HNO_2 and many organic compounds (aldehydes, phenol, benzene, toluene, xylenes, styrene and cresol). The method is appropriate for obtaining the average concentrations of a pollutant across an urban area or along the length of an industrial plant boundary.

Scattering LIDAR. The Lidar and DIAL systems described above measure the backscattered radiation at the wavelength at which it was emitted. The radiation is scattered elastically, and changes in flux density are not necessarily due to the target species alone. By using other techniques such as Raman scattering, a signal is obtained at different wavelengths which is more directly attributable to the target species.

Example 3 – Fourier transform infrared (FTIR) absorption spectroscopy

This is the most optically sophisticated remote sensing technique, using a scanning Michelson interferometer to detect the entire spectral region at once, and therefore capable of measuring many gases simultaneously. The wavebands used are within the atmospheric absorption windows of 8.3–13.3 or 3.3–4.2 μm. However, sensitivity is generally limited to a few tens or hundreds of ppb (depending on the gas) over a 200 m pathlength; the technique is therefore more appropriate to perimeter monitoring for gas escapes from an industrial site than to general ambient monitoring.

4.4.4   Gas chromatography

Chromatography involves the passage of chemical species in a liquid or gaseous mobile phase through a liquid or solid stationary phase. In gas chromatography (GC), the mobile or carrier phase is a gas (commonly nitrogen, helium or argon). In gas–solid chromatography, the stationary phase is a porous polymer. In gas–liquid chromatography, the stationary phase is a liquid (such as methyl gum or an ethylene glycol ester). The liquid coats a solid supporting substrate such as silica or alumina which is packed inside a long thin coiled glass or stainless steel column. The sample is either dissolved into a volatile solvent such as acetone which is injected into the carrier gas, or injected directly in the gaseous phase after thermal desorption (Figure 4.7).

As the carrier gas flows along the column, different volatilised pollutant compounds in the gas have different affinities for the stationary phase, and migrate along the column at different rates so that they separate. With capillary columns, the stationary phase substrate is bonded to the inside surface of a column that might be only 0.1 mm internal diameter and over 100 m long. The exit gas is passed to a detector. The time of passage of the component through the system identifies the material, and the time-integral of the output signal gives the amount and hence the concentration. The most commonly used detectors are the flame ionisation detector (FID), and the electron capture detector (ECD). Both of these detectors have a detection limit of about 1 pg (picogram = 10^{-12} g).

- In the FID, organic compounds, when injected into H_2-in-air flames, are pyrolysed to produce ionic intermediates and electrons. The current generated when an electric field is applied is a measure of the initial concentration.

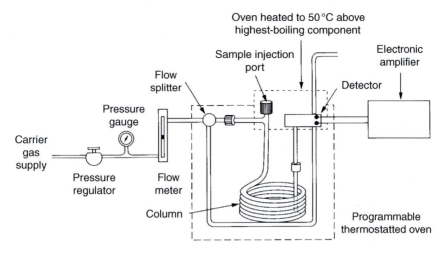

Figure 4.7 Gas chromatograph.

- In the ECD, the eluent is passed over a radioactive electron source such as ^{63}Ni, which ionises the carrier gas

$$N_2 + \beta^- \Leftrightarrow N_2^+ + 2e^-$$

  This generates a steady current between two electrodes. When molecules that capture electrons are present in the carrier gas, the current decreases. Hence ECD is particularly sensitive for compounds with electronegative functional groups, such as halogens, nitro groups and carbonyls, chlorinated pesticides and polychlorinated biphenyls.
- Flame photometers can be used for S- and P-containing compounds. When these are combusted in a hydrogen-rich flame, they emit at 394 and 526 nm respectively.
- Katharometers, which respond to the thermal conductivity of the gas stream, can be used for detection of very low concentrations of H_2, He, H_2O, N_2 and CO_2.

4.4.5 Liquid chromatography

The principles of high pressure liquid chromatography (HPLC) are just the same as those of GC, except that the mobile phase is a liquid solvent such as ethanol, methanol or hexane that is forced at pressures of around 40 MPa through a 30 cm long, 5–10 mm diameter stainless steel column packed with 1–10 μm particles. The stationary phase may be in several forms – a liquid which may be retained on the column either by physical absorption or by the formation of covalent bonds with the support, or a finely divided polar solid such as alumina or silica. This technique is used for the determination of high-molecular weight insoluble organic compounds. With ion exchange chromatography, the stationary phase is 5 μm beads of an ion exchange resin formed from high molecular weight copolymers of styrene and vinylbenzene with functional groups attached to the aromatic ring. As with other chromatography, the polarity of the solvent and substrate may affect the performance of the system dramatically. The most common detectors use UV absorption, fluorescence (especially for organic molecules with organic rings), or electrical conductivity, depending on the molecule concerned. There may be difficulties due to the overlap of the solvent carrier absorbance with that of the analyte. If the analyte does not fluoresce, it may be induced to do so by reaction with a reagent after it has been through the column.

4.4.6   Chromatography with mass spectroscopy

A powerful combination of instruments is increasingly being used to discriminate and quantify environmental pollutants. First, either a gas or liquid chromatograph is used to separate a mixture of compounds. The eluate is then delivered to a mass spectrometer for quantitative determination of each component. The mass spectrometer ionises the constituent atoms and measures their mass by firing them into a magnetic and/or electrostatic field and observing the deviation, which depends on their speed, time of flight and charge to mass ratio. As well as the chemical determinations, it is an important technique for distinguishing stable isotopes such as ^{15}N and ^{18}O, which are not radioactive but which are chemically identical.

4.4.7   Inductively-coupled plasma (ICP) spectroscopies

The ICP is a very high temperature radiation source, created by passing an electric current through an inert gas such as argon. This generates an ionised region, known as a plasma, which radiates strongly due mainly to collisions between the carrier gas ions. Solid particles or liquid droplets of sample are vaporised and ionised in the ICP. When used in conjunction with atomic absorption spectroscopy (AAS), light from a normal source is passed through the plasma and the absorption lines determined. There is also some radiation from the analyte which is present at low concentration in the carrier gas, and this emission can be analysed by conventional AAS. Alternatively, the ionised material from the ICP can be passed to a mass spectrometer.

4.4.8   Optical spectroscopy

Spectroscopy uses the different energy levels of electronic or vibrational states in atoms and molecules to identify the presence of that atom or molecule. In general, the energy needed to move between levels is lowest for vibrations, higher for the outer electrons and highest for the inner electrons. Since photon energy increases as frequency increases, the vibrational states can be excited by IR, whereas outer electrons need visible or UV and the inner electrons need UV or even X-ray.

A spectrometer needs a stable and tuneable source of radiation, a transparent sample container, a wavelength selector and a photodetector. The radiation source may be a tuneable laser, which replaces the combination of thermal radiation source, collimation optics, wavelength selector and focussing optics. However, the tuneable range of many lasers, especially in the UV-visible, is small. In the IR the CO_2 laser, tuneable over 80 lines in the range 9–11 μm, is used. Commercial instruments still largely depend on a range of tried and tested thermal sources such as the Nernst Glower (rare earth oxide), Globar (silicon carbide), incandescent wire (Nichrome), tungsten filament and high-pressure mercury arc. An interference filter can be used to select one line (1–5 nm wide) from a broadband source, or a scanning monochromator with holographic gratings will do the same job over a continuous wavelength range. Photon detection is by photosensitive detector (e.g. mercury cadmium telluride) in the IR or photomultiplier tube in the UV-visible.

Molecular spectroscopy includes absorption, emission, fluorescence and scattering processes. In the UV-visible wavelengths, the transitions are between electronic energy levels, whereas IR radiation causes changes in the vibrational energy.

Absorption

When the electrons that are bound to atoms change their energy level, the difference in energy is emitted or absorbed at wavelengths determined by the energy change and given by Planck's formula (E = hν). The wavelengths are related to the atomic configuration, and are therefore specific to the element concerned. For molecules, as well as the energy levels of the constituent atoms there is a whole new set of energy levels for molecular orbitals of electrons shared by the atoms plus vibrational states.

Absorbance is defined by $A = \log_{10}(1/T)$ where $T = I/I_0 = \exp(-\mu L)$, I, I_0 are the transmitted and incident flux density, respectively, μ is the attenuation coefficient of the medium, L is the pathlength through the medium. If the molecules of interest are the only significant absorber, then we can write

$$T = \exp(-\alpha CL)$$

where α is the *absorption cross section* (usually expressed as cm^2 when the concentration C is in molecules cm^{-3} and L is in cm), or the *absorptivity* (1 g^{-1} cm^{-1}) when C is in g l^{-1}, or the *molar absorptivity* (1 mol^{-1} cm^{-1}) when C is in mol l^{-1}.

Hence absorbance $A = 0.434\,\alpha CL$, so that if the output from the spectrometer is presented in terms of A, it is directly proportional to concentration provided that the concentration is sufficiently dilute (typically <0.01 M).

Ultraviolet-visible molecular absorption mainly involves unsaturated double bonds at wavelengths between 200 and 700 nm. Infrared molecular absorption mainly involves transitions between successive levels of the fundamental stretching vibrational modes, stimulated by radiation in the 2.5–15 μm spectral region.

If the excited molecule decays back to an intermediate excited state, rather than to the ground state, then the emitted photon will have less energy (i.e. longer wavelength) than the incident photon. This is called Raman scattering, and each molecular species generates a set of narrow Raman absorption lines in the UV-visible that correspond to the various vibrational modes observed in infrared absorption spectroscopy.

If the molecule is raised to the lowest vibrational state of the excited electronic level, then radiative transitions to the various vibrational states of the ground electronic state will occur after a relatively long period (10^{-8} s instead of 10^{-12} s). This is called fluorescence and is characterised by having an emitted wavelength longer than the exciting wavelength.

The concepts in atomic spectroscopy are very similar to those in molecular spectroscopy. The material is reduced to its atomic components if not already, then the absorption, emission and fluorescence of radiation are studied. Instead of dealing with the molecular excited states we are now involving the electron energy levels of multi-electron atoms. The atomisation is achieved either with a simple flame (temperatures of around 2000 °C for natural gas in air, 3000 °C for acetylene in oxygen), or with ICP or electric arc–spark (temperatures of 4–6000 °C).

In atomic absorption spectrophotometry (AAS, Figure 4.8), the material for analysis is atomised at high temperature (either injected into an air–acetylene flame at 3000 °C, or vaporised in a graphite furnace). A beam of radiation is passed through the vapour, so that photons knock electrons from a low (usually ground) energy state up to a higher state; hence photons are absorbed and the radiation flux is depleted. The radiation source is a hollow cathode lamp constructed from the element to be analysed, so that it emits narrow spectral lines appropriate to that element rather than a continuous spectrum. This greatly reduces the stray light transmission and increases the sensitivity. The total amount transmitted at different wavelengths is measured photoelectrically, and the absorbance determined. From the Beer–Lambert law, $I_t = I_0 \exp(-kL)$, where I_t, I_0 are the transmitted and incident flux densities, k the extinction coefficient and L the optical path length. Also $k = \epsilon C$, where ϵ is a constant for the substance and wavelength, and C is the vapour concentration. Hence C can be calculated. Atomic absorption spectroscopy is used almost exclusively for determinations of metallic elements.

In atomic emission spectroscopy (AES), atomic collisions due to the high temperature of the sample knock electrons into higher energy states from which they subsequently decay with the emission of a photon (Figure 4.9).

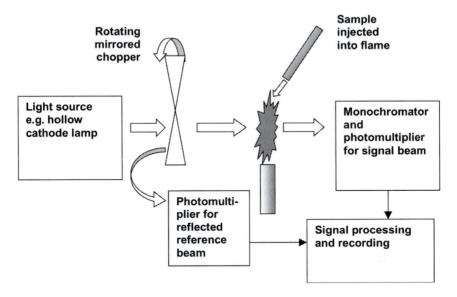

Figure 4.8 Atomic absorption spectrophotometer.

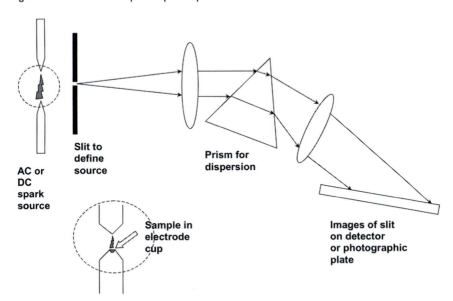

Figure 4.9 Atomic emission spectrometer.

The excitation energy is provided by a flame or spark, which also vaporises the sample from solution. The range of elements that can be excited (and therefore detected and measured) depends on the temperature of the source. Low temperature propane flames have been used for the alkali metals that have

Table 4.3 Elements determined by low-temperature AES

Element	Wavelength/nm	Detection limit/ppm
Li	671	0.005
Na	589	0.0003
K	766	0.003
Rb	780	0.004
Cs	852	0.004
Mg	285	0.008
Ca	423	0.005
Sr	461	0.001
Ba	422	0.004

emission lines near the low-energy red end of the spectrum (Table 4.3). The typical solution concentration range that can be used is 0–20 ppm. At higher concentrations the response is non-linear.

Nitrous oxide–acetylene flames enable the detection of a similar range of metals to AAS. Inductively-coupled plasma sources achieve temperatures of *c.* 6000 K and make it possible to analyse elements such as Ce, U and Zr that form very stable refractory oxides. If higher excitation energy is provided by a spark rather than a flame, every element above helium can be detected, and the sample can be solid as well as liquid.

4.5   QUALITY CONTROL

All measurement devices eventually produce an output (voltage, weight, absorbance etc.) that has to be translated into a gas concentration. An effective quality assurance and quality control (QA/QC) programme is an important part of gaining confidence in the truth of the output value and in the errors associated with it. A minimum QA/QC programme will include blanks, duplicates and calibration standards; for more demanding assessments, it may be necessary, for example, to have independent laboratories carry out duplicate analyses.

4.5.1   Blanks

If a measurement method such as a bubbler or diffusion tube is being used, then contamination of the absorbent is possible. To evaluate this effect, sample units are prepared and analysed following exactly the same procedure as those used for the measurements themselves, but are kept sealed. The concentration measured is then a baseline value which should be subtracted from all actual measurements.

In an extended sampling programme, it would be normal to retain at least 10% of sampling units for construction of this baseline.

4.5.2   Calibration

Even the highest specification gas analysers are not yet as stable or repeatable as other more common electronic devices such as computers and radios. Over a period of time that may be as short as days, the output signal due to a constant input gas concentration can vary significantly. Such variations might be due to drift in the value of electronic components, dirt penetrating through to the optics, or temperature changes. Large sudden changes can be caused by replacement of failed components. For these reasons it is essential to have good calibration facilities available.

Calibration involves supplying the analyser with a known concentration or range of concentrations of the gas to be measured, together with a zero air sample containing none of the gas. Several methods can be used to obtain these supplies.

4.5.2.1   Permeation tubes

Small stainless steel tubes are sealed at one end and capped at the other by a membrane that is permeable to the gas. The tube is filled under pressure with the liquid phase of the calibration gas. If the tube is kept in a constant-temperature oven (typically 50 ± 0.1 °C) then gas leaks out through the membrane at a very stable rate. The gas is added to a known volume flow rate of clean air to produce a known gas concentration. Tubes are available for hundreds of gases including SO_2 and NO_2 (but not O_3). The release rate from the tube is individually certified by the manufacturer, and can be confirmed by the user by weighing the tube at intervals during use. These sources can be incorporated into a self-contained calibrator, or built into an analyser to provide a calibration signal on demand.

4.5.2.2   Gas bottles

Specialist gas suppliers can provide pressurised bottles of calibration gases – a known concentration of the gas in an inert carrier. If the concentration is too low, the concentration may change in the bottle because of reactions with the surfaces or with other gas molecules, and also the bottle will be used up very quickly. If it is too high, then an impractical dilution factor will be needed to bring the concentration down within the analyser range. Hence it is normal to store, say, 50 ppm of NO in nitrogen, and to use a precision dilution system to generate concentrations of between 50 and 200 ppb when calibrating an NO analyser that has a 200 ppb range.

4.5.2.3   UV O_3 generator

Ozone cannot be stored, so that any calibration method must create it on the spot. In this method, a proportion of oxygen molecules irradiated with UV light is converted to O_3; by arranging for clean air to be so irradiated by a very stable UV lamp, stable O_3 concentrations can be generated. By mixing NO with O_3, known concentrations of NO_2 can also be made (gas phase titration).

4.5.2.4 Zero air

All of these calibration methods rely on the availability of a clean air supply, known as zero-air. It is not a trivial task to generate such a supply, particularly if calibrations in the low- or sub-ppb range are required. In order to make zero-air on site, the basic principle is to pass ambient air through activated charcoal to adsorb pollutant molecules, but there are some difficulties. First, the air must be passed through cooling coils to condense out excess water vapour. Second, NO is present at significant concentrations but is not adsorbed by charcoal. Hence the air is first irradiated with UV. This generates O_3 that reacts with the NO, converting it to NO_2 (which *is* efficiently adsorbed by the charcoal). Provided that the charcoal is renewed frequently, very low trace gas concentrations can be achieved. Alternatively, compressed bottles of zero-air can be purchased from the usual suppliers.

The weakest link in the calibration chain is frequently the measurement of volume flow rate. On a calibrator, for example, the volume flow rate of the clean air used to dilute the output from the permeation tube is often measured with rotameters that are only accurate to 5%. This determines the best overall accuracy of calibration, no matter how accurately the permeation rate is known. The precision can be increased – at a price – by the use of electronic mass flow meters instead of the rotameters.

A comprehensive QA/QC programme for a network of continuous analysers might involve:

- daily automatic calibration checks of the analysers
- site checks every one or two months
- regular analyser servicing by trained instrument engineers
- regular intercalibrations (either taking several analysers to the same calibrator, or taking one calibrator around all the analysers in succession)
- periodic checking of the calibrator against primary standards
- detailed manual and automatic scrutiny of the data to eliminate false values caused by instrument malfunction or transmission faults. It is much better to have holes in the data set than to have spurious values
- comprehensive review of period data sets
- telemetry of the data to the processing centre, so that faults can be spotted as soon as they occur

The automatic urban and rural monitoring networks operated by DEFRA and described in Chapter 5 aim to achieve overall error bands of ±5 ppb on SO_2, NO_x and O_3, ±0.5 ppm on CO_2, and ±5 $\mu g\ m^{-3}$ on PM_{10}. The actual time for which the network generates valid data, expressed as a percentage of the possible time, is called the data capture. The EU Directive on monitoring ambient NO_2 specifies 75%, the USEPA requires 80%, and the UK DoE network achieves 95%.

4.6   PARTICLE SAMPLING

The task of sampling is to get a representative sample of particles from the air into the measuring device without changing the concentration or the characteristics on the way. Thus we must make sure that we do capture all the particles from a known volume of air in the first place, and then that we do not lose any on to tubing walls etc., or break up large particles and aggregates into small particles, or lose particles by the evaporation of volatile materials, or gain them by condensation.

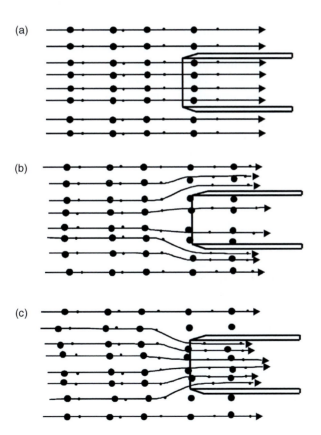

Figure 4.10 Flow streamlines of air into a sampling inlet: (a) isokinetic flow; (b) inlet speed less than the free stream speed; (c) inlet speed greater than the free stream speed.

4.6.1   Isokinetic sampling

As soon as we extract an air sample into the inlet of a measuring device, we start to corrupt the sample. We can think of the air as moving along streamlines that are parallel if the air is not flowing round an obstacle. If the air deviates round an obstacle, the streamlines bend. For each particle, there is then an imbalance between the inertial forces that tend to keep its movement in a straight line, and pressure gradient forces that tend to make it follow the streamlines. The heavier the particle, the more it will tend to come off the streamlines. When we sample air through an inlet, the streamlines will almost always be deviated to some degree. In the ideal case, the inlet is faced directly into the prevailing wind, and the sampling flow-rate is chosen so that the inlet speed is the same as the windspeed. This condition is said to be isokinetic (Figure 4.10(a)), and results in the least possible distortion of the particle sample.

Even when this condition is achieved on average, however, the turbulence that is always present in wind outdoors means that the instantaneous speed and direction at the inlet will usually differ from isokinetic. When the average inlet speed is too low (Figure 4.10(b)), the streamlines diverge around the inlet. Some particles will enter the inlet from air that has not been sampled; the larger the particle, the less likely it is to follow the streamlines, so that the sampled air becomes enriched with larger particles. Conversely, when the average inlet speed is too high (Figure 4.10(c)), the streamlines converge on the inlet so that the sampled air is depleted of larger particles.

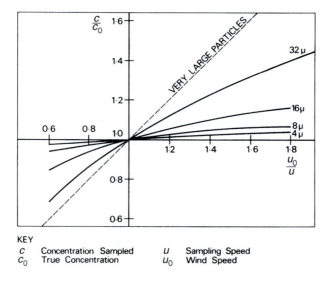

KEY

c	Concentration Sampled	u	Sampling Speed
c_0	True Concentration	u_0	Wind Speed

Figure 4.11 Variation of sampling efficiency (c/c_0) with degree of isokineticity (u_0/u) for particles of different sizes.
Source: Pio, C. A. (1986) In: R. M. Harrison and R. Perry (eds) *Handbook of Air Pollution Analysis*, Kluwer Academic Publishers, Dordrecht, The Netherlands.

In Figure 4.11, the ratio of windspeed u_0 to inlet speed u is plotted against the ratio of sampled concentration c to true concentration c_0 for different diameters of particles having unit specific gravity. For the smallest particles (the fine particle fraction <2.5 µm, for example), the concentration ratio is within a few percent at all speed ratios. As the diameter increases, so does the distortion of the sample, up to the limit for very large particles at which the ratios are equal (that is, a 20% error in the sampling speed gives a 20% error in the concentration).

This description holds for an inlet that is oriented directly into the wind. If the inlet is at an angle to the wind, then even if the speed is correct, the whole set of streamlines will have to bend on their way into the inlet, so that the sample will always be depleted of heavier particles. Figure 4.12 shows that this depletion could be 5%, 10% and 15% for 4, 12 and 37 µm particles (respectively) entering an inlet aligned at 30° away from the mean wind direction.

These issues are especially severe when the sampling is to be undertaken from an aircraft. The aircraft will be flying at typical speeds of 40–200 m s^{-1}, which requires a very small inlet diameter if a reasonable volume flow rate is used. There may be very high wall deposition losses just inside the inlet due to severe turbulence. Furthermore, the air must then be rapidly decelerated before the particles are filtered out. This raises the air temperature by 5–20 °C, evaporating water and volatile species and changing the size distribution.

A different set of problems arises if the sampling is to be undertaken from still air. Now there is no wind direction to be oriented into, and no windspeed to be matched isokinetically. The errors are due both to the particles' settling velocities and to their inertia. For example, suppose the inlet tube is facing vertically upward. Then particles that were not originally present in the sampled air volume

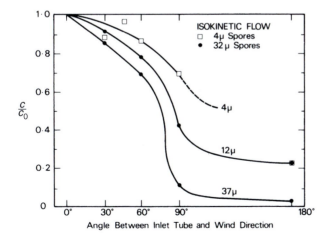

Figure 4.12 Variation of sampling efficiency (c/c_0) with the angle of the inlet to the wind direction, when sampling isokinetically.
Source: Pio, C. A. (1986) In: R. M. Harrison and R. Perry (eds) *Handbook of Air Pollution Analysis*, Kluwer Academic Publishers, Dordrecht, The Netherlands.

will fall into the inlet, so increasing the apparent concentration. The situation is reversed for an inlet that points downward. In principle there should be no error from this cause if the inlet is horizontal. Provided that the inlet velocity is sufficiently high (>25 V_s, where V_s is the terminal velocity of the largest particle of interest) then settling errors should be negligible.

It is not so easy to eliminate inertial errors. If the air is still then the flowlines in the sampled air must converge into the inlet, so that particles will have to follow them. The larger the particles, the faster the flow or the smaller the inlet, the more will be lost. Inertial errors always result in an underestimate of the concentration. The general criterion for negligible errors is for $D_s > 4 (Q\tau)^{1/3}$, where D_s is the diameter of the sample tube inlet, Q is the volume flow rate and τ the relaxation time of the largest particles. Taking the relaxation times given in Chapter 2 for particles of 0.1, 1, 10 and 100 μm, sampled at a flow rate of 20 l per minute, we would need inlet diameters of 2, 6, 28 and 129 mm respectively. Thus the inertial and settling criteria set upper and lower bounds for the inlet diameter at a specified volume flow rate.

Once particles have been coaxed into the inlet in the correct proportions, it follows that they should be measured as soon as possible to reduce risks of corruption. This involves the shortest physical path lengths, least severe bends, shortest practical residence time and closest approach to ambient temperature, pressure and humidity.

4.7   PARTICLE MEASUREMENT METHODS

There are many particle measurement methods, each with its own merits and disadvantages. Selection of a particular method will depend on the objectives of the measurement and the resources available.

4.7.1   Filtration

Probably the oldest method of extracting particles from the air in which they are suspended is to pass them through a filter. Yet even this primitive technology is full of subtleties that have to be appreciated if we are to understand how the collected sample differs from the original population of particles in the air. A particle sampling filter consists either of a mat of tightly-packed, randomly-oriented fibres, or a membrane with a network of randomly-arranged air passages through it, or a membrane punctured by microscopic cylindrical pores. Particle collection mechanisms are complex, and vary with filter type. For fibrous filters, the particles are collected by a mixture of impaction, interception, Brownian diffusion and electrostatic attraction, with the significance of each depending on the flow rate, face velocity, porosity, fibre and particle composition. Filtration is not at all like sieving, because although a nominal pore size is quoted for a filter material, particles much smaller than that will be retained. Also, collection of particles will itself change the airflow through the filter and the collection efficiency. There are many different filter types, and selection will depend on the properties of both the particles and gases in the air and the ultimate purpose of the sample collection.

Among the critical factors are:

Particle collection efficiency and pressure drop. For normal applications, filters should remove >99% of the particles presented to them. However, the efficiency is increased by reducing the pore size, which also increases the pressure drop; this in turn reduces the flow rate or increases the pump size required. The pressure drop across the filter must not clog the filter even when the highest expected particle loading is sampled – otherwise the flow rate will fall and the measurement of the once-in-20-years pollution episode will be corrupted. Glass fibre filters have been the mainstay of particle sampling for decades. They are cheap and rugged and offer high collection efficiencies at low pressure drops. Particles are trapped on fibres through the depth of the filter, rather than on the surface alone. Hence the rate of increase of pressure drop as the sample collects is low. Membrane filters made from cellulose nitrate or cellulose triacetate are suitable if the sample needs to be recovered for chemical analysis, because the substrates dissolve easily. Polyvinylchloride, which is widely used for quartz particles, suffers less from weight change due to water absorption. Membrane filters typically have higher flow resistances than fibrous ones, so are used for low flow rate applications. Increasing the face velocity can increase the collection efficiency, because more of the particles will be collected by impaction on fibres. On the other hand, the smallest particles will be less likely to be collected by diffusion. On a more exotic level, nuclepore filters are

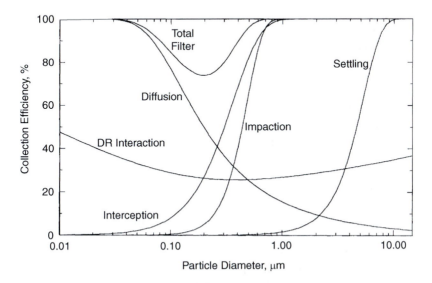

Figure 4.13 Filter efficiency due to different mechanisms.
Source: Hinds, W. C. (1999) *Aerosol Technology: Properties, Behaviour and Measurement of Airborne Particles*, Wiley, New York.

made by piercing holes through a membrane with sub-atomic particles, then etching the holes to enlarge them to repeatable and known diameters. Particles are deposited on the flat smooth surface of the membrane. Costs and pressure drops are both high.

Figure 4.13 shows how the different particle collection mechanisms contribute to the overall filter efficiency of a fibre filter, with different mechanisms having quite different effectiveness in different size ranges, based on the aerosol mechanics that were discussed in Chapter 2.

Chemical stability and purity. Filters must be inert to gases present in the air sample, and not react with the deposited particles even in the presence of extraction solvents. Glass fibres absorb sulphur dioxide, quartz fibres absorb organic vapours, and cellulose fibres absorb water. If chemical analysis of the deposits is needed, then the filter must not contain significant and/or variable levels of the elements of interest. These properties must be determined *before* the sampling programme.

For a basic determination of particle mass concentration, the filter will need to be conditioned by being kept at $50 \pm 3\%$ relative humidity and 20 ± 0.5 °C until constant weight is achieved before weighing clean; this conditioning is repeated after use to minimise errors due to water vapour uptake or loss by the filter. It is also done at a fairly low temperature to reduce any effect of volatilisation of ammonium nitrate and other salts. The microbalance used for the weighing should be kept in a cabinet at the same controlled conditions.

4.7.1.1   British Standard (BS) smoke method

This technique was devised in the 1920s to measure rather high urban concentrations that were known to be made up mainly of black carbonaceous smoke particles; the method was widely adopted, both for the British National Survey and elsewhere. Ambient air is drawn through a Whatman No. 1 (cellulose) filter paper at around 1.5 l min^{-1} for 24 h. The standard inlet arrangements involve passing the air down small-bore tubing, giving a size cut of around 5 μm. There is no attempt to sample isokinetically or even to point the sample inlet into the wind. The volume of air sampled is measured by gas meter. The method was originally devised to assess the 'blackness' due to smoking chimneys, and the amount of material collected is impracticably small for routine mass determination by weighing (e.g. a concentration of 50 μg m^{-3} would give a deposit of about 100 μg). Hence the standard method of analysis is based on reflectometry, which is much more sensitive at low particle loadings. As urban particle concentrations have declined from the hundreds of μg m^{-3} that were common in the 1960s to the tens that are more typical today, and the components have become less influenced by black smoke and more by other materials such as ammonium sulphate, the reliability of this method has decreased. Although results are issued in gravimetric units, conversion factors must be used to compare these with other, truly gravimetric, methods.

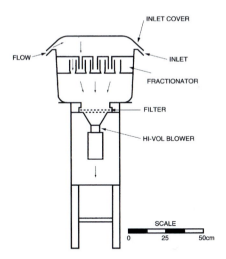

Figure 4.14 Schematic of the high-volume sampler, including PM$_{10}$ head.
Source: Quality of Urban Air Review Group (1996) *Airborne Particulate Matter in the United Kingdom*, Third Report of the UKQUARG, Department of the Environment, London.

4.7.1.2   *High-volume sampler*

The 'Hi-Vol' (Figure 4.14) was developed in North America, where it has been widely used. This device produces the equivalent result over the same time period as the filter above, but gravimetrically. Air is drawn in below the eaves of a protective roof, then vertically down through a rectangular 254 × 203 mm (10″ × 8″) glass-fibre filter paper. A separator may be included before the filter, as in Figure 4.14, to confine the measurement to PM$_{10}$, for example. The airflow rate is about 1 m^3 min^{-1}, so that over a 24-h period about 1400 m^3 of air are sampled, which would contain 70 mg of particles if the mass loading was 50 μg m^{-3}. The filter papers themselves weigh around 2.7 g, and with careful procedures such as the equilibration described above, the mass can be determined within a few percent. The filter material must collect >99% of 0.3 μm particles. If the filters are to be analysed for chemical composition, then various grades of filter that have reduced or more stable background concentrations of certain elements are available. Glass or quartz-fibre filters can also be treated by high-temperature soaking or by washing with distilled water to remove volatile organics or ionic species. Although glass-fibre filters are widely used, care must be taken because conversion of gaseous SO$_2$ to particulate sulphate on the filter surface itself can artificially elevate apparent particle loadings. The USEPA specifies an alkalinity of less than 25 microequivalents per gram, and that the filter itself should not gain or lose more weight during a 24-h sample than is equivalent to 5 μg m^{-3}. Additional errors are possible due to reactions of gases with particles on the filter surface, or volatilisation of particles after they have been collected. The supporting measurement needed in order to calculate particle concentration is the volume of air sampled. This is obtained by measuring the pressure drop across orifice plates in the flow path. For basic models the flow rate is measured before use, and again at the end because the build-up of particles reduces the flow; more sophisticated (and therefore more expensive) models have electronic flow controllers to keep the sample rate within ±5% throughout the 24 h.

4.7.2   Optical methods

Some of the electromagnetic radiation shining on a particle is scattered away from its original direction. The amount scattered depends on various parameters including the particle's size, and several instruments use this fact to measure properties of the particle population without collecting the particles (which means, of course, that they are not available for chemical analysis either). First we describe methods for the overall mass loading. Methods which are mainly used to give information on the size distribution are described in Section 4.7.5.

Nephelometers. This family of instruments employs a large view volume to measure scattering from many particles simultaneously. The air sample is passed through a cylinder (typically 2 m long and 15 cm diameter) that is illuminated from one side, and light scattered between 5° (almost forward) and 175° (almost backward) is detected by photomultiplier at one end of the cylinder. Single-wavelength instruments are principally used for measurement of atmospheric optical properties such as extinction coefficient or visibility. Multiple wavelength nephelometers can provide additional information on aerosol size distributions.

The *aethalometer* is used to measure real-time total concentrations of black carbon (BC) particles. The particles are sampled at around 10 l min^{-1} onto a quartz fibre filter, through which a broad-spectrum light source is shone. The beam is attenuated, and the attenuation converted into a mass concentration of BC on the filter via a calibration factor (expressed in m^2 g^{-1}). The calibration factor itself varies from site to site, since it will depend on the detailed properties of the BC particles; different workers have found values ranging from 20 in urban areas down to five in remote regions. These values compare with a figure of around 0.05 for mineral dusts such as Saharan sand or road quartz. There is a problem with converting the measurements directly into the equivalent scattering of those particles in the atmosphere, because the close packing of the particles on the filter results in multiple scattering. The same principle has been used in the US to measure coefficient of haze (COH).

A related handheld device known as the direct-reading aerosol monitor (DRAM) is used to measure ambient and workplace particle concentrations. The PM$_{10}$ fraction is selected from an inlet stream that has a flowrate of 2 l min^{-1} and is passed to an optical cell. Near-infrared light is forward scattered from the sample, then the scattered signal is converted to mass concentration via a calibration factor. No information is given on the size distribution. The device is useful for fast semi-quantitative determinations or assessment of changes, but the calibration factor varies with aerosol characteristics.

4.7.3   Beta-attenuation

The air sample (possibly after being drawn through a size-selective inlet for PM_{10} or $PM_{2.5}$) is pulled through a filter to deposit particles. A beam of ~0.1 MeV electrons from a radioactive source (a few MBq of krypton 85 or carbon 14) is directed through the filter paper and the flux measured by a Geiger–Müller counter or silicon semi-conductor on the other side. There is a constant attenuation of the beam due to the filter itself, and an increasing attenuation, described by the Beer–Lambert law, as particles accumulate on the filter. The method is most commonly employed as an automated sampler, with separate samples taken on a continuous reel of filter tape. The mass loading is measured intermittently, with a typical minimum time resolution of 30 min, and a detection limit of 5 μg m^{-3} for a 1-h average. System calibration can be automated by inserting a membrane of known mass density into the beam, although this does not allow for the effects of different sample properties. One attraction is that the attenuation is quasi-gravi-metric – it should only be affected by the mass of material collected, and not by the size distribution, particle shape or chemical composition. In fact attenuation is due to scattering of the beta particles by atomic electrons in the filter media and deposited particles, and therefore depends on the areal density of these electrons. Different elements have different ratios of atomic number (i.e. number of electrons) to mass. The ratio varies between 0.4 and 0.5, except for hydrogen for which it is one. In practice, for the mixtures of elements found in particle samples, the ratio is in the range 0.47–0.50, and the error is acceptable. Ammonium compounds such as ammonium nitrate and ammonium sulphate, which contain a high proportion of hydrogen atoms, may require local calibration.

4.7.4   Resonating microbalance

There are two main variants of this principle. In the first, particles are impacted inertially or electrostatically onto a piezoelectric quartz crystal that oscillates at its natural resonant frequency of around 10 MHz. As the weight of deposited material accumulates, the resonant frequency decreases according to

$$\Delta m = K\left(\frac{1}{f^2} - \frac{1}{f_0^2}\right) \quad \text{or} \quad \Delta m \cong \frac{2}{f_0^3}\Delta f$$

so that the change in frequency Δf is directly proportional to the change in mass Δm, provided that Δf is small compared to f_0. The frequency is continuously compared to that of a reference crystal placed out of the particle flow, and the frequency difference is directly proportional to the collected mass. The device is sensitive, with a minimum time resolution of one minute, but requires the crystal to be manually cleaned periodically.

The other variant of this principle is known as the tapered element oscillating microbalance (TEOM, commonly pronounced tee-om, Figure 4.15). The particles are collected continuously on a filter mounted on the tip of a glass element that oscillates in an applied electric field. Again the natural frequency falls as mass accumulates on the filter. This device has been chosen for PM_{10} measurements in the UK Automated Urban Network, largely because it was the only gravimetric method capable of the required 1-h time resolution when the network was set up in the early 1990s. Although the method is specific to mass

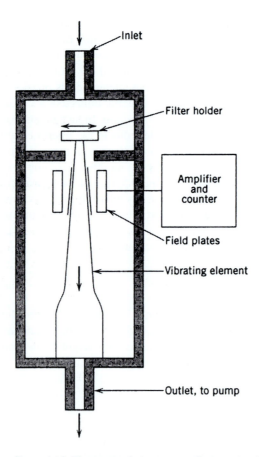

Figure 4.15 The tapered element oscillating microbalance (TEOM).
Source: Hinds, W. C. (1999) *Aerosol Technology: Properties, Behaviour and Measurement of Airborne Particles*, Wiley, New York.

loading, there is a significant systematic error. In order to maintain the constancy of the base resonant frequency, the oscillating element is maintained at a temperature of 50 °C. This temperature is high enough to evaporate volatiles from some particles, resulting in a long-term gradual loss of mass. In specific experiments in which pure ammonium nitrate aerosol was sampled, serious volatilisation losses were experienced which were greatly reduced when the substrate was maintained at 30 °C rather than 50 °C. Unfortunately this lower temperature creates even larger errors due to variable water content in the sample. In field trials where the TEOM has been compared with co-located high or low volume gravimetric samplers, it has been found that the TEOM under-measures daily mean PM_{10} concentrations by about 30%, and that this proportion will also vary with particle composition. Since attainment of concentration limits is framed in terms of the number of days on which the limit is exceeded, significant errors in attainment may be generated by this under-reading. For the UK network, it is now recommended that the indicated mass loadings be increased by 30% to correct for this effect, and it is necessary to specify the measurement method when reporting readings (i.e. $PM_{10\,(grav)}$ or $PM_{10\,(TEOM)}$). Despite this imperfection, the TEOMs in the AUN have certainly shown the power of rapid-response continuous monitors for understanding the behaviour of urban particulates.

4.7.5   Size fractionation

The methods above give only the total mass loading of particles after arbitrary changes to the ambient distribution have occurred in the sampling system. The methods were widely used in the first generation particle networks – high-volume samplers in the US, smoke shade in the UK. Subsequently, it has been realised that we need more specific information about the mass loading in specific size fractions in order to assess potential health effects: the very minimum information required is the mass concentration of particles less than 10 μm (PM_{10}); we are now moving towards a need to know the mass of particles less than 2.5 μm, and for some research activities a complete breakdown of mass loadings in each of five or ten diameter sub-ranges below 10 μm. The range of sizes that we might be interested in stretches for five decades – from 0.001 μm up to 100 μm. No available particle sizer will cover that range, and at least three quite different techniques would be needed. Each technique will be based on a different particle property such as inertia or mobility, each of them will have been calibrated using a precisely controlled laboratory aerosol quite unlike that being measured, and the results from the different methods may not be directly comparable. Laboratories need to use a bank of different instruments to build up an overall particle size distribution. This requirement has become more pressing as it has become recognised that vehicle exhausts generate particles below 1 μm, which are not measurable by the most common (and cheapest) methods.

4.7.5.1   Optical sizers

Single particle counter. A small view volume (a few mm^3) is illuminated by a laser, and the light scattered from it, at an angle to the beam direction, is measured by photomultiplier. When there is no particle in the volume, no light is scattered towards the photomultiplier and no signal results. A pump draws an air sample through the view volume at such a rate that there is usually only one particle in view at a time. Each particle traversing the volume scatters a pulse of light which is recorded. When enough particles have been sampled to give statistically meaningful results (a minute or two at typical ambient air particle concentrations) the pulse heights are analysed in terms of the number of particles within different diameter ranges. The minimum particle diameter that is measurable with commercial instruments is around 300 nm. Although more powerful lasers can in principle detect smaller particles, there may be a problem with particle heating. The technique has a very large dynamic range, so that one device can be used for a wide range of concentrations, from clean-room conditions with 10 particles m^{-3} up to polluted ambient air with millions of particles m^{-3}. The lower limit is set by the length of time it takes to build up a statistical description of the concentration distribution, and the upper by the increasing probability of having more than one particle in the view volume at a time. The manufacturer's calibration is usually achieved by using homogeneous spherical particles of latex or similar material. Natural particle populations are neither spherical nor chemically homogeneous, and it will often be necessary to recalibrate the device gravimetrically against a sample of the aerosol of interest.

4.7.5.2   Aerosol spectrometer

With this technique, light is used to detect the particles, but not to size them. The sample air stream in which the particles are suspended is accelerated rapidly. The acceleration of the particles lags behind that of the air because of their inertia, and the lag is larger for aerodynamically larger particles. The lag is measured by the transit time between parallel laser beams. Hence the true aero-dynamic diameter (rather than the mass or physical diameter), which is the parameter that defines particle inhalation and deposition, is determined directly.

4.7.5.3   Inertial impactors

Ironically, this technique relies on exactly the inertial properties of particles that make them so difficult to sample accurately. Air is drawn through an orifice (or jet) towards an obstruction (or particle sampling substrate). The orifice diameter, jet speed and separation of the jet tip from the substrate combine to determine the cut diameter of the stage – that is, the particle diameter above which 50% of particles will be deposited on to the substrate, and below which 50% will remain airborne. The efficiency of each stage depends on the Stokes number, which in this instance is the ratio of the particle stop distance to the half-width of the nozzle aperture.

$$\text{Stk} = \frac{C\rho_p U d_p^2}{18\mu d}$$

The cut diameter is then given by

$$d_p^* = \left[\frac{18\mu d \text{Stk}_{50\%}}{C\rho_p U}\right]^{1/2}$$

$\text{Stk}_{50\%}$ is around 0.25 for round jets, 0.5 for rectangular jets. In practice the impaction stage will be designed approximately using these criteria, then calibrated using monodisperse aerosols.

This principle is used in a variety of devices. First, there are size-selective PM_{10} inlets that fit above the filter of the standard high-volume sampler in place of the gabled roof (see Figure 4.14). Air is drawn in along a curved path which, at the normal sampler flow rate, leaves most particles of less than 10 μm diameter still in the airstream *en route* to the filter, but separates out the larger particles. Although a cut diameter of 10 ± 0.5 μm is specified, it is very hard to achieve this level of precision in practice, because the cut diameter is affected by the flow rate and inlet cleanliness. When properly carried out, the method should yield a precision within ± 5 μg m^{-3} for PM_{10} concentrations less than 80 μg m^{-3} or $\pm 7\%$ for higher concentrations. This is one of the reference methods used in the US for measuring compliance with the National Ambient Air Quality Standards (NAAQS).

Second, with the so-called dichotomous sampler (Figure 4.16), air from a PM_{10} preseparator is passed to a virtual impactor. The latter involves a very slow flow rate through filter C, so that particles greater than 2.5 μm are 'impacted', whereas those less than 2.5 μm follow the flow to filter F. This provides both respirable and fine-particle fractions, usually averaged over a 24-h period. The sample rate is much lower than that of the high volume sampler, at around 17 l min^{-1}. This volume flow rate is not arbitrary – it is designed to simulate the typical adult average breathing rate.

Finally, various specialised size discriminators, called cascade impactors, are available both as inlets for high-volume samplers and as stand-alone devices. Figure 4.17 shows the basic principle. The sample air is passed first through Jet 1, which has the largest diameter and greatest separation from the substrate below it. Hence the larger particles are deposited on Substrate 1 but smaller particles follow the streamlines around it. Jet 2 is smaller in diameter (higher air speed) and closer to Substrate 2, so that the smaller particles are retained. By correct arrangement of a sequence of stages having smaller jets, higher speeds and closer substrates, up to ten different cut diameters can be accommodated. The performance of each stage is determined by the Stokes number, and characterised by the cut diameter for which the collection efficiency is 50%. The smallest particles which escape deposition on the final stage may be collected on a back-up filter.

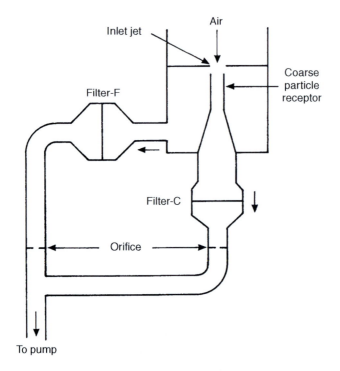

Figure 4.16 Dichotomous sampler.
Source: Harrison, R. M. (1999) 'Measurements of the concentrations of air pollutants', In: S. T. Holgate, J. M. Samet, H. S. Koren and R. L. Maynard (eds) *Air Pollution and Health*, Academic Press, London.

Each substrate can be weighed individually if the variation of mass loading with diameter is required, and the collected material is available for analysis to show the variation of chemical properties with particle size. Of course, as with most things in life, there are trade-offs. The large number of stages gives a high pressure drop and uses more power. High jet speeds in the final stages mean a greater risk of breaking up large particles into smaller ones, and also of re-entraining deposited particles back into the air stream. The very small amounts collected on some stages may make it difficult to weigh or analyse them with the required accuracy.

The standard cascade impactor cannot normally be used to measure particles smaller than about 0.3 μm because the high air speeds needed for deposition cause unacceptable pressure drops and impact forces. The size range has been extended with the electrostatic low pressure impactor (ELPI), in which reduced air pressure (and hence a larger value for C) enables deposition of particles down to around 30 nm. Even this is not small enough to include the mode of the number distribution produced by engines running on low-sulphur fuel. One disadvantage is that the low air pressure (about 8 kPa at the final stage) increases the

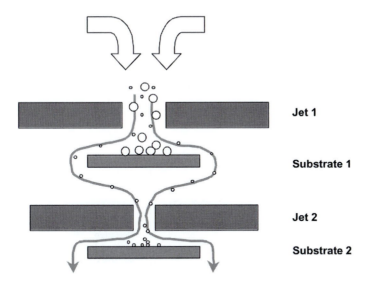

Figure 4.17 The principle of a two-stage cascade impactor.

evaporation from semi-volatile particles. The ELPI also charges the particles electrostatically and measures the rate of particle deposition from the current flow to each stage. Hence it can be used in real time, rather than having to wait for a weighable deposit to accumulate on each stage. However, these image and space charges themselves generate a new force on the particles which may change their deposition characteristics. In another version of this principle, the micro-orifice uniform deposit impactor (MOUDI), the particles are deposited on rotating substrates to avoid re-entrainment, and the mass determined gravimetrically.

4.7.5.4 Electrical mobility analysers

In the differential mobility analyser (DMA)/spectrometer, electric fields are used to separate particles according to their mobility (which is a function of diameter). The transfer function is the particle penetration as a function of the electrical mobility for a given voltage and flow setting. The DMA has also been used to measure both cloud droplets and the aerosols that accompany them. If an individual droplet is first charged and then evaporated to dryness, the electric charge on the residue is a function of the original droplet diameter. Detection is via either condensation of water vapour on each particle and subsequent light scattering, or deposition of electric charge to an electrometer. In instruments such as the scanning mobility particle sizer (SMPS) and electrical aerosol analyser (EAA), the aerosol is first passed through a charger (e.g. Krypton 85) that applies a bipolar electrostatic charge to the particles (Figure 4.18). The

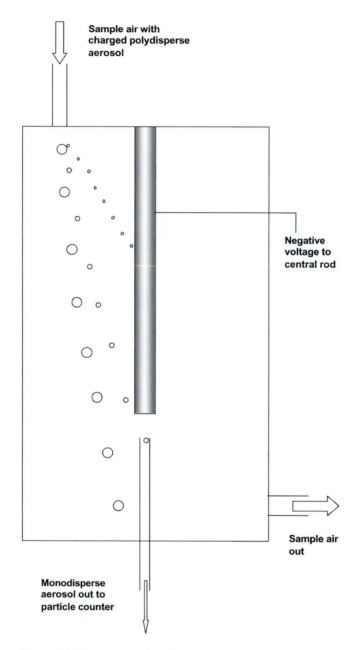

Sample air with charged polydisperse aerosol

Negative voltage to central rod

Sample air out

Monodisperse aerosol out to particle counter

Figure 4.18 Electrostatic classifier.

mean particle charge is close to zero, but a fraction of the particles carry charges of ±1, ±2 etc.

The system is designed so that the charge distribution is independent of the chemical composition and morphology of the particles. Then the aerosol flows through the annular region between a cylindrical steel housing and a coaxial central rod. A variable voltage difference is applied between the two. Positively charged particles are attracted towards the negative central rod, while negatively charged particles are attracted towards the outer cylinder. Each particle takes up an equilibrium radial drift velocity v_{drift} towards the appropriate plate such that

$$v_{drift} = \frac{EeC}{3\pi\mu d_p}$$

where E is the electric field strength, e is the electron charge, C is the Cunningham slip correction factor, μ is the air viscosity, d_p is the particle diameter.

The particles will have a residence time in the cylinder which is determined by the axial velocity v_{axial} and the length. At any given voltage setting, the smallest particles with the highest mobilities will be deposited on the rod. The largest will leave without attaining significant radial velocities. Hence a small diameter range will be selected by the outlet, where they can be counted by a condensation particle counter (or the current measured). By varying the field strength and residence time, particles of different sizes may be allowed to pass through. The measured size range is typically 10 nm–0.5 μm, and 2–3 min are needed to accumulate the measurements across the sequence of cut diameters. This makes the technique too slow, for example, to determine variations in particle emissions from engines operating in a transient mode.

4.7.5.5 Diffusion batteries

The sample air passes in laminar flow between closely-spaced parallel plates or along cylindrical pipes. Particles of different sizes diffuse with different speeds to the walls, with the smaller particles more likely to be retained. By varying the flow parameters, particles above a cut diameter will penetrate and be counted (by a CNC). Size ranges of 5–500 nm have been determined.

4.7.5.6 Condensation particle counter

The condensation particle counter (CPC, also known as the condensation nucleus counter (CNC) or Aitken nucleus counter) is used to measure the total number concentration of particles in the diameter range from a few nanometres to 1 μm. The particles are introduced into a chamber where they are exposed to a water-or alcohol-based supersaturated vapour. The particles act as condensation nuclei for droplets that grow to around 10 μm diameter, when they are large enough to be detected and counted, usually by optical methods. Since the growth is diffusion-limited, all the droplets are of similar diameter even if the original particles were highly polydisperse. Because the supersaturation is very high, it makes no difference what the original particle was made of – it simply serves as a condensation centre. In some designs the supersaturation is created cyclically by adiabatic expansion, and in some it is created continuously by conduction cooling. Individual particles can be counted if the concentration is low enough. At higher concentrations an optical method based on attenuation of a beam by the droplet cloud is used. Problems may be caused by excessive concentrations of nanoparticles (<20 nm), which are undercounted because the vapour has been depleted.

An adaptation of this technique is used to count cloud CCN particles. These are activated by the low supersaturations (0.01–1%) found in updrafts and clouds. The controlled supersaturation is achieved by maintaining two wetted surfaces at different temperatures.

4.7.5.7　Calibration

Whatever size classifier method is used, it will need calibrating during and after the design stage, as well as by the user to confirm correct operation. It is not possible to generate a polydiperse aerosol with known standard deviation and median diameters. Instead, calibration is achieved with the use of a series of quasi-monodisperse aerosols with very narrow size distributions. A range of methods exists for generating these.

The process of condensing a vapour on to a nucleus, which is used to count aerosol particles in the CNC, can also be used to create almost monodisperse aerosol. The aerosol material, which is typically a high-boiling substance such as dioctyl phthalate or oleic/steric esters, is atomised and evaporated in an electrically-heated glass tube. Residual non-volatile impurities provide the condensation nuclei for droplet formation. Although the nuclei may vary in size initially, the rate of condensation growth slows as the droplet diameter increases, so that after a suitable time all the droplets have achieved almost the same diameter. The original solution is diluted with alcohol, which evaporates to produce smaller droplets. By controlling system parameters such as dilution, temperature and residence time, particles in the size range 0.04–$1.0\,\mu m$ can be produced at number concentrations of around 10^7 particles cm^{-3}.

Bottles of spherical polystyrene latex beads in aqueous suspension are available, with diameters in the range 0.02–$100\,\mu m$. Aerosols can be generated by nebulising a diluted suspension and mixing the droplet suspension with dry air to evaporate the liquid component. The distribution is extremely monodisperse, and the particles are almost the same density as water so that the aerodynamic diameters are nearly the same as the physical diameters. Hence these particles are favoured for instrument calibration. There may be problems with chain formation if multiple particles are present in one drying droplet, and with formation of nanoparticles from stabilisers if there are no particles present in a drying droplet.

With a spinning disc generator, a small (3 cm diameter) conical disc with a flat top is suspended on the airflow leaving a vertical pipe. Fluting on the underside of the disc forces it to rotate in the airstream, achieving up to 60 000 rpm. Liquid is fed onto the top centre of the disc, and moves radially across the surface under centripetal force before being projected from the edge in a thin sheet. Hydrodynamic forces break the liquid sheet up into droplets. Most of the droplets are in one larger size class, while about 10% of the liquid forms satellite droplets which are about 1/4 of that diameter.

$$d = \frac{0.47}{\omega} \left(\frac{\gamma}{\rho_1 R} \right)^{1/2}$$

where ω is the speed of rotation (revs s^{-1}), γ and ρ_1 are the surface tension and density, respectively, of the liquid, and R is the disc radius.

The two size groups separate because of their different stopping distances and both sets can be used. The droplet diameter is changed by adjusting the rotation speed of the disc. Further control can be achieved by air-drying solutions of different concentrations to leave a solid residue. By using both of these techniques, aerosol particles in the range 0.6–3000 μm can be generated.

Highly monodisperse droplets can be produced from a vibrating orifice. For example, a liquid is forced under pressure through a 5–50 μm orifice, which is vibrated in the direction of the jet by the application of a sinusoidal force from a piezoelectric crystal. The Rayleigh instability which breaks up the jet into droplets occurs at the imposed frequency, and the droplet diameter can be calculated directly from

$$d = \left[\frac{6}{\pi} \frac{Q}{f} \right]^{1/3}$$

where Q is the liquid flow rate and f the applied frequency. The initial droplet diameters are 15–100 μm. As with other liquid-based particle generators, solvent dilution and evaporation can be used to create solid particles and to further extend the diameter range. Rapid dilution with clean air is essential to prevent the droplets from coagulating.

In a fluidised-bed aerosol generator, a chamber is filled with bronze beads about 200 μm diameter, to a depth of 15 mm. An upward air current suspends the beads above the floor of the container. Dust from a reservoir is fed into the chamber, ground up by the bronze beads, and entrained into the updraft leaving the chamber. This technique is used to disperse powders such as ground Arizona road dust. It generates a broad but quasi-reproducible size distribution rather than a monodisperse aerosol.

4.8   CHEMICAL COMPOSITION OF AEROSOL

The primary component of the accumulation and ultrafine aerosol includes elemental carbon and high-molecular weight organic compounds from fossil fuel combustion and biomass burning. The secondary component includes photochemical reaction products and organic condensates. Some of the particles will include significant water if they are hygroscopic and the relative humidity is high (over about 60%). Other substances include short-lived intermediates, such as peroxides and free radicals, of gas- and aerosol-phase chemical reactions. Many of the measurements made of airborne particles have focussed on the mass loading and size distribution regardless of the chemical composition, and this has been reflected in legislation and air quality guidelines concerning particles. We are just beginning to address this lack of understanding, and it is possible that in the future it will be seen to be just as naive to discuss total particle concentration as it is to discuss total gas concentration today. A range of techniques has been used to achieve our current understanding:

- Wet chemistry of bulk samples. Particles are accumulated on a filter, typically over a period of at least a few hours, and analysed by the full range of chemical laboratory techniques for the materials of interest. This method has been used to investigate the photochemical formation of ammonium sulphate and nitrate, the occurrence of sea salt particles from ocean droplets, fluoride pollution from point sources. Increased sophistication is available from other devices. Atomic absorption is routinely used for metal analysis; ion chromatography for ions such as sodium and calcium; inductively-coupled plasma spectroscopy (ICP) can give 20 metals simultaneously, neutron activation analysis (NAA) can give up to 35 elements. Carbon content (by solvent extraction or thermal decomposition) can also be determined.
- For semi-volatile compounds such as ammonium nitrate, volatilisation of the collected particles in the presence of variable concentrations of gaseous component before analysis is a serious source of error. In this case the diffusion denuder can be used to separate the gas from the particles.
- Optical microscopy. Studies of individual particles on a substrate can be useful for source attribution. Collections of photographs of known materials are available for comparison. Standard microscopy can be enhanced with techniques such as polarising filters.
- Transmission electron microscopy (TEM). Again, collections of micrographs are available for comparison.
- Scanning electron microscopy (SEM). Not only does this give 3-D images of particles, which makes visual identification more straightforward, but the instrument can be used with a number of bolt-on analytical devices, such as energy-dispersive analysis of X-rays (EDAX), which give a powerful method of performing chemical analyses on individual particles once they have been collected on a substrate.

- X-ray fluorescence is a common method for multi-element particle analysis. Particles are first collected on a membrane filter (so that they remain on the surface, rather than getting buried within the filter as they would with glass fibre). Photons (X-rays) are used to knock an inner orbital electron out of an atom. The vacancy is filled by an outer electron, and the excess binding energy released as an X-ray. The energy of the X-ray is determined by the difference between the energies of the two orbits, and hence spectroscopic analysis gives the element involved, with determination of all elements simultaneously. Elements with atomic numbers less than 12 cannot be determined by this method, which rules out many environmentally-important elements such as sodium and magnesium. The methods for determining the energy of the emitted X-ray are either wavelength dispersive or energy dispersive, with the latter being most common because it is cheaper. Detection limits for both methods are in the $1-10 \mu g \ g^{-1}$ range. The vacuum environment of the EM, together with local heating from the electron beam, may give problems with volatile components.
- In the *PIXE* method, heavy charged particles, such as $1-4$ MeV protons, are used to generate the X-ray fluorescence instead of photons. The X-ray spectrum is measured with an energy-dispersive detector. The minimum sample mass is lower, and the detection limit an order of magnitude smaller, than for EDXRF (down to ng g^{-1}), although the inability to sample light elements remains.
- Neutron activation analysis involves irradiating the sample with thermal neutrons from a nuclear reactor. This makes a small proportion of the stable atomic nuclei in the sample radioactive. The original elements present can be determined by measuring the radioactivity. Normally, the emitted γ-rays are measured with a spectrometer and the elements identified from the occurrence and strength of the lines in the γ-ray spectrum.
- Laser microprobe mass spectrometry involves the irradiation with a high power laser of particles that have been collected on a substrate. The ejected ion fragments are then analysed by a mass spectrometer. Trace levels of metals can be detected at the ppm level in individual particles, inorganic compounds such as nitrates and sulphates can be speciated, and surface species can be differentiated from those in the body of the particle.

These methods will only give the elemental composition, rather than compounds, although much can be inferred from the ratios of components. A further limitation is that examination of a statistically significant number of individual particles can be both laborious and time-consuming.

Clearly there is no simple single method that will give the information needed for a complete characterisation of particle chemistry. Use of complementary methods is needed in order to build up as full a picture as possible. Even when that has been done, it must be remembered that repeat sampling at a different time of year or in a different location will probably give different results.

4.8.1   Particle microstructure

With advances in instrumental analysis techniques it has become feasible, although not yet routine, to investigate the microstructure of individual particles on spatial scales of ~0.01 μm. Very few particles are physically or chemically homogeneous. It is more normal for any particle to have, for example, a core of one material such as carbon, with a surface shell of another material such as condensed volatiles. Elements of anthropogenic origin such as tin, titanium, chromium and lead are mainly released into the atmosphere in gaseous form during fossil fuel combustion. They eventually form a thin surface film on large mineral particles. Fly-ash particles from coal burning can also have very non-uniform element distributions, with surface concentrations of toxic elements such as lead which can be thousands of times the average concentration in the particle. These particles have been much studied due to their large emissions and widespread prevalence in the troposphere. Studies downwind of smoke plumes from oil well fires have shown that particle surface sulphate deposits increase with distance, indicating a gas-to-particle conversion process. Urban aerosol particles have been found to have a shell structure, with an inner core of soil-based material, a layer of carbonate and organic carbon compounds, and a surface layer only 60 nm thick of nitrate and ammonium sulphate.

4.8.2   Real-time chemical analysis

Some of the methods normally used for laboratory analysis have been automated to give *in situ* measurements in near-real time, usually on a cycle that integrates over about 1 h. Methods are currently being developed to give true real-time chemical properties of individual particles, although the smallest particle size for which they have been used so far is around 1 μm. The most developed technique is particle analysis by mass spectrometry (PAMS). Typically, a few hundred particles per minute are first sized by a time-of-flight method. The particles are then individually volatilised and ionised by a high-energy (e.g. excimer) laser, which is itself triggered by the passage of the particle through the sizing laser. The ionised fractions then enter a time-of-flight mass spectrometer.

4.9   MEASUREMENT OF COARSE PARTICLE DEPOSITION

Large particles (having aerodynamic diameters greater than about 20 μm) have sedimentation speeds that are significant compared to vertical turbulent speeds in the atmosphere and therefore settle out onto surfaces. When industrial processes such as crushing, grinding, ploughing, quarrying or mining generate large emissions of these large particles, there can be a local nuisance problem with deposition to the surrounding area, making cars, paintwork and laundry dirty. There is unlikely to be a health issue, since the particles are too large to be inhaled. Standards have been difficult to lay down, since what is clean to one person may be dirty to another, and the effect will vary depending on the combination of surface colour and dust colour – coal dust will be noticed much more quickly on white windowsills than would limestone quarry dust. A guide value of 200 mg m^{-2} day^{-1} has been suggested, but is certainly not accepted universally. These particles are not in suspension in the air and cannot be measured by the active sampling/filtration methods discussed above for aerosol. Particle deposition has instead been measured for many years by a variety of deposit gauges. Their common feature is a collector which presents a horizontal circular area into which the particles fall.

For the BS deposit gauge, the circular area is the upper rim of a glass or plastic funnel, as in Figure 4.19. Particles that fall into the funnel are washed by rain into a collecting bottle during a period of one month, after which they are filtered from the liquid, dried and weighed. The mass of particles is then divided by the funnel area and the collection period to calculate the deposition rate in mg m^{-2} day^{-1}. The sample can also be analysed for soluble and insoluble fractions or other chemical components which might give additional information on the source of the dust. The collection process sounds straightforward but is far from it. First, although the particles are sedimenting, their terminal speeds are much lower, typically by factors of 10 or 100, than typical horizontal windspeeds in the lower atmosphere. Hence they are 'falling' almost horizontally, and the deposit gauge will present a very thin ellipse rather than an open circle. Edge effects caused by turbulence as the wind blows over the rim of the gauge also interfere with the smooth deposition process. Wind-tunnel tests have shown the collection

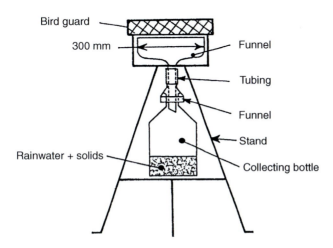

Figure 4.19 BS deposit gauge.

efficiency to be very low at normal combinations of particle diameter and wind-speed. Second, rainfall is an intermittent process – it only rains for 7% of the time in the UK. Hence dust will be deposited into a dry container most of the time, and can get blown out again, or re-entrained, by atmospheric turbulence. The gauges are usually left out for many days or weeks, so significant losses can occur during long dry spells.

Other deposit gauges have been designed on the same basic principles. The International Standards Organisation (ISO) gauge is simply an upright vertical cylinder, designed to allow particles to fall into stagnant air well below the rim and reduce particle re-entrainment. The flow disturbance and re-entrainment problems of deep gauges have prompted the development of very shallow dish collectors with a low rim, often referred to as frisbee gauges (Figure 4.20). They are cheaper and lighter, which also makes them attractive for short-term surveys where vandalism might be a problem. Their low profile reduces both the disturbance to the flow and the development of the internal vortex that causes re-entrainment, and their collection efficiency is less sensitive to windspeed. Various modifications have been tested to improve their performance, such as fitting a nylon mesh across the top and a foam liner to the inside to reduce splash. These changes will all affect performance by uncertain amounts. Petri dishes have also been used for measurements taken over short periods (a day or two) under calm conditions.

In the middle part of the twentieth century these deposit gauges were the main method of measuring atmospheric particulate pollution in urban areas, and

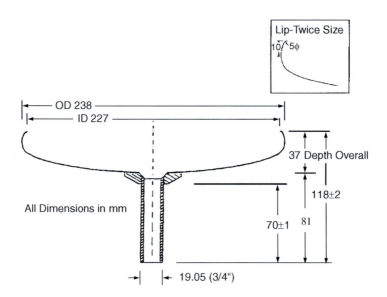

Figure 4.20 Frisbee deposit gauge.
Source: Mark, D. (1998) In: R. M. Harrison and R. van Grieken (eds) *Atmospheric Particles*, Wiley, Chichester.

formed a part of the UK National Survey. As coarse dust emissions have been reduced, urban deposition values have fallen from around 500 mg m^{-2} day^{-1} to below 100. Attention has switched to smaller suspended particles, and deposit gauges have fallen into disuse for general air pollution measurements, although they are still appropriate for specific source investigations.

The fact that most particles are falling with almost horizontal trajectories has been used to advantage with the directional dust gauge. In this design (Figure 4.21), there are four cylindrical collectors, each of which has a vertical slit facing outward. When the wind is blowing towards a particular slit, the streamlines diverge around the cylinder and the particles are deposited inertially into the slit, after

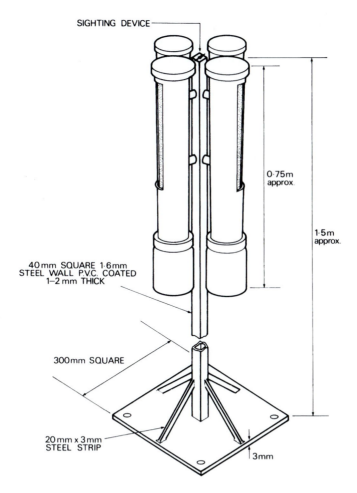

Figure 4.21 BS directional deposit gauge.
Source: Pio, C. A. (1986) In: R. M. Harrison and R. Perry (eds) *Handbook of Air Pollution Analysis*, Kluwer Academic Publishers, Dordrecht, The Netherlands.

which they settle into removable bottles below. The directional gauge can be placed, for example, on the perimeter of a source such as an open-cast coal mine or quarry. After exposure for a few days or weeks, examination of the relative amounts collected in the four bottles, after correction for the proportion of time for which the wind has been blowing from those directions, will give an indication of the source direction. This might be confirmed by microscopic or chemical analysis of the samples. There are uncertainties connected with swirl around the cluster of cylinders causing deposition in the 'wrong' collector, and the variability of collection efficiency with windspeed.

The size distributions of coarse particles can be determined, but much care is needed with sampling conditions in order retain the larger particles. Size fractions can be separated by means of an elutriator, in which a vertical updraft of air carries away particles having less than the designed sedimentation speed while retaining the remainder. Centrifugal forces generated by relatively gentle bends can be used to discriminate particles in the 10–100 μm diameter range.

Other simpler semi-quantitative methods are also in use to assess nuisance from deposited dust. First, white sticky tape can be exposed and the results expressed as percent area coverage by measuring the percent reduction in reflectance (per day). Second, glass microscope slides can be exposed horizontally, and the resulting reduction in reflectance expressed in soiling units, where one soiling unit is a 1% reflectance reduction.

4.10 EMISSION MEASUREMENT FROM STATIONARY SOURCES

Emissions from stationary sources may be divided initially into confined and unconfined. Confined emissions are those from chimneys and other organised release arrangements, usually as a result of a specific ventilation process and driven by a fan or other prime mover. Unconfined emissions are those from roof vents and other gaps, often driven by natural convection due to buoyancy. Confined emissions are much easier to quantify. For both types of emission, two measurements are often of interest – the concentration of the pollutant and its mass flow rate. Calculation of the latter requires knowledge of the gas flow rate.

Both the temperatures and pressures of process gas emissions are likely to be different to ambient values. Gravimetric concentrations are therefore converted to mass per Normal m^3 (Nm^3, at 20 °C and 1 atmosphere), by the method given in Chapter 1, before reporting.

4.10.1 Confined emissions

These are usually chimney stacks from continuous or intermittent sources such as power stations, chemical works and many other processes.

4.10.1.1   Measurement of volume flow rate

For various reasons it may be useful to measure either the distribution of gas velocities across the cross-section of a duct or stack, or the gas volume flow rate. First, correct particle sampling involves establishing isokinetic flow through the sample inlet – that is, an inlet speed equal to the local gas speed. Second, process emission limits may be expressed in mass of pollutant per day, for example, as well as or instead of pollutant concentration.

The classical method for measuring the local value of gas speed is by the velocity head. In a single-point fixed installation, a small-bore tube points upstream (there is no flow through it) and measures the total head, which is the combined static and dynamic pressure. A second tube just penetrates the duct wall and measures the static pressure alone. The difference between them, which can be measured on one manometer, is the velocity head. This is equal to $1/2\ \rho v^2$, where ρ is the gas density at the flow conditions (composition, temperature and pressure). If the gas speed has to be measured at different points across a section of the duct, then a Pitot-static probe is used, in which the total and static heads are combined at the end of a stainless steel probe which may be several metres long. For low particulate loadings, the standard type is preferred since its calibration is within 1% of the theoretical value. For high particulate loadings the small upstream hole of the standard type tends to get blocked, and the S-type should be used, although the calibration of this may be as much as 15% away from theoretical and corrections must be applied. The pressure differential is measured by a slant oil-filled manometer or magnehelic gauge. There are also available electronic digital instruments which measure the differential pressure, temperature and static pressure, take averages and compute the flow.

Volume flow rate is determined as the average gas speed over a cross section of the chimney or a duct leading to it, multiplied by the cross-sectional area. The volume flow rate may then be multiplied by the gas concentration (which is usually assumed uniform) to obtain the mass emission rate. It is important to take the measurement in a section of the duct where the flow is as uniform as possible – guidelines are at least five duct diameters downstream, and two diameters upstream, of any disturbance or obstruction such as a damper or elbow. Many flows are rather non-uniform across the width of the duct, especially if the measurement has to be taken close to a bend, and the average value will need to be constructed from many point measurements. These are taken according to a standard protocol (Figure 4.22), with the Pitot probe marked up beforehand so that it is inserted to the correct depth. In a rectangular duct, sample points should be chosen to sample equal areas, while in circular ducts they should sample equal annular areas. For example, in the configuration shown, the radii of the two annuli will be 0.58 and 0.82 of the duct radius.

The dry bulb temperature, wet bulb temperature and static pressure will all be needed as well in order to convert the flow rate to standard conditions. Basic methods for these have been the mercury in glass thermometer and water U-tube manometer respectively, although these are being replaced by thermocouple

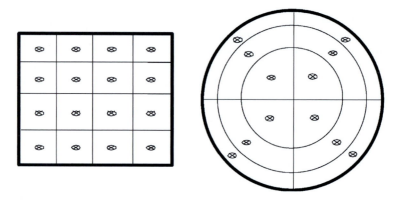

Figure 4.22 Measurement locations in rectangular and circular ducts.

probes and electronic pressure sensors. The internal cross-sectional area of the duct or stack can often be obtained from drawings without needing direct measurement. If the duct is under significant suction (if the main extraction fan is downstream) then the sample access hole will need to be sealed to prevent the inward flow of air from affecting the readings.

Gas volume flow rate from combustion processes can also be calculated from a knowledge of fuel combustion rate and composition, together with either oxygen or carbon dioxide content of the flue gases. This is more likely to be done for efficiency optimisation than for pollutant emission calculations.

4.10.1.2 *Measurement of gas concentrations*

A wide range of methods is in use, corresponding to those described for ambient air. The measurements may be either extractive, in which a sample is drawn from the flue gases and presented to the analyser, or *in situ*, in which the measurement is made directly in the flue gases.

The standard manual extractive method (e.g. US EPA Method 6 for sulphur dioxide) is to reduce the particle interference problem by plugging the sample inlet with a quartz wool filter before drawing the sample gas at a flow rate of about 1 l min^{-1} through a heated probe into an appropriate absorber for the gas of interest. The absorbant is taken back to the laboratory for analysis. A wide range of absorbers and systems has been developed to cope with the wide range of gases. For SO_2, for example, it is hydrogen peroxide in midget impingers, whereas for NO_x it is sulphuric acid and hydrogen peroxide in evacuated flasks. The flue gases are often hot, acidic, saturated with moisture and laden with particles, so getting the sample to the analyser can be difficult. A variety of solutions to this problem is commercially available.

First, hot/wet systems keep the flue gas sample at source conditions by heating the sample line and pump to a higher temperature than the dew point and delivering the gas to an analyser with a heated sample cell. This method is appropriate for water-soluble gases such as SO_2, NH_3 and NO_2 which would be lost or reduced in condensate.

Second, cool/dry systems cool the sample under controlled conditions to remove moisture while minimising absorption of the gas of interest. The sample is then measured by an analyser at ambient conditions. The sample stream is extracted through a sintered stainless steel or ceramic filter to remove most of the particles. The filter will often be blown back intermittently with pressurised air or gas to clear blockages. The sample is then drawn down a 6 mm internal diameter heated line made of PTFE or stainless steel, at a flow rate of a few 1 min^{-1}, to the moisture removal system. This sample line may be a hundred metres long, and there must be no condensation anywhere along its length. In a properly engineered system designed for long life with minimum maintenance, the sample line may be integrated into an 'umbilical' assembly which also carries power cables, an air line for cleaning the filter and a calibration gas line, all mounted inside the heating elements (Figure 4.23). Moisture is then removed by passing the sample stream through either a condensation drier in which the gas is cooled below the water dew point, or a permeation drier in which the gas flows through ion exchange membranes which only allow the passage of water vapour.

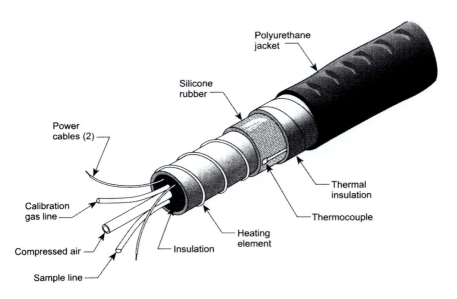

Figure 4.23 Integrated heated sample, power and calibration line.
    Source: Jahnke, J. A. (2000) *Continuous Emissions Monitoring*, Wiley, New York.

Both the hot/wet and cool/dry extractive systems can be close-coupled to the extraction point by removing the sample line and putting the pump and analyser right by the extraction point. This eliminates sample absorption in the cooler and elsewhere, which may improve the accuracy of VOC measurements. It also eliminates a long length of expensive sampling line, although the downside is that the analyser may be at an inaccessible location halfway up a tall stack.

With a dilution system, a much lower sample flow rate (20–500 ml min^{-1}) is extracted, which reduces the particle and moisture contamination problems. The sample gas is immediately diluted with a known flow rate of clean dry air (a few 1 min^{-1}, giving dilution ratios between 50 and 300:1). The dew point is thus lowered to below ambient temperatures, and the sample can be brought to the pump and analyser (which can be an ordinary ambient model) in an unheated sample line. In one common design of dilution probe (Figure 4.24) the dilution air supply drives an ejector pump which draws the sample from the flue gases. The quartz wool filters out particles, and the enclosed design means that calibration air and zero air can be injected to displace the sample when required. Dilution systems automatically give a concentration in terms of the volume including moisture, which may be a useful advantage if that is the way an emission limit is expressed.

The extractive methods described above deliver a sample of the gas through a pipe to an analyser, and the analysis method can be any of those described earlier for the component of interest. A further range of techniques is available by which

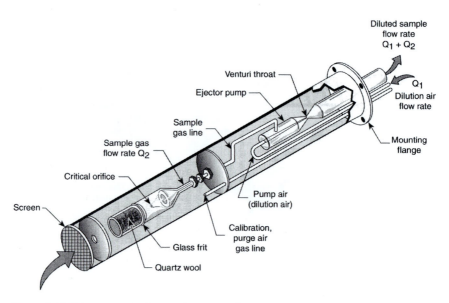

Figure 4.24 Dilution probe for stack gas sampling.
Source: Jahnke, J. A. (2000) *Continuous Emissions Monitoring*, Wiley, New York.

the gas concentration is measured *in situ*, by optical techniques based on some form of the Beer–Lambert law. The Beer–Lambert law is the simplest description of the change in flux density of a light beam as it passes through a medium.

$$\tau = \frac{F}{F_0} = e^{-\alpha(\lambda)CL}$$

where τ is the transmittance (dimensionless), F (W m^{-2}) is the flux density after distance L (m), F_0 is the initial flux density (W m^{-2}), $\alpha(\lambda)$ is the molecular absorption coefficient (m^2 kg^{-1}) as a function of wavelength λ, C is the concentration of the gas (kg m^{-3}).

Optical measurements may be made in the IR, visible or ultraviolet regions of the spectrum. Infrared sources are either Globars (rods of zirconium and yttrium oxides, or of silicon carbide, heated to around 1500 °C) or solid state lasers. Visible sources include conventional tungsten and quartz-halogen lamps, light-emitting diodes and lasers. Ultraviolet sources include hollow cathode gas discharge tubes, xenon arc lamps and mercury discharge lamps. A narrow region of the emission spectrum, within which the gas of interest absorbs strongly, is selected by an interference filter or diffraction grating. There is a clear trade-off between specificity (which needs a narrow window but that reduces the total flux density) and sensitivity (which is *vice versa*). The transmitted flux density is detected by solid-state quantum detectors, photomultiplier tubes or photodiode arrays.

The optical arrangement can be single pass, with the transmitter and receiver on opposite sides of the stack, or double pass (Figure 4.25), with the transmitter and receiver co-located. Some instruments, known as *point in situ* have the optical path (i.e. the source and detector) fixed on a probe in the gas stream, with a shorter path length of 1–100 cm.

One important issue is calibration. It will often be a requirement to automatically zero and calibrate the system every 24 h by inserting either clean air or air containing a known concentration of the gas of interest (from a certified cylinder) into the optical path. This is not usually possible with single-pass systems, possible with difficulty with double-pass systems, and more straightforward with point *in situ* arrangements. The latter have a compact sample measurement volume which is often protected from particle deposition by enclosure in a sintered filter which allows penetration of the flue gases but keeps the particles out. The filter will retain zero or span gas when it is injected into the view volume, driving out the flue gas.

Two main measurement methods are in use:

Differential optical absorption spectroscopy has been described previously for the long path measurement of ambient concentrations, and the same technique is applied for stack gas concentrations. Commercial DOAS systems can measure NO, NO$_2$, SO$_2$, CO, CO$_2$, NH$_3$, VOCs, HCl and HF.

Derivative spectroscopy involves scanning a spectral absorption peak to obtain its second derivative at the maximum. This is accomplished in practice by

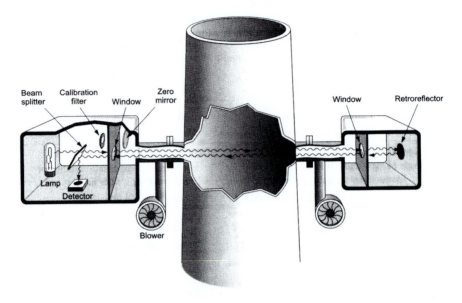

Figure 4.25 A double-pass transmissometer.
Source: Jahnke, J. A. (2000) *Continuous Emissions Monitoring*, Wiley, New York.

modulating the output from a tunable diode laser so that it sweeps sinusoidally through a range of wavelengths. Stack concentrations of HF, NH_3, HCl, O_2, CO, H_2S and NO have been measured with such systems.

4.10.1.3   *Measurement of particle concentrations*

Standard gravimetric sampling

Standard methods have been developed to sample particles from ducts and stacks. The gases are often hot and humid, which poses problems for repeatable measurement. The US EPA has developed Method 5 for out-of-stack measurements of low particulate loadings, and Method 17 for in-stack measurements of high loadings which might block the sample probe used for Method 5. The requirements are simple in principle – to separate the particles from a known volume of gas. The essential features are a stainless steel sample probe, a heated filter for coarse particles, a set of impingers in iced water to trap fine particles and condensates, a vacuum pump and a dry gas meter. In addition to the particles trapped in the filter and impingers, the sample probe will be washed out with acetone and the deposited particles collected. Although particle size analysis can be done on all the collected material, it will not usually be known how that compares with the *in situ* size distribution.

Opacity

The most obvious sign of high particle emissions from a chimney is thick black smoke, and this has been the basis of pollutant observations for many years. It is no longer acceptable for a chimney to discharge such emissions continuously, although there may be circumstances when it is allowed for a short period for a specific purpose, such as lighting up or soot blowing in a large boiler. The severity of such emissions has been measured by trained observers, and put onto a semi-quantitative basis by comparing the appearance of the smoke with a standard series of increasing shades of grey (the Ringelmann chart).

The instrumental equivalent of the trained observer is the opacity monitor. A light is shone single- or dual-pass across the stack and the transmittance measured. Then opacity (%) = 100 − transmittance (%). The spectrum of the lamp must peak between 500 and 600 nm, and not extend into the UV or IR, so that the results can be compared with those of the human eye. A zero value is created by inserting a mirror into the optical path close to the transmitter. A span value is created by inserting a filter of known attenuation in front of the zero mirror. The optical density of the particles depends on factors such as particle size distribution as well as number concentration, so this method is strictly for opacity only.

Accurate measurement of process particle emissions is much more difficult than for gas concentrations. The types of technique used include:

- *Manual in-stack*: Isokinetic sampling procedures must be followed strictly if a representative sample is to be obtained. Large particles cannot be extracted since they do not remain suspended in the gas sample, so the particle capture must occur right at the inlet of a probe inserted into the flow. Since the gas flow speed varies across the duct, multiple samples need to be taken at different points, with the sample gas flow rate (and hence the sample inlet speed) adjusted to maintain isokinetic sampling at each location. Bends and other flow disruptions produce turbulence and inhomogeneous particle distributions; hence particle emission measurements are often made halfway up the chimney. Commercial devices are available either to give total particulate matter or a size distribution from a cascade impactor.
- *Manual extractive*: If the particle emission is downstream of an efficient control device such as a fabric filter or an electrostatic precipitator, the particle diameters will usually be small enough to extract without serious losses. Isokinetic samples can then be extracted through a heated pipe to a filter for determination by weighing.
- *Instrumented in-stack*: Most commonly, optical transmissometry is used to determine total concentration from the reduction in flux density of a beam passing across the width of the duct. This is similar to the opacity technique discussed above but requires a calibration of the transmittance against

gravimetric concentration determined by a manual isokinetic method. The transmittance τ is related to particle properties by Bouguer's Law, which is just a rephrased version of the Beer–Lambert law discussed previously for gases.

$$\tau = e^{-Na\kappa L}$$

where N is the particle number concentration, a is the particle projected area, κ is the particle extinction coefficient, L is the path length of the light beam.

In practice it is more likely that optical density will be used. Optical density is defined by

$$D = log_{10} \frac{1}{1 - \text{opacity}} = log_{10} \frac{1}{\tau}$$

Also since $\ln \tau = -Na\kappa L$, then $\ln (1/\tau) = Na\kappa L = 2.303$ D, since $\ln 10 = 2.303$. Hence $D = Na\kappa L/2.303$. This is framed in terms of the particle number concentration N, whereas it is usually the mass concentration C which is measured, and in terms of which the regulations are framed.

We write κ_S = specific mass extinction coefficient = $\pi r^2 \kappa/m$. This is a property of the given particle composition and size distribution, and will be effectively constant for an individual source. Then $D = \kappa_S CL/2.303$, so that the optical density is directly proportional to the mass concentration and pathlength.

An electrodynamic method is available in which an electrical current, which is measured directly by an electrometer, is generated by charge induction on a probe inserted into the gas stream. Although the technique is useful for detecting gross changes such as bag failure in a baghouse, the many factors such as particle charges and resistivity, probe surface condition and humidity which affect the results render it only semi-quantitative. A more quantitative version of the method relies on AC detection of the variable voltage induced by particles as they move past the probe.

4.10.2   Unconfined emissions

Often intermittent industrial processes, such as steel manufacture when hundreds of tonnes of molten metal are poured from one container to another, generate heavy particulate and gaseous emissions. Most of these may be collected by an extraction system for subsequent cleaning and discharge from a chimney. Part of the emissions, however, is not collected and leaves the building by whatever route presents itself. These are referred to as fugitive emissions, because they have escaped the control system. Measurement of fugitive emissions follows the same principles of concentration and volume flow measurement as described above for confined emissions. Obtaining representative values is harder, however, because the releases are often sporadic, and both the source strength and the ventilation are variable. Quantitative manual assessment involves multiple repeat sampling at many locations. Long-path optical transmissometry may be able to provide an average concentration along the length of a vent.

FURTHER READING

Ahmad, R., Cartwright, M. and Taylor, F. (eds) (2001) *Analytical Methods for Environmental Monitoring*, Pearson Education, Harlow.

Clarke, A. G. (1998) (ed.) *Industrial Air Pollution Monitoring*, Chapman and Hall, London.

Conti, M. E. and Cecchetti, G. (2001) 'Biological monitoring: lichens as bioindicators of air pollution assessment – a review', *Environmental Pollution* 114: 471–492.

Harrison, R. M. and Perry, R. (eds) (1986) *Handbook of Air Pollution Analysis*, Chapman and Hall, London, UK.

Hewitt, C. N. (ed.) (1991) *Instrumental Analysis of Air Pollutants*, Elsevier, Barking, UK.

Jahnke, J. A. (2000) *Continuous Emissions Monitoring*, Wiley, New York.

Chapter 5

Concentrations and deposition

Air pollution concentrations vary greatly from place to place at any one time, and with time of day and from year to year at any one place. Before we can understand where and when the concentrations of hazardous air pollutants rise to unacceptable levels, we need to have an extended period of data available from a network of pollution monitors having appropriate spatial distribution and response time. In this chapter we will look at measurement networks in the UK and elsewhere that have given us our current understanding of the occurrence of air pollutants, and present some results that demonstrate key aspects of the conclusions so far. We will also look in some detail at the factors that determine both the short and long-term airborne concentrations and the rates of dry and wet deposition of gaseous pollutants. We will relate much of the discussion of gases to sulphur and nitrogen oxides and to ozone, since they have been the subjects of the longest-established and most systematic measurement programmes.

We will also highlight the difference between regional-scale measurements of concentrations and deposition across the country at large, and local-scale measurements that are more appropriate to urban areas. At the present time 90% of the population of the UK, and half the people in the world, live in urban areas. For several hundred years, air pollution in towns was dominated by domestic emissions from wood and coal combustion, so that sulphur dioxide and smoke were the key issues. During the last 50 years these pollutants have declined, to be replaced by new ones due mainly to motor vehicles. Concerns have shifted away from acute health problems due to specific episodes, and towards the more insidious threat posed by low levels of pollutants that may have significant impact yet be hard to detect. For example, there has been a steady increase in the number of asthma cases reported in England and Wales. This increase is certainly one health impact that has become associated in the public's mind with motor vehicle pollution, and the two may well be related, yet there is no scientific proof of causality. For example, New Zealand has low population density, much of its electricity is generated from hydro and its air quality is high, yet it has the highest asthma incidence in the world.

5.1   GASEOUS POLLUTANTS

Sulphur and nitrogen species are predominantly emitted as gases. They have life-times in the atmosphere of days, and return to the surface either as gases (dry deposition), or as acidified rain (wet deposition) or as acidified cloud droplets (droplet deposition). All of these processes have different characteristics and different implications for different organisms. Many of the implications are not yet fully understood. Since ozone is formed in the atmosphere from precursors rather than being emitted directly, its occurrence is quite different in character.

5.1.1   Measurement networks in the UK

Measurements of airborne concentrations are derived from one of several extensive networks of sampling and measurement stations that have been established to achieve various objectives:

- assessing population exposure and health impact
- identifying threats to natural ecosystems
- determining compliance with national or international standards
- informing or alerting the public
- providing input to air-quality management schemes
- identifying and apportioning sources.

Assessment of health effects is usually the first reason for air pollution monitoring. There are then choices that are related mainly to the budget available. A single advanced monitoring station to measure the pollutants for which there are agreed health guidelines might cost £100 000. Although concentrations might vary widely across a city, there would not normally be such a monitoring station in a large town or small city, one in a city of a few hundred thousand population, and 2–4 in the largest cities such as Birmingham. These will be used as representative of the concentrations in that urban area. It may also be useful to divide the population into groups according to the amounts of time spent on different activities such as home, commuting, city work etc. There might be one group that lives and works in the suburbs, one that lives in the suburbs but commutes into the city, one that lives and works in the city. Then contributions could be calculated for each group provided that modelled or measured concentrations were available from each of the area types.

Another issue is exposure dynamics. The health impact of some pollutants, such as carcinogens, depends on the integrated lifetime exposure. For these pollutants, there is no safe lower limit, simply a reduction in the statistical probability of an adverse health outcome. Other pollutants, such as photochemical smog, have health effects that depend on the concentration and duration. The body can accommodate steady low levels but some people, especially those who are predisposed to react for any reason, may have acute responses if the concentration exceeds a threshold value. Concentration limits are usually written in terms that recognise this relationship, so that both short-term (1 h or 1 day) and long-term (1 year) targets have to be met. Also, the response may be acute or chronic. Cancer may take years to develop, continues to develop once the exposure has stopped, and is then irreversible without treatment, whereas carbon monoxide poisoning may have acute effects which go completely once the exposure stops.

5.1.1.1   The UK national survey

Following the severe smogs and loss of life in London in 1952, a National Survey of smoke and SO_2 concentrations was set up in 1961, to monitor progress following the implementation of pollution control measures. All the stations use identical standard equipment, known as an 8-port sampler, to take 24-h average samples of black smoke and SO_2. Smoke is captured on a filter paper and measured by the reflectance method described in Chapter 4. SO_2 is absorbed in hydrogen peroxide and the concentration calculated by acid titration. In the 8-port sampler, eight pairs of smoke filter/wash bottle combinations are connected to a common pump and air volume flow recorder. An automatic timer switches the sample flow to a new pair at midnight, so that the system will accumulate daily averages while being left unattended for a week. The National Survey started with 500 sites, grew to a peak of 1200 stations in 1966 and clearly revealed the marked decline in annual average urban smoke and sulphur dioxide (Figure 5.1). The pattern of concentration decline has not been only a British phenomenon. Figure 5.2 shows that SO_2 concentrations in five US cities increased steadily from 1880 to 1930, and after falling since 1950, by 1980 were lower than they had been 100 years previously.

In 1981 the National Survey was reorganised in response to these falling urban concentrations. The basic urban network (BUN) currently involves 154 sites selected

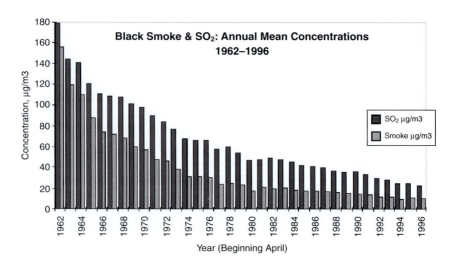

Figure 5.1 The decline in UK SO_2 and smoke concentrations between 1962 and 1996.

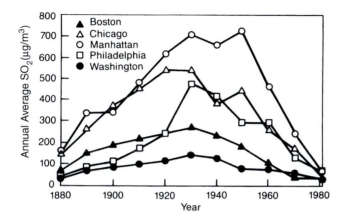

Figure 5.2 The rise and fall of SO_2 concentrations in five US cities, 1880–1980. Source: NAPAP (1990) *Integrated Assessment Report*, National Acid Precipitation Assessment Program, Washington DC.

to provide information on long-term changes across the whole of the UK. The EC Directive Network involves sites in areas of relatively high concentration, monitoring compliance with EC Directive 80/779/EEC on smoke and SO_2. There were initially 370 sites in the network, but a number has been withdrawn as the concentrations have declined. This network is by no means representative of the country as a whole, with pollutant information being confined to areas of high population density such as London and/or industrial activity such as the North-west of England. If the concentrations at an individual site do not exceed the Directive limit values for three consecutive years the site can be taken out of service, so the numbers of operational sites should continue to decline. There is some overlap between the Basic Urban and EC Directive networks, giving a current total of about 225 sites.

5.1.1.2 Secondary networks

Although the National Survey had measured high urban concentrations successfully, some major defects became apparent in the 1980s.

- Urban smoke and SO_2 concentrations had declined to such low levels that the relatively crude methods of the National Survey no longer offered adequate sensitivity.
- The network had naturally been concentrated in urban areas so that there were relatively few rural measurements.
- There was no capability at all to measure ozone or its precursors, which had by that time been identified as major factors in poor air quality.
- There was no capability to measure the most toxic organic compounds, which were by then known to be carcinogenic at very low concentrations.

- Although the existing National Survey results were appropriate to show compliance with the Commission of the European Communities (CEC) Directive on SO_2 and TSP, there was no corresponding capability for the Directive on NO_2 (see Chapter 13).

To meet these needs, new networks of monitoring sites based on high-sensitivity continuous analysers were established by the UK DoE from the mid 1980s to measure O_3 and other pollutants in rural areas, and a range of pollutants in urban areas. The first sites of the rural monitoring network (known as the automated rural network (ARN)) were established in 1986, and there is now a total of 15 sites. Ozone is measured at all of them, and SO_2 and NO_x at three of them. In 1987 five new sites, known as the NO_2D network, were set up to monitor compliance with the CEC Directive on NO_2. Site locations were based on a national survey of NO_2 concentrations, using diffusion tubes, that had been carried out at over 360 sites the previous year. Three of the sites were in the urban centres of West London, Glasgow and Manchester, and the remaining two in the industrial areas of Walsall and Billingham. There are now seven sites in the NO_2D network.

The first six sites of the automated urban network (AUN) started operation in January 1992, and there are now (2001) around 115 sites. The following pollutants (at least) are measured at each site:

O_3 – by ultraviolet absorption
NO_x – by chemiluminescence
SO_2 – by pulsed fluorescence
CO – by IR absorption
PM_{10} – by tapered element oscillating microbalance (TEOM).

The NO_2D network has been incorporated into the AUN, and hydrocarbons and toxic organic micro-pollutants (TOMPS) are measured continuously at a small number of sites. Hourly measurements of 27 species of hydrocarbons, including two known carcinogens – benzene and 1,3 butadiene – are taken by automatic gas chromatographs.

5.1.1.3   Data processing

For all the continuous data, average concentration values over periods of 15 and 60 min are compiled from instantaneous readings taken every 10 s, and telemetered to a central data processing station at the National Environmental Technology Centre (NETCen), Culham. The measurements from these networks are compared with a broad range of international criteria, including EU Directives, WHO Guidelines and United Nations Economic Council for Europe (UNECE) critical levels. Since the data are available almost in real time, it has been possible since 1990 to make daily summaries of the data available to the public on television information systems and by telephone. The full data set of hourly observations from all the stations is available on the web.

The objectives of these networks have been summarised as follows:

- to provide a broad picture of the spatial and temporal distribution of photo-chemical pollutants throughout the country
- to aid the assessment of the impact of these pollutants on sensitive areas, particularly in relation to the exposure of crops and the natural environment
- to permit trend analysis as the data accumulate over a period of years
- to improve understanding of the sources, sinks and national-scale transport of these pollutants
- to allow validation and evaluation of the performance of computer-based models.

5.1.1.4   Other surveys

These continuous sites have been supported by intermittent surveys of NO_2 taken with diffusion tubes. Although diffusion tubes have rather poor time resolution – typically one month – their low cost means that spatial variations of average concentration can be investigated. Figure 5.3, for example, shows the results of a NO_2 survey of Madrid.

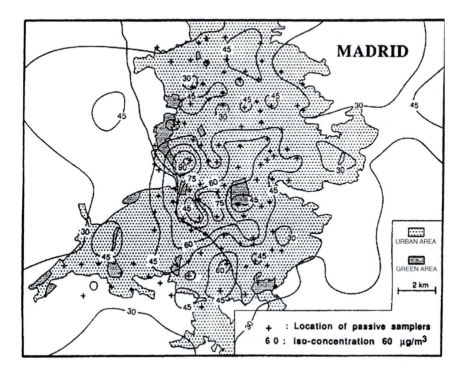

Figure 5.3 Results of a diffusion tube survey of NO_2 taken in Madrid.
Source: *http://europa.eu.int/comm/environment/air*

Diffusion tubes were placed at the locations marked by crosses. Contour calculation and plotting software was then used to generate the spatial concentration distribution. These results could not have been obtained with a small number of continuous analysers at fixed sites, or with a mobile instrumented vehicle.

National UK diffusion tube surveys in 1986 and 1991 showed that NO_2 concentrations were increasing in urban areas, primarily due to increased densities of motor vehicles. In response to this finding, an enlarged long-term diffusion tube survey was started in 1993. Three hundred local authorities are now operating diffusion tubes at over 1200 sites. The tubes are placed at three different types of site by each local authority – one kerbside, two background and one in between. Monthly average results are again compiled by NETCen.

Rural sulphur dioxide samples are taken at 38 sites in the UK. The sample is captured in a peroxide bubbler as with the urban surveys; the low rural concentrations and risk of interference from ammonia require sulphate analysis by ion chromatography in place of acid titration. The measurements of national SO_2 levels are subsequently input to a dry deposition model in order to calculate sulphur inputs to sensitive ecosystems.

Wet deposition of rain and snow is sampled at 32 sites in order to assess pollutant inputs by this route. The samples are analysed for conductivity, pH, five cations (NH_4^+, Na^+, K^+, Ca^{2+} and Mg^{2+}) and four anions (NO_3^-, Cl^-, SO_4^{2-} and PO_4^{3-}). The total deposition (whether it is raining or not) is measured at all the sites, and at five of the sites there are wet-only collectors that are covered when it is not actually raining.

5.1.2   Other measurement networks

There are extensive air quality monitoring networks in all the developed nations. The 15 EU countries between them have around 2200 SO_2 monitors, 1000 NO_x monitors, 800 O_3 and CO monitors and 1000 particle monitors (plus a further 1000 smoke monitors). These monitors have generally been put in place during rather haphazard and uncoordinated monitoring campaigns by the constituent countries of the EU, rather than by planned EU monitoring programme; hence they are by no means uniformly distributed. For example, Figure 5.4 shows the ozone monitors that were known to be operating in Europe in 1999.

The network has been analysed in terms of spatial representivity, and found to cover only 20–40% of forests and 30–50% of crops. Only 25% of EU city dwellers, and 12% of total EU residents, were within the area that was covered effectively by the monitors. The European Environment Agency (EEA) coordinates Europe-wide monitoring through its European Topic Centre on Air Emission (ETC/AE) and European Topic Centre on Air Quality (ETC/AQ). Both of these Topic Centres are part of the European Environmental Information and Observation network (EIONET), although this organisation has a much broader remit than air quality. European Topic Centre on Air Quality has set up the European Air Quality Monitoring Network (EUROAIRNET) to

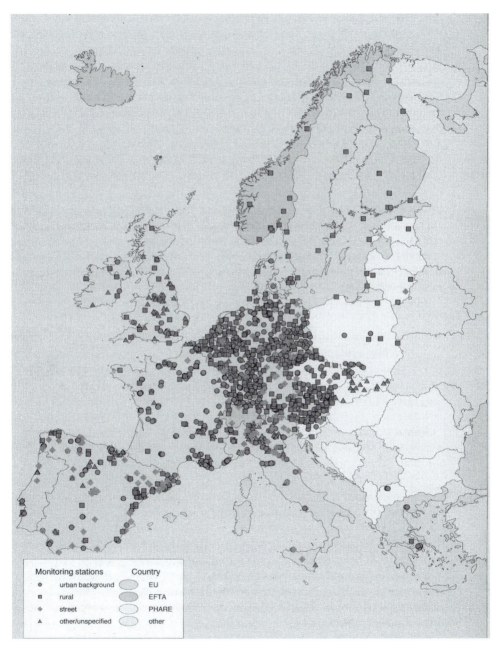

Figure 5.4 EU ozone analyser network.
Source: de Leeuw, Sluyter and Camu (2000) *Air Pollution by Ozone in Europe in 1998 and Summer 1999*, European Environment Agency, Copenhagen.

annually report on and assess European air quality. At the end of 1999, EURO-AIRNET was taking data from 1451 stations in 29 of 32 possible partner countries, covering 355 cities, 46 industrial areas and 218 rural or remote locations. The network is responsible for annually evaluating exceedances of the EU Ozone Directive (92/72/EEC). Data from selected stations of this network is also input annually to the air quality database, AIRBASE. European Topic Centre on Air Quality sets standards for how representative the monitors are of the country's urban or rural populations, QA/QC protocols and data reporting formats. As at September 1999, 566 stations in 20 countries were reporting to AIRBASE. These included SO_2 and NO_2 from 300 stations, O_3 from 250 stations, PM_{10} and CO from 150 stations. One of the major purposes of all this monitoring and reporting of data is to work out whether air quality across Europe is getting better or worse, and whether the air quality limits and guidelines are being met. For example, a rather broad-brush empirical modelling approach has been used to evaluate air quality in 200 European urban agglomerations for 1995 and 2010. The proportion of the population of these urban areas that is exposed to background concentrations in excess of the limit values was calculated (see Chapter 13). Hence appropriate emission reductions can be targeted effectively.

In the US, there are four major network categories.

- State and local air monitoring stations (SLAMS) have around 4000 stations (typically with multiple monitors at each one) which are designed to confirm compliance with State Implementation Plans.
- National air monitoring stations (NAMS) are a subset of around 1100 SLAMS stations that are chosen to focus on urban and multi-source areas of high population density where maximum concentrations are anticipated.
- Special purpose monitoring stations (SPMS) make up a flexible and relocatable network to conduct special studies of particular issues.
- Photochemical assessment monitoring stations (PAMS) are required in each O_3 non-attainment area that is designated serious, severe or extreme. O_3 precursors (about 60 VOCs and carbonyl) are measured.

There are hundreds more sites in North America within national air pollution surveillance (NAPS) in Canada, and Metropolitan Networks of Mexico. The stations at these sites generally measure O_3, NO, NO_x, CO, SO_2, TSP and PM_{10}. More recently, $PM_{2.5}$ is being measured at 1500 sites. The focus of these networks is to measure air quality in or near large urban centres. In addition, there are many other smaller networks set up to monitor air quality for more specific purposes – for example, in National Parks.

These measured data are interpreted with the help of models which predict ozone concentrations from precursor emissions, photochemistry and meteorology. Figure 5.5 shows the output of such a model for global ozone concentrations

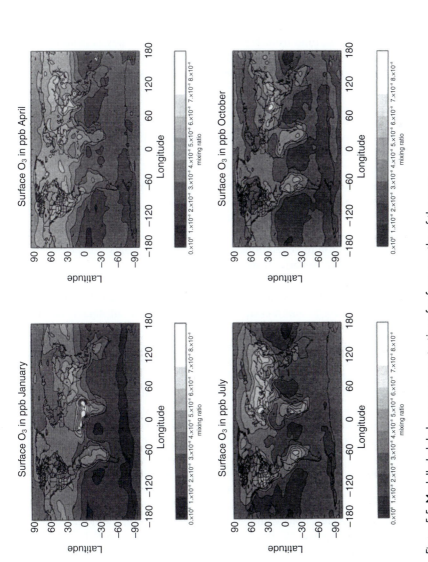

Figure 5.5 Modelled global ozone concentrations for four months of the year.
Source: Collins, W. J., Stevenson, D. S., Johnson, C. E. and Derwent, R. G. (2000) 'The impact of human activities on the photochemical production and destruction of tropospheric ozone', *Quarterly Journal of the Royal Meteorological Society* 126: 1925–1951.

at four seasons of the year, clearly demonstrating the general dependence of concentration on sunlight and precursors. This dependence causes the peak ozone regions to move around the globe with the seasons (the paler the shade, the higher the concentration).

5.2 PATTERNS OF OCCURRENCE

Even when spatial variations in concentration appear to be smooth, they may conceal large variations with time from hour to hour and day to day depending on wind direction and other variables. Nevertheless, when averages are taken over a sufficient number of individual days, patterns can often be seen.

5.2.1 Ozone

The instantaneous concentration of tropospheric ozone is influenced by five main factors:

- photochemical production (which depends in turn on the supply of precursor materials and the availability of sunlight)
- chemical destruction (largely by reaction with NO)
- atmospheric transport (which may generate a net increase or reduction)
- surface dry deposition to vegetation, water or other materials
- folding down of stratospheric ozone.

In Europe, the resulting average concentrations show a broad pattern of increase from north-west to south-east, although this masks substantial annual variability. In the summer months, there is a general gradient from lower concentrations in the north-west to higher in the south-east. This is effectively the outer limb of a pan-European gradient which generates the highest concentrations in Italy and Greece due to the combination of solar radiation and precursors. In the winter the relative situation is reversed, because without the solar radiation the NO emissions (which continue all year round) remove the ozone most effectively over mainland Europe, leaving the concentrations higher in the north-west.

Globally, summer ozone peaks are created over five large areas of the world as seen in Figure 5.6. Fossil fuels and solvents are largely responsible for the northern hemisphere production over North America, Europe and Asia, biomass combustion in the southern hemisphere. These results were generated by the UK Meteorological Office model, which is a Lagrangian 3-D model incorporating 50 chemical species and 90 reactions within 50 000 cells that represent the Earth's atmosphere up to the tropopause. Surface emissions are added at a spatial resolution of $5° \times 5°$, and stratospheric exchange is included. Some coarse approximations currently have to made in such models, either to reduce computing time or simply because our understanding of processes is incomplete. Nevertheless many useful

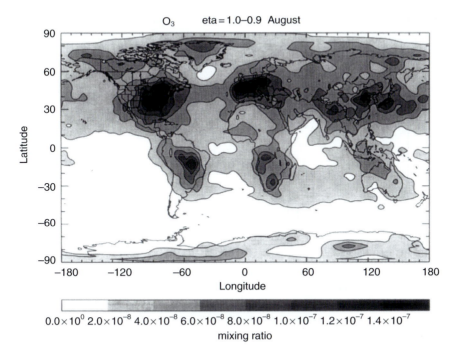

Figure 5.6 Modelled global ozone concentrations in August.
Source: Collins, W. J., Stevenson, D. S., Johnson, C. E. and Derwent, R. G. (1997)
'Tropospheric ozone in a global-scale three-dimensional Lagrangian model and
its response to NO_x emission controls', *J Atmospheric Chemistry* 26: 223–274.

results can be generated, and 'what if' experiments enable the consequences of
changes to be predicted. For example, when the model was run with European NO
emissions reduced to half their current value, ozone concentrations fell by 17% at
most (Figure 5.7).

The same factors that influence the north-south ozone gradient also influence
the diurnal variation. Typically, ground-level ozone concentration decreases at
night, because there is no UV to photolyse NO_2, and the stable atmospheric
temperature profile allows ozone to deposit to the surface without replenishment
from the atmosphere above. In the morning, emissions from motor vehicles and
other sources into the stable atmosphere increase concentrations of precursors;
the increasing flux density of UV radiation produces more ozone, so that the
concentration peaks in the middle of the afternoon. As the sun sets, concentration
decreases once again. Figure 5.8 shows diurnal profiles taken in the same month
of the same year at different sites in Europe. All the profiles show a night-time
minimum and a peak at about 3 pm solar time, but there are differences between
sites due to their local conditions. The French site (FR) has very low windspeeds

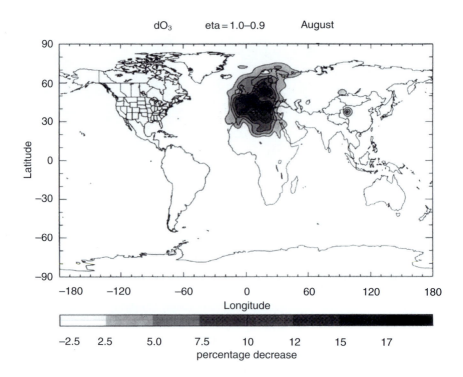

Figure 5.7 The response of global ozone concentrations to a 50% reduction in NO
emissions from Europe.
Source: Collins, W. J., Stevenson, D. S., Johnson, C. E. and Derwent, R. G. (1997)
'Tropospheric ozone in a global-scale three-dimensional Lagrangian model and
its response to NO$_x$ emission controls', *J Atmospheric Chemistry* 26: 223–274.

and high solar radiation, so that concentrations go very low at night and high during the day. The Nottingham site (GBN) has higher windspeeds that keep the night-
time concentration up, and lower average solar radiation that suppresses the
afternoon peak. There are also higher concentrations of NO in the English Midlands
that sweep up the O$_3$. The Swiss profile (CHZ) is characteristic of a high-altitude
(900 m) site, above the stable inversion so that night-time ozone deposition can be
replaced from the upper troposphere and maintain the concentration. Although this
can result in high 24-h average concentrations, it is not necessarily damaging to
plants because their gas exchange is minimised at night. On an annual basis, tropospheric ozone production depends on the availability of precursors and solar UV,
and on atmospheric conditions that suppress mixing. The optimum period tends to
be April–June. During the remainder of the year, production is driven mainly by the
level of UV flux. Occasionally, ozone from the stratospheric ozone layer can be
injected into the troposphere by a weather system, resulting in a short-lived spike at
any time of the year.

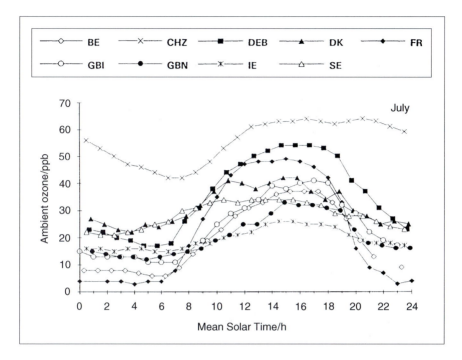

Figure 5.8 Diurnal ozone profiles from different sites in Europe. BE = Belgium, CHZ = Switzerland, DEB = Germany, DK = Denmark, FR = France, GBI = London, GBN = Nottingham, IE = Ireland, SE = Sweden.

Studies of ozone episodes in Middle Europe have shown that exceedances are characterised by certain meteorological situations, namely:

- anticyclonic conditions
- a ridge of high pressure across middle Europe
- a wind direction that brought air from regions expected to have high concentrations of ozone precursors
- high daily maximum surface temperatures
- low surface wind speeds.

Although measures are in place in Europe and North America to reduce emissions of ozone precursors and hence reduce ozone concentrations, the same is not true in developing countries. Figure 5.9 shows diurnal average ozone concentrations during April or June from an analyser in Beijing, China. The peak concentration in the early afternoon has increased by a factor of three, from 40 to 120 ppb, between 1982 and 1997. As will be discussed in Chapters 9 and 10 in this book, the latter concentration is high enough to damage both people and plants.

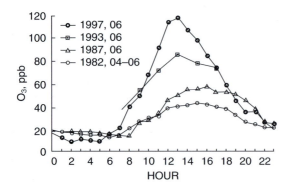

Figure 5.9 Spring ozone concentrations in Beijing, China.
   Source: Redrawn from Zhang, Y., Shao, K., Tang, X. and Li, J. (1998) 'The study of urban photochemical smog pollution in China', *Acta Scientiarum Naturalarum Universitatis Pekinensis* 34: 392–400.

It has been quite a challenge to work out long-term trends of global average ozone concentrations because of the high variability of the concentrations at any one location. Concentrations at remote sites appeared to increase through the 1970s and 1980s, both on the surface and higher in the troposphere. Since 1990, concentrations at some of those background sites have levelled off.

5.2.1.1   The general picture

In the decade 1990–2000, it has been estimated that at least $600 million were spent in Europe and North America conducting over 30 major field studies on the precursor emissions, chemistry and meteorology of ozone. These studies have confirmed the general role of anthropogenic emissions in ozone formation, with biogenic emissions such as isoprene also being important in some areas such as the south-eastern US. Methane and CO may also be important, despite their lower reactivities, because of their widespread abundance in the troposphere. Meteorological processes drive the patterns of ozone occurrence, generating the substantial variations that are found on daily, seasonal and annual timescales. High concentrations of ozone are most likely to be observed during an episode of air stagnation lasting a few days, when there are high temperatures, low wind speeds, strong sunlight and an ample supply of either natural or anthropogenic precursors. Local recirculation due to temperature differentials caused by mountain–valley or water–land opposition may also be important. In contrast to short-lived pollutants such as SO_2, O_3 pollution is a regional-scale phenomenon, with areas of elevated concentration extending over many hundreds of km, and atmospheric transport moving ozone over similar scales. Most rural areas have been found to be VOC-rich, so that O_3 concentrations could be most effectively reduced by reducing NO_x emissions. Conversely, VOC reductions would be most effective in most urban areas.

5.2.2   Nitrogen dioxide

In the 1990s, NO_2 has replaced smoke and sulphur dioxide as the prime indicator of poor urban air quality. Nitric oxide (NO) and nitrogen dioxide (NO_2), collectively referred to as NO_x, are generally the most important nitrogen compounds in urban locations, although N_2O_5, HNO_3 and HNO_2 may play a significant role in the formation of urban photochemical pollutants. Around 90% of the emissions from combustion sources are of NO rather than NO_2; however, since the NO can all potentially be converted to NO_2, it is usual to express all of the NO_x as NO_2 when making mass emission estimates. Long-term average concentrations of NO_2 in UK city centres are typically 30–50 ppb, with NO being similar to, or greater than, NO_2. Winter NO is around twice summer NO, but NO_2 does not vary much with season.

Since exhaust-pipe concentrations of NO and NO_2 can both be several hundred ppm, we should expect peak kerb-side concentrations to be extremely high in comparison to background levels. In fact the record-high kerb-side concentrations of these gases in London are around 2000 ppb. Annual mean concentrations of NO are around 4.5 times those of NO_2, reflecting their higher concentrations in the exhaust gases. Concentrations should also be expected to fall off fairly rapidly with distance from the road – measurements of NO_2 in London, for example, have shown 80–90 ppb at the centre of a road; 50–60 ppb on the kerb; 40–50 ppb at the back of the pavement, decreasing to background within about 20 m of the kerb.

Diurnal variations in NO_x concentration are also to be expected because of variations in traffic flow and atmospheric dispersion. Data for London (Figure 5.10) show a minimum at about 4 am, a strong peak between 8 and 10 am during the weekday morning rush-hour when atmospheric stability is high, and a gradual decline during the remainder of the day. Although there must be an equivalent peak in emissions during the evening rush-hour, increased advective and convective dilution, together with reactions with the O_3 created during the afternoon, prevent a recurrence of the concentration peak. On Sundays the variation is quite different – the average concentration is lower because of the general reduction in traffic, and the peak occurs in the early morning as people return home from Saturday night activities.

Short-term episodes of high NO_2, lasting from a few hours to a few days, can occur in either summer or winter via different mechanisms. In the summer, high NO_2 levels are brought about by rapid photochemistry, and are therefore associated with low NO levels. In the winter, trapping of emissions beneath temperature inversions is responsible, so that NO concentrations are also high. A particularly severe NO_x smog occurred in London between 12 and 15 December 1991. Figure 5.11 displays the NO_2 record for the whole of 1991 from one London monitoring site, and shows how strongly the episode, with a peak concentration over 400 ppb, stood out from the typical NO_2 concentrations that year of 30–50 ppb. The

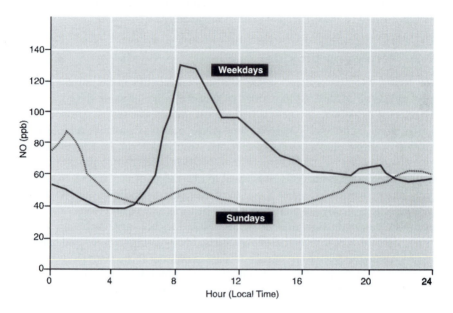

Figure 5.10 Diurnal variation of NO concentration on weekdays and Sundays at Earls
Court in London, 1987.
Source: United Kingdom Quality of Urban Air Review Group (1993a) *Urban
Air Quality in the United Kingdom*, DoE/HMSO, London, UK.

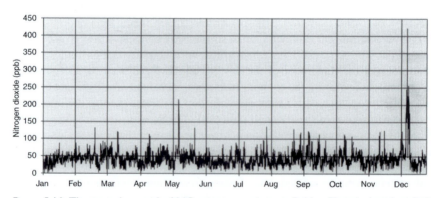

Figure 5.11 The annual record of NO₂ concentrations at Bridge Place in London during
1991.
Source: United Kingdom Quality of Urban Air Review Group (1993a) *Urban
Air Quality in the United Kingdom*, DoE/HMSO, London, UK.

episode extended over four days, with average NO_2 concentrations of around
150 ppb and the peak concentration rising to 423 ppb; this was associated with
NO concentrations (not shown) of around 1200 ppb. Comparison of mortality
data for the period indicated that about 160 deaths could be attributed to the

incident. Direct NO_2 emissions, which have been shown from measurements in road tunnels to make up only 5–7% of the NO_x in exhaust gases, cannot have been responsible because the NO_2/NO_x ratio in the smog was 20–30%. Furthermore, NO_2 concentrations at three different sites in London were much more similar than the NO concentrations at those sites, indicating that the NO_2 had been formed in the atmosphere. It has been proposed that certain very specific hydrocarbons might enhance the production of NO_2 from NO. If this is the case, then the reduction of NO_x emissions by TWCs may have unpredictable effects on NO_2 concentrations. It is also ironic that, despite this being the worst pollution episode in London since the combustion smogs thirty years previously, the EC limit value for NO_2 was not broken. The limit is framed in terms of a 98th percentile exceedance, so the concentration would have had to remain high for seven days. In future years, as more petrol cars become fitted with TWCs, these pollution episodes should be controlled.

5.2.3   Sulphur dioxide

The diurnal and annual cycles of SO_2 used to be clear when ground-level concentration was dominated by low-level sources. Both coal burning and the likelihood of stable atmospheric conditions increased in the winter, and concentrations rose. However, with the decrease in importance of domestic emissions in Western nations, temporal cycles have become of less significance. In China, in contrast, 75% of the energy consumption, 93% of the SO_2 emissions and 62% of the particle emissions are from coal, and the majority of that is domestic. There is a strong annual cycle in the concentrations of these pollutants, and a strong diurnal cycle with peaks at 0800 and 2000 associated with domestic cooking.

The situation for urban SO_2 is quite different to that for NO_x. The sulphur content of petrol is low (150 ppm) and falling, though it is still rather higher for diesel (0.2%); hence vehicle emissions are quite small. Domestic and commercial (non-industrial) emissions from coal and oil combustion have now been replaced by those from natural gas, which are essentially free of sulphur at the point of use. The consequence has been the steady decline in urban SO_2 concentrations described above. Even in towns that have never been heavily industrialised, such as Lincoln, the concentration in the centre has fallen since the early 1960s and by 1990, at around 10 ppb, was below that in the surrounding countryside.

5.2.4   Carbon monoxide

About 90% of the UK emission of CO is from road vehicles, having increased from 61% in 1970. The rural ambient background of less than 50 ppb may be increased to several 10s of ppm in urban conditions in the short term, although annual mean urban concentrations are likely to be below 1 ppm. Maximum hourly mean concentrations during the London smog episode in December 1991 were 13–18 ppm, indicating that concentrations in urban areas can certainly come close to those having recognised health effects (WHO guideline for CO is an 8-h average of 10 ppm).

5.2.5   Volatile organic compounds

As we have seen in Section 1.4.5, this phrase covers a wide range of chemical compounds, largely associated with vehicles. Effects are covered elsewhere, and here we will look at particular features of the urban environment. Urban VOC levels can be expected to be high because of the concentration of traffic; although the VOCs may take part in photochemical ozone production in the downwind urban plume, it is for their toxicity *per se* that they are important within the city limits. A comprehensive UK NMVOC emission inventory has been published by NETCen. It shows that some 37% of emissions from stationary sources are solvents (mainly alcohols, acetates, and chlorinated compounds), whereas over-all, transport emissions (mainly alkanes, alkenes, aromatics and aldehydes), con-tribute 33% of the total. These percentages will be higher in urban areas because of the absence of other sources such as vegetation.

Concentrations of VOCs in urban air have annual averages that lie in the range of a few ppb and show the expected diurnal variation due to human activity. Volatile Organic Compounds are unlikely to represent a health hazard from their immediate toxicity, since the WHO guidelines, even allowing for large and arbitrary safety margins, are a few ppm. However, there is also the much more difficult aspect of carcinogenicity, particularly of aromatics, to consider. It is generally not possible to specify a safe concentration of a carcinogen, since any exposure, however small, increases the risk of cancer by a finite amount.

5.2.6   Air quality information from the UK Government

In the UK, DEFRA publishes a wide range of information and results connected with the various measurement networks described above. First, there is a free-phone Air Quality Helpline on 0800 55 66 77 which accesses recorded messages giving current and predicted air quality for specific pollutants on both a regional and local (for a small number of sites only) basis. Second, DEFRA has a home page on the Internet at *http://www.open.gov.uk/defra*. There are associated pages allocated to air quality bulletins, networks, forecasting, acid deposition informa-tion, the National Atmospheric Emissions Inventory and the chemistry of atmos-pheric pollutants. Each of these in turn can be expanded into further pages giving both data and maps of individual pollutant distribution. The telephone number and Internet address were correct in October 2001.

5.3 PARTICULATE MATTER

Black smoke from coal combustion was the first form of suspended particulate matter recognised as being a hazard in the urban environment. As with SO_2, the combination of regulations such as those for smokeless zones, together with social/energy changes such as the move from open coal fires to gas-fired central heating, led to a steady decline in particle emissions and concentrations. Later measurements of black smoke from the much wider range of instruments in the National Survey confirmed the influence of the UK Clean Air Act in reducing concentrations (Figure 5.1). By the end of the twentieth century concentrations had come down to only a few percent of those at the start. As with gaseous pollutants, Britain was not alone in improving the atmospheric environment, and similar trends were apparent in other countries. Measurements from the UK Automated Network show that typical urban PM_{10} concentrations now lie in the range 10–50 μg m^{-3}, and cumulative frequency distributions show 1% exceedances of around 100 μg m^{-3}. Four issues remain of concern: (1) the long-term chronic effects on health of very low concentrations of very small particles (<2.5 μm in diameter), whatever their chemical composition; (2) the presence among these small particles of especially toxic materials such as heavy metals and PCBs; (3) the soiling properties of smoke particles on buildings and other surfaces; (4) the loss of amenity caused by light attenuation and reduction in visual range.

Determination of the chemical composition of particles, and of its variation with particle size, is a challenge. Many research programmes have concentrated on one aspect such as nitrogen or sulphate partitioning without any sizing. Many other programmes have concentrated on measuring size distributions, for example with a cascade impactor, with little or no chemical analysis. Identification of the components of an aerosol requires use of several complementary techniques such as wet chemistry, atomic absorption, electron microscopy with EDAX, mass spectrometry. It is rarely possible to conduct an investigation with this full suite of analytical methods. For example, cascade impactor samples at low flow rates may not yield enough sample on each substrate to permit chemical analysis at all. Other effects may also interfere, such as the change in diameter of hygroscopic particles when the humidity changes.

Many sources contribute to the particles found in urban air. A generic description of the chemical composition in different size fractions, formed by compounding the results from different sampling campaigns, is shown in Figure 5.12. The proportion (by mass) of the fine fraction is most heavily influenced by smoke emissions, whereas that of the coarse fraction is largely due to wind-blown dust and other insoluble minerals. The carbonaceous fraction is made up mainly of organic carbon, plus typically 20–40% of particulate elemental carbon (PEC). PEC is effective at both scattering and absorbing light, being responsible for 25–45% of visual range reduction, and at soiling surfaces.

(a) Total

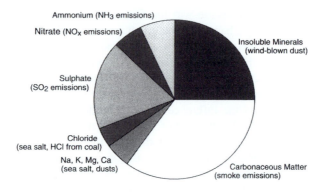

(b) Fine Fraction

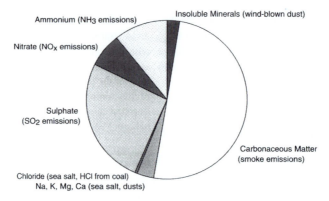

(c) Coarse Fraction

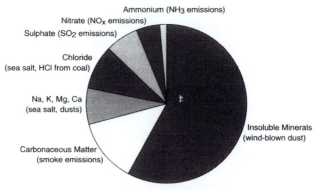

Figure 5.12 Particle chemical composition in different size fractions.
Source: United Kingdom Quality of Urban Air Review Group (1993a) *Urban Air Quality in the United Kingdom*, DoE/HMSO, London, UK.

Diesel engines are major and increasing sources of PEC; even in 1986 they were thought to be responsible for 91% of total PEC emissions in the UK, and diesel use has increased greatly since then. The preponderance of diesel is due to its high emission factor − 0.2% of diesel fuel burned is emitted as PEC, compared to 0.002% for petrol.

Anthropogenic particulate metals are derived from activities such as fossil fuel combustion (including vehicles), metal processing industries and waste incineration. They may occur as the elemental metal, as inorganic oxides or chlorides, or as organic compounds. Reactivity, absorptivity and toxicity all vary from compound to compound, and there are no EC standards for metals other than lead. Both the emission and airborne concentrations of lead have fallen steadily since the mid-1980s due to the introduction of unleaded petrol. Unsystematic *ad hoc* measurements of the concentration of other heavy metals in UK suspended particulate over the last 10 years have indicated that WHO guideline values are not generally approached, let alone exceeded. The concentrations lie in the range 10–30 ng m^{-3}, and do not present an environmental hazard.

Finally, we should expect correlations between gaseous and particulate pollution, since to some extent they derive from the same sources and activities. Figure 5.13 gives an example of such a relationship, between NO_x and $PM_{2.5}$.

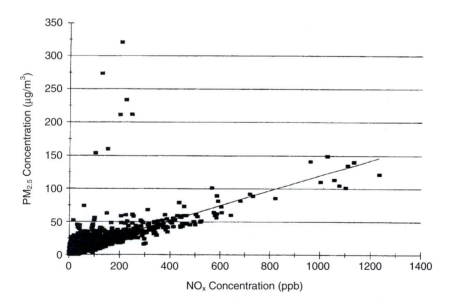

Figure 5.13 Co-occurrence of NO_x and $PM_{2.5}$.
Source: Williams, M. L. (1999) 'Patterns of air pollution in developed countries', In: S. T. Holgate, J. M. Samet, H. S. Koren and R. L. Maynard (eds) *Air Pollution and Health*, Academic Press, London.

5.4 DRY DEPOSITION OF GASES

The main factor that determines deposition of a particular gas species to a surface is the concentration gradient between the atmosphere and the surface. Three processes contribute to the movement of gases down the gradient. First, turbulent diffusion brings the gas close to the surface (within a few mm); second, molecular diffusion takes the gas across the laminar boundary layer adjacent to the surface; third, the gas molecules must adhere to, dissolve in, or react with, the surface itself. If any of these three processes is absent or inhibited, then the surface concentration will tend towards the atmospheric concentration, and both the gradient and the flux will be reduced.

The gas flux can be measured by a variety of methods. The most straightforward is to measure the concentration gradient between one point close to the surface and another point at a reference height such as 2 m. Then by using Fick's law and a measured value for the turbulent diffusivity, the flux can be calculated. Usually the measurement is averaged over a 30-minute period to smooth out short-term fluctuations. A more sophisticated technique, known as conditional sampling, is shown in Figure 5.14. The principle of the device is to measure directly the upward-going and downward-going fluxes, and subtract them to give the net flux. An ultrasonic anemometer measures the windspeed and direction with very high time

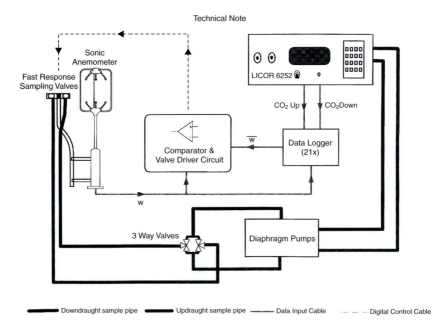

Figure 5.14 Conditional sampling system for flux measurement.
Source: Beverland, I. J., Scott, S. L., ÓNéill, D. H., Moncreiff, J. B. and Hargreaves, K. J. (1997) *Atmospheric Environment* 31: 277–281.

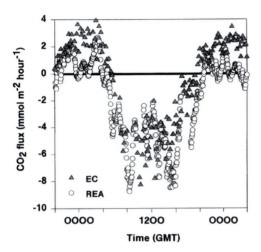

Figure 5.15 CO$_2$ flux measured by the conditional sampler.
Source: Beverland, I. J., Scott, S. L., ÓNéill, D. H., Moncreiff, J. B. and Hargreaves, K. J. (1997) *Atmospheric Environment* 31: 277–281.

resolution. When the vertical component of the wind vector is upward, the air is switched by high-speed solenoid valve into one sample line. When the wind is downward, the sample air is switched into a second line. Either the samples can be accumulated in bags for later analysis, or the difference measured real-time by an in-line gas analyser (for CO$_2$ in this example, but in principle any gas can be measured this way). A sample of the results from this system is shown in Figure 5.15. The CO$_2$ flux is small and upwards (respiration) during the night, large and downwards (photosynthesis) by day.

Two concepts are commonly used to parameterise gas deposition: the deposition velocity and the transfer resistance.

5.4.1   Deposition velocity

Estimation of deposition velocity assumes that the flux is due to a gradient between atmospheric concentration measured at a reference height (usually 1 or 2 m) and zero concentration at the surface (because of absorption by the surface). This variation of any parameter above the surface is called the profile. The basic relationship is the windspeed profile, which often follows a logarithmic variation with height which is due to the loss of air momentum by turbulent interaction with roughness elements at the surface. This is expressed by the standard wind profile equation, which in differential form becomes

$$\frac{\partial u}{\partial z} = \frac{u_*}{k(z - d)}$$

where
   u_* = the friction velocity (the larger the value of u_*, the larger the turbulent velocity variations),
   k = von Karman's constant = 0.41,
   d is the zero plane displacement.

Under neutral atmospheric stability conditions the turbulent 'eddies' can be thought of as circular, with equal components from horizontal and vertical velocity fluctuations. When there is surface heating with associated instability the vertical variations are enhanced, and the wind speed increases rapidly going away from the surface. Conversely, in stable conditions the vertical eddy component is suppressed, and the windspeed increases more slowly going away from the surface.
   Then

$$\frac{\partial u}{\partial z} = \frac{u_*}{k(z - d)} \Phi_M$$

where Φ_M is a dimensionless stability function which has a value larger than unity in stable conditions and less than unity in unstable conditions.
   The flux of momentum τ from the atmosphere to the surface is given by

$$\tau = pu_*^2 = K_M \frac{\partial(pu)}{\partial z}$$

where K_M is the turbulent diffusivity for momentum
   and

$$K_M = \frac{ku_*(z - d)}{\Phi_M}$$

Essentially the same process drives the fluxes of molecules such as water vapour and carbon dioxide, and these can be written generally as

$$K_S = -\frac{F}{\partial S/\partial z} \quad \text{and} \quad K_S = \frac{ku_*(z - d)}{\Phi_S}$$

where F is the flux of material and S is its concentration at height z.
   Corresponding stability functions Φ_S are used, and there has been much discussion in the literature about whether the values of K and Φ are really the same for heat, water and other molecules when the conditions are non-neutral. The relationships should also apply to particles if they are small enough for their sedimentation speeds to be insignificant.

Then

$$v_g = \frac{F_g}{C(z_r)}$$

where
F_g = flux to surface (kg m^{-2} s^{-1}),
$C(z_r)$ = atmospheric concentration (kg m^{-3}) at reference height z_r.

Hence the SI units of v_g are m s^{-1}, as we would expect for a velocity (Table 5.1). The area that is measured in m^2 is normally the land or sea surface, although in some cases it can be the leaf area (one side or two, or projected). It should be remembered that this single parameter of deposition velocity, while being convenient to use, does combine the overall effects of three processes – the turbulent diffusion that brings molecules and particles close to a surface, the molecular or Brownian diffusion that takes them across the final few millimetres to the surface itself, and the absorption/dissolution/adhesion which results.

Note that the roughness lengths are not simply the physical height of the surface roughness elements, but have been determined experimentally as the effective length of the surface elements for turbulence generation (Table 5.2). The

Table 5.1 Typical values of v_g/mm s^{-1}

Gas	Surface		
	Soil	Grass	Sea
SO$_2$	2 for acid soil 10 for basic soil	8	10
O$_3$	5 to wet soil 15 to dry soil	6	0.4
PAN	2.5	2.5	<0.2
NO$_2$	5	5	
CO$_2$	–	0–0.2	0.1
NH$_3$	–	–	–

Table 5.2 Values of roughness length for different surfaces

Surface	Roughness length/m
Still water or smooth ice	0.00001
Smooth snow or smooth sand	0.0003
Soil	0.003
Short grass	0.005
Long grass	0.05
Uniform crops	0.1
Open countryside	0.4
Suburbs	1.0
Forests or urban areas	2.0

roughness length of some surfaces varies with wind speed – water develops ripples and waves, long grass blows flat, trees bend over.

5.4.2   Resistance

Another way of looking at the idea of v_g is as a measure of the conductance of the atmosphere – surface combination for the gas. Resistance is the inverse of conductance, and R_t ($= 1/v_g$, with units of s m^{-1}) is then the total transfer resistance for the pollutant between the reference height and the surface. Now, since

$$R = \frac{1}{v_g} = \frac{C(z_r)}{F_g}$$

we can add and subtract the surface concentration C_s. Hence

$$R = \frac{C(z_r) - C_s}{F_g} + \frac{C_s}{F_g} = R_{atm} + R_{surface}$$

The resistance R_{atm} can further be separated into two resistances in series, R_a to describe turbulent transport by eddies to within a few mm of the surface, and R_b to describe molecular diffusion of the gas through the laminar boundary layer to the surface itself. When the surface roughness is high (as in a forest or an urban area), or the windspeed is strong, R_a will be low, and *vice versa*. Typically, R_a is around 200 s m^{-1} for windspeeds <1 m s^{-1} over vegetation 10 cm tall; 20 s m^{-1} for windspeeds >4 m s^{-1} over 1 m vegetation; <10 s m^{-1} over a forest canopy. These atmospheric resistances are then subtracted from the total resistance R to give the surface resistance R_s.

Figure 5.16 shows the application of this concept to the complex surface of a vegetation canopy. After the gas molecule has penetrated close to the surface, there will be several possible paths by which it can be taken up. In the schematic shown here, the paths involve entry through the stomata, deposition to leaf surfaces, transport and deposition to the lower canopy, and deposition to the soil. Each of these paths involves its own resistance and flux. In this case we have

$$\frac{1}{R_d} = \left[\frac{1}{R_a} + \frac{1}{R_b} + \frac{1}{R_c} \right]$$

where R_d is the dry deposition resistance, R_a is the aerodynamic resistance of the atmospheric surface layer (above the canopy), R_b is the resistance due to molecular diffusion across the quasi-laminar boundary layer, R_c is the canopy resistance. Also

$$\frac{1}{R_c} = \left[\frac{1}{R_{stom} + R_{meso}} + \frac{1}{R_{cut}} + \frac{1}{R_{lowercanopy}} + \frac{1}{R_{soil}} \right]$$

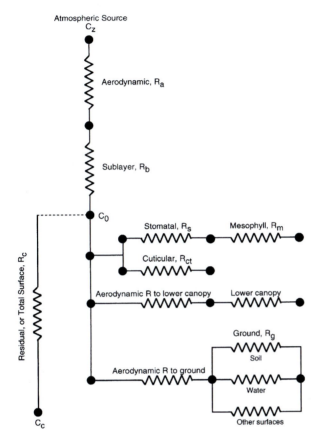

Figure 5.16 A schematic diagram of the series and parallel resistances that control the flux of pollutant to a vegetated surface.
Source: Wesely, M. L. and Hicks, B. B. (2000) 'A review of the current state of knowledge on dry deposition', *Atmospheric Environment* 34: 2261–2282.

where R_{stom} is the resistance to pollutant flux through the stomata, R_{meso} is the resistance of the internal organs of the plant, R_{cut} is the resistance of the leaf cuticle, R_{soil} is the soil resistance.

As mentioned above, the behaviour of gases will also be influenced by factors such as their reactivity (very high for HF, moderate for SO_2, O_3, NO_2, PAN), or their biological activity (high for CO_2), or solubility (high for SO_2).

This scheme is highly simplified compared to the real world. A real vegetated surface will have components of resistance due to atmospheric transport to the soil (depends on canopy structure, density and turbulence), across the soil boundary layer and into the soil (depends on roughness, wetness, organic matter content); atmospheric transport to the plant, into the non-leaf plant parts, into the leaves through cuticles and stomata, and eventually into the intercellular spaces. Equivalent models can be constructed for other surfaces such as water bodies, buildings and animals. In practice, canopy resistance typically makes up around 70% of the total, so that deposition rate is controlled by changes in r_c rather than in roughness or windspeed, even over forests.

5.5   WET DEPOSITION

5.5.1   Rainfall

As well as being deposited in the gaseous form, S and N are eventually incorporated into cloud droplets, raindrops and snow flakes, and deposited back to land and water bodies via cloud droplet deposition, rainfall and snowfall. If the gas molecules or particles are incorporated into cloud droplets which then turn into raindrops, it is called rainout. If they are collected by the raindrops as the latter fall through the atmosphere, it is called washout. It may be hard to distinguish the two routes in practice. A particle may form the CCN on which a cloud droplet first condenses, or it may be collected by a raindrop as it is falling to the surface. Equally, gases may dissolve into cloud droplets or into the falling raindrop. Droplets may have lifetimes of a few minutes in convective shower clouds or a few hours in frontal rain systems. Also, the raindrops themselves do not have a simple history. They may start as frozen hail or snow but melt in the warmer air they fall through. They may undergo several lifting and falling cycles in convective motions in the cloud, and they may evaporate substantially in the drier air below the cloud. Many clouds generate rain which falls out of the cloudbase but never reaches the ground. During this process all the particles and gases that were taken up into the raindrops will be converted into quite different forms.

5.5.2   Rainout

Cloud droplet formation involves the generation of air supersaturated with water vapour during convective or orographic uplift. Droplet formation is a complex process involving competition between several different mechanisms. As the air rises and cools, the saturated vapour pressure decreases towards the actual vapour pressure. Eventually the vapour becomes supersaturated, and condensation is possible. When an air mass that is completely free of particles is cooled, no condensation occurs until extremely high supersaturations are reached, several times the saturated vapour pressure. This is because saturated vapour pressure over a curved water surface follows the Kelvin relationship:

$$\ln\frac{e_d}{e_\infty} = \frac{4\sigma v_B}{RT} \cdot \frac{1}{d}$$

where e_d, e_∞ are the SVPs above a surface of diameter d and a flat surface, respectively, σ and ν_B are the surface tension and molar volume, respectively, of the liquid, R is the Gas Constant and T the absolute temperature.

When $d = 0$, the required supersaturation is infinite. When there are no particle surfaces present, condensation can only occur onto chance clusters of a few molecules. The latter have values of d of a few $\times 10^{-10}$ m so the supersaturation needed is extremely high. This Kelvin effect also controls natural aerosol production from vapours and the manufacture of aerosol particles for instrument calibration. When very high values of supersaturation occur, it is possible to get homogeneous nucleation – the formation of droplets from combining clusters of molecules.

Although supersaturations in the real atmosphere rarely exceed 1%, cloud droplets do form and grow, due to the presence of CCN. When there are CCN present, they will usually be either hydrophobic, insoluble materials such as silica or carbonates from soil particles, or hygroscopic soluble salts such as chloride. In maritime air, oceanic sources of secondary sulphate particulate from planktonic DMS, and primary sea salt particles, contribute to the CCN. When the air is of continental origin, the CCN are from anthropogenic combustion and soil entrainment.

The hydrophobic nuclei simply present surfaces onto which condensation can begin when the supersaturation is sufficiently elevated, although this represents an order of magnitude increase on the activation supersaturation for hygroscopic CCN. The latter are often salts such as sulphates, nitrates or chlorides, and can be of natural or human origin. One of the main natural production routes is sea spray. On average, a few percent of the surface area of the world's oceans is covered by breaking wave crests. Each of these crests generates millions of breaking bubbles, and each of the breaking bubbles typically generates a few larger droplets (up to a few hundred μm in diameter) and many smaller droplets (5–20 μm in diameter). Some of these droplets will fall back into the sea, but many will remain airborne long enough for their water to evaporate and leave behind a residual salt particle. For example, sea water contains about 35% salts, of which 29% or 29 g l^{-1} is NaCl. If a sea water droplet of 50 μm diameter is ejected from spray and evaporates, it will leave a salt CCN having a mass of 2.2 ng and a diameter of 15 μm. In addition, the droplets and particles are formed from sea water in the very near surface layer, where there are enriched concentrations of many materials. These CCN then remain unchanged in the airmass until the latter is involved in convective uplift. As the relative humidity increases in the uplifted air, the hygroscopic CCN absorb water vapour. Eventually they deliquesce, or dissolve into their own water content, to form a small liquid droplet of highly concentrated salt solution. There is a hysteresis effect here – if the relative humidity is then decreased, the droplet evaporates water but does not recrystallise until a relative humidity of 40% (in the case of sodium chloride) is reached. Low solubility salts such as NaCl absorb a lot of water before they dissolve, and hence make large droplets. High solubility salts, such as those in the ammonium family, form smaller initial droplets of more concentrated solutions. If the relative humidity were constant, the droplet could exist at a stable radius depending on the mass of the initial CCN. As uplift continues and reduces the saturated vapour pressure the droplet grows, the salt concentration declines but so does the curvature, and growth can continue. Hence the supersaturation within the cloud is kept below about 0.5%.

A hygroscopic CCN promotes the formation of a cloud droplet in two ways. First, it provides a surface with a larger radius of curvature onto which new water vapour molecules can condense at lower supersaturations. Second, the presence of the dissolved salt reduces the surface vapour pressure below that of pure water, and hence reduces the supersaturation needed for the uptake of new water molecules from the vapour. The Kelvin equation becomes

$$\ln \frac{e_d}{e_\infty} = \frac{a}{d} - \frac{b}{d^3}$$

where a and b are constants.

For droplets that have just dissolved and for which the solution is very concentrated, the reduction in vapour pressure due to this effect (the second term on the right hand side) outweighs the increase due to curvature, and condensation can be initiated even while the relative humidity is as low as 70 or 80%. The new cloud droplet can grow at low supersaturations, with the two effects competing: as the radius increases, the salt concentration decreases with the droplet volume (cube of radius) while the curvature of the surface decreases with the radius. The resultant of these opposing tendencies is the set of Köhler curves shown in Figure 5.17.

This figure repays some study. First, note that the supersaturation scale is expanded by a factor of 100 compared to the relative humidity scale. This accounts for the strange discontinuity at the boundary. It further emphasises the very small supersaturations that are present in clouds. Second, the supersaturation over a pure water surface (the outer curve) is almost zero for very large droplets, which have almost flat surfaces. As the droplet diameter falls, the surface curvature increases, and the supersaturation goes off to infinity. Third, note that each of the curves is for a different initial CCN mass. The smallest CCN at the left gives the least benefit from solute reduction of vapour pressure, and the saturated vapour pressure needed for growth becomes very high while the droplet is still very small. Fourth, on the left sides of the curves, the droplets are in equilibrium – if the humidity increases, the droplet will grow, but if it decreases, then the droplet will shrink again.

An NaCl CCN which grows from activation at 75% relative humidity to equilibrium at 90% will take on about six times its initial mass of water vapour molecules, by which point its diameter will have increased by a factor of 1.8. This is a confusing situation for aerosol sampling, because if the sample is based on the aerodynamic diameter at ambient humidity, the adsorbed water will determine the cut diameter, whereas when the water evaporates, the smaller CCN will

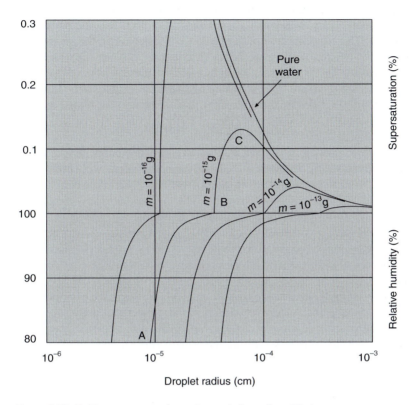

Figure 5.17 Köhler curves to show the variation of equilibrium water vapour concentration with droplet radius and mass of CCN.
Source: Wells, N. (1997) *The Atmosphere and Ocean*, Wiley, Chichester.

remain. Hence a sampling protocol should involve bringing the particles to equilibrium at an agreed low relative humidity – the World Meteorological Organisation (WMO) suggestion is an RH of 40%. However, if the droplet radius exceeds the value corresponding to the peak of the curve, then the 'activated' droplet will grow continuously even if the humidity falls (provided it remains above saturation). The rate of growth decreases as the diameter increases, and is extremely slow for droplets above about 40 μm diameter. The increase in diameter necessary to reach even drizzle-drop size of about 100–200 μm is achieved by collision and coalescence with other droplets. Hence in an air mass that contains a size distribution of CCN, together with turbulence which causes variations in humidity, a complex selection process operates which determines which nuclei get activated (scavenged) and eventually contribute to the growing droplet population. Cloud condensation nuclei may participate in 10–20 condensation/evaporation cycles before eventually becoming incorporated in a raindrop which reaches the ground. Cloud condensation nuclei concentrations are rather low (tens to hundreds cm^{-3}) in maritime air, and higher (hundreds to thousands cm^{-3}) in continental air. Of these, 20–30% become activated in an updraught. The concentration of activated CCN depends on the supersaturation, going as cS^k where c is the total CCN concentration, S is the supersaturation and k is a constant. The supersaturation achieved is a function both of the number of CCN (the more there are to be activated, the faster they will sweep up the excess water vapour generated by the updraught) and of the updraught (the faster the air rises, the faster it will cool and release water vapour). Hence activated CCN concentration $N \propto c^{2/(k+2)}U^{3k/(2k+4)}$.

The rate of growth of diameter slows as the diameter increases, so that the diameter distribution becomes increasingly monodisperse with time. Since the initial CCN particles may vary in mass from 10^{-21} kg (0.01 μm) to 10^{-5} kg (1 μm), the concentration of sulphate and other ions can also vary by many orders of magnitude from droplet to droplet. Also, the initial population of CCN in the updraft will have a wide range of masses. The larger ones will activate first (because they make the greatest initial reduction in vapour pressure above the droplet surface) and sweep up all the excess water vapour as the air cools. Hence smaller nuclei may never be activated to form droplets. The CCN are not restricted to soluble salt particles. Other insoluble particles, for example entrained dust, can provide the surface on which initial condensation can occur. As the droplets grow, their volumes increase as the cube of their diameters but the rate of vapour uptake increases only as the square (i.e. surface area). The rate of diameter growth declines. Eventually collision processes between drops, which are unimportant for small drops because of low impaction efficiency, become the dominant way in which droplet diameter increases. All this depends on the droplet remaining suspended in the updraft. As the diameters are increasing, the sedimentation speeds increase, and eventually the old cloud droplet or new raindrop will acquire a fall speed greater than the updraft and fall out of the cloud.

Growth by condensation can be dramatically accelerated by the presence of ice crystals. When the air temperature is less than 0 °C, some of the condensation will occur as ice onto ice CCN. At temperatures between −40 and 0 °C, there will be a mixture of ice crystals and water droplets. However, the SVP over ice is lower than the SVP over water at the same temperature, so the supersaturation with respect to ice may be 10% while that with respect to water is zero. The ice crystals will then grow extremely rapidly. When the large ice crystal falls out of the cloud, it will probably melt back to water on its way down to the ground.

5.5.2.1   Washout

As this raindrop is falling out of the cloud, it has an opportunity to collect raindrops, particles and gases within the cloud, and particles and gases below. We can use Stokes Number to get an idea about the efficiency of particle collection. For example, a raindrop of 1 mm diameter will have a fall speed of around 4 m s^{-1}. At that speed a 10 μm diameter particle will have a stopping distance of 1 mm and a Stokes No. of 2, while a 3 μm particle will have a stopping distance of 0.1 μm and a Stokes No. of 0.2. The 10 μm particles will be collected at 55% efficiency by the falling raindrop, while the 3 μm particles will hardly be collected at all with an efficiency of only 4%.

We can also define a time constant λ for particle concentration decay. If the rainfall rate is R m s^{-1}, and the raindrop diameter is D, then the raindrop rate N is $R/(\pi D^3/6)$ raindrops m^{-2} s^{-1}. Each of these drops sweeps out a cylinder of area $\pi D^2/4$, collecting particles from it with efficiency E. Hence the time constant for particle removal is $(\pi D^2/4)NE$. The particle concentration will decline exponentially in continuing rain according to $\exp(-\lambda t)$. For 1 mm h^{-1} of rain due to drops 1 mm in diameter, half of 10 μm particles would be washed out in 50 minutes, whereas it would take 10 h for the 3 μm particles to be reduced by the same proportion.

A washout ratio W can be defined

$$W = \frac{\text{concentration per kg of rainwater}}{\text{concentration per kg of air}}$$

A small proportion of the S and N dissolves directly into the water. However, the concentrations of S and N in rainwater are far higher than equilibrium solution concentrations would allow, and can only be achieved by oxidation processes that form sulphate and nitrate ions. The processes involved are complex, involving not only the S and N species themselves but other gaseous compounds such as O_3 and H_2O_2 and reactions with catalysts such as manganese and iron that take place on the surface of aerosol particles. The timescale of these reactions is typically days, so that the S and N from emitted gases may be deposited thousands of km from their source.

5.5.3 Dry reactions

An important requirement is the availability of the OH radical through the decomposition of O_3 and the subsequent reaction of the free excited oxygen atom with water vapour

$$O_3 + h\nu \rightarrow O^* + O_2$$

$$O^* + H_2O \rightarrow 2OH$$

and hence

$$SO_2 + OH + M \rightarrow HSO_3 + M$$

$$HSO_3 + O_2 \rightarrow HO_2 + SO_3$$

$$SO_3 + H_2O \rightarrow H_2SO_4$$

where M is a passive reaction site such as a nitrogen molecule. Measurements in dispersing plumes indicate that the conversion rate can rise as high as 4% per hour on a sunny summer afternoon, with a typical overall rate of about 1% per hour.

The OH radical is also involved in the dry production of nitrate

$$NO + O_3 \rightarrow NO_2 + O_2$$

$$NO_2 + OH + M \rightarrow HNO_3 + M$$

This conversion occurs an order of magnitude faster than for sulphate.

At night, a second route is available which involves the nitrate radical NO_3.

$$NO_2 + O_3 \rightarrow NO_3 + O_2$$

$$NO_2 + NO_3 \rightarrow N_2O_5$$

$$N_2O_5 + H_2O \rightarrow 2HNO_3$$

This process is unimportant in the daytime because the radical is photolytically unstable.

After 24 h, a dispersing plume with an average windspeed of 10 m s^{-1} would travel about 1000 km (up to the Highlands of Scotland from Northern Europe, or across the North Sea to Sweden from the UK). In this time around 30% of the sulphur dioxide would be dry-converted to sulphuric acid, and all of the nitric oxide to nitric acid.

These acid molecules may condense on to existing particles, or they may self-nucleate to form pure acid droplets. Such droplets will absorb ammonia rapidly to form hygroscopic ammonium sulphate or ammonium nitrate aerosol particles, which will eventually act as CCN and be returned to the surface in precipitation.

5.5.4   Wet reactions

Within cloud droplets the production of sulphuric acid is much faster. Sulphur dioxide dissolves in water to form mainly the bisulphite ion

$$2SO_2 + 2H_2O \rightarrow SO_3^- + HSO_3^- + 3H^+$$

There are then two main routes for oxidation to sulphuric acid, depending on the SO_2 concentration. At the low SO_2 concentrations found in remote areas, the bisulphite is oxidised by hydrogen peroxide, i.e.

$$HSO_3^- + H_2O_2 \rightarrow HSO_4^- + H_2O$$

A key role is played by the hydrogen peroxide, which is regenerated slowly from reactions involving HO_2 and carbonyl compounds such as CO and HCO. The availability of H_2O_2, and of other pollutants and their by-products, can therefore control the rate of sulphuric acid production. In more polluted regions, where the SO_2 concentration is higher, the peroxide supply is inadequate. Then oxidation by ozone, which is highly non-linear with SO_2 concentration, becomes significant. Hence turbulent entrainment of new supplies of oxidants may play an important part in determining the overall rate of sulphate or nitrate production. Carbon monoxide, which is usually thought of for its toxicity at high concentrations in urban areas, is also important in atmospheric photochemistry at low concentrations.

Both nitric oxide and nitrogen dioxide have low solubilities in water, and the rate of wet oxidation is less significant.

These conversion processes have several impacts on the quality of precipitation. First, they change the acidity. The acidity is measured by the pH, which is the negative logarithm of the molar hydrogen ion concentration

$$pH = -\log_{10} [H^+]$$

For pure water, $[H^+] = 10^{-7}$ M, giving a pH of 7.0. A reduction of one pH unit, from 7.0 to 6.0 for example, indicates that the H^+ ion concentration has increased by a factor of ten. In the clean atmosphere, water is in equilibrium with 360 ppm CO_2, giving a pH of 5.6. In remote temperate areas of the world, average rain pH is about 5.0. In the UK, precipitation quantity and quality are measured at 68 sites in two Precipitation Composition Monitoring Networks. In the secondary network, precipitation is sampled every week or fortnight at 59 sites by bulk collectors that remain open to the atmosphere all the time and hence collect dry deposition from gases and particles as well as the rain, snow or hail. In lowland areas it only rains for 7% of the time, so that dry deposition of sulphate and nitrate can make a significant contribution (perhaps as much as one-third) to the total. The secondary network is therefore supplemented by a primary network of nine wet-only collectors. These are uncovered automatically when rain is actually falling, and store the rainfall in a different container for each 24-h period. There are several penalties for this improvement. First, the first part of each precipitation event, which is the period when solute concentrations are often the highest, is lost because the collector is still covered. Second, the rainfall measurement itself is affected by the aerodynamics of the collector. Since wet-only collectors are a different shape to their less sophisticated counterparts, the results are not automatically comparable. Third, the wet-only collectors normally need an electricity supply and cannot necessarily be sited in the optimum position. Across Europe, a further 90 sites in the European Monitoring and Evaluation Programme (EMEP) also record precipitation quality. Standard analyses include H^+, SO_4^{2-}, NO_3^-, NH_4^+ and a range of other ions. The results have shown that the annual average pH of European precipitation ranges between 4.1 and 4.9. The record low pH from a single rain event in the UK is 2.4, which occurred at Pitlochry in 1974. This amounts to a hydrogen ion concentration which is over 1500 times that found in clean rain, and represents a similar acidity to that of lemon juice. In fact there are naturally acidified lakes around the globe that sustain ecosystems at similar pH values – it is the short-term shock of relatively sudden changes in pH that causes stress.

The second impact of the acidity lies in the quantities of sulphur and nitrogen that are returned to the terrestrial environment by precipitation. Two factors determine the total amounts of S and N that are deposited – the concentration of the ions in the rain, and the total amount of rain. It is thought that some biological processes may be sensitive to the former, and others to the latter. The spatial distributions of the two factors are normally quite different. For example, the annual mean nitrate *concentration* in UK rainfall falls strongly from east to west, because it is influenced by European sources to the east of the UK. On the other hand, total nitrate *deposition* is influenced heavily by topography, with the total deposition being increased by the high amounts of rainfall over the uplands of Wales, north-west England and Scotland. The maximum annual values of wet deposition in the UK are over 6 kg N ha^{-1} and over 14 kg S ha^{-1}. The effects of this deposition on terrestrial environments vary greatly with the mode of deposition and the chemical properties of the particular ecosystem. For example, although the precipitation in the English Midlands has low pH, it falls mostly as rain, is evenly distributed during the year, and falls on to alkaline soils. In contrast, in Scandinavia the winter precipitation falls as snow on to acid soils. During the spring thaw, all the stored winter acidity is released over a short period, generating an 'acid pulse' with potentially severe consequences for aquatic life.

The networks show that the lowest pH in the UK (<4.1) is found in the East Midlands while the highest total deposits are found in the upland regions where the rainfall is highest. In mainland Europe, the lowest pH values are found within a belt stretching from southern Sweden down into eastern Germany and Poland.

The concentrations of ions in precipitation are known from analyses of material collected largely from raingauges in lowland areas. These concentrations are not necessarily the same as those experienced in the higher rainfall upland areas, because of a process called the seeder–feeder mechanism. An orographic feeder cloud forms when advected air cools as it rises over hills. Droplets are washed out of the cloud by raindrops falling from a seeder cloud above. Since the ionic concentrations in the small feeder cloud droplets are several times as high as those in the larger seeder raindrops, this process enhances the ionic concentrations in the rain. The higher the altitude, the greater the orographic cloud cover, and the greater the total wet deposition. Wet deposition values calculated from raingauge data must be corrected for this altitude effect.

During the period that the old industrialised nations have been controlling emissions, the chemical balance of precipitation has been changing. By 1993, sulphur emissions in Western Europe had decreased by around 50% from their value in 1980, and both the acidity and sulphur content of precipitation had fallen by 35%. Emissions and deposition of nitrogen, however, remained about the same, so that N now generates around 60% of the acidity in European precipitation.

5.5.5   Cloudwater deposition

The processes of droplet formation from CCN, together with the deposition mechanisms that are common for all particles, mean that significant fluxes of salts and acidity can occur to vegetation by the deposition of droplets when it is not actually raining. This is most likely to occur from orographic cloud which is formed when air blows up the windward side of a range of hills. Some upland areas can be in cloud for a high proportion of the year. At such sites, the deposition of cloud droplets to vegetation – a process also known as occult precipitation, because it is not detected by a standard raingauge – can be a significant S and N supply in itself, and the high concentrations (up to 50× those in raindrops) may have important effects on the impacted vegetation. For example, experiments have shown that the freezing resistance of conifers is reduced by acid mists, making the trees more liable to cold damage in the winter. Altitude effects on precipitation chemistry are likely to be significant above 200 m, and above 400 m cloudwater deposition can be more important than the direct input from precipitation. Although direct measurements of droplet deposition are technically difficult and infrequently made, values that have been taken over moorland and forests show that the deposition velocity is similar to that for momentum. This has provided the basis for calculations of deposition across the country from known average windspeeds, land-use data and cloud water ion concentrations. Such calculations have shown that in upland areas of Wales and Scotland, droplet deposition can increase wet nitrate deposition by 30%. The deposition rate is strongly influenced by the aerodynamic roughness of the surface, and is therefore higher over forests than over grass or moorland.

5.6   TOTAL DEPOSITION AND BUDGETS

The total deposition at any site is the sum of wet (rain/snow or droplets) and dry deposition. The relative proportions of these three vary greatly with the location. In the south of England, rainfall and altitude are both low, while proximity to urban and industrial sources is high, so that dry deposition is high, rainfall and droplet deposition both low. In the uplands of Wales and Scotland, on the other hand, dry deposition is small and wet deposition high. Since windspeeds also increase with altitude, the efficiency of droplet impaction on to vegetation increases, so that droplet deposition becomes a more significant fraction of the wet proportion. There has been considerable debate on the influence of vegetation type on deposition – for example, when an upland area is afforested, the trees act as efficient droplet and gas collectors and can significantly increase the total N and S fluxes. These fluxes are very significant for the nutrient balance of the vegetation – total upland N fluxes can be around 30 kg ha^{-1} a^{-1}, to systems that normally receive no fertiliser nitrogen.

Once all the dry and wet emission and deposition fluxes have been estimated for a particular element, they can be combined into a budget for an area, country or region. Table 5.3 shows one such calculation for nitrogen for the UK, for fluxes averaged over the period 1988–1992. The total deposition on any one country or area consists of some material that was emitted from the country itself, and some that has been 'imported' from other countries. Similarly, some of the total national emissions will be exported, and some deposited internally. The balance between these import and export terms can be very different for different countries. The UK is a relatively small, intensively industrialised landmass surrounded by water, so it is bound to be a net exporter. Scandinavian countries recognised in the 1970s that they were importing S, N and acidity on the prevailing winds from the high emitters in the UK, Germany and Central Europe, while their own emissions were low because of their dependence on hydroelectric power. For Europe as a whole, we have the balance given in Table 5.4.

Estimates are given in Figure 5.18 and Figure 5.19 of the corresponding global budgets for sulphur and nitrogen, including natural emissions. Both budgets confirm that anthropogenic emissions (large bold font in Figure 5.18, bracketed values in Figure 5.19) dominate the totals.

Table 5.3 UK nitrogen budget

Source	Direction	Flux/kt N a^{-1}
Stationary sources of NO_x	Emission	425
Mobile sources of NO_x	Emission	421
Agricultural NO_x	Emission	350
Dry dep of NO_2	Deposition	100
Dry dep of NH_3	Deposition	100
Wet dep of NO_3^-, NH_4^+	Deposition	108, 131
Dry dep of HNO_3	Deposition	15
	Net export of NH_x[*]	119
	Net export of NO_y[*]	623

Notes
[*] $NH_x = NH_3 + NH_4^+$
$NO_y = NO + NO_2 + HNO_3 + HONO + NO_3 + N_2O_5 + NO_3^-$ aerosol + PAN

Table 5.4 European emissions and deposition of S and N

Emitted from Europe	Deposited in Europe
7 Mt NO_x-N	3.6 Mt NO_3-N
22.3 Mt SO_2-S	13.9 Mt SO_x-S
7.6 Mt NH_3-N	5.9 Mt NH_x-N

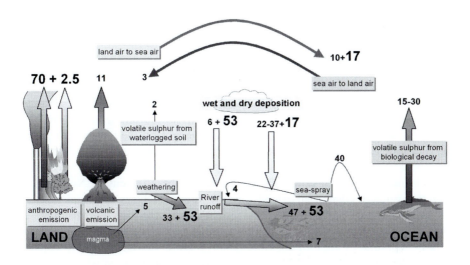

Figure 5.18 Global sulphur budget.
        Source: National Expert Group on Transboundary Air Pollution (2001)
        *Transboundary Air Pollution: Acidification, Eutrophication and Ground-level Ozone
        in the UK*, Centre for Ecology and Hydrology, Edinburgh. Based on Graedel,
        T. E. and Crutzen, P. J. (1993) *Atmospheric Change – an Earth system perspect-
        ive*, W. H. Freeman, New York.

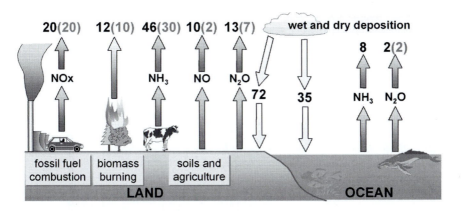

Figure 5.19 Global nitrogen budget.
        Source: From Fowler, D. (1999) In: Macaulay Land Use Research Institute
        Annual Report 1998.

5.7   ANALYSIS OF AN AIR POLLUTION EPISODE

On 2 September 1998, an unusual air pollution episode occurred in the
English Midlands. The meteorological situation was that, on the day before, an

occluded cold front had moved across the UK before stalling over the North Sea due to a blocking High over Scandinavia. The resulting pressure field over the UK was very slack, with weak gradients and low windspeeds. Several hours of calm (zero windspeed) were recorded at the Watnall meteorological station near Nottingham. Nevertheless the presence of cloud inhibited thermal convection during the day. Radiosonde ascents at 1115 and 1715 from Watnall showed an inversion at around 300 m above ground, which probably persisted all day. During the afternoon of the 2 September, the wind flows were light and variable, with a lot of shear (wind direction varied with height) in the lowest 1000 m of the atmosphere. The overall effect of these conditions was to hold and perhaps pool local emissions to the north of Nottingham in the English East Midlands, and to bring this material slowly south-west late on the 2 September. Nottingham is on the southern edge of an extended coalfield on which are sited many of the UK's coal-burning power stations, and it appears that the emissions from these accumulated in the stagnant air before drifting over Nottingham. The records from the continuous monitors in Nottingham (Figure 5.20), show sharp coincident peaks in both SO_2 and PM_{10} concentrations, a combination which is characteristic of emissions from coal-burning power stations. The concentration of ozone, which is not emitted from power stations, peaks just before the others, but then decreases sharply and then varies widely. This was probably due to a combination of variable weather conditions and reactions with the plume.

The peak 15-min mean SO_2 concentration was 653 ppb, two orders of magnitude greater than the long-term mean, and the peak hourly-mean PM_{10}

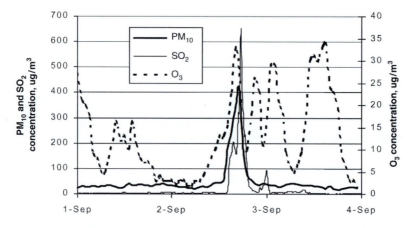

Figure 5.20 Concentrations of SO_2 and PM_{10} at Nottingham during the pollution episode of 2 September 1998.
Source: Environment Agency (2000) *Report into an Air Pollution Episode*, Environment Agency, Bristol.

concentration was around 420 μg m^{-3}, one order of magnitude larger. Both levels greatly exceed the EPAQS/NAQS health guidelines. The episode was modelled during an intensive analysis by the EA (UK). They found that the usual Gaussian-based models were unsuitable because they are not designed to handle the situation of pollutant accumulation and long residence times in calm air. Instead they turned to a model called nuclear accident model (NAME) which had been developed by the UK Met. Office to predict the dispersion of radionuclides after a fire or similar event at a nuclear power plant. Nuclear accident model predicted the movement and concentration build-up rather well, and enabled the attribution that just three power stations lying close together on the River Trent had been responsible for 95% of the SO$_2$ peak.

FURTHER READING

Brasseur, P. Orlando, J. L. and Tyndall, G. S. (eds) (1999) *Atmospheric Chemistry and Global Change*, Oxford University Press, New York.

Demerjian, K. (2000) 'A review of national monitoring networks in North America', *Atmospheric Environment* 34: 1861–1884.

Environment Agency (2000) *Report into an Air Pollution Episode*, Environment Agency, Bristol.

Holland, D. M., Principe, P. P. and Sickles II, J. E. (1999) 'Trends in atmospheric sulphur and nitrogen species in the eastern United States for 1989–1995', *Atmospheric Environment* 33: 37–49.

Hornung, M. and Skeffington, R. A. (eds) (1993) *Critical Loads: Concepts and Applications*, NERC/HMSO, London, UK.

Jones, K. C. and de Voogt, P. (1999) 'Persistent organic pollutants (POPs): state of the science', *Environmental Pollution* 100: 209–221.

Mason, B. J. (1992) *Acid Rain: Its Causes and Its Effects on Inland Waters*, OUP, Oxford, UK.

National Acid Precipitation Assessment Program (1990) *1990 Integrated Assessment Report*, NAPAP Office of the Director, Washington, DC, USA.

National Acid Precipitation Assessment Program (1998) *NAPAP Biennial Report to Congress: an Integrated Assessment*, NAPAP, Silver Spring, MD.

National Expert Group on Transboundary Air Pollutants (2001) *Transboundary Air Pollution*, Centre for Ecology and Hydrology, Edinburgh.

National Expert Group on Transboundary Air Pollution (NEGTAP) (2001) *Transboundary Air Pollution: Acidification, Eutrophication and Ground-Level Ozone in the United Kingdom*, CEH, Edinburgh.

Solomon, P. S., Cowling, E., Hidy, G. and Furiness, C. (2000) 'Comparison of scientific findings from major ozone field studies in North America and Europe', *Atmospheric Environment* 34: 1885–1920.

UNEP/WHO (1992) *Urban Air Pollution in Megacities of the World*, Blackwell, Oxford, UK.

United Kingdom Impacts of Atmospheric Nitrogen Review Group (1994) *Impacts of Nitrogen Deposition in Terrestrial Ecosystems*, DoE, London, UK.

United Kingdom Photochemical Oxidants Review Group (1987) *Ozone in the United Kingdom*, DoE/HMSO, London, UK.

United Kingdom Photochemical Oxidants Review Group (1990) *Oxides of Nitrogen in the United Kingdom*, DoE, London, UK.

United Kingdom Photochemical Oxidants Review Group (1993) *Ozone in the United Kingdom 1993*, DoE/HMSO, London, UK.

United Kingdom Photochemical Oxidants Review Group (1997) *Ozone in the United Kingdom 1997*, DoE/HMSO, London, UK.

United Nations Environment Programme (1991) *Urban Air Pollution*, UNEP, Nairobi, Kenya.

Wayne, R. P. (2000) *Chemistry of Atmospheres*, Oxford University Press, Oxford.

Chapter 6

Meteorology and modelling

Measurements tell us what the concentrations are (or have been) at a particular location. They cannot tell us what the concentration is going to be in the future, or what it is now at locations where no measurements are being made. Air pollution models help us to understand the way air pollutants behave in the environment. In principle, a perfect model would enable the spatial and temporal variations in pollutant concentration to be predicted to sufficient accuracy for all practical purposes, and would make measurements unnecessary. We are a very long way from this ideal. Also, it should be remembered that a model is no use unless it has been validated to show that it works, so that the process of model development goes hand in hand with developments in measurement. There are many reasons for using models, such as working out which sources are responsible for what proportion of concentration at any receptor; estimating population exposure on a higher spatial or temporal resolution than is practicable by measurement; targeting emission reductions on the highest contributors; predicting concentration changes over time.

There are four main families of model:

- Dispersion models, which are based on a detailed understanding of physical, chemical and fluid dynamical processes in the atmosphere. They enable the concentration at any place and time to be predicted if the emissions and other controlling parameters are known;
- Receptor models, which are based on the relationships between a data set of measured concentrations at the receptor and a data set of emissions that might affect those concentrations;
- Stochastic models, which are based on semi-empirical mathematical relationships between the pollutant concentrations and any factors that might affect them, regardless of the atmospheric physical processes;
- Compartment or box models, in which inputs to, and outputs from, a defined volume of the atmosphere are used to calculate the mean concentration within that volume.

All modelling and validation need some understanding of relevant meteorology, so we will cover that before looking at the models themselves.

6.1   METEOROLOGICAL FACTORS

The main meteorological factors that affect dispersion are wind direction, windspeed and atmospheric turbulence (which is closely linked with the concept of stability).

6.1.1   Wind direction

Wind direction is conventionally specified as the direction from which the wind is blowing, because we have been more interested in what the wind has collected before it reaches us than in where it will go afterwards. Wind direction is commonly identified with one of 16 (or sometimes 32) points of the compass (e.g. a south westerly wind is a wind blowing from the south-west), or more scientifically as an angle in degrees clockwise from north (so the south-west wind will be blowing from between $213.75°$ and $236.25°$). What is spoken of colloquially as the wind direction is rarely constant in either the short or long term. The average value over periods of between one and a few hours is determined by meteorology – the passage of a frontal system, or the diurnal cycle of sea breezes near a coast, for example. This medium-term average value of wind direction is of fundamental importance in determining the area of ground that can be exposed to the emission from a particular source. Short-term variations (between seconds and minutes), due to turbulence, are superimposed on this medium-term average – they can be seen on a flag or wind-vane. As well as these short-term horizontal variations, there are also short-term variations in the vertical component of the wind that affect turbulent dispersion. The magnitudes of both horizontal and vertical variations are influenced by the atmospheric stability, which in turn depends on the balance between the adiabatic lapse rate and the environmental lapse rate. These concepts are described in Section 6.1.3.

6.1.2   Windspeed

Windspeed is measured in m s^{-1} or knots (one knot is one nautical mile per hour; one nautical mile is 6080 feet, or 1.15 statute miles). Although the use of SI units is encouraged in all scientific work, some professions have stuck with earlier systems. Thus mariners, pilots and meteorologists are all comfortable with knots. Mariners also use the Beaufort Scale, which relates windspeed to its effects on the sea.

Windspeed is important for atmospheric dispersion in three distinct ways. First, any emission is diluted by a factor proportional to the windspeed past the source. Second, mechanical turbulence, which increases mixing and dilution, is created by the wind. Third, a buoyant source (hot or cold) is 'bent over' more at higher windspeeds, keeping it closer to its release height. Friction with the ground reduces the windspeed near the surface, so that the speed at the top of an industrial chimney (such as that of a power station, which might be 200 m tall)

Table 6.1 Variation of the windspeed exponent p with stability category

Pasquill stability Class	Exponent p for rough terrain	Exponent p for smooth terrain
A – the most unstable	0.15	0.07
B	0.15	0.07
C	0.20	0.10
D	0.25	0.15
E	0.40	0.35
F – the most stable	0.60	0.55

will be substantially greater than at the bottom. The change with height can be approximated by a power law such as

$$u(z) = u_0 \, (z/z_0)^p \tag{6.1}$$

where $u(z)$ is the windspeed at height z, u_0 is the windspeed measured by an anemometer at height z_0, p is an exponent that varies with atmospheric stability.

Table 6.1 gives the values of the exponent p appropriate to the Pasquill stability categories which are described in 6.1.3.5. This supports the intuitive idea that the best-mixed air (Class A) has the smallest variation of speed with height.

6.1.3   Atmospheric stability

6.1.3.1   Dry adiabatic lapse rate

Atmospheric pressure decreases exponentially with height. Hence, as an air parcel moves up (or down) in the atmosphere, it must expand (compress) and cool (warm). For a dry atmosphere (containing water vapour, but not liquid water droplets) the rate of decrease of temperature with height caused by this type of displacement is called the dry adiabatic lapse rate (Γ, or DALR). Adiabatic simply means that the air parcel's total energy content is conserved during the displacement, not exchanged with the surrounding air. The value of Γ is fixed by physical constants, and can be calculated from

$$\Gamma = g/c_p$$

where g is the acceleration due to gravity $= 9.81$ m s^{-2} and c_p is the specific heat of air at constant pressure $= 1010$ J kg^{-1} K^{-1}.

Hence $\Gamma = 9.8$ °C km^{-1}. This value applies to sufficient accuracy in the lower 20 km of the atmosphere. Above that height, changes in the molecular composition (which affect the molar mass) and in g (due to increasing distance from the centre of the Earth) start to affect the value. This is the reason for the steady decrease in temperature with height – up mountains, for example.

6.1.3.2   Saturated adiabatic lapse rate

If the air temperature falls below the water vapour dew point while the parcel is being lifted, then the excess water vapour will start to condense. This will release latent heat of vaporisation which will reduce the rate of cooling. Conversely, if a descending parcel contains liquid water droplets, then the process is reversed – the droplets evaporate as the parcel warms, extracting sensible heat from the air and reducing the rate of warming.

The variation of temperature with height when there is liquid water present is known as the saturated adiabatic lapse rate (SALR), or Γ_{sat}.

$$\Gamma_{sat} = \frac{dT}{dz} = \frac{-g}{c_p} \left\{ \frac{1}{1 + (L/c_p)(dq_s/dT)} \right\} \tag{6.2}$$

where L is the latent heat of evaporation of water, q_s is the saturation mixing ratio (kg water/kg air).

Since the saturated vapour pressure of water depends on both the temperature and pressure, the SALR is also variable (Table 6.2). The minimum values are found for surface pressures and the warmest temperatures. Since the air also has to be saturated, this is more likely to be in the tropics than above the Sahara. Maximum values are found where the air is very cold, because then the water vapour content is low. The average lapse rate over the Earth is about $6.5\ °C\ km^{-1}$.

6.1.3.3   Atmospheric stability and temperature profile

The adiabatic lapse rates describe the temperature changes expected in a parcel of air when it is displaced vertically. This is not usually the same as the vertical

Table 6.2 Variation of saturated lapse rate ($°C\ km^{-1}$) with temperature and pressure

Pressure/mb	Temperature/°C				
	−40	−20	0	20	40
1000	9.5	8.6	6.4	4.3	3.0
800	9.4	8.3	6.0	3.9	
600	9.3	7.9	5.4		
400	9.1	7.3			
200	8.6				

temperature profile of the air, as would be measured, for example, by recording the temperature transmitted by a sounding balloon as it rose through the atmosphere. This environmental lapse rate (ELR) is the vertical variation, or profile, of air temperature with height that exists at any particular time and place. It may be equal to the adiabatic lapse rate over at least part of the height range of interest (up to, say, 1 km for pollutant dispersion), but it may be substantially different. The local balance between the two lapse rates gives an insight into the concept known as stability.

First, consider the different physical situations shown in Figure 6.1. In Figure 6.1(a), a marble is at rest in a bowl. Any small displacement of the marble results in a restoring force – a force that moves the marble back towards its initial position. Such a system is said to be stable. In Figure 6.1(b), the marble is on a flat surface. No force in any direction is generated by a displacement, and the system is described as neutral. Finally, in Figure 6.1(c), the marble is poised on the top of the upturned bowl. Now any displacement results in a force away from the initial position, and the system is unstable.

These ideas can be applied to air parcels in the atmosphere. Consider, for example, the different situations shown in Figure 6.2(a–c). In each Figure, the prevailing ELR is shown, and the DALR is given for comparison. A parcel of air that starts at A in Figure 6.2(a) and moves upward will cool at the DALR, reaching a lower temperature at B. However, the air around the parcel at the same height will be at C on the ELR. The parcel has become cooler and denser than its surroundings and will tend to sink back towards its starting height. If its initial displacement is downward, it will become warmer and less dense than its surroundings and tend to rise back up. The ELR is therefore stable, because small disturbances are damped out. In Figure 6.2(b), a rising parcel will again cool at the DALR. Since this is equal to the ELR, the parcel will be in air of the same temperature and density after displacement. The ELR is therefore neutral, because vertical motions are neither accelerated nor damped. Examination of vertical displacements for Figure 6.2(c) shows that they are accelerated, and that such an ELR is unstable.

If we look at the forces acting on a parcel displaced upwards in a stable atmosphere, an equation of motion can be written for which the solution is an

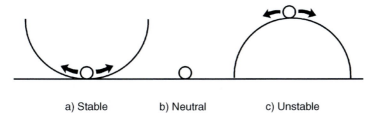

a) Stable   b) Neutral   c) Unstable

Figure 6.1 Stable, neutral and unstable systems.

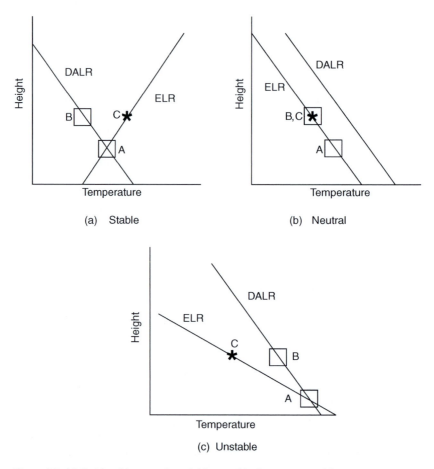

Figure 6.2 (a) Stable; (b) neutral; and (c) unstable Environmental lapse rates.

oscillation about the starting height. That is, the parcel experiences a restoring force that accelerates it back towards its original position. Since by then it is moving, it overshoots downwards, and a restoring force accelerates it upwards again, and so on. The frequency of this oscillation, which is called the buoyancy frequency, corresponds to a natural period of a few minutes in the lower atmosphere.

In practice, real temperature profiles in the atmosphere often consist of a mixture of different ELRs, so that vertical dispersion will be different at different heights. Consider the Environmental Lapse Rate shown in Figure 6.3. Between A and B, strong solar heating of the ground has warmed the lowest layers of air; the middle layer BC is close to DALR, while the layer CD is showing an increase of temperature with height (this is known as an inversion of the temperature profile).

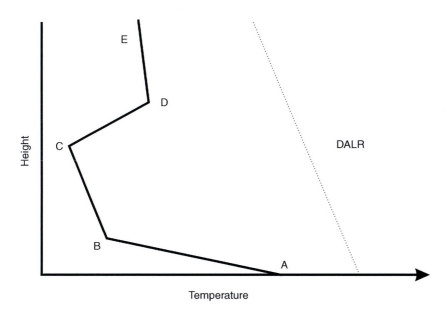

Figure 6.3 An atmospheric temperature profile involving a range of stabilities.

This profile will be very unstable (strong dispersion) in AB, close to neutral in BC, and very stable (poor dispersion) in CD. When a persistent anticyclonic high pressure region is present, there is often an area in the centre where high-level air is sinking. The sinking raises the temperature, generating an elevated inversion which may act as a lid to vertical plume dispersion and increase ground-level pollutant concentrations. Since the pressure gradients and hence windspeeds are low under these meteorological conditions, a serious pollutant episode may result.

We saw in Section 6.1.3.2 that the lapse rate for saturated air is lower than for dry air. Hence the same air mass may be stable or unstable, depending on whether it is saturated or not. As an unsaturated air parcel starts to rise, it will cool at the dry lapse rate of 9.8 °C km^{-1}. If the ELR is, say 8 °C km^{-1}, then the movement is stable. But if the upward movement cools the air enough to cause condensation of water vapour, then the lapse rate will fall to the saturated value of, say 7 °C km^{-1}, and the displacement is now unstable. This is called conditional instability. The condition is not rare – since the average global lapse rate of 6.5 °C falls between the dry adiabatic rate and the average moist lapse rate of 6 °C, conditional instability is the norm. Furthermore, a humidity gradient can change the stability. This is most often seen in the warm atmosphere near the surface above the tropical ocean. The sea surface evaporates water into the air. The molar mass of water is 18 compared with 29 for the air, so the air density decreases and promotes instability.

6.1.3.4   Potential temperature

A useful additional concept is that of potential temperature, θ. The DALR shows a steady reduction of temperature with height, and it is not obvious at first glance whether a particular rate of reduction is greater than or less than the DALR. This divergence is important for the assessment of atmospheric stability and pollutant dispersion, as we have seen. We can define the potential temperature of an air parcel at any pressure (i.e. height) as the temperature that it would have if the parcel were moved adiabatically to a standard or reference pressure. This reference pressure is normally taken to be 100 kPa, which is very close to the standard global average atmospheric pressure at sea level of 101.325 kPa. The potential temperature, then, is the temperature that the parcel would have if it was moved down to sea level adiabatically. By definition, θ is constant for an air parcel that is moved upwards adiabatically from sea level. An actual temperature profile such as that in Figure 6.4 is transformed into a potential temperature profile, where parts with neutral stability are vertical; in stable atmospheres θ increases with height, while in unstable atmospheres it decreases with height.

6.1.3.5   Pasquill stability classes

The general effect of environmental variables on stability can be summarised as follows:

- On cloud-free days, solar radiation will warm the ground during the day, making the lowest air layers unstable.
- On cloud-free nights, net upwards longwave thermal radiation cools the ground, making the lowest air layers stable, and creating a ground-level inversion.
- As windspeed rises, vigorous vertical exchange tends to generate a neutral ELR equal to the DALR.
- As cloud cover increases, daytime heating and nighttime cooling of the ground are both reduced, again making the ELR closer to the DALR.
- A persistent high pressure region has a downward air movement at the centre, which often creates an elevated inversion and traps pollutants near the ground. This is known as 'anticyclonic gloom'.

The influence of solar radiation and windspeed on stability led Pasquill to identify stability classes on a scale from A (very unstable) to G (very stable), and to relate them to the meteorology in the simple way shown in Table 6.3.

First, look at the stability classes during the daytime. When the sun is strong and the winds light, we would expect maximum ground heating, raising the temperature of the lowest layers of air and creating unstable conditions. With less sunshine, this effect is reduced and the instability is less marked. As the windspeed increases, vertical mechanical mixing starts to override the buoyancy effects, and frequent exchanges of air parcels at DALR generate a neutral ELR (class D). At night, net radiation losses and ground cooling are most marked under clear

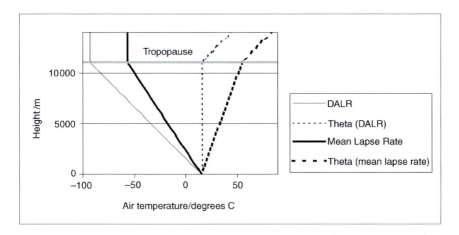

Figure 6.4 Actual and potential temperature profiles.

Table 6.3 Dependence of stability on meteorological parameters

Surface windspeed ms^{-1}	Day time sun (flux density in W m^{-2})			Night time (cloud amount in oktas)		
	Strong (>590)	Moderate (300–590)	Weak (<290)	8	4–7	0–3
<2	A	A–B	B	D	G	G
2–3	A–B	B	C	D	E	F
3–5	B	B–C	C	D	D	E
5–6	C	C–D	D	D	D	D
>6	C	D	D	D	D	D

skies. As the ground cools, the lowest air layers become stable (class G). The cloudier or the windier it becomes, the more probable class D becomes again.

Occurrence of Pasquill classes

Table 6.4 shows the average occurrence of different stability classes in the UK. As would be expected, our temperate maritime climate gives us high average windspeeds and cloud cover, so that categories C and D are the most likely. The very high values for ELR given here, equivalent to up to 40 °C km^{-1}, can only be sustained within the lowest few metres of the atmosphere.

6.1.3.6   Other stability estimators

The Pasquill scheme described in Section 6.1.2 is semi-quantitative, in that it ascribes categories to ranges of the driving variables wind speed and solar radiation, but does

Table 6.4 Occurrence of different stability classes in the UK

Stability class	Occurrence in UK/%	ELR °C (100 m)$^{-1}$	u at 10 m ms^{-1}	Weather
A: Very unstable	1	<-1.9	1.0–2.5	Very sunny
B: Moderately unstable	5	$-1.9 - -1.7$	1.0–5.0	Sunny
C: Slightly unstable	15	$-1.7 - -1.5$	2.0–6.0	Part cloud (day)
D: Neutral	65	$-1.5 - -0.5$	2.0–>10.0	Overcast
E: Stable	6	$-0.5 - +1.5$	2.0–5.0	Part cloud (night)
F: Very stable	6	$+1.5 - +4.0$	2.0–3.0	Clear night
G: Even more stable than F	2		Calm	

not involve a formula with explicit values. There are two more quantitative parameters in common use – the Richardson Number and the Monin–Obukhov length.

The Richardson Number

The Richardson Number Ri is calculated from gradients of temperature and windspeed

$$\mathrm{Ri} = \frac{\dfrac{g}{T}\dfrac{\partial T}{\partial z}}{\left(\dfrac{\partial u}{\partial z}\right)^2}$$

where T is the absolute temperature.

The Richardson Number is dimensionless. We can see by inspection that if temperature increases strongly with height, which is the stable condition that we have discussed in Section 6.1.3.3, then Ri will tend to be large and positive. If T decreases with height at the adiabatic lapse rate, Ri will be negative. If there is a large gradient of windspeed with height, Ri will be small.

Ri < -1	free convection
$-1 <$ Ri < -0.01	unstable mixed convection
$-0.01 <$ Ri $< +0.01$	forced convection
$+0.01 <$ Ri $< +0.1$	damped forced convection
Ri > 0.1	increasingly stable

The Richardson Number is mainly used by meteorologists to describe events in the free atmosphere, rather than by air pollution scientists discussing dispersion closer to the ground.

The Monin–Obukhov Length

The Monin–Obukhov Length, L is a function of heat and momentum fluxes.

$$L = -\frac{\rho c_p T u_*^3}{kgC}$$

where T is the absolute temperature and C is the vertical sensible heat flux.

The formula for L recognises the two contrasting processes that drive atmospheric turbulence and dispersion. Mechanical turbulence is generated by wind blowing over surface roughness elements. It is therefore strongest at the surface, and reduces upward depending on the stability. Convective turbulence is generated by heat flux from the surface, and will increase upwards in unstable conditions but be damped out in stable conditions (Table 6.5). The result has units of metres, and represents the scale of the turbulence. Since C may be positive (up) or negative (down), L may be positive or negative. When conditions are unstable or convective, L is negative, and the magnitude indicates the height above which convective turbulence outweighs mechanical turbulence. When conditions are stable, L is positive, and indicates the height above which vertical turbulence is inhibited by stable stratification. In practice, values range between infinity (in neutral conditions when C is zero) and -100 m. Atmospheric stability is related to the ratio h/L, where h is the depth of the boundary layer. L is increasingly being used in dispersion models as a more sophisticated estimator of turbulence than the Pasquill category.

For practical use in the lowest few hundred metres of the atmosphere, we can write

$$L = \left(\frac{\rho c_p T}{kg}\right)\frac{u_*^3}{C} = \text{const.}\frac{u_*^3}{C}. \tag{6.3}$$

Note that although this expression does not look like a length at first sight, the dimensions do confirm that it is
i.e.

(kg m^{-3})(J kg^{-1} K^{-1})(K)(m s^{-1})^3/(m s^{-2})(W m^{-2}) has dimensions of length and units of metres.

Table 6.5 Stability as a function of L and h

Stability	Range of h/L
Stable	$h/L \geq 1$
Neutral	$-0.3 \leq h/L < 1$
Convective	$h/L < -0.3$

The values of ρ, c_p, T, k and g are then taken as constant. With

$$\rho = 1.2 \text{ kg m}^{-3}$$
$$c_p = 1010 \text{ J kg}^{-1} \text{ K}^{-1}$$
$$T = 288 \text{ K (15 °C)}$$
$$k = 0.41$$
$$g = 9.81 \text{ m s}^{-2},$$

we have $L = 8.7 \times 10^4 \, u_*^3/C$

Other stability estimators have been used, such as the standard deviation of the fluctuations in the horizontal wind direction, and the ratio of the windspeeds at two heights. Comparisons of the predictions made by different estimators from the same data set have shown that the Monin–Obukhov length method correlates best with the Pasquill classes.

Boundary layer height

The Monin–Obukhov formulation can also be used to estimate the height of the boundary layer in stable-to-neutral conditions, from

$$\frac{h}{L} = \frac{0.3u_*/|f|L}{1+1.9h/L}$$

The situation in unstable conditions is more complex, because the boundary layer grows from an initial value of zero when the surface heat flux becomes positive, at a rate governed by the surface heat flux, the temperature jump across the top of the layer, and the friction velocity (Table 6.6).

Table 6.6 Relationships between stability estimators

Windspeed u/m s^{-1}	Pasquill category	Boundary layer height h/m	Monin–Obukhov Length/m	h/L
1	A	1300	−2	−650
2	B	900	−10	−90
5	C	850	−100	−8.5
5	D	800	∞	0
3	E	400	100	4
2	F	100	20	5
1	G	100	5	20

6.1.4   Meteorological data

In the UK the main provider of meteorological data is the Meteorological Office. Typically a user, such as a Local Authority, will need to use a commercial dispersion model to calculate ground-level concentrations of a pollutant from a particular source. If this is done hour-by-hour for one year, then 8760 sets of meteorological data will be needed. Each set will include values of wind speed, wind direction, boundary layer height and an atmospheric stability criterion such as Pasquill class or Monin–Obukhov length scale. Wind speed and direction may be measured locally, but the other parameters are derived from other values and different methods are in use. Trinity Consultants are the main provider of nationwide data in the US, and this company also supplies compilations of UK data in the UK.

6.2   DISPERSION MODELS

The qualitative aspect of dispersion theory is to describe the fate of an emission to atmosphere from a point, area or line source. Quantitatively, dispersion theory provides a means of estimating the concentration of a substance in the atmosphere, given specific information about meteorological factors and the geometry and strength of the source. Dispersion models include:

- Eulerian models, which numerically solve the atmospheric diffusion equation. Eulerian models work on the measurement of the properties of the atmosphere as it moves past a fixed point. A windvane or cup anemometer are Eulerian sensors.
- Gaussian models, which are built on the Gaussian (normal) probability distribution of wind vector (and hence pollutant concentration) fluctuations. Strictly these are a subset of Eulerian models but are usually treated as a group in their own right.
- Lagrangian models, which treat processes in a moving airmass, or represent the processes by the dispersion of fictitious particles. Tracking a neutral-density balloon as it moves downwind is a Lagrangian measurement.

Dispersion is essentially due to turbulence, and turbulence occurs with a great range of length scales in the atmosphere. Hence there are further families of model depending on the length (or time) scale of interest:

- Macro scale (~1000 km, or days). Atmospheric flow is driven by synoptic phenomena such as high/low pressure areas. For example, the long range transport of air pollution from Central Europe to the UK or from the UK to Scandinavia.
- Meso scale (10s–100s of km, or hours). Air movement is synoptically driven, but modified by local effects such as surface roughness and obstacles. For

example, the dispersion of emissions from a power station chimney over the surrounding countryside.

- Microscale (<1 km, or minutes). Air flow depends mainly on local features. For example, urban flows in the labyrinth of street canyons.

6.3 GAUSSIAN DISPERSION THEORY

Eulerian models are based on the concept of a fixed reference point, past which the air flows. Lagrangian models, in contrast, are based on the concept of a reference point that travels with the mean flow. In the commonly-used Gaussian method, the simplest realisation of this idea is to think of an instantaneous release of a pollutant from a point source. This 'puff' moves downwind along the average wind direction. As it does so it expands in volume, incorporating dilution air from around it and reducing its concentration. The puff also experiences small random movements, caused by turbulence, away from the mean direction. A continuous emission is then described as an infinitely rapid series of small individual puffs. Gaussian dispersion theory enables the concentration due to such a trajectory to be calculated at any location downwind.

We are now going to outline the system of equations that we need in order to predict what concentration of a substance will be found at any point in the atmosphere, if that substance is being released at a known rate from a source. We will show how the practical equations are related to physical processes in the atmosphere, but we will not derive the equations from first principles.

6.3.1 Coordinate system

First, we have to be able to describe the position of the place at which we wish to estimate concentration, relative both to the source and the ground. A standard Cartesian (x, y, z) coordinate system is used (Figure 6.5).

In this coordinate system:

the physical source (e.g. the base of a chimney) is located at the origin $(0, 0, 0)$;
the x axis lies along the mean wind direction;
x is the distance downwind from the source;
y is the lateral distance from the mean wind direction;
z is the height above ground level;
h is the physical height of the chimney;
dh is the additional height by which the plume rises due to its buoyancy and/or momentum;
$H = h + dh$ is the effective height of the release;
$\bar{u}$ is the mean windspeed at plume height.

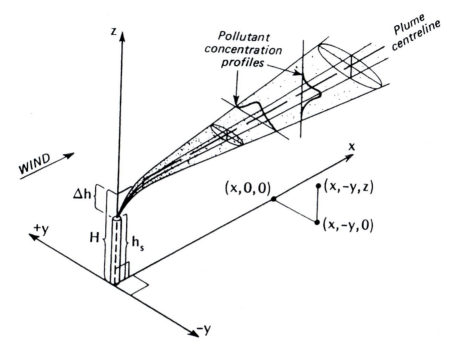

Figure 6.5 The Cartesian coordinate system used to specify dispersion geometry.

6.3.2   Fickian diffusion

The starting point for dispersion theory is to consider the diffusion of a cloud of material released into the atmosphere instantaneously from a point source. In one dimension, turbulent transfer in the atmosphere can be described by a Fickian equation of the form

$$\frac{d\bar{q}}{dt} = K\frac{\partial^2 \bar{q}}{\partial x^2}$$

where q is the concentration of material and K is the turbulent diffusivity. This equation has an analytical solution

$$\bar{q} = \frac{Q}{(4Kt)^{0.5}}\exp\left(\frac{-x^2}{4Kt}\right)$$

where Q is the source strength (e.g. in kg s^{-1}). The concentration is a maximum at the point of release, and decays away in both positive and negative directions following the Gaussian bell-shaped distribution (Figure 6.6). Hence the value of

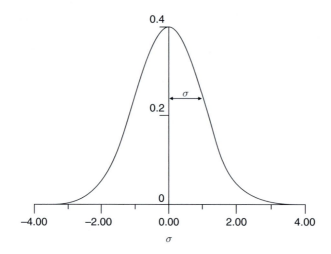

Figure 6.6 The standard Gaussian (normal) distribution, showing the physical signifi-cance of sigma.

σ at any distance from the source is a distance in metres which represents the lateral or vertical half 'width' of the plume.

Now imagine that this instantaneous point source of material is diffusing isotropically in three dimensions (puff model) and at the same time drifting away from the point of release at the average windspeed $\bar{u}$. The solution becomes

$$q(x, y, z, t) = \frac{Q}{(2\pi\sigma^2)^{3/2}} \exp\left(\frac{-r^2}{2\sigma^2}\right)$$

where σ is the standard deviation of the puff concentration at any time t, and $r^2 = (x - \bar{u}t)^2 + y^2 + z^2$.

6.3.3 The Gaussian plume equation

Following the analysis given for puff dispersion, emission from a continuous source can be considered as a continuous series of overlapping puffs, giving

$$q(x, y, z) = \frac{Q}{2\pi\bar{u}\sigma_y\sigma_z} \exp\frac{-y^2}{2\sigma_y^2} \exp\left\{\frac{-(z-H)^2}{2\sigma_z^2}\right\} \tag{6.4}$$

where σ_y and σ_z are the standard deviations in the y and z directions, respect-ively; and H is the total height of release. The first exponential term describes lateral dispersion, and the second describes vertical dispersion. Equation 6.4 is the basic Gaussian dispersion equation which is at the heart of many air quality prediction models.

It is important to realise that σ_y and σ_z describe the width of the concentration distribution, not of the plume itself. This aspect is clarified in Figure 6.7. The plan view shows a snapshot of the plume at any one time as the plume wanders either side of the mean wind direction. The concentration within the plume itself is approximately uniform, as shown by the 'top hat' shape of the

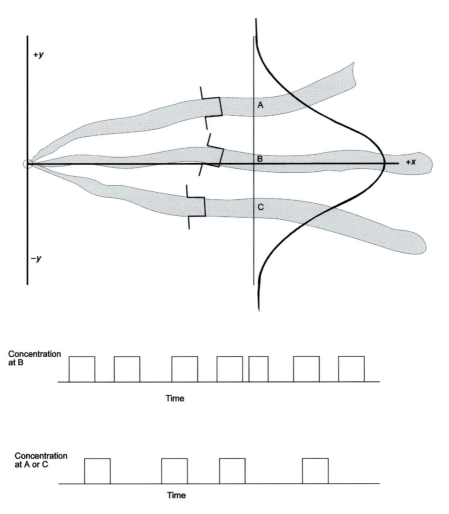

Figure 6.7 A plan view showing how the Gaussian distribution is formed from the differing frequency of exposure to the plume as it wanders about the mean wind direction.

distribution. The plume moves across to locations A and C only rarely – although the peak concentration may be high for a short time, the average is low because most of the time the plume is elsewhere. Point B is on the centre line of the plume along the mean wind direction. By definition, the plume spends a higher proportion of the time in this direction than any other, so the average concentration is highest. The frequency distribution of direction variations about the mean is Gaussian – hence the lateral Gaussian profile is built up. Similar variations in the vertical plane also give a Gaussian profile.

6.3.4   Assumptions of the Gaussian model

- Release and sampling times are long compared to the travel time from source to receptor. This means that the release is effectively steady state and that diffusion along the mean wind direction is negligible compared to advection (movement with the mean wind). Measurement time scales of hours rather than minutes are implied.
- The material is chemically stable and is not deposited to the ground. This means that gases must be unreactive, and particles must be <20 μm in diameter so that they do not sediment out. The equation of continuity will then apply – the integral of the concentration over all space at any time is equal to the total mass of material emitted. In practice, most gases are deposited to some extent; this can be allowed for by, for example, an additional exponential decay factor in the concentration with distance from the source.
- The lateral and vertical variations of the material concentration can both be described by Gaussian distributions, which are functions of x only.
- The windspeed is constant with height. This is never true in practice, as has already been seen. Windspeed variation with height can often be described by a logarithmic profile. More advanced versions of the Gaussian formulation divide the atmosphere up into layers, each layer having a specified set of characteristics such as windspeed and stability.
- The wind direction is constant with height. Again, this is rarely true. The most common form of the variation is the Ekman spiral, in which the direction tends towards the geostrophic (parallel with the isobars) as height increases, over the first few hundred metres.

6.3.5   Plume rise

Most plumes have either some exit velocity that carries them up into the air, or some degree of buoyancy due to temperature or density difference with the surrounding air, or both. Hence the effective plume height is likely to be different (usually, greater) than the physical height of the chimney. The dispersion equation shows that concentration is a function of the square of the release height, and hence much research has been done into the dependence of plume rise on other factors.

In principle it is possible to solve the conservation equations for mass, momentum, enthalpy and amount of emitted material, and hence predict the trajectory of the plume and the rate of entrainment into it of surrounding air. A simplified, more empirical method has often been used which is less mathematically and computationally intensive, although increased computing power is making full solution more practical.

The standard theory assumes that a rising buoyant plume entrains ambient air at a rate proportional to both its cross-sectional area and its speed relative to the surrounding air. The driving force is expressed in terms of an initial buoyancy flux F_b, where

$$F_b = w_0 R_0^2 \frac{g}{T_{p0}} (T_{p0} - T_{a0})$$

with w_0 the initial plume speed, R_0 the inside stack radius, T_{p0} the initial plume temperature (K), and T_{a0} the ambient temperature at stack height. Hence the SI units of F_b are $m^4 \, s^{-3}$.

Briggs solved the equations of motion for a bent-over plume in a neutral atmosphere to show that the height z at any distance x downwind was given by

$$z = \frac{C_1 \left(\dfrac{F_b}{\pi} \right)^{\frac{1}{3}} x^{\frac{2}{3}}}{u} \left(1 + \frac{u\tau}{x} \right)^{\frac{1}{3}} = C_1 Br(F_b, x, u)$$

where $Br(F_b, x, u)$ is the 'Briggs variable'
and

$$\text{and } C_1 = \left(\frac{3}{2\beta^2} \right)^{\frac{1}{3}}, \text{ with plume radius } r = \beta z.$$

The most important value for use in dispersion calculations is the final plume rise dh. This has usually been found by fitting curves to observed data from a range of sources, and Briggs recommended the following relationships

Buoyant rise in unstable – neutral atmospheres

In Pasquill categories A–D, different equations are to be used depending on whether F_b is greater or less than $55 \, m^4 \, s^{-3}$.

$$\text{For } F_b < 55 \text{ use } dh = 21 \, F^{0.75}/u_h \tag{6.5}$$

For $F_b \geq 55$ use $dh = 39\,F^{0.6}/u_h$ $\qquad\qquad$ (6.6)

where u_h is the windspeed at height h.

Buoyant rise in stable atmospheres

In a stable atmosphere, Briggs found that the most reliable expression for the final plume rise dh was given by

$$dh = 2.6\left(\frac{F_b}{uS}\right)^{1/3}$$

Here S, the ambient stability factor $= \dfrac{g}{T}\left(\dfrac{\partial T}{\partial z} + \Gamma\right)$, and is a function of atmospheric conditions only.

Clearly, these estimates of plume rise need quite detailed information about the properties of the source. If less is known about the source, dh can be estimated from

$$dh = \frac{515 P_T^{0.25}}{\bar{u}}$$

which gives the plume rise in metres when P_T is the thermal power of the chimney gases in MW – i.e. the product of mass flow rate, the specific heat of the flue gases and their temperature difference from that of the ambient air. For example, a large power station will have an electrical output at full load of 2000 MW, and a thermal output in the flue gases of about 280 MW. Hence a plume rise of 220 m would be expected in a windspeed of 10 m s^{-1}. The chimney height for such a power station in the UK is around 200 m, so plume rise contributes significantly to dispersion, especially since ground level concentrations are inversely proportional to the square of the effective source height.

6.3.6   The effect of vertical constraints

The simplest Gaussian solution (Eqn 6.4) assumes that the plume is free to expand in all directions without constraint. This may be true in practice during a limited period of time when an emission is moving away from an elevated source and has not yet moved close to the ground. For real sources, the pollutant cannot disperse in all directions without constraint. In the usual situation of an elevated source (chimney + plume) at some height above the ground, downward dispersion is always limited by the presence of the ground, while upward dispersion may be limited by an elevated inversion.

6.3.6.1   Reflection at the ground

Assuming that no pollutant is absorbed by the ground, any pollutant that reaches the ground is available for upward dispersion into the atmosphere again. This is accounted for theoretically by adding an 'image source', equal in magnitude to the actual source but located at $(0, 0, -H)$, i.e. a distance $2H$ vertically below the actual source and distance H *beneath* the ground surface (Figure 6.8). Although this is a physically meaningless mathematical contrivance, it does the required job of effectively reflecting the plume away from the ground.

The dispersion equation therefore becomes

$$q(x, y, z) = \frac{Q}{2\pi\bar{u}\sigma_y\sigma_z}\exp\frac{-y^2}{2\sigma_y^2}\left\{\exp\frac{-(z-H)^2}{2\sigma_z^2} + \exp\frac{-(z+H)^2}{2\sigma_z^2}\right\} \quad (6.7)$$

where the second exponential term inside the curly brackets represents 'additional' material from the image source located at $(0, 0, -H)$. This is shown as Image Source A on Figure 6.9.

6.3.6.2   Reflection at an inversion

Dispersion downwards will always be constrained by a physical surface such as water or vegetation. Dispersion upwards may also be constrained – not by a physical surface but by the temperature structure of the atmosphere itself. If there is an elevated inversion present, as was described in Figure 6.3, then this can act as a barrier to free upward dispersion. Buoyant air parcels penetrating into an elevated inversion from below will have buoyancy removed because the temperature of the surrounding air will increase with height while that of the parcel will decrease. The worst pollution episodes, such as those that occurred in London in

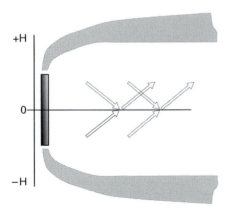

Figure 6.8 Side view to show how the image source allows for reflection of plume material at the ground.

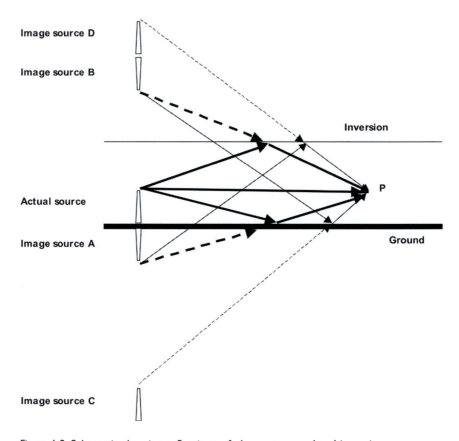

Figure 6.9 Schematic showing reflections of plume at ground and inversion.

1952 and 1991, are often associated with the presence of ground-level or low-altitude elevated inversions that trap ground-level emissions within a thin layer of the atmosphere. This is allowed for in the solution to the diffusion equation by the addition of further reflection terms, which are generated by an image source as far above the inversion as the actual source is below it (i.e. at a height of $(2H_{bl} - H)$). Hence

$$
\begin{aligned}
q(x, y, z) = {} & \frac{Q}{2\pi\bar{u}\sigma_y\sigma_z}\, \exp\frac{-y^2}{2\sigma_y^2}\left\{\exp\frac{-(z-H)^2}{2\sigma_z^2} + \exp\frac{-(z+H)^2}{2\sigma_z^2}\right. \\
& + \exp\frac{-(z-2H_{bl}+H)^2}{2\sigma_z^2} + \exp\frac{-(z+2H_{bl}-H)^2}{2\sigma_z^2} \\
& \left. + \exp\frac{-(z-2H_{bl}-H)^2}{2\sigma_z^2}\right\}
\end{aligned}
\tag{6.8}
$$

where H_{bl} is the height of the inversion or the top of the boundary layer. In this equation, the first term in the curly brackets is the direct component, and the second is due to the image source discussed above. Figure 6.9 shows how these and the other components contribute to the concentration at any point P. The third exponential term in the curly brackets represents reflection of the source at the inversion (Image source B, Figure 6.9), the fourth term (Image source C) is reflection of Image source B at the ground, and the fifth (Image source D) is reflection of Image source A at the inversion.

Far enough downwind, the plume has expanded so far vertically that it fills the boundary layer completely. This occurs for $\sigma_z > H_{bl}$. After that it expands as a horizontal wedge, as though it had been released from a vertical line source the height of the boundary layer. The plume is well mixed throughout the boundary layer, so there is no z dependence and the concentration is given by

$$C(x, y) = \frac{Q}{\sqrt{2\pi}\sigma_y h U_{h/2}} \exp\left(\frac{-y^2}{2\sigma_y^2}\right) \qquad (6.9)$$

The Gaussian formulation described above applies to stable and neutral boundary layers. There is increasing evidence that the distribution of vertical velocities, and hence of concentrations, is non-Gaussian (skewed) when the boundary layer is convective. The result of this skewness is that for elevated sources, the height within the plume at which the concentration is a maximum descends as the plume moves downwind, while the mean height of the plume rises.

6.3.7   Estimating stability

Figure 6.10 shows the idealised effect of the different ELRs on plume behaviour. Each diagram shows a temperature profile in the atmosphere, together with the expected plume behaviour. The most unstable conditions (a, Pasquill A) can result intermittently in very high concentrations of poorly-diluted plume close to the stack. This is called looping. Neutral conditions (b, Pasquill C–D) disperse the plume fairly equally in both the vertical and horizontal planes. This is referred to as coning. A strong inversion of the profile (c, Pasquill E–F) prevents the plume from mixing vertically, although it can still spread horizontally. If the plume is released above a ground-based inversion (d), then it is prevented from mixing down to the ground and ground-level concentrations are kept low. Conversely, if it is released below an elevated inversion (e) then it is prevented from dispersing upwards and ground-level concentrations are enhanced. In practice, complex ELR structures interact with stack and plume heights to create a much more variable picture.

The effect of stability on dispersion pattern can also be seen in Figure 6.11. Pasquill A generates the highest maximum concentration and the closest to the stack, although the narrow peak means that the integrated downwind concentration is low. As stability increases, the peak concentration decreases and moves downwind.

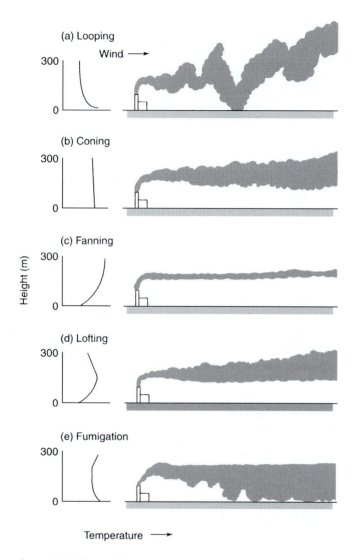

Figure 6.10 Effect of lapse rate on plume dispersion.
Source: Oke, T. R. (1987) *Boundary Layer Climates*, Methuen, London.

6.3.7.1   *Estimates of σ_y, σ_z*

Expressions for σ_y and σ_z have been derived in terms of the distance travelled by the plume downwind from the source under different Pasquill stability classes. They are available both as graphs and as empirical equations that are valid in the range 100 m $< x <$ 10 km. Figure 6.12(a) and (b) show the variations of standard

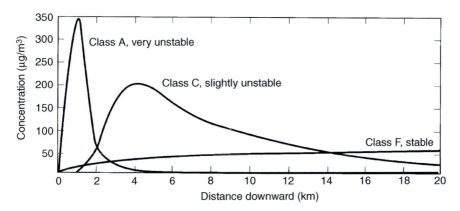

Figure 6.11 Variation of ground level concentration with distance downwind, under different atmospheric stabilities.
Source: Masters, G. M. (1998) *Environmental Engineering and Science*, Prentice Hall, New Jersey.

deviation with distance from the source for the different Pasquill stability categories. Note that the x scale runs from 100 m to 100 km – this dispersion methodology is designed for medium physical scales, neither small enough to be influenced by individual topographic features nor large enough to be controlled by synoptic windflows. Note also that σ_z, the vertical standard deviation, is much

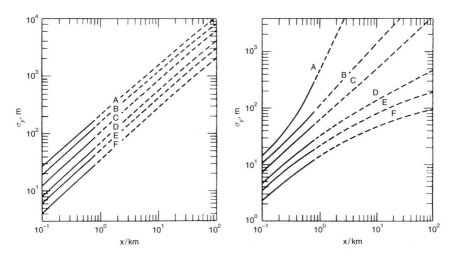

Figure 6.12 Variations of plume standard deviation in the y and z directions with distance x from the source. A, B, C, D, E and F are Pasquill stability categories.

Table 6.7 Equations for the variation of σ_y and σ_z with stability class

Pasquill category	σ_y/m	σ_z/m
Open-country		
A	$0.22x \, (1 + 0.0001x)^{-0.5}$	$0.20x$
B	$0.16x \, (1 + 0.0001x)^{-0.5}$	$0.12x$
C	$0.11x \, (1 + 0.0001x)^{-0.5}$	$0.08x \, (1 + 0.0002x)^{-0.5}$
D	$0.08x \, (1 + 0.0001x)^{-0.5}$	$0.06x \, (1 + 0.0015x)^{-0.5}$
E	$0.06x \, (1 + 0.0001x)^{-0.5}$	$0.03x \, (1 + 0.0003x)^{-1}$
F	$0.04x \, (1 + 0.0001x)^{-0.5}$	$0.016x \, (1 + 0.0003x)^{-1}$
Urban		
A–B	$0.32x \, (1 + 0.0004x)^{-0.5}$	$0.024x \, (1 + 0.001x)^{0.5}$
C	$0.22x \, (1 + 0.0004x)^{-0.5}$	$0.20x$
D	$0.16x \, (1 + 0.0004x)^{-0.5}$	$0.14x \, (1 + 0.0003x)^{-0.5}$
E–F	$0.11x \, (1 + 0.0004x)^{-0.5}$	$0.08x \, (1 + 0.0015x)^{-0.5}$

more influenced by stability than is σ_y, the horizontal standard deviation, owing to the influence of buoyancy forces.

Table 6.7 gives corresponding equations for open-country and urban conditions. The main difference is that the greater surface roughness in built-up areas generates greater turbulence.

6.3.8   Versions of the Gaussian equation

The full Gaussian equation (6.7) describes the concentration field everywhere in the free atmosphere due to a point source. More usually, we need to know the concentration at ground level, or from a ground level source, or directly downwind of the source. In these cases various simplifying assumptions can be made.

Ground-level concentration due to an elevated source, bounded by the ground surface

Set $z = 0$ everywhere in equation (6.7)

$$q(x, y, 0) = \frac{Q}{\pi \bar{u} \sigma_y \sigma_z} \exp - \left[\frac{y^2}{2\sigma_y^2} + \frac{H^2}{2\sigma_z^2} \right]$$

Ground-level concentration due to an elevated source, bounded by the ground surface, directly downwind of the source at ground level

Set $y = z = 0$ everywhere in equation (6.7).

$$q(x, 0, 0) = \frac{Q}{\pi \bar{u} \sigma_y \sigma_z} \exp \left[\frac{-H^2}{2\sigma_z^2} \right] \tag{6.10}$$

It can also be shown that the maximum downwind concentration, q_{max}, is given by

$$q_{max} = \frac{2Q}{\pi \bar{u} e H^2} \frac{\sigma_z}{\sigma_y} \tag{6.11}$$

at the distance x_{max} for which $\sigma_z = \dfrac{H}{\sqrt{2}}$.

In equation (6.11), $e = 2.71828$, the base of natural logarithms. First, use the known value of H to calculate σ_z, and read off this value on the vertical axis of Figure 6.12(b). Now move across to the curve corresponding to the known stability class, and read down to the horizontal axis to find x_{max}. Set this value on the x-axis of Figure 6.12(a), and read up to the same stability class and across to the y-axis. This gives the value of σ_y at x_{max}. Finally, use these values of σ_y and σ_z, together with Q, u and H, to find q_{max} from equation (6.11).

Note that the maximum ground-level concentration is inversely proportional to the square of the effective release height H, so that building taller chimneys is a very cost-effective method of reducing ground-level pollution.

Ground-level concentration due to a ground level source, directly downwind of the source (e.g. bonfire)
Set $y = z = H = 0$
Hence

$$q(x, 0, 0) = \frac{Q}{\pi \bar{u} \sigma_y \sigma_z} \tag{6.12}$$

Line source. When a continuous series of point sources moves along a line, as vehicles do along a road, they are equivalent to a continuous line source. If the x-axis is perpendicular to the road, then there is no variation of concentration in the y direction, and we have

$$q(x, z) = \sqrt{\frac{2}{\pi}} \frac{Q}{\bar{u} \sigma_z} \exp\left[\frac{-z^2}{2\sigma_z^2}\right] \tag{6.13}$$

where Q now has units of kg m^{-1} s^{-1}.

6.4   DISPERSION THEORY IN PRACTICE

Do not be misled into thinking that the neat-looking equations given above provide us with an infallible method of predicting dispersion – they fall well short of this ideal. Many field measurement programmes have been undertaken to validate both the plume rise and dispersion equations. A favourite method for evaluating plume dispersion is to release sulphur hexafluoride (SF_6) into the flue gases where they

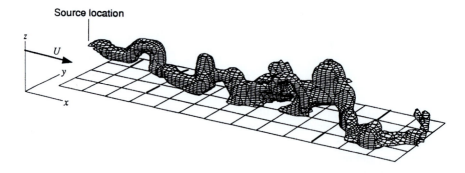

Source location

Figure 6.13 A modelled surface of constant concentration downwind of a continuous release.
Source: Henn, D. S. and Sykes, R. I. (1992) 'Large-eddy simulation of dispersion in the convective boundary layer', *Atmospheric Environment* 26A(17): 3145–3159.

enter the stack. SF$_6$ is a passive conservative tracer – it is very stable and there is no ambient background, so one can be sure that the concentration measured in the ambient air was emitted from the source under investigation and not from somewhere else altogether. Alternatively, the sulphur dioxide emissions from major commercial sources such as power stations have been monitored. The underlying problem in validation is the large spatial and temporal variability of the plume concentrations caused by the random nature of atmospheric turbulence. This is graphically illustrated by Figure 6.13, which is a computer simulation of plume geometry – the wiggly tube represents a snapshot of a surface of constant concentration. Dispersion theory must predict the average concentration at a point in space due to the highly variable exposure of that point to this fluctuating plume.

6.4.1 Worked example of Gaussian dispersion

We will apply the concepts that have been described above to predict the down-wind ground-level concentration of sulphur dioxide that will be produced from a chimney of known height. We are given the following information:

Source characteristics
Height 100 m
Internal radius 5 m
Exit speed 20 m s^{-1}
Exit temperature 80 °C = 353 K
Coal burn = 3000 t day^{-1}: S content = 1.4%

Environmental conditions
Windspeed at 10 m = 8 m s^{-1}

Weather – cloudy
Air temperature at stack height $= 10\,°C = 283$ K

Receptor location
6000 m downwind of the source, at ground level on flat terrain.

The data given above have been invented for this example, but are representative of a medium-sized coal-fired facility. The same Gaussian equations can be applied to a wide range of possible source configurations and weather conditions.
   Step 1. Find the stability category
   Use Table 6.3 to find the stability from the environmental conditions. Because it is windy and cloudy, we have Category D.
   Step 2. Calculate σ_y and σ_z
   Now go to Table 6.7. In open country, with stability category D and $x = 6000$ m, we have

$$\sigma_y = 0.08 \times 6000\,(1 + 0.0001 \times 6000)^{-0.5} = 379 \text{ m; and}$$
$$\sigma_z = 0.06 \times 6000\,(1 + 0.0015 \times 6000)^{-0.5} = 113 \text{ m.}$$

Step 3. Calculate the windspeed at the release height
   Use Equation (6.1) to find the windspeed at 100 m. From Table 6.1 for Category D, $p = 0.15$. Hence $u(100) = 8(100/10)^{0.15} = 11.3$ m s^{-1}
   Step 4. Calculate the plume rise and effective release height
From

$$F_b = w_0 R_0^2 \frac{g}{T_{p0}} (T_{p0} - T_{a0}),$$

the buoyancy flux $F_b = 20 \times 5^2 \times 9.81 \times (353-283)/353 = 973$ m^4 s^{-3}. Following Section 3.3.1, with Category D and $F_b > 55$, we use equation (6.6) to find the plume rise dh. Hence $dh = 39 \times 973^{0.6}/11.3 = 214$ m and $H = 100 + 214 = 314$ m.
   Step 5. Calculate the SO_2 release rate
   Coal burn $= 3000$ t day^{-1}, S content $= 1.4\%$
   Sulphur emission $= 0.014 \times 3000/(24 \times 60 \times 60)$ t s^{-1}
$$= 4.86 \times 10^{-4} \text{ t s}^{-1} = 4.86 \times 10^2 \text{ g S s}^{-1}$$
   Sulphur dioxide emission $= 2 \times$ S emission (because the molecular weight of SO_2 is twice that of S) $= 9.72 \times 10^2$ g s^{-1}.

Step 6. Solve the dispersion equation
For ground-level concentration on the plume centre-line, use equation (6.10).

$$q(x, 0, 0) = \frac{Q}{\pi \bar{u} \sigma_y \sigma_z} \exp\left[\frac{-H^2}{2\sigma_z^2}\right]$$

with $Q = 9.72 \times 10^2$ g s^{-1}, $u = 11.3$ m s^{-1}, $\sigma_y = 379$ m, $\sigma_z = 113$ m and $H = 314$ m.

Hence

$$q(6000, 0, 0) = \frac{9.72 \times 10^2}{\pi \times 11.3 \times 379 \times 113} \exp\left[-\frac{314^2}{2 \times 113^2}\right]$$

$$= (6.39 \times 10^{-4}) \times 0.021$$

$$= 13.4 \times 10^{-6}\,\text{g m}^{-3} = 13.4\,\mu\text{gm}^{-3}\ (\text{about 5 ppb})$$

6.4.2   Predictions for planning purposes

When a new industrial development involving pollutant emissions is proposed, calculation of pollutant concentrations can become an important component of the environmental impact assessment. Long-term exposures due to a point source are determined by accumulating average values for the 8760 hourly means of wind direction, windspeed, atmospheric stability and other environmental variables, and then running the dispersion model for each hour. The most important factor is the wind rose – the proportion of the time for which the wind blows from each direction (Figure 6.14).

Commonly, the 360° of the compass will be divided up into 16 sectors of 22.5° or 32 sectors of 11.25°, and the pollutant concentration determined while the wind is blowing from that sector. The resulting angular distribution is then called the pollutant rose. Since the statistical distribution of the concentrations follows

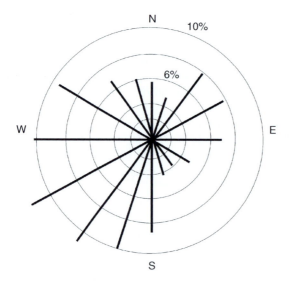

Figure 6.14 A typical windrose. The length of the bar shows the proportion of the time for which the wind was blowing from the sector centred on that direction.

a log-normal pattern, the correct description of concentrations is in terms of the median (the value exceeded for half the time, or 4380 h), together with the 10% (876 h) and 1% (88 h) exceedances. Very often organisms, including people, respond to the highest values that occur for the shortest periods, so that this knowledge of statistical distribution is important for predicting effects. Measurement of the pollutant rose also suggests attribution of the concentrations to a particular source. In many areas of the UK, winds come most often from the south-west to west sector and from the north-east, and this is reflected in the geographical distribution of concentrations around a source.

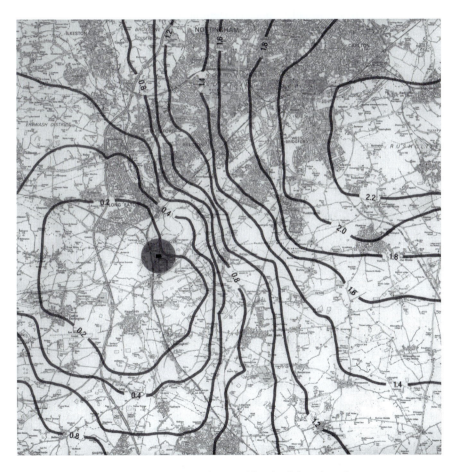

Figure 6.15 Annual average contours of ground-level sulphur dioxide concentration pre-dicted around the PowerGen power station at Ratcliffe-on-Soar. The power station is the rectangle within the dark circle. The urban area to the north-east is Nottingham. The contours are at intervals of 0.2 μg m^{-3}.
Source: PowerGen *Environment Statement for the Proposed Flue Gas Desulphurisation Plant, Ratcliffe-on-Soar Power Station*, PowerGen, UK.

Figure 6.15 shows an example of calculations made at the planning stage of a new Flue Gas Desulphurisation plant being designed for the PowerGen power station at Ratcliffe-on-Soar, Nottinghamshire, UK. Average dispersion conditions were calculated for each hour of the year, and a Gaussian model used to calculate the pollution contours for that hour. All the individual hours were then combined to give the annual average contours shown. The modelled contours of sulphur dioxide concentration show one peak to the north-east due to winds from the south-west, and a second, much lower, peak to the south-west due to the much lower proportion of wind from the north-east. Other, less obvious, factors also influence concentrations; for example, windspeed is correlated with wind direction, and higher windspeeds will usually reduce concentrations.

There may be complications in matching the predicted concentrations to guidelines and limit values. For example, the UK air quality standard for SO_2 specifies a 15-min mean of 266 $\mu g\ m^{-3}$ as the 99.9th percentile (equivalent to no more than 35 exceedances per year), whereas the Chimney Heights Memorandum (which must be used by Local Authorities to calculate stack heights for planning purposes) specifies 452 $\mu g\ m^{-3}$ as a 3-min average. In such cases, empirical factors derived from more complete data sets can be used to convert. For example, the factor 1.34 is recommended to convert the 99.9th percentile 15-min mean from short stacks to the 99.9th percentile 1-h mean. Unfortunately the factor itself varies with parameters such as stack height, adding another level of uncertainty to the predictions.

6.4.3   Effects of topography

The simplest dispersion theory deals with a source situated in the middle of a flat plain. Although such sources do exist, most landscapes involve hills of some sort, and for some landscapes the hills seriously affect dispersion.

Valleys tend to deviate and concentrate the wind flow, so that the average windspeed is higher than, and the wind blows along the valley for a higher proportion of the time than, would be expected in the open. There will also be thermal effects due to differential heating of the valley sides. For example, the north side of an east-west oriented valley will receive stronger solar radiation during the day than the south side. This asymmetry will generate a convective circulation which involves air rising up the relatively warm northern slope and descending down the relatively cool southern slope. At night, cool air may accumulate in the valley floor, producing a stable temperature inversion in which pollutants emitted from activities on the valley floor (which is where towns, roads and factories tend to be located) will accumulate.

The actual flow pattern of air over non-uniform terrain is not easy to predict or measure. For an isolated obstacle, the upwind streamlines are gradually compressed as the flow approaches the obstacle. At some point near the obstacle, the flow separates, with a turbulent and possibly recirculating region in the downstream wake where the wind direction will be opposite to the initial direction. At some further distance downstream the flows converge again. The wake region must be allowed for in pollutant release calculations; if an emission is released into such a recirculation zone, it can build up to much higher concentrations than simple dispersion theory would indicate. Air flow over hilly terrain will involve a mixture of processes. In smoother areas, the streamlines will follow the contours. In rougher areas there may be flow separation and turbulent wake formation under some combinations of wind direction and speed but not under others.

6.4.4 Effects of buildings

It would be unusual for any emission to be released from an isolated chimney, standing alone in the middle of a flat plain. More often, the chimney will be part of a complex of other buildings which disturb the flow in the boundary layer and change the dispersion of the chimney gases. The flow patterns are very complex, and only the main features will be indicated here. Furthermore, the predictions are part-theoretical, part empirical and part based on wind-tunnel studies which often do not reliably predict full-scale field results. The main features of the disturbed flow for air impinging directly onto the upwind face of a cuboidal building are seen in Figure 6.16.

- a stagnation point (SP) where streamlines impinge on the upwind face
- a recirculating flow region, separated from the main flow, immediately downwind of the downstream face. This recirculation may separate from the top edge of the upwind face, or it may attach to the roof and separate from the top edge of the downstream face. Concentrations are assumed to be uniform in this well-mixed region
- a downstream wake where the turbulence is enhanced and the average flow vector is down toward the ground
- a long trail of more structured (i.e. less randomised) vortices downwind.

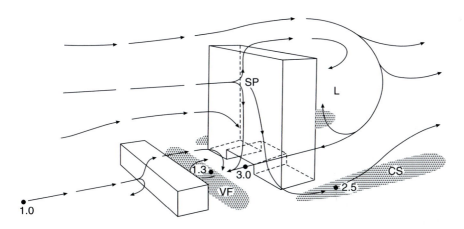

Figure 6.16 Idealised flow pattern round a cuboidal building. Values are relative wind speeds; SP = stagnation point; CS = corner stream; VF = vortex flow; L = lee eddy.

The effects of these disturbances on the dispersion of emissions from the building, or from nearby sources, depend sensitively on the exact position and characteristics of the source. For example, a weak (in the jet sense) source close to the roof may be entrained into the recirculation, and thence into the wake, or it may mix into the wake, or it may avoid both and mix into the undisturbed flow. Variable proportions of the plume may also experience these histories. If a roof level emission does not escape the recirculation, then its effective emission height will be zero. This can clearly have serious consequences for downwind pollution concentrations.

6.4.5   Recent developments

The Pasquill categorisation scheme was the first method to be widely adopted for relating plume standard deviation to atmospheric conditions, and has become the most generally used. Although it has proved its worth in many trials, it has certainly got important deficiencies. The method works best under stable and neutral conditions, but tends to fail under the more fluctuating conditions typical of an unstable atmosphere. In order to improve the predictions, changes have been made to the model. First, the values of σ_y and σ_z can be derived from fundamental characteristics of turbulence. If we measure the lateral and vertical components of windspeed (v and w respectively), we find that there is a normal distribution of speeds above and below the mean. The widths of these distributions can be characterised by their standard deviations, σ_v and σ_w, which can then be related in turn to the standard deviation of the concentration.

The Gaussian model is in widespread use because of its comparative simplicity, but it should be clear by now that it can only offer rather coarse predictions of dispersion in practice. More powerful models have been developed based on more rigorous turbulence theory. Although a detailed treatment is beyond the scope of this book, we can see the potential improvements. Figure 6.17 gives an example of how an SF_6 release experiment was used to test the differences in prediction performance of two dispersion models used in the US. In each graph, the measured 1-h average concentration c_o is plotted against the concentration predicted by the particular model, c_p. The 'pdf' model (Figure 6.17(a)) is based on the probability distribution function describing the relative abundance of convective updraughts and downdraughts in the advecting plume, while the CRSTER model (Figure 6.17(b)) is a conventional Gaussian dispersion model as described above. In this instance the pdf model clearly offered improved prediction power over the standard Gaussian. In particular, the concentration c_p predicted by the Gaussian model was zero on a large number of occasions when a concentration was actually observed (the column of points up the y axis of Figure 6.17(b)). The correlation coefficient of the pdf model was substantially higher than that of the Gaussian ($r^2 = 0.34$, 0.02 respectively).

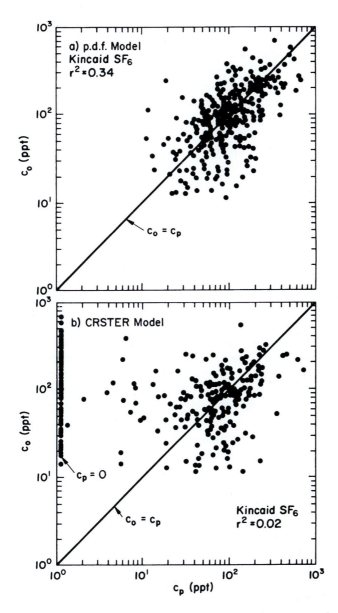

Figure 6.17 Comparisons of measured ground level concentrations with those predicted by: (a) the 'pdf' model; and (b) a standard Gaussian model.
Source: Weil, J. C., Sykes, R. I. and Venkatram, A. (1992) 'Evaluating air-quality models: review and outlook', *Journal of Applied Meteorology* 31(10): 1121–1145.

6.5   DISPERSION OF VEHICLE EMISSIONS

In Chapter 2, we discussed the complicated set of procedures that has to be undertaken in order to compile an estimate of vehicle emissions. After these source strengths (in g m^{-1} s^{-1}, for example) have been calculated, ambient concentrations can be found from dispersion theory. Away from built-up areas, for motorways, main roads and other roads in the open, the Gaussian method described above can be used. There are many implementations of the method in common use; for example, a model called CALINE was developed in California, and has evolved into CALINE4. The basic principles are quite straightforward. An urban network of roads is divided into links, where each link is a straight length of road having known emission characteristics. That link is treated as a line source for the calculation of concentrations at receptors (the points at which a concentration estimate is required). The total concentration at the receptor is the sum of the contributions due to all the individual links, plus any background concentration due to the unmapped parts of the area. There are certain features of the dispersion that are not catered for in the simple Gaussian method. For example, the effective source is regarded as the roadway itself, after initial dilution of the exhaust gases by mechanical (wake) and thermal (due to waste vehicle heat) turbulence. These effects are used to modify the dispersion parameters such as stability class.

In built-up urban areas, simple Gaussian dispersion theory does not apply – the sources are often at the bottom of 'street canyons' which reduce windspeeds below the ambient level, especially when the wind direction is almost perpendicular to that of the road. Under those conditions, a recirculating vortex can keep emissions confined within the canyon, so that road-level concentrations build up to several times the values they would have in the open. Several methods of analysis have been developed for this situation. One example is the Canyon Plume Box (CPB) model originally developed for the German EPA and subsequently refined by the US Federal Highway Administration. CPB models the vehicle-induced initial turbulence, plume transport and dispersion (Gaussian) on the lee side of the canyon, advective and turbulent exchange at the top of the canyon, pollutant recirculation, and clean air injection on the downwind side of the canyon. Allowance can also be made for the effects of canyon porosity – i.e. gaps in the canyon wall due to intersections, missing buildings or other irregularities.

The UK Meteorological Office has produced a simple model called Assessing the Environment of Locations In Urban Streets (AEOLIUS) to screen for worst case conditions in urban areas, on the assumption that the highest concentrations are produced by only two conditions – wind direction parallel or perpendicular to the road. As a result of this simplification they were able to reduce the complex computer modelling to a paper nomogram. First an uncorrected concentration is calculated for the prevailing windspeed and unit emissions. This is then multiplied by four correction factors estimated graphically from values for canyon height and width, and traffic flow and speed.

Modelling pollutant concentrations due to vehicles involves calculation of source strengths for a series of roadlinks, then dispersion in an urban area that usually consists of a mixture of open and canyon environments. At Nottingham, we developed a general-purpose suite of vehicle emission and dispersion models, called SBLINE. SBLINE uses a very detailed evaluation of emissions based on local measurements (if available) or default estimates. It is crucial to estimate this source strength as accurately as possible for the particular circumstances, making due provision for the components of the vehicle fleet (types of vehicle, year of legislation under which manufactured, or operating mode). If a receptor lies in a canyon, then CPB is used to find the contribution due to sources on that link, and NOTLINE for the contributions due to all other links. If the receptor lies on open ground, then NOTLINE is used throughout. This model has been validated against measured carbon monoxide concentrations in Leicester, UK. A specific part of Leicester was chosen within the City's Urban Traffic Control scheme, so that information on traffic flow was available. Three continuous carbon monoxide analysers were sited on London Road. Detailed surveys were carried out to build up profiles of traffic flows, vehicle age structure (from number plates), vehicle types, and periods spent in different operational modes. These data were input to SBLINE to predict the CO concentrations at the three receptor sites. Figure 6.18 shows the capability of the model in predicting both the average level and temporal variations in CO.

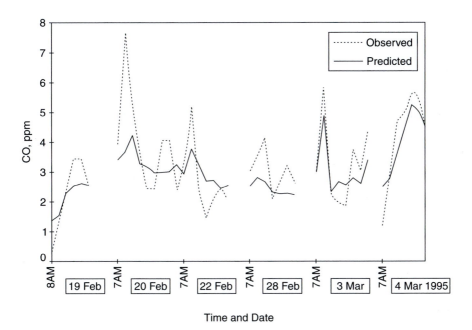

Figure 6.18 Comparison of the measured and modelled CO concentrations over the period 19 February–4 March 1995.
Source: A. Namdeo. (Pers. Com.)

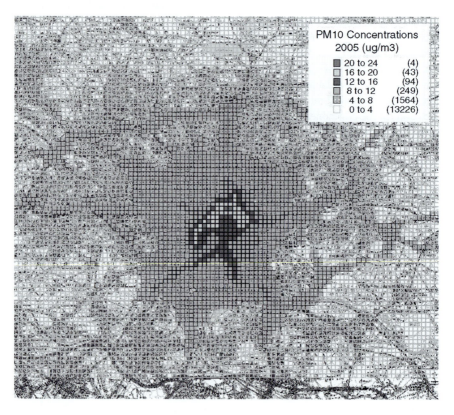

Figure 6.19 Predicted PM$_{10}$ concentrations around Leeds in 2005.
Source: A. Namdeo. (Pers. Com.)

This approach can be developed into a model of a whole urban area, such as that shown in Figure 6.19 for Leeds, UK. Having made predictions for 2005 of the traffic flows on the network of roads around Leeds, and allowing for factors such as the change in vehicle emission control by then, the variation of PM$_{10}$ concentrations across the city has been calculated. This kind of output can then be used to improve urban planning and meet air quality guidelines.

6.6   RECEPTOR MODELS

Receptor models are used to estimate the contribution of different sources to the concentration at a location (the receptor) using statistical evaluation of the data collected at that receptor. This process is called source attribution or source apportionment. Receptor models have been used most intensively for the attribution of sources to explain particle concentrations, because the distinctive chemical fingerprints of particles enable them to be identified with specific emissions. In some ways the process is the reverse of dispersion modelling, because the source is worked out from the concentration, rather than *vice versa*. The first step is to collect a large data set of concentration measurements at the receptor, using as comprehensive analysis as possible. Typically elemental, organic and gaseous concentrations would be measured, as well as particle concentration and possibly size distributions. Intensive and sophisticated analytical methods would often be used, such as X-ray fluorescence, ICP-MS, PIXE and neutron activation analysis. Certain elemental tracers may be used to identify sources, such as lead for petrol, but in that case the link needs to be unambiguous. Hydrocarbons such as ethane and propane may be used as tracers for exhaust or petrol storage tank emissions, PAH for wood combustion and diesel smoke, toluene for solvents. In chemical mass balance methods, data on chemical composition of the ambient measurements is related to that of known emissions. It is assumed that the elements or compounds are not reactive in the atmosphere. Clearly the chemical and physical characteristics of the measured pollutants emitted from different potential sources must be different, and those characteristics must be preserved during transit through the atmosphere to the receptor. By using more powerful statistical methods, such as principal component analysis (PCA) with multiple linear regression, the sample properties alone can be used to work out the number of source categories, the chemical composition of their emissions and their relative contributions to the measured concentrations. This is necessary if some sources have not been identified or if their emission characteristics are poorly defined – as with roads.

6.6.1   Chemical mass balance models

Chemical mass balance was the first receptor modelling method developed for source attribution. It uses two data sets: (1) the chemical composition of the atmospheric aerosol at a measurement site; and (2) the chemical composition of the aerosol emitted from the principal sources in the region. The material collected, for example on a high-volume filter paper, is composed of contributions from various sources

$$\rho = \sum_{j}^{p} m_j \, (j = 1, 2, \ldots p)$$

where ρ is the local aerosol mass concentration, m_j is the contribution from source j, and there are p sources altogether. In practice, there may be some discrete individual sources if large local ones exist, and other smaller sources will be lumped and treated collectively as domestic, vehicular etc. It is also normally assumed that the overall chemical balance of the sources is the same as that at the receptor – i.e. that there have been no changes due to selective deposition or gas/particle conversion.

The concentration ρ_i of the ith chemical component in the aerosol sample is related to the source contributions by

$$\rho_i = \sum_j^p c_{ij} m_j \, (i = 1, 2, \ldots n)$$

where c_{ij}, which is the mass fraction of component i in m_j, is the source concentration matrix, and there are n chemical components. The main task is to invert this equation to give the source contributions m_j. This could be done by varying the source contributions to minimise the weighted mean square deviations. The method has been applied to a wide variety of source attribution problems, from single point sources to complex urban and regional airsheds and Arctic air quality.

6.7 BOX MODELS

Box models are the simplest ones in use. As the name implies, the principle is to identify an area of the ground, usually rectangular, as the lower face of a cuboid which extends upward into the atmosphere. A budget is then made on the box over a specified timestep, for the pollutant(s) of interest. This budget will involve the initial contents of the box due to the prevailing pollutant concentration, emissions into the box from sources, deposition from the box to surfaces, additions/losses to/from the box due to wind and concentration gradients, and gains/losses due to chemical reactions. At the end of the timestep, a new average concentration within the box is calculated, and the process is repeated to track temporal variation of the pollutant concentration. Spatial variability is calculated by having a grid of adjacent boxes of appropriate size, and vertical variations by extending the boxes upwards using the same principles.

6.8 STATISTICAL MODELS

Statistical models do not involve emissions at all. Data on concentrations at a location are collected, together with data on local factors that might affect those concentrations. Such factors might include wind direction and speed, solar radiation, air temperature, time of day, day of the week, season. Statistical correlations are made to suggest what proportion of the variability of the concentration about the mean value is due to each of the factors. Then if a value for each of the factors is known, an estimate can be made of the likely concentration. Because it is based on historical data sets, the prediction assumes that the influence of the factors on concentration will be the same in the future. This is likely to be true on a few-day timescale; of course, each new day's data set may be input to the model so that any gradual changes in the effects of the different factors are incorporated.

FURTHER READING

Arya, S. P. (1999) *Air Pollution Meteorology and Dispersion*, Oxford University Press, New York.

Lyons, T. J. and Scott, W. D. (1990) *Principles of Air Pollution Meteorology*, Belhaven Press, London, UK.

Pasquill, F. and Smith, F. B. (1983) *Atmospheric Diffusion*, Ellis Horwood, Chichester, UK.

Turner, D. B. (1994) *Workbook of Atmospheric Dispersion Estimates*, CRC Press, Boca Raton, Florida.

UK Department of the Environment, Transport and the Regions (1998) *Selection and Use of Dispersion Models*, Stationery Office, London.

Chapter 7

Analysis of an air quality data set

In this Chapter we will make a detailed analysis of a comprehensive set of air pollution concentrations and their associated meteorological measurements. I would like to thank Professor David Fowler and Dr Robert Storeton-West of the Centre for Ecology and Hydrology (CEH) for their help in supplying the data. The measurements were taken over the period 1 January–31 December 1993 at an automatic monitoring station operated by CEH at their Bush Estate research station, Penicuik, Midlothian, Scotland. Three gas analysers provided measurements of O_3, SO_2, NO, and NO_x; NO_2 was obtained as the difference between NO_x and NO. Windspeed, wind direction, air temperature and solar radiation were measured by a small weather station. The signals from the instruments were sampled every 5s by a data logger, and hourly average values calculated and stored.

7.1   THE RAW DATA SET

There were 8760 hours in the year 1993. Hence a full data set would involve 8760 means of each of the nine quantities, or 78 840 data values in all. An initial inspection of the data set showed that it was incomplete. This is quite normal for air quality data – there are many reasons for loss of data, such as instrument or power failures or planned calibration periods. Any missing values have to be catered for in subsequent data processing. The data availability for this particular data set is given in Table 7.1.

Many of these lost values were due to random faults and were uniformly distributed throughout the year. The reliability of gas analysers depends largely on the simplicity of their design, and the ranking of the analysers in this example is quite typical. Ozone analysers based on UV absorption are very straightforward instruments that rarely fail provided that regular servicing is carried out. UV fluorescence sulphur dioxide analysers are rather more complicated, and NO_x analysers even more so. The lower availability for the nitrogen oxides in this case was in fact due to an extended period at the start of the year when problems were being experienced with the analyser.

Table 7.1 Data availability for the 1993 CEH data set

Measurement	Number of hours available	Percentage of 8760
Ozone	8569	97.8
Sulphur dioxide	8251	94.2
Nitric oxide	7010	80.2
Nitrogen oxides	7010	80.2
Nitrogen dioxide	7010	80.2
Windspeed	8443	96.4
Wind direction	8459	96.6
Air temperature	8650	98.7
Solar radiation	8642	98.7

The raw data for gas concentrations, wind speed and wind direction are shown as time sequences in Figure 7.1. Plotting the data in this form is an excellent rapid check on whether there are serious outliers (data values lying so far outside the normal range that they are probably spurious). Differences in the general trends of the concentrations through the year are also apparent. The O_3 concentration (Figure 7.1(a)) rose to a general peak in April–May before declining steadily through to November, and the most common values at any time were around half the maxima of the hourly means. Sulphur dioxide concentrations (Figure 7.1(b)) were typically much smaller, although the maxima were nearly as great as for ozone. Typical NO concentrations (Figure 7.1(c)) were low throughout the year, although the intermittent maxima were higher than for the other gases. NO_2 concentrations (Figure 7.1(d)) were high around May and November, and particularly low in June/July. There were occasions when the concentrations of any of these gases increased and declined very rapidly – they appear as vertical lines of points on the time series. Superficially, there does not appear to be any systematic relationship between the timing of these occasions for the different gases. The windspeed (Figure 7.1(e)) declined during the first half of the year and then remained low. The wind direction (Figure 7.1(f)) was very unevenly distributed, being mainly from around either 200° (just west of south), or 0° (north).

7.2   PERIOD AVERAGES

The plots shown in Figure 7.1 give an immediate impression of the variations during the year, but are not of much use for summarising the values – for comparison with other sites or with legislated standards, for example. We therefore need to undertake further data processing. The most straightforward approach is to explore the time variations by averaging the hourly means over different periods. First, we need to decide how to handle those missing values.

(a)

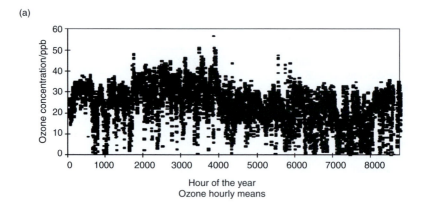

Hour of the year
Ozone hourly means

(b)

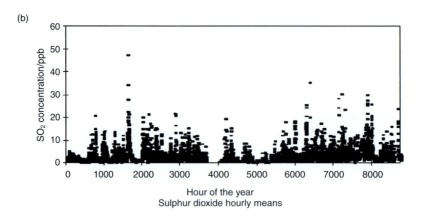

Hour of the year
Sulphur dioxide hourly means

(c)

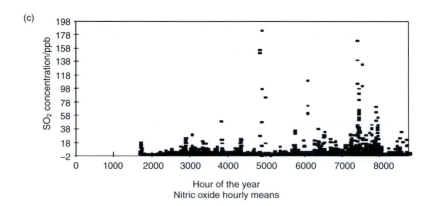

Hour of the year
Nitric oxide hourly means

Figure 7.1 Time series of hourly means for the 1993 CEH data set.

(d)

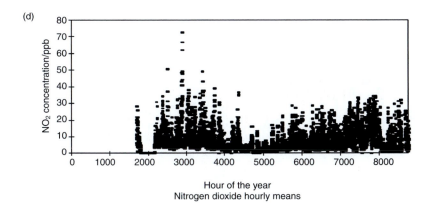

Hour of the year
Nitrogen dioxide hourly means

(e)

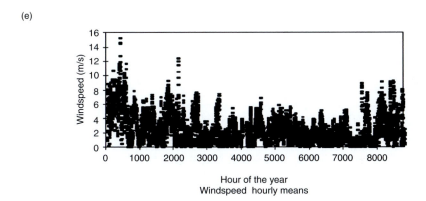

Hour of the year
Windspeed hourly means

(f)

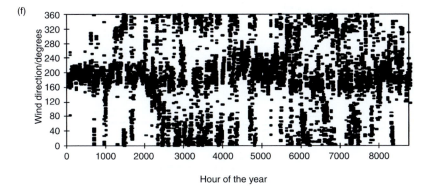

Hour of the year

Figure 7.1 Continued.

(a)

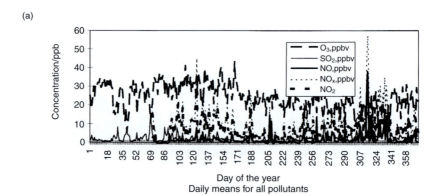

Day of the year
Daily means for all pollutants

(b)

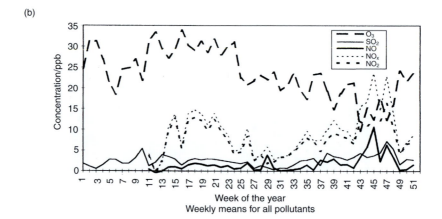

Week of the year
Weekly means for all pollutants

(c)

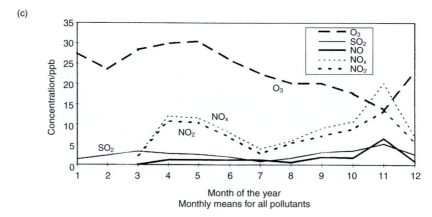

Month of the year
Monthly means for all pollutants

Figure 7.2 Time series of: (a) daily, (b) weekly and (c) monthly means for the 1993 CEH dataset.

When we plotted the (nominal) 8760 h means, a few missing values did not have a big effect on the appearance. As we average over longer periods, the number of values decreases to 365 daily means, 52 weekly means and only 12 monthly means, so that the effect of missing values becomes proportionately greater. Before we start averaging, we must decide on a protocol that includes as much of the data as possible, but does not create an average value when there is simply not enough data to justify it. For example, consider the calculation of a daily average from the 24 individual hourly averages that contribute to it. If one value is missing, and the sequence of values before and after is varying smoothly, then it is legitimate to substitute the missing value with the average of the adjacent values. If the sequence varies erratically, this procedure serves little purpose. Instead, we can ignore the missing value, and calculate the daily mean as the average of the remaining 23 h, arguing that the whole day was still well repre-sented. If only 12 or 8 h remain (as might happen if a faulty instrument was reinstated in the early afternoon), then this argument loses credibility and the whole day should be discarded. The same idea can be applied to the calculation of weekly, monthly and annual averages, with a requirement that, say, 75% of the contributing values be present if the average is to be calculated. This philosophy must particularly be adhered to when the data is missing in blocks, so that no measurements have been taken over significant proportions of the averaging period. For example, it is clear from Figure 7.1(d) that the annual average NO_2 concentration would not include any of January or February, and therefore might not be representative of the year as a whole.

In Figure 7.2(a–c) the 1993 data for gas concentrations are presented as daily, weekly and monthly averages respectively. The short-term variations are succes-sively reduced by the longer averaging periods, and the trends that we originally estimated from the raw data become clearer.

7.3   ROSES

In Section 6.1.1 we discussed the influence of wind direction on the pollutant concentration at a point. We have analysed the CEH dataset specifically to highlight any such dependencies. The 360° of the compass were divided into 16 sectors of 22.5° each. The hourly means taken when the wind was from each sector were then averaged, and the values plotted in the form shown in Figure 7.3. These diagrams are known as roses or rosettes. Figure 7.3(a) shows that the wind direction was usually from between South and South-west, and Figure 7.3(b) that these were also the winds with the highest average speeds. Figure 7.3(c) gives the ozone rose for the year – the almost circular pattern is expected because ozone is formed in the atmosphere on a geographical scale of tens of km, rather than being emitted from local sources. Hence the concentration should be relatively free of directional dependence. Sulphur dioxide, on the other hand, is a primary pollutant which will influence

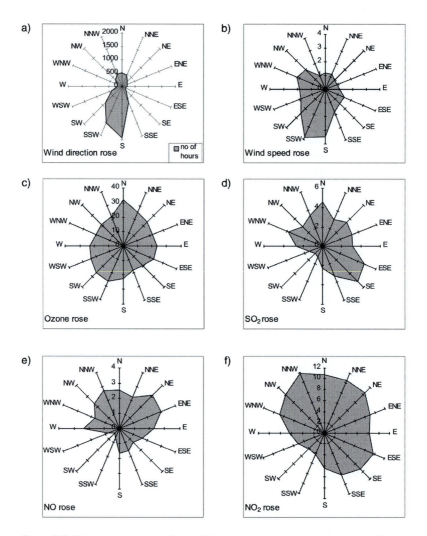

Figure 7.3 Direction roses of wind frequency, wind speed and pollutant gas concentration

concentrations downwind of specific sources. Figure 7.3(d) indicates possible sources to the North, South-east and West-north-west of the measurement site. Concentrations from the sector between South and South-west – the predominant wind direction – are the lowest. The roses for NO (Figure 7.3(e)) and NO_2 (Figure 7.3(f)) are not so well defined. These gases are both primary pollutants (dominated by NO) and secondary pollutants (dominated by NO_2). Hence we can see both patterns of directional dependence, with NO behaving more like SO_2, and NO_2 more like O_3.

Figure 7.4 shows the location of the monitoring site in relation to local topographical features and pollution sources, knowledge of which can help to understand the pollution data. Two factors strongly influence the wind rose – the Pentland Hills run north-east–south-west, and the Firth of Forth estuary generates north–south sea breezes. The combined effect of both factors produces the strongly south to south-west wind rose which was seen in Figure 7.3(a). The main sources responsible for the primary pollutants are urban areas and roads. The city of Edinburgh, which lies 10 km to the north, generates the northerly SO_2 peak seen on Figure 7.3(d). Although the small town of Penicuik lies close to the south, there is apparently no SO_2 peak from that direction, nor is there an identifiable source responsible for the south-east SO_2 peak. Detailed interpretation of such data cannot be made without a detailed source inventory, since weak low sources close to the monitor can produce similar signals to those from stronger higher more distant sources.

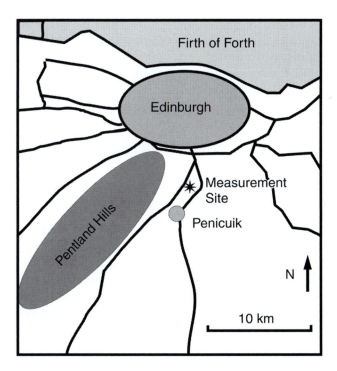

Figure 7.4 The topography, urban areas and roads around the CEH measurement site at Penicuik.

7.4   DIURNAL VARIATIONS

Another way of gaining insight into pollutant occurrence is to average the concentrations according to the hour of the day. We must be careful, though, to allow for the ways in which the characteristics of the days themselves change during the year. In Figures 7.5 and 7.6, we have carried out diurnal analyses for the months of June and December respectively.

In June, the period of daylight is long and the peak solar radiation high (Figure 7.5(a)). The air temperature is warm (Figure 7.5(b)) and shows a pronounced afternoon increase in response to the solar radiation. Average windspeeds are low, with convective winds increasing during daylight hours (Figure 7.5(c)). The ozone concentration (Figure 7.5(e)) shows a background level of about 23 ppb, with photochemical production increasing this concentration to a peak of 30 ppb at around 1600. The diurnal variation of SO_2 (Figure 7.5(f)) is quite different – there are sharp peaks centred on 1000 and 1700 which result from local emissions, and no clear dependence on solar radiation. As with the pollutant roses, the diurnal variations of NO and NO_2 are a blend of these two behaviours.

Figure 7.6(a–h) show the corresponding variations during December. Now, the days are short and solar energy input is a minimum. Air temperatures are low and barely respond to the sun, windspeeds are higher and almost constant through the day. As a consequence of these changes, ozone shows no photochemical production in the afternoon. Somewhat surprisingly, SO_2 has lost all trace of the 1000 and 1700 spikes, although these remain very clear for NO. The pattern for NO_2 is very similar in December and June.

7.5   SHORT-TERM EVENTS

So far in this chapter, we have grouped data in different ways specifically to smooth out short-term variations and clarify patterns. We can also benefit from a detailed look at shorter periods of measurements. In Figure 7.7 are shown the time series for one particular period of 300 h (between 4100 and 4400 h, or roughly from the 20 June to the 3 July). The wind direction was generally southerly, except for two periods of about 50 h each when it swung to the north and back several times. When the wind direction changed, there were bursts of higher concentrations of NO, NO_2 and SO_2, and the O_3 background concentration was disturbed. These changes were probably associated with emissions from a local source that was only upwind of the measurement site when the wind was from one particular narrow range of directions. This would not only bring the primary pollutants, but also excess NO to react with the O_3 and reduce the concentration of the latter.

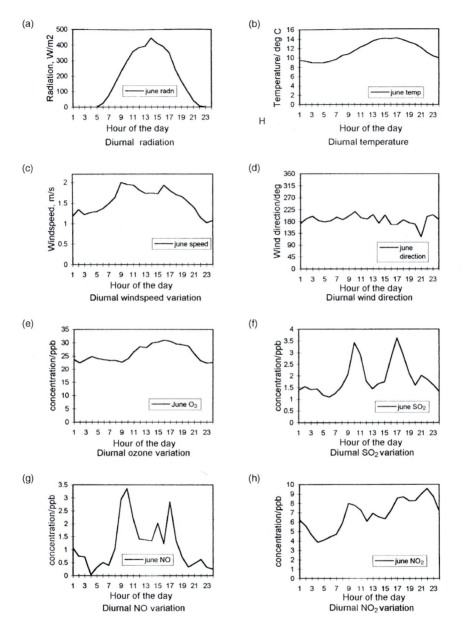

Figure 7.5 Average diurnal variations in June.

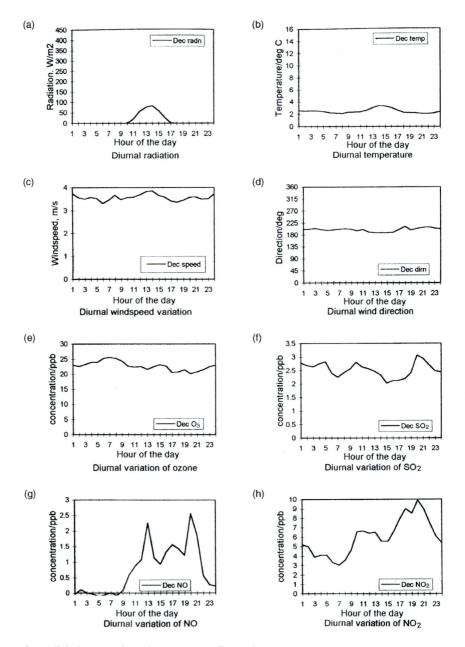

Figure 7.6 Average diurnal variations in December.

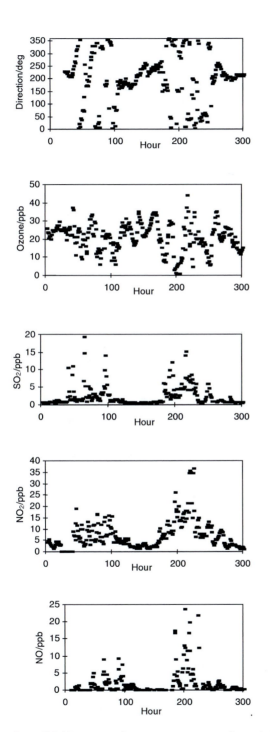

Figure 7.7 Variations of gas concentration with wind direction over a single period of 300 h.

7.6   FREQUENCY DISTRIBUTIONS

As discussed in Chapter 4, the concentrations of air pollutants often show a log-normal frequency distribution – i.e., the logarithms of the concentrations are distributed normally. We have analysed the hourly and daily means from the Centre for Ecology and Hydrology data set to confirm this. The overall range of the concentrations that occurred over the year was divided into subranges, and the number of values that fell within each subrange was counted. This is the frequency distribution. These counts were then expressed as a percentage of the total number, and summed by subrange to give the cumulative frequency distribution. If the frequency distribution is log-normal, then the cumulative distribution plots as a straight line on log-probability axes. In Figures 7.8 and 7.9, the distributions for the hourly and daily means of the five gases are shown. Those for SO_2, NO and NO_x are quite linear, NO_2 less so, and O_3 not at all. The O_3 distribution is characteristic

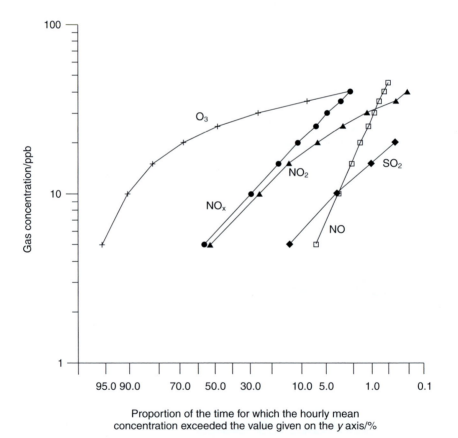

Figure 7.8 Cumulative frequency distributions of hourly-mean pollutant concentrations.

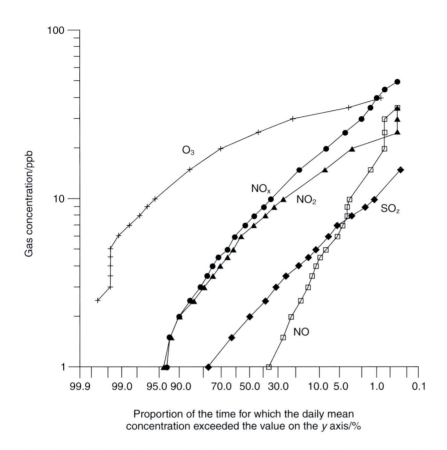

Figure 7.9 Cumulative frequency distributions of daily-mean pollutant concentrations.

of a variable that has a significant background component – the concentration was not low for as high a proportion of the time as the log-normal form requires.

Certain statistical parameters can be derived from the log-probability curves. Commonly quoted are the concentrations below which the value falls for 50% (median), 90%, 98% and 99% of the time. The 98% value of daily means is used in European Union Directives on air quality – it is equivalent to stating that the concentration should not be exceeded for more than seven days in the year. The values indicated by Figures 7.8 and 7.9 are extracted in Table 7.2. It is clear that one such parameter alone is not sufficient to define the distribution. If the distribution is linear, we can measure the gradient, which is equivalent to the standard geometric deviation of the sample. Then the median and gradient completely define the population distribution. There are more complex formulations of the log-normal distribution that can be used to describe non-linear data sets.

Table 7.2 Percentiles of hourly and daily means

Pollutant	Hourly means				Daily means			
	50	90	98	99	50	90	98	99
		(per cent)				(per cent)		
O_3	24	34	40	42	24	32	36	38
SO_2	1.5	6	13	15	2	5	9	11
NO	<1	3	16	27	<1	4	13	18
NO_x	5	21	41	56	7	18	30	39
NO_2	5	17	27	31	7	14	22	23

7.7   FURTHER STATISTICAL ANALYSES

Other standard statistical parameters can be used to describe the data. The summary statistics for the 1993 data are given in Table 7.3.

Finally, we can apply the ideas on the relationships between the period maxima that were outlined in Chapter 4. If the maximum 1-h concentration is $C_{max,1\,h}$, and the maximum over any other period t is $C_{max,t}$, then we should find that $C_{max,t} = C_{max,1\,h}\, t^q$, where q is an exponent for the particular gas. For the

Table 7.3 Summary statistics for the 1993 CEH data set

Gas	Hourly means/ppb			Daily means/ppb		Weekly means/ppb	
	Mean	Median	Standard deviation	Median	Standard deviation	Median	Standard deviation
O_3	23.6	24.7	9.2	23.8	7.4	23.4	5.8
SO_2	2.6	1.6	3.1	2.0	2.2	2.7	1.4
NO	1.7	0.3	7.5	0.7	3.7	1.3	2.0
NO_x	9.2	5.8	11.6	7.5	7.8	8.9	5.3
NO_2	7.5	5.3	6.9	6.7	5.2	7.8	27.2

Table 7.4 Values of $C_{max,t}$ for the different pollutants

$C_{max,t}$	Pollutant				
	O_3	SO_2	NO	NO_x	NO_2
$C_{max,\,1\,h}$	57	47	185	186	72
$C_{max,\,1\,day}$	43	16	38	57	37
$C_{max,\,1\,week}$	34	7	11	24	16
$C_{max,\,1\,month}$	31	5	7	20	13
q	−0.095	−0.35	−0.51	−0.35	−0.27

1993 data set, the maximum values over the different averaging periods are shown in Table 7.4. Plotting log $C_{\mathrm{max},t}$ against log t gives the results shown in Figure 7.10, in which the gradients of the lines give the values of q for the different gases.

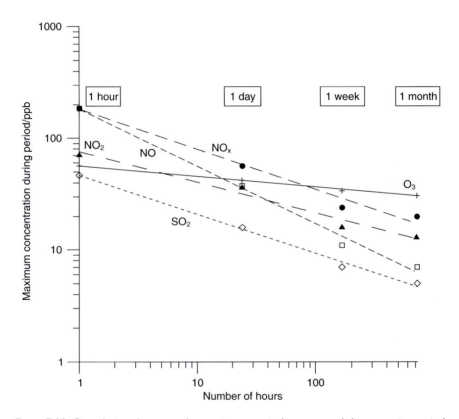

Figure 7.10 Correlations between the maximum period average and the averaging period.

Chapter 8

Indoor air quality

Not only do most people in the most polluted regions of the world live in urban areas, but they also spend nearly 90% of their time inside buildings, a further few per cent in vehicles and only around 6% outdoors, so in this Chapter we shall look briefly at special features of indoor air quality. Despite its importance and complexity, the indoor environment has been much less thoroughly studied than that outdoors. The general perception is that one is 'safe' from air pollutants indoors, and many who are today beware of high NO_2 concentrations near roads will not worry about cooking with a gas hob that potentially gives them an even higher NO_2 exposure. The sources of air pollution that have been discussed so far have been characterised by their release into the atmosphere at large. In general, the concentrations of these pollutants inside buildings will be lower than the concentration outside, because the pollutants will be deposited to internal surfaces during the residence time of air in the building. However, there are many additional sources inside buildings, and the same reduced air turnover that protects from external pollutants may cause the concentrations from internal sources to rise to unacceptable levels. Furthermore, the species involved include not only the same gases and particles that penetrate in from the outside, but a whole range of new pollutants that only increase to significant concentrations in the confines of the indoor environment. The sources include people and domestic animals, tobacco smoke, cooking (especially wood-burning stoves), heating and lighting, building and decorating materials, aerosol sprays, micro-organisms, moulds and fungi. The concentration of any particular pollutant at any time will be the resultant of the external and internal source strengths, deposition velocities to different surfaces, air turbulence and building ventilation rate.

Correct measurement of personal exposure is a science in itself. Often the pollutant will be very non-uniformly distributed both in space and time. Many of the instruments that are used for outdoor ambient pollutant monitoring will therefore be unsuitable. The main uptake routes for airborne pollutants are through the nose and mouth, so this is where the concentrations need to be measured. Most real-time analysers are bench-based, mains-powered items which cannot be carried around. Although devices such as diffusion tubes are highly portable and take no power, they offer only time-integrated measurements which give no information about short-term peaks. There is a major need for real-time, highly portable gas and particle instrumentation which can be used to assess personal exposure. As an added complication, it has also been found that personal exposure to many pollutants can exceed that predicted from the ambient concentration – this has been called the personal cloud effect.

8.1   BUILDING VENTILATION

Ventilation of buildings controls the rate of exchange of inside air with outside air, and hence not only the ingress of outside pollutants into the building, but the release of indoor pollutants from the building. Controlled ventilation of buildings is a compromise between energy costs, maintenance of comfort and air quality. If a building is well sealed, energy efficiency will be high but emissions from occupants and other sources will quickly become unacceptable. In many buildings, there are significant routes for air infiltration (i.e. leaks, as distinct from ventilation) other than through doors and windows. This infiltration is either advective or convective. Advective flow is driven by pressure gradients across the building, so that it increases with wind speed. Convective flow is driven by the density difference between the building air and ambient air, so that it increases with temperature difference.

The most basic model of building air exchange involves an enclosure with two vents (Figure 8.1). One vent represents the sum of all the areas (whether specific controlled ventilation devices or simply gaps round doors and windows) through which at any one time air is leaking inwards. The other vent is a corresponding one for air leaking outwards. The air venting inwards has the properties of the outdoor ambient air – temperature, humidity, pollutant concentrations etc. The air venting outward has the corresponding properties of the room air. The mass and volume flow rates of air through the two vents must be equal, or the room would become pressurised with respect to ambient.

During a short time interval dt, the mass of pollutant in the room increases by an amount dm, where dm = inward ventilation mass flow + internal production − outward ventilation mass flow.

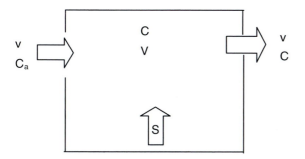

Figure 8.1 Basic building venting.

Hence

$$dm = VdC = vC_a dt + Sdt - vCdt$$

where V is the room volume, dC is the small change in concentration C during time dt, v is the ventilation rate, C_a is the ambient (outside) concentration and S is the internal source strength of that pollutant. It is assumed here that the compound is stable (does not decay or react) and does not deposit to surfaces, and that the air in the room is well mixed so that the concentration is C everywhere. Hence

$$\frac{dC}{C_a + \dfrac{S}{v} - C} = \frac{v}{V} dt$$

which integrates to give

$$C_t = C_a + \frac{S}{v} - \left(C_a + \frac{S}{v} - C_0 \right) e^{-\frac{vt}{V}}$$

where C_t is the concentration at time t, and C_0 is the initial concentration at $t = 0$. Inspect the behaviour of this equation: if $C_a = S = 0$ (outside concentration negligible and no internal sources), then we have

$$C_t = C_0\, e^{-\frac{vt}{V}}$$

and the indoor concentration decays exponentially from its initial value. This equation is used for measuring the ventilation rate of a room, provided the air is well mixed. For example, the concentration of an inert tracer such as CO_2 or N_2O is raised by releasing some from a bottle into the room. When the supply is turned off, the decay in concentration is measured over a period of an hour or so.

Since $\log_e C_t = \log_e C_0 - (v/V)t$, a graph of $\log_e C_t$ against time will have a gradient of $-v/V$ from which v can be determined. In fact the result is often left as v/V, which has units of s^{-1}. This represents the air turnover rate in the room, and is usually expressed in air changes per hour (ACH).

If $S \neq 0$ in the ventilation equation, an equilibrium concentration C_{equ} will be achieved, at which there is as much pollutant leaving in the ventilation air as there is entering and being generated. Then $C_{equ} = C_a + S/v$. Note that this equilibrium concentration is independent of the room volume.

Additional pollutant behaviour can be added to this basic method. If a gas such as ozone, which deposits readily to surfaces, is involved, then a term would be included which specified the rate of loss in terms of appropriate deposition velocities to the carpet, paintwork and other materials. If particles were involved that were large enough to have significant sedimentation velocities, then corresponding loss terms would be used. For example, if a particle of diameter d has a deposition velocity of $v_{d,t}$ and a concentration at time t of $C_{d,t}$, then the exponential decay will occur with a time constant of $(v + Av_{d,t})/V$, where A is the surface area to which deposition can occur. Hence we could use the decay rate of gas concentration to measure the ventilation rate v, and the decay rate of $C_{d,t}$ to measure $v_{d,t}$.

Figure 8.2 shows the results of an informal test of these relationships. The particle concentration in a kitchen was artificially increased by burning some toast (deliberately, for once). Then the ventilation rate was set at a high constant value by turning on an extractor fan, and the concentration measured by an optical particle counter which generates size distributions as one-minute averages. The three particle size ranges shown in Figure 8.2(a) are inhalable (all particles less than 50 μm), thoracic (roughly equivalent to PM_{10}), and alveolar (about PM_4). The concentrations increased to a sharp peak at around 78 min, then decayed roughly exponentially between the dashed vertical time limits. The increases in concentration before the peak and after 110 min were due to moving around in the kitchen, rather than burning the toast; this mechanical process generated inhalable and thoracic fractions but little alveolar material.

Inhalable includes thoracic and alveolar, while thoracic includes alveolar. Hence the size classes were subtracted to generate three narrower ranges, which are referred to here as coarse (inhalable minus thoracic), medium (thoracic minus alveolar) and fine (alveolar). These three ranges are plotted in Figure 8.2(b), which shows much more clearly that the burning toast was generating fine particles. Natural logs of the three size classes were then plotted (Figure 8.2(c)), for the exponential decay period. The gradients are in the order that we would expect if sedimentation was increasing the loss of larger particles above the rate due to ventilation alone. The gradient of the fine particle regression line (which is assumed to be by ventilation alone) gives a ventilation rate of 0.12 min^{-1}, or 7 ACH. Using a floor area of 16 m^2 and a room volume of 40 m^2 gives deposition velocities for the medium and coarse fractions of 2.7 and 8.9 mm s^{-1}, respectively.

The indoor concentrations of some pollutants will be affected mainly by their penetration into the room from outdoors and their subsequent deposition to internal surfaces. For other pollutants there will be significant indoor sources and concentrations will be determined by the rate at which emissions from those sources are diluted by the outdoor air that ventilates the room. The most effective ventilation rate for indoor pollution control will vary with the circumstances. For example, a high ventilation rate will prevent the build-up of VOC concentrations from new paintwork or carpet, but encourage the penetration of particles from outside.

Personal exposure of an individual is computed as the sum of the exposures in different microenvironments, each of which has a characteristic concentration of the pollutant or pollutants:

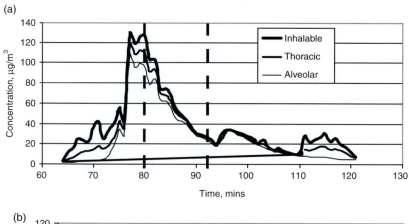

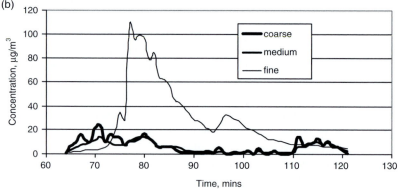

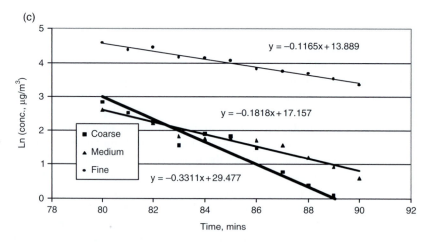

Figure 8.2 Particle concentration decay.

$$E_i = \sum_{j=i}^{m} E_{ij} = \sum_{j=1}^{m} f_{ij} \times C_{ij}$$

where E_{ij} is the exposure of individual i who spends a fraction of their time f_{ij} in microenvironment j, in which the pollutant concentration while they are there is C_{ij}. Certain severe microenvironments will contribute disproportionately to the total exposure – for example if someone in the room is smoking, or if there is cooking, or if new paint or other VOC-emitting substances are in the room. The number of microenvironments should be kept to a minimum, subject to adequately representing the exposure of the individual. National studies have shown that in industrialised countries, people average around 87% of their time indoors, 7% in transit and only 6% outdoors. Hence full representation of the indoor environment is vital if the overall exposure estimate is to be meaningful.

The particle Total Exposure Assessment Methodology (TEAM) study in the US calculated the indoor particle concentration as a function of ventilation parameters and contributions from outdoor air, environmental tobacco smoke, cooking and other unaccounted sources.

$$C_{in} = \frac{P \times a}{a + k} C_{out} + \frac{N_{cig} S_{smoke} + t_{cook} S_{cook}}{(a + k) V t_{sample}} + \frac{S_{other}}{(a + k)V}$$

where

C_{in} = pollutant concentration indoors ($\mu g\ m^{-3}$),
C_{out} = pollutant concentration outdoors ($\mu g\ m^{-3}$),
P = penetration fraction (no units),
a = air exchange rate (h^{-1}),
k = pollutant decay or deposition rate (h^{-1}),
N_{cig} = number of cigarettes smoked during the period,
V = house volume (m^3),
t_{cook} = cooking time in hours,
t_{sample} = sampling time in hours,
S_{smoke} = source strength for cigarette smoking (μg/cigarette),
S_{cook} = source strength for cooking ($\mu g\ h^{-1}$),
S_{other} = source strength for other indoor sources, including resuspension ($\mu g\ h^{-1}$).

Note that the units of the three terms on the right hand side are all $\mu g\ m^{-3}$, as they have to be if we are going to add them together. Factors in the numerator, such as source strengths and penetration, act to increase the concentration. The air exchange rate acts to increase the concentration due to outdoor particles but to decrease it if the sources are internal. The higher the deposition rate k, or house volume V, the lower the concentration.

Relationships like this have been used to show that indoor/outdoor ratios for PM_{10} are about 0.5, rising to 0.7 for $PM_{2.5}$. The value for $PM_{2.5}$ is higher because the deposition rate is lower (around 0.4 h^{-1} against 0.65 h^{-1} for PM_{10}). Indoor

NO_2 concentrations are greatly influenced by whether there is a permanent pilot light on the gas cooker, which can add 5–10 ppb to the average concentration. NO_2 is not a highly reactive gas, and indoor concentrations will be around 0.6 of those outdoors if there are no internal sources. In contrast, O_3 concentrations are greatly reduced when outdoor air comes indoors, because O_3 is deposited rapidly, with a decay rate of around 2.8 h^{-1}, and indoor concentrations in houses with the windows closed might only be 0.1–0.2 of those outdoors. Indoor VOC concentrations, on the other hand, depend on which VOCs are included. For a new building, within which paint solvents, furnishings and carpet glue are strong sources, the concentration may be a hundred times ambient. Many other indoor sources contribute to VOC concentrations, including smoking, dry-cleaned clothes, air fresheners, house cleaning materials, chlorinated water, paints, adhesives and deodorisers.

Building regulations are framed in terms of a target minimum number of ACH – typically 0.35–0.7 – in association with a guide value for volume flow rate per person of around 12 l s^{-1} (when smoking is permitted) down to 2.5 (when it isn't). The values found in practice range from 0.1 ACH (modern well-sealed office building) to 3 ACH (old drafty accommodation with sash windows and open fireplaces). For housing used by people or other animals, the carbon dioxide concentration is the basic measure of ventilation effectiveness. The outdoor concentration is usually between 300 and 600 ppm. A person doing light work exhales 0.5–1 l of air s^{-1}, corresponding to about 25 l h^{-1} of CO_2. This would need a ventilation rate of outside air of 10–30 l s^{-1} person^{-1} to reduce the concentration to 1000–500 ppm. Of course, ventilation may act either as a net importer of an ambient pollutant or as a net exporter of an internal pollutant, depending on the balance between indoor and outdoor concentrations. As energy costs for space heating and air-conditioning continue to increase, there is also increasing financial pressure to restrict building ventilation rates.

8.2   COMBUSTION

Any combustion source (e.g. gas, wood or coal fires, gas ovens and hobs) emits a wide range of pollutant gases – NO, NO_2, CO, hydrocarbons – and particles. The particles may comprise mainly soot (if combustion conditions are poor). High-temperature combustion may produce carcinogenic PAH.

8.2.1   Solid-fuel combustion

In developed countries, domestic solid-fuel combustion usually takes place in a grate with a chimney, and provided that ventilation is adequate the increase of indoor pollutant concentrations should be small. In developing countries, women who spend long periods in poorly-ventilated cooking areas with open unflued stoves are at the greatest risk. The most common traditional fuels are wood, charcoal, agricultural residues and animal wastes. Around 2 billion people rely on these fuels to meet most of their energy needs, amounting to around 25% of the total energy consumption in these countries. Appallingly high peak concentrations of TSP have been reported – in the range 5000–20 000 $\mu g\ m^{-3}$. Even daily average TSP concentrations of over 1000 $\mu g\ m^{-3}$ have been found for women and children (who are liable to be carried on their mother's back while she is cooking) in a rural area of India. In the more developed nations, concentrations have generally decreased with improved ventilation and displacement of solid-fuel stoves by gas-fired central heating, although wood-burning stoves can produce serious emissions during lighting and refuelling. There is some evidence of a higher incidence of respiratory illness and reduction of lung function, especially amongst children, in homes with such stoves.

8.2.2   Gas combustion

Combustion of natural gas (CH_4) always produces some CO (from incomplete combustion to CO_2) and NO_x (from fuel and thermal routes). Personal exposure is highly erratic, and depends on factors such as room ventilation, time of operation and the movements of a person in the vicinity of the source. For example, someone cooking may be exposed to intermittent peaks of several hundred ppb of NO_2 when they bend over a gas hob. If measurements of personal exposure are to be meaningful, they will need to be taken with a personal dosimeter worn near the face, otherwise they are likely to be erroneous. Also, we need to measure peak short-term exposures, since personal response is likely to be dominated by the concentration exceeded for 1% or 10% of the time, rather than by the long-term arithmetic mean.

8.2.3   Tobacco combustion

A large number of gaseous and particulate air pollutants are produced by tobacco combustion. Of course, the exposure of a smoker is very high compared to a non-smoker, but a smoker's exposure is voluntary, whereas a non-smoker's is not. Of the more than 4500 compounds found in tobacco smoke, 50 are known or suspected to be carcinogenic. Conventionally, the emissions are classified into mainstream (MTS), sidestream (STS) and environmental (ETS) tobacco smoke.

Mainstream emissions, formed during active puffing, are sucked directly from the end of the cigarette into the mouth; burn temperatures are up to 1000 °C, so thermal NO_x is formed, as well as HCN (hydrogen cyanide) and *N*-nitrosamines (carcinogenic). Most of the components are generated by a combination of pyrolysis and distillation in an oxygen-deficient atmosphere. Very high concentrations (around 10^{10} per cm^3 of evolved gas) of particles (mode diameter around 0.5 μm) are produced. Although these are serious pollutants, they are only inhaled by the smoker.

Sidestream tobacco smoke refers to emissions from cigarettes that are alight but not being actively puffed; the burn temperature then falls to 400 °C. Such a low combustion temperature can lead to the formation of more toxics than in MTS. Concentrations can be extremely high (many ppm of formaldehyde, for example), before they are diluted by air turbulence. Some compounds, such as the carcinogen N-nitrosodimethylamine, are emitted much more strongly in STS than in MTS, so that an adjacent non-smoker can have a higher exposure than the smoker. About half of the cigarette tobacco is consumed during the production of MTS, half during STS, but puffing only takes about 4% of the total lifetime of the cigarette.

Environmental tobacco smoke is the mixture present in a room, consisting of a variable proportion of MTS (after inhalation and subsequent exhalation) and STS. It is not possible to measure and specify the whole range of possible toxics. Various investigators have identified and measured different fractions, such as carbon monoxide, nicotine, NO_x, hydrocarbons, acrolein, acetone, benzene, benzo-a-pyrene (BaP) and particles.

Concentrations of PM_{10} in ETS can be up to 1000 μg m^{-3} (this compares with 1–5 μg m^{-3} in clean ambient air, 100 μg m^{-3} in polluted ambient air, 1000–3000 μg m^{-3} in the air during the London smogs). At the highest concentrations, such as those found in crowded and poorly ventilated pubs and clubs, the nicotine concentration alone can reach hundreds of μg m^{-3}. About 1 μg m^{-3} of PM_{10} is added to long-term indoor concentrations for each cigarette smoked per day (e.g. two heavy smokers would add several times the ambient background). Homes with smokers have been found to have around three times the mass concentration of suspended particulate as homes without smokers. Inhalation of ETS is colloquially known as passive smoking; it can induce effects ranging from eye watering (tearing), increased coughing and phlegm production, allergic reactions through lung cancer to death. There is also evidence for reduced lung function development in children, increased asthma incidence, increased rates of hospitalisation from bronchitis and pneumonia. One study showed that around 17% of lung cancers among non-smokers may be attributed to ETS exposure during childhood. About 100 000 people die each year in the UK through smoking-related illness, and extrapolation from the mortality statistics of these active smokers indicates that an additional several thousand people die from passive smoking. The general effects of cigarette smoke include:

- bronchoconstriction
- depletion of intracellular antioxidants
- stimulation of excess mucus production
- increased permeability of airway epithelial linings to macromolecules
- reduced mucociliary function

- suppression of immune response (alveolar macrophages have impaired capacity to kill bacteria)
- inflammation of mucus-secreting glands.

8.3   INDOOR ORGANICS SOURCES

Hundreds of volatile organic compounds are emitted by many processes and/or materials in buildings. They are compounds that mainly exist in the vapour phase at the normal range of air temperatures and pressures. There are also semi-volatile VOCs, that are present both as liquids and solids, but which also evaporate. Many compounds are outgassed over a long period of time from the increasing range of synthetic materials, particularly polymeric ones. For example, just one popular polymer, PVC, can emit a wide range of substances including dibutylphthalate, vinyl chloride, mesethylene, toluene, cyclohexanol and butyl alcohol. A single source may dominate, such as a damp-proof membrane, or many separate sources such as carpet glue, paint, chipboard and plastics may contribute. Flooring materials may involve thermoplastic polymers compounded with process aids such as stabilisers, viscosity modifiers and co-solvents. Some materials are produced from monomers and other low-molecular weight reactive organic compounds and involve curing processes such as cross-linking, vulcanisation or oxidation. Additives such as plasticisers, antioxidants, surface-treating agents and colorisers may be used. Natural materials such as wood and cork contain many volatile compounds. The release process is for diffusion of the compounds through the body of the material to the surface, followed by evaporation from the surface. Conversely, some flooring materials such as foam-backed carpet can adsorb materials such as tobacco combustion products from the atmosphere and release them later, possibly creating an odour problem. Varnished or laquered wooden furniture is coated in a material containing binders, solvents, light stabilisers, antioxidants, radical starters and other additives. The binders might be polyurethane, unsaturated polyesters or cellulose nitrate. Acid-curing laquers are notable for their formaldehyde emissions.

Emissions have been measured from a great range of everyday products found around the home or office – newspapers, books, electronic devices, air freshener, adhesives, cleaning fluids, furniture polish, wax and hair spray. Some of the most commonly found VOCs are toluene, benzene, xylenes, decane and undecane. Many VOCs are bioeffluents – ethanol, acetone, methanol, butyric acid, acetic acid, phenol and toluene are examples. Others are quasi-bioeffluents – components of deodorants or make-up are examples. Exhaled breath contains around 2000 $\mu g\,m^{-3}$ acetone and ethanol, several hundred $\mu g\,m^{-3}$ isoprene, methanol and 2-propanol. These are endogenous metabolism by-products originating inside the body. Smoking naturally raises the exhaled concentrations of many substances. One smoker, even when not smoking, can raise the room concentration of a compound such as benzene significantly. Microbial VOCs have been used as indicators –– 3-ethylfuran, dimethyldisulphide and 1-octen-3-ol in combination have been shown to be strongly correlated with microbial activity. One survey of 100 homes found that the average total concentration of the 50–300 VOC species that are normally found in indoor air was 553 μg m^{-3}, with 2% (i.e. 2 homes) above 1777 μg m^{-3}. The average ambient concentration during the period of the survey was 32 μg m^{-3}. Another survey showed that total VOC concentrations might be several thousand μg m^{-3} in new homes, but fall to around 1000 μg m^{-3} in homes more than 3-months old. The EU has issued a list of 50 VOCs relevant to indoor air exposure, although the proportional representation of each varies from case to case. In general, as one might expect, the emissions from all building products tend to be greatest when new and to decay fairly rapidly (within days to weeks). Exceptions to this decay are the actions of heat and moisture which can accelerate oxidative reactions.

The measurement of these organic compounds indoors is often a trade-off between time resolution and sensitivity. Real-time methods, such as flame ion-isation detectors, do not have the necessary sensitivity in many situations, so the sample has to be preconcentrated by adsorbing it over a period of time. Active samplers pass the sample air through the adsorbent with a pump. Often 30 min will be sufficient, and will generate a desorbed concentration which is 1000–100 000 times the original ambient value. Passive samplers rely on molecular diffusion over periods of hours to days. This is convenient for low-maintenance integrated measurements, although errors can accumulate if the adsorbed gas self-desorbs during the sampling period. Solid adsorbents (Table 8.1) may be inorganic, porous materials based on carbon and organic polymers. They vary in specific surface area, polarity, thermal stability and affinity for water; hence each tends to have preferred applications.

Carbon-based adsorbents with high specific surface areas can be used to trap very low-boiling compounds, but high-boiling compounds may be difficult to desorb once trapped. If they are reactive, they may even decompose before desorption. Although porous polymers have lower specific surface areas, they can be used effectively for high-boiling compounds such as glycols and phthalates, but not for low-boiling materials such as alkanes.

The sample may be removed from the adsorbent by solvent extraction or thermal desorption. In the former method, an organic solvent such as dimethyl-formamide, carbon disulphide or dichloromethane, having a high affinity for the sample gas and low response on the detector, is flushed through the adsorbent. With thermal desorption, the adsorbent is heated to at least 20 °C above the boiling point of the adsorbed sample to drive it out of the adsorbent.

Formaldehyde (HCHO) is used in many products – in urea-formaldehyde resins for making chipboard, as a preservative in medicines and cosmetics, and as a carrier solvent for fabrics and carpets. Emissions from chipboard and similar wood products are aggravated by dampness, which accelerates the reversible polymerising reaction. Formaldehyde is also given off by most combustion sources – solid-fuel, gas, petrol, diesel and cigarettes (up to 40 ppm). But HCHO is a pungent, irritating toxic gas. One key indoor air pollution problem in the 1970s was the release of formaldehyde from urea-formaldehyde foams used for cavity wall insulation. The formaldehyde gas leaked out of cracks in walls and

Table 8.1 Solid adsorbents

Type	Structure	Specific surface area ($m^2\,g^{-1}$)	Tradenames	Water affinity	Application
Inorganic	Silica gel Molecular sieve	1–30	Volasphere		PCBs, pesticides Permanent gases
	aluminium oxide	300	Alumina F1		Hydrocarbons
Carbon-based	Activated charcoal	800–1200		High	Non-polar and slightly polar high-boiling VOCs
	Carbon molecular sieve	400–1000	Carbosieve	Low	Non-polar and slightly polar low-boiling VOCs
	Graphitised carbon blacks	12–100	Carbotrap	Low	Non-polar high-boiling VOCs
Porous polymers	Styrene	300–800	Porapak Chromosorb	Low	Non-polar and moderately polar high-boiling VOCs
	Phenyl-phenylene oxide	12–35	Tenax	Low	Non-polar high-boiling VOCs
	Polyurethane foams			Low	Pesticides

gaps at door and window frames, and accumulated in the indoor air. If the rate of leakage was high, and the ventilation rate low, then sensitive individuals could experience a serious level of skin and respiratory irritation. Substitution by other materials such as rockwool has decreased the significance of this issue. The ambient concentration of formaldehyde is rarely above 10 ppb, but concentrations of several hundred ppb have been reported in buildings that contain formaldehyde-generating materials. Formaldehyde can also be photodissociated by the UVA light emitted from fluorescent sources ($\lambda < 330$ nm) and oxidised to the hydroperoxy radical.

8.4 BIOAEROSOLS

Although not conventionally regarded as air pollutants, many biological materials can accumulate in buildings to higher concentrations than are normally found outdoors, and be irritating or toxic to occupants. This category includes airborne particles, large molecules or volatile compounds that are either living or were released from a living organism. Some bioaerosol such as bacteria and viruses may replicate, others such as pollen and mite faeces may simply be an irritant *per se*. In the limit, a single bacterium or virus can be fatal. The particles can vary in size from 100 nm to 100 μm (Figure 8.3).

As with other air pollutants, direct measurement of personal exposure may be very difficult. Measurements of airborne concentration, combined with collection of samples from carpets, mattresses and other surfaces, will provide an indication of presence or absence and relative abundance, but not of personal exposure.

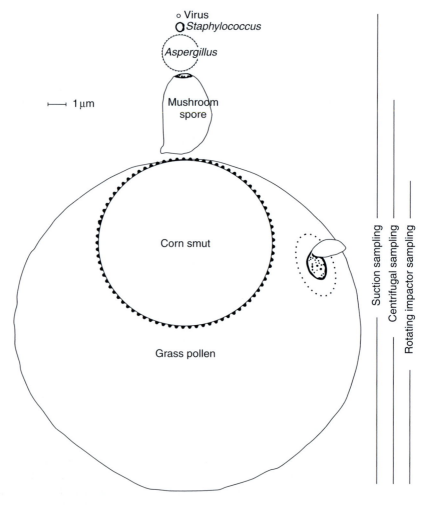

Figure 8.3 Size ranges of bioaerosol.
Source: Yocom, J. E. and McCarthy, S. M. (1991) *Measuring Indoor Air Quality*, Wiley, Chichester.

8.4.1   Categories

The major categories of bioaerosol are bacteria, endotoxins, fungi, viruses, antigens and pollen. Pollen can stimulate allergic reactions. Viruses, bacteria and fungi spread disease. Skin scales (dander) and lice excreta can cause allergic reactions (for example, allergic rhinitis due to the house dust mite).

8.4.1.1   Bacteria

Bacteria are prokaryotic (no distinct cell nucleus) organisms that have a cell membrane, DNA and some sub-cellular components. Some bacteria such as *Legionella pneumonophila, Streptococcus* and *Meningococcus* are associated with specific diseases, others such as *Thermoactinomyces* and *Enterobacter* are non-infectious (i.e. not pathogenic by inhalation). Drinking water contaminated with *Legionella* does not cause Legionnaire's disease; the aerosolised water droplets must be inhaled. Other toxic materials such as endotoxins and exotoxins are associated with them. Their rate of multiplication and dissemination can depend strongly on certain aspects of the indoor environment, such as humidity. The size range of bacteria is 0.3–30 μm, while that of bacterial spores is 0.3–10 μm. Microbial growth needs a surface relative humidity above 25%, and then increases with Rh up to Rh values of 70%.

8.4.1.2   Endotoxins

Endotoxins are highly toxic lipopolysaccharide materials contained in the outer membranes of Gram-negative bacteria. They stimulate the immune systems of a small proportion of sensitive individuals, which may produce fever, malaise, airway obstruction and acute pulmonary infection (humidifier fever). In the airborne environment, they are usually associated with dust particles or liquid aerosol having a wide distribution of sizes.

8.4.1.3   Fungi

Fungi are eukaryotic (having a nucleus). Some, such as yeasts are unicellular while others such as moulds are multicellular. Fungal growth usually needs a relative humidity >70%, so bathrooms and kitchens are favoured. Although fungi live on surfaces or in liquids, many produce spores which are often airborne. Apart from the production of infectious diseases such as by *Aspergillus fumigatus*, most fungi generate metabolic products such as proteins that can induce an antibody response and hypersensitivity symptoms. Fungi also achieve reproductive success by releasing mycotoxins to kill off competitor organisms. *Penicillium* is the classic case which has been turned to human advantage, but airborne mycotoxins are not generally good news. Commonly problems are experienced by people handling damp (i.e. mouldy) paper, furniture and textiles. The most publicised mycotoxin is aflatoxin, which is a known human carcinogen, although the predominant uptake route for that is ingestion of mouldy foodstuffs such as groundnut rather than inhalation. The fungal spore size range is 0.5–30 μm.

8.4.1.4   Viruses

These are the smallest life forms, parasitic replicating organisms that are dependent on their host cell for reproduction. Size range 0.02–0.3 μm, although most airborne viruses are attached to other particles.

8.4.1.5   Allergens

Allergy, which is any adverse reaction resulting from immunologic sensitisation towards any specific substance (the allergen), is an increasing problem in society, and airborne uptake of allergens plays a significant part. Allergens are macro-molecules, produced by living organisms, that produce a detectable response from the immune system. The symptoms are characteristic, such as allergic rhinitis (runny nose) and allergic aspergillus. Human sensitivity may be three orders of magnitude greater than for the classic air pollutants such as ozone, with serious responses to concentrations in the $ng\, m^{-3}$ range. The allergens are produced from a wide variety of sources such as fungal and bacterial spores, cat and rodent dander and hairs, house dust mites and cockroach excreta. Responses to allergens vary widely between individuals. Although certain allergens such as cat dander and house dust mites affect many people and have therefore had a lot of publicity, many other allergens affect smaller numbers. For example, allergic responses to the rubber tree latex used in surgical gloves is topical at the time of writing.

Each person can shed as much as 1 g of skin cells each day, and this constitutes the organic matter for the domestic micro-ecology, regardless of the additional supply from pets, foodstuffs etc. The skin scales accumulate in furniture, carpets and bedding, where they are colonised by saprophytic fungi, and in turn by mites. The latter 300 μm arthropods (*Dermatophagoides*, which means skin-eater) enjoy conditions of 25 °C and 70–80% relative humidity, and grow from egg to adult in 25 days. Hence their abundance has been favoured over the last 50 years by increased use of central heating (which raises temperatures), combined with reduced ventilation (which increases humidity) and use of fitted deep-pile carpets (which reduces cleanability). Floor or mattress dust samples typically contain 100–1000 mites (dead or alive) g^{-1}. Mite faeces, which are sticky pellets about 20 μm in diameter, constitute a large proportion of the allergen in house dust. The gastric enzymes produced by the mites, which are in their faeces and are strong human allergens, become airborne and are eventually inhaled. The concentrations of the two specific allergenic proteins are in the range 2–10 μg (g settled dust)$^{-1}$ or 5–50 ng m^{-3} air in the houses for which house dust mite allergy has been implicated in triggering an asthma attack. Corresponding allergens are produced by cats (*Fel d*), dogs (*Can f*) and cockroaches (*Bla g*). Again, the concentrations and inhalation rates are increased by the use of tighter buildings with better draught-proofing and reduced ventilation. The allergic response may be a gentle 'runny nose' of little concern, but it may also be associated with a violent or even fatal asthma attack. Although specific materials are related to sources – cat allergen concentrations are naturally higher in homes with cats, for example – there may be high enough concentrations in adjacent properties, or in trains and other public places, to cause reactions in sensitive individuals.

8.4.1.6   Pollen

Pollen grains are the largest bioaerosol particles, having diameters from 10 up to 170 μm (predominant range 30–50 μm). They are produced by plants to transport genetic material to the flowers of other plants. Pollen may be either windborne or insect borne. Windborne pollen is produced on the end of protuding anthers: as with many of the bioaerosols, airborne dispersion is an evolutionary tool for reproduction, so we can expect the pollen to become airborne easily and to remain so for long periods. Both number and mass concentrations are generally small; even a high pollen episode involving hundreds of pollen grains per m^3 would correspond to only around 10 μg m^{-3}. However, sensitivity to pollen as allergens can be extremely high for a small proportion of the population. For most people this amounts only to an inconvenience such as allergic rhinitis (hay fever), but at worst it can exacerbate chronic obstructive pulmonary disease and possibly be fatal. Pollen is not uniformly irritant – maple, birch and sycamore are generally the worst tree pollens, while bluegrass and Timothy are the worst grasses.

8.4.2   Sampling bioaerosols

Bacteria, fungi and viruses are all particles, and in principle they can all be sampled by the standard methods described in Chapter 4. In practice a specialised subset of equipment has been developed over the years for sampling these organisms.

Viruses are in practice not sampled physically, but identified by their effects on people. For bacteria and fungi, there are sampling methods based on gravity, inertial impaction, filtration and liquid impingement. Cascade impactors have their cut-diameters optimised for bacteria or fungi, which are relatively coarse compared to suspended particles in general. For organisms of a few μm or smaller, the acceleration and impact forces needed to separate them from the airstream are too violent, and other methods must be used. Filtration and electrostatic precipitation, or impingement in a solution, are the preferred techniques. If the morphology of the organisms is known, direct microscopy on the substrate can be used for identification. More sophisticated laboratory techniques, such as analysis for guanine to identify house dust mite excreta, are also used. Often the collection substrate will be coated in a growth medium such as agar (a nutrient jelly prepared from seaweeds, blood, soy and other materials) so that the microorganisms can be identified from the colonies formed, and the areal density quantified. The plated colonies are cultured over periods of a few days to a week at constant temperature, with the media and growth conditions optimised for the species of interest. Often this agar plate will be the substrate of an impactor stage so that the bioaerosol is deposited directly on to it. Multi-jet impactors may have up to 500 jets arranged in this way, spreading out the collected particles to avoid overloading. This is quantitatively important because if two bacteria are deposited close to each other, only one colony will develop. The most complex collectors ramp up the volume flow rate and/or decrease the jet-substrate spacing in order to change the cut-off diameter. There are imperfections in this capture and identification process. Some fungi can repress the growth of other organisms, which would lower the predicted count. The sampling process itself is quite violent, with impaction speeds of up to several hundred m s^{-1} needed to separate the smallest microorganisms from their streamlines.

Not all important bioaerosol particles are culturable – indeed, up to 95% of viable bioaerosol particles are not. For such material, other techniques come to the forefront for identification after capture. A non-specific count of total bacteria can be made by staining them with a fluorescent dye such as a monomeric or dimeric cyanine and identifying them with scanning confocal microscopy. Species can be identified by polymerase chain reaction (PCR) techniques to amplify their DNA. Pollen grains can be impacted on to a rotating adhesive square transparent rod for microscopic counting and identification. Liquid impingers trap the microorganisms by impinging the airstream onto the agitated surface of the collection liquid, which is usually a dilute nutrient solution. Many interactions such as asthma and allergic responses are due to dead material. Techniques such as enzyme-linked immunosorbent assay (ELISA) are used to identify such material, by binding an antibody with specificity for the aeroallergen. This primary antibody is then reacted with an enzyme-linked second antibody. The enzyme in turn reacts with a substrate to produce a spectral shift which can be quantified. These techniques are only semi-quantitative or indicative at present.

8.5   SICK BUILDING SYNDROME

In controlled-environment office buildings, there has sometimes been reported a high incidence of minor but discomforting illnesses such as headache, sore throat, runny nose and coughing. This phenomenon has come to be known as Sick (or Tight) Building Syndrome (SBS). It is probably a compound response which is partly due to air quality, and also due to factors such as isolation from the natural environment, thermal discomfort, light flicker and other unsettling aspects of working conditions. In early investigations, VOCs were thought to be respons-ible, but this finding has not been supported by later work. Ventilation rate has been investigated but not conclusively shown to be a cause. Although there have been many investigations of alleged SBS, the complexity of the problem has meant that many of these have only measured a small subset of the possible relevant parameters. Building-related illness is distinguished from SBS because it is a specific response due to a single identifiable factor such as bio-allergens from air conditioning or formaldehyde from insulation.

8.6 ODOUR AND VENTILATION

Odour is one of the most exacting problems to deal with. Some chemicals such as ammonia (from the body) and mercaptans (from tobacco smoke) have extremely low odour thresholds; a ventilation rate that is entirely adequate for health and thermal comfort may still leave a perceptible smell in the air. For example, in one study, it was found that an air exchange rate of $50 \, l \, s^{-1}$ per smoking occupant was still insufficient to eliminate a smell problem. A measurement unit has been proposed, but not yet widely adopted – one olf (=olfactory unit) is the emission rate of bioeffluents from a standard sedentary person in thermal comfort. A smoker has been reported as being about six olf even when not actually smoking, and in practice the carpets and other fittings can contribute at least as much as the occupants. A further complication is caused by acclimation – a person who finds a smell noticeable or even offensive when they first walk into a room will usually adapt after a short time and reduce their sensitivity to that particular compound. Ammonia has a very low odour threshold and is given off by the human body, especially when sweaty. Surveys of ammonia which have been carried out with diffusion tubes in the kitchens, living rooms and bedrooms of a small sample of UK homes have shown average concentrations, in the range 30–40 $\mu g \, m^{-3}$, which were surprisingly similar in the three environments.

8.7 CLEAN ROOMS

Extremely high standards of particle control are required in some activities, for example electronic chip manufacture, to reduce the risk of product contamination (Table 8.2). United States Federal standards are based on particle counts per volume of circulated air, with the performance graded into Classes.

Table 8.2 Extreme US clean-room standards

Particle concentration ft^{-3} not to be exceeded	
Class 100 000	Class 1
100 000 > 0.5 μm	1 > 0.5 μm
700 > 5 μm	3 > 0.3 μm
	7.5 > 0.2 μm
	35 > 0.1 μm

Table 8.3 Equivalence between US and SI clean-room standards

US	SI
100 000	M 6.5
10 000	M 5.5
1000	M 4.5
100	M 3.5
10	M 2.5
1	M 1.5

The Class number corresponds to the number of particles larger than 0.5 μm being less than that value per ft^3. Corresponding to these are SI classes, in which the cleanness is indicated by an exponent p, where the number of particles larger than 0.5 μm is less than 10^p m^{-3}.

For example, M 6.5 on the SI system can contain no more than $10^{6.5} = 3.53 \times 10^6$ particles per m^3 greater than 0.5 μm, which corresponds to 100 000 ft^{-3} (Table 8.3).

Obtaining these very low concentrations requires a special branch of control technology. Specialist methods are used called high efficiency particulate air (HEPA) or ultra low penetration air (ULPA) filtration which obtain efficiencies of 99.97% for 0.3 μm particles, and 99.999% for 0.2–0.3 μm particles, respectively. Very large airflows may also be needed in order to remove particles emitted from within the controlled volume. An M 6.5 clean room may have 30 ACH, which can rise to 600 for an M 2.5 room.

FURTHER READING

Bardana, E. J. Jr. and Montanaro, A. (1997) *Indoor Air Pollution and Health*, Marcel Dekker, New York.

Gammage, R. B. and Kaye, S. V. (eds) (1990) *Indoor Air and Human Health*, Lewis Publishers, Chelsea, Michigan, USA.

Hansen, D. L. and David, L. (1999) *Indoor Air Quality Issues*, Taylor & Francis, London.

Jones, A. P. (1999) 'Indoor air quality and health', *Atmospheric Environment* 33: 4535–4564.

Leslie, G. B. and Lunau, F. W. (1992) *Indoor Air Pollution: Problems and Priorities*, Cambridge University Press, Cambridge, UK.

Perry, R. and Kirk, P. W. (eds) (1988) *Indoor and Ambient Air Quality*, Selper Ltd, London, UK.

Samet, J. M. and Spengler, J. D. (1991) *Indoor Air Pollution: A Health Perspective*, John Hopkins, Baltimore, USA.

Salthammer, T. (1999) *Organic Indoor Air Pollutants: Occurrence – Measurement – Evaluation*, Wiley-VCH, Weinheim.

Yocom, J. E. and McCarthy, S. M. (1991) *Measuring Indoor Air Quality*, Wiley, Chichester.

Chapter 9

Effects on plants, visual range and materials

Air pollution can have both long-term (chronic) and short-term (acute) effects on plants, which are economically significant. They also affect visual range by absorbing and scattering light, and they damage materials by chemical action.

9.1 EFFECTS ON PLANTS

9.1.1 Historical evidence

9.1.1.1 Early experiments in Leeds

Before the First World War, Cohen and Ruston carried out a series of experiments in Leeds to determine the effects of urban air pollution on a range of plants. At that time Leeds was typical of a thriving urban/industrial complex. There was a profusion of small industrial coal-fired premises releasing smoke and sulphur dioxide from rather inadequate chimneys, set amongst large areas of closely packed terraced housing which was again dependant on coal for energy. Cohen and Ruston set experimental plants out in pots of standard soil in various parts of Leeds, chosen because of their different levels of air pollution. They found severe growth reductions to both roots and shoots of a range of plants, from radish and lettuce to wallflower. There were no sophisticated gas analysers available at the time, and the growth reductions were attributed to a variety of possible causes, from acid rain and sulphurous fumes to the reduction in flux density of solar radiation caused by the pall of smoke particles.

9.1.1.2 Canadian smelters

Between the First and Second War, there was rapid expansion of the smelting industry in parts of Canada. Ore deposits at Copper Cliff, Trail, Sudbury and Wa-Wa were rich in iron, tin, zinc and copper. Sulphur is associated both with the ores and with the coal used for smelting, so that large volumes of sulphur dioxide at high concentration were released, often from undersized stacks. For example, the iron smelter at Wa-Wa in Ontario released its emissions into the south-west prevailing wind, so that ground-level concentrations were confined to the north-east sector. Large tracts of native boreal forest (spruce, cedar, larch and pine) to the north-east were destroyed, in a belt stretching some 30 km from the smelter. Furthermore, the damage pattern clearly reflected the interaction between two competing influences: (1) the decline of average concentration with increasing distance away from the source; (2) the different exposure of different vegetation types. The trees are aerodynamically rough and, being tall, experience the highest windspeeds and lowest boundary layer resistances. They therefore had the highest potential gas deposition fluxes. The understory of bushes was not only lower but was protected to some degree by the trees. Finally, there were lichens, mosses and other low-lying species that survived to within a few km of the source. Investigators described the vegetation as having been peeled off in layers. Other pollutants, such as copper and arsenic, contributed to the effects of these emissions. Although these smelter emissions have now been controlled, there are still parts of the world, such as the 'Black Triangle' in Eastern Europe referred to below, where forests are being killed by intense sulphur dioxide concentrations.

9.1.1.3   Urban/industrial pollution in the Pennines

The Pennine Hills in England lie in a scenically attractive region of relatively low population density, much used today as a recreational area. However, the southern Pennines are surrounded by urban/industrial towns such as Manchester, Leeds, Sheffield and Stoke, so that throughout at least two centuries of industrial development the area has been exposed to emissions of sulphur dioxide and other phytotoxic gases. Since there have been sources in all directions, changes in wind direction have brought no respite. Research in the 1960s showed that the area had been denuded of Scots Pine, and this was thought to be due principally to air pollution.

9.1.1.4   The Black Triangle

The Black Triangle is the area where the borders of Germany, Poland and the Czech Republic meet. It is a region of managed forests which is also the location of substantial brown coal deposits. The latter have been strip-mined and burned for electricity production, resulting in high sulphur emissions and deposition. Sulphur deposition in the Erzgebirge mountains was up to 125 kg ha^{-1} a^{-1} in 1989, but is now down to around 40 (for comparison, the UK maximum is around 15). Damage to the mainly Norway spruce forest (reddening of the needles followed by defoliation) spread gradually during the 1950s and 1960s, and worsened suddenly in 1979 after an unusually sudden fall in air temperature. Many square kilometres of forest have now been devastated, particularly above an altitude of 750 m.

9.1.1.5   Lichen

Lichen are widespread symbiotic associations of fungi and algae that grow slowly on rocks, roofs and other mineral surfaces. Surveys of lichen abundance and species diversity across the UK have shown a strong correlation with the distribution of SO_2 sources and therefore with ground-level SO_2 concentrations. Parts of the UK have become 'lichen-deserts', although the reduction of SO_2 concentrations is expected to foreshadow a recovery. Certain areas from which a particular species was found to be absent during the 1970s have been recolonised during the 1980s and 1990s, and resurveys of specific sites where air pollutant concentrations are known to have decreased have shown a general reduction in the proportion of grid squares from which species are absent. Of course, long-term acidification of the growth substrate may mean that the recovery takes much longer than the initial response. For example, oak trees do not shed their bark, so once the bark has been given a reduced pH, it tends to stay that way.

9.1.1.6   Diatoms

Diatoms are microscopic (a few tens of μm), single celled algae which occur in great abundance in all aquatic habitats. Individual species of diatom have been found to have quite different sensitivities to the pH of the water that they inhabit. When they die, each species leaves behind a characteristic silica 'skeleton' which is preserved in lake sediments. Hence, changes in the relative abundance of the various species can be used to track changes in pH. The sediments accumulate at about 1 mm per year, and if a true chronology can be assigned to the sediment record, then the change of pH with time will also be known. Over the last two decades, intensive sampling of Scottish fresh-water lochs and lakes in Europe and North America has shown the impact of fossil-fuel emissions on the acidity of rainfall and its consequences for life in fresh waters.

9.1.2   Experimental methods

The observations above provide strong evidence that air pollution damages plants, and it is not hard to demonstrate experimentally the sensitivity of plants to high concentrations of air pollutants. In the early days of such research, plants would be exposed, for example, to a few ppm of SO_2, a concentration within the range found during the worst episodes in urban areas: acute damage was caused to the plant within hours or days. Nowadays the situation is rather different. Although acute damage does still occur, it is usually localised and the source is identifiable. Control is then a matter of policy and economics rather than science. In addition to these short-range problems, large areas of the world are experiencing low concentrations of various gaseous and particulate air pollutants that may or may not be causing chronic 'invisible' plant damage. How do plants respond to these low pollutant levels? Is there damage that is economically or ecologically significant? A range of different experimental techniques has been developed to address these questions. They all have their own advantages and limitations. The great majority of experiments has been undertaken on single agricultural species such as grasses, cereals and legumes; although this gives important economic information, it tells us very little about the responses of complex communities of mixed natural species.

9.1.2.1   Laboratory chambers

This is the most straightforward experimental approach. Plants are grown in 'growth cabinets', and exposed to the pollutant(s) of interest. There are many factors that have to be considered in assessing the results of such an exposure: was the plant grown for the whole of its life cycle (e.g. all the way to final yield for a cereal); what are the changes in response brought about by environmental differences in lighting (quality and quantity), temperature, humidity, windspeed, soil conditions or plant density? For example, many early experiments were undertaken with rather stagnant air in the cabinets. It is now recognised that the leaf boundary layer resistances were artificially high, pollutant transport to the leaf surface reduced, and responses to a particular concentration of that pollutant suppressed. The cure in this case was simple – to stir the air in the chamber with fans – but many other aspects of plant growth need to be optimised if useful results are to be obtained. In general, it is now accepted that laboratory chambers cannot be used to obtain *quantitative* data on responses to pollutants that can be applied to plants grown outdoors. However, such chambers are much more useful than field facilities for short-term qualitative experiments – to elucidate biochemical mechanisms, for example.

9.1.2.2   Field release systems

Field release systems lie at the logical extreme from laboratory chambers. The principle is to expose an area of forest, ecosystem or commercial crop to an elevated pollutant concentration without changing any other environmental parameter. In one system, developed at the University of Nottingham to investigate the responses of winter cereals to sulphur dioxide (Figure 9.1), a square of gas release pipes, 20 m on each side, was supported about 10 cm above the growing crop. Each pipe was pierced with small holes along its length and acted as a line source. Gas release was controlled by computer in the following sequence:

- check the wind direction, and operate solenoid valves to release gas only from the upwind pipe (or from two adjacent pipes if the wind direction was blowing close to a diagonal);
- measure the gas concentration upwind of the square (ambient value);
- measure the gas concentration within the square;

- adjust mass flow controllers to increase the gas release rate if the concentration is below target, and *vice versa*;
- repeat the loop.

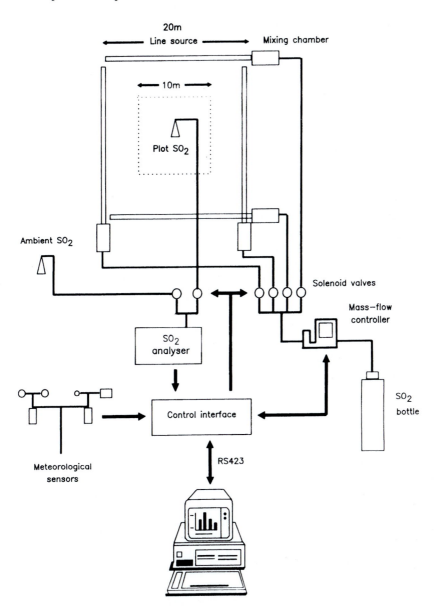

Figure 9.1 A field release system for sulphur dioxide.
Source: Colls, J. J., Geissler, P. A. and Baker, C. K. (1991) 'Long term performance of a field release system for the intermittent exposure of agricultural crops to sulphur dioxide', *Atmospheric Environment* 25A: 2895–2901.

In a typical experiment, the equipment was set up on a commercial field of winter wheat or winter barley soon after crop emergence in the autumn, and operated continuously until harvest in July or August. The plants were thereby exposed to elevated SO_2 concentrations for almost their whole lives, growing under completely natural conditions with no modification of temperature, windspeed or other environmental variables.

This principle has also been adapted for studying responses to carbon dioxide. In the Free Air Carbon Dioxide Enrichment (FACE) system developed by the Brookhaven National Laboratory, a circle of release pipes, around 25 m in diameter, is used; the circular pattern increases the uniformity of exposure above that possible with the square design. The main practical difference is that CO_2 enrichment involves raising the concentration by several hundred ppm instead of the tens or hundreds of ppb typical of ozone or SO_2, so that very high rates of gas supply are needed. If possible, an experiment involving CO_2 release is sited close to a brewery, fertiliser factory, natural CO_2 well or other low-cost source to reduce the operating cost.

There have been other designs of field release system. In one, an array of vertical pipes hanging down between rows of vines has been used to study responses to low concentrations of fluoride. In another, ozone release pipes have been distributed within a canopy of mature trees, with the release rate automatically adjusted to keep the local concentration approximately constant.

The limitations of field release systems include:

- There is no possibility of reducing pollutant concentrations below ambient – they cannot answer the question 'what is the prevailing level of air pollution doing to plants?' Of course, one solution might be to grow the plants in an area that has very low background concentrations of the pollutant of interest (such as the south of England for SO_2). Then additions can be used to raise the concentration and simulate the polluted environment at the original location. This procedure raises other issues – the soil, climate and background levels of other pollutants will be different at the new site, and these changes may themselves affect the growth of the plants. This problem has been tackled with field filtration systems (see Section 9.2.3).
- In addition, whatever burden of other pollutants that exists in the ambient air has to be accepted, and may interact in an unknown way with the experimental gas. There are mixed views on this. We certainly need to understand the responses of plants to individual pollutants. On the other hand, many locations such as the English Midlands only experience mixtures of pollutants. Changing the concentration of one gas against a steady background of the others is then a realistic experiment.
- It is difficult in practice to achieve a uniform exposure over the experimental area. The gas must be released at a rather high concentration from a distributed array of point sources (small holes in the pipe). It diffuses rapidly within the first few meters, so that a guard area of plants has to be sacrificed where

it is known that the concentration field is unacceptable. With the square geometry described above, there were also significant variations of exposure with wind direction.

*   A large ground area is required, within which the plant cover has to be reasonably uniform. Associated with this aspect of size is the large gas release rate that an open air system requires. This is especially true with forests because of the depth of atmosphere over which the gas concentration has to be increased.

9.1.2.3   Field filtration

Field filtration systems have been used to retain the natural environmental qualities of field release, while enabling the reduction of concentrations below ambient. Pipes are laid between rows of plants, and charcoal-filtered air blown out amongst the plants, so that pollutant concentrations are greatly reduced. Experience has shown that the pipes have to be quite large in diameter and quite close together in order to blow clean air out at a sufficiently high rate: this means that a normal field planting density cannot be used and one of the attractions of a field-based system is therefore reduced.

9.1.2.4   Open-top chambers

Open-top chambers (OTCs) were designed in America in the early 1970s as a way of controlling the air quality (both adding and subtracting pollutants) to which field-grown plants were exposed without creating unacceptable environmental changes. Typically, an OTC consists of an upright cylindrical aluminium framework covered in transparent plastic film. In one design, the lower half of the film is double-skinned, and there are several hundred holes on the inside wall (Figure 9.2). Alongside the OTC is a fan unit, which draws in ambient air and blows it into the double skin and out through the holes to the inside of the chamber. The air can be passed through a charcoal filter to reduce concentrations of O_3, SO_2 and NO_2, or it can have controlled concentrations of these gases added to it. Hence the plants growing inside the chamber are exposed to the controlled air supply. The ventilation rate is kept high (3–4 air changes per minute). This both stirs the air inside the OTC to keep the boundary layer resistance down, and keeps the temperature close to ambient. Other designs of gas distribution have been used – from holes in a toroidal plenum suspended above the canopy, or from grilles in the chamber floor. Chamber sizes have ranged from portable units 1.25 m in diameter to designs 5 m in diameter, 8 m tall for use on trees.

Open-top chambers also have their limitations:

*   Although the environmental conditions are significantly closer to ambient than they are in laboratory chambers, they are by no means perfect. Temperatures may be 4 °C higher than ambient in the middle of a sunny

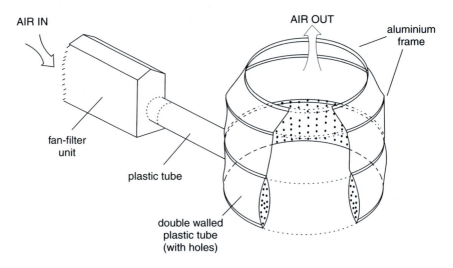

Figure 9.2 An open-top chamber.

afternoon, and average 0.5–1 °C higher over the year. Apart from the short-term impact of these excursions on photosynthesis and transpiration, for example, thermal time will accumulate faster in OTCs than outside so that experimental plants go through their growth stages faster. Rainfall inside OTCs is significantly reduced, since most rain falls at an angle to the vertical and the shelter effect is substantial. The wind environment is quite unnatural, with rather high windspeed at the edges of the circular area, decreasing towards the centre.

- The small environmental changes may be tolerated for the annual plants typical of most agricultural crops. The situation is quite different for trees, however. Very small proportional changes in growth, maintained over tens or hundreds of years, can have large cumulative effects. A new design of chamber at Nottingham has been used to minimise these differences. The chambers have been constructed with roll-down side panels for use in experiments involving intermittent ozone exposures. For most of the year, the panels are rolled up and the trees inside are effectively in the open air. For about 30 days spread over the spring and summer, the panels are rolled down to form a complete chamber while ozone exposures are carried out. In this way the environmental changes due to the chambers themselves are minimised.
- Scientists like to find out what is happening to plants during the course of a growing season rather than waiting until the end, and will want to take destructive plant harvests for measurements in the laboratory. Like any chamber, OTCs only offer small areas of plant material for measurements.

Furthermore, destructive harvests disrupt the vegetation canopy and may cause the remaining plants to grow differently and to have different exposures to the pollutant.

9.1.2.5   Ventilated glasshouses

A wide variety of controlled-environment buildings has been used for pollution control studies. In an early example, rye-grass was studied in glasshouses near Manchester. The ventilation air was scrubbed with water sprays to reduce the sulphur dioxide concentration before it entered the glasshouse. In the UK, small hemispherical forced-ventilation glasshouses known as 'solardomes' have been popular. Such systems are cheaper and more permanent than OTCs, and usually located closer to a research laboratory for more intensive supervision of experiments. Their main disadvantage is that they are one stage further removed from the outdoor environment, particularly since plants are grown in pots rather than in the ground.

9.1.2.6   Branch bags

The size and longevity of mature trees has usually made it unrealistic to enclose them with chambers of any sort for a significant proportion of their lives. Instead, parts of the tree have been enclosed. A mini-chamber is constructed around a single branch, with the appropriate ventilation to prevent the temperature and humidity increasing unacceptably. A gas such as ozone can then be added to the air supply. The technique has the advantage that it can be applied to mature trees growing in their natural environment. However, since only part of the tree is exposed to the stress, it begs the question of whether the tree is able to mitigate the effects by importing resources from, or exporting toxic products to, other parts of the tree.

9.1.2.7   Ethylene diurea

One specific experimental method is available for ozone that is quite dissimilar to the methods discussed above. It has been found that the chemical ethylene diurea (EDU), when applied to plants as either a soil drench or to the foliage, can protect the plant against elevated ozone concentrations. Hence the responses of EDU-treated and untreated plants can be compared to derive dose–response functions. This relationship has been used as the basis of a European programme on ozone effects research co-ordinated by the UNECE.

9.1.2.8   Bioindicators

Some species of plant have fairly high specificity towards individual pollutants. For example, gladioli are sensitive to fluoride, tobacco plants to ozone and lichen to sulphur dioxide. Such species (or, more usually, a particularly sensitive variety of the species) can be used to show variations in pollutant concentration, either with time at one place, or from place to place. An unambiguous response is needed (for example, formation of brown upper-surface speckle by ozone) which can be at least approximately linked to pollutant concentration. Tobacco plants have been used in surveys of responses to ozone in the UK, while the Netherlands has had a long-term bio-indicator programme to monitor HF and O_3. In Figure 9.3, leaf-tip necrosis on gladioli was shown to be strongly correlated with proximity to industrial sources of HF in the south-west Netherlands.

Figure 9.3 The degree of damage to gladioli was dependent on their proximity to fluoride sources in the south-west of the Netherlands.
Source: Postumus, A. C. (1982) 'Biological indicators of air pollution', In: M. H. Unsworth and D. P. Ormrod (eds) *Effects of Gaseous Air Pollution in Agriculture and Horticulture,* Butterworths, Sevenoaks, UK.

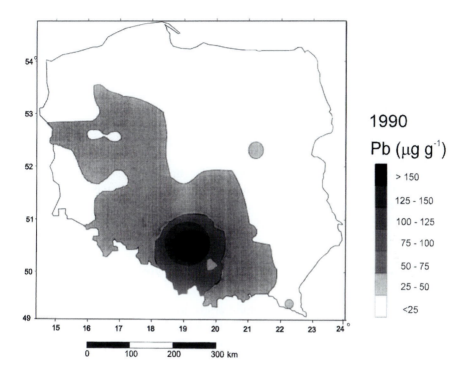

Figure 9.4 Contours of moss lead content in Poland.
Source: Grodzinska, K. and Szarek-Lukaszewska, G. (2001) 'Response of mosses to the heavy metal pollution in Poland', *Environmental Pollution* 114: 443–451.

In another version of this idea, mosses have been used extensively to measure the deposition of heavy metals from the atmosphere. Mosses have no root system to take up minerals from the substrate on which they grow, and accumulate materials such as heavy metals from wet and dry deposition. Either indigenous mosses are harvested, or a network of mossbags is set up, with the moss being collected and analysed after a period. Figure 9.4 shows the results of lead analyses of mosses in Poland, with the highest concentrations associated with the most heavily industrialised part of the country.

9.1.3   Plant responses to specific air pollutants

Summary data on plant responses to air pollutants, particularly at low levels, show complex and often inconsistent results. We shall describe them from two directions: the observed responses regardless of mechanism; and the mechanisms themselves.

9.1.3.1   Visible injury

Visible injury to leaves is the most obvious indicator of air pollution damage when the concentrations have been high enough. Different gases produce different symptoms:

- ozone – a speckle of brown spots, which appear on the flat areas of leaf between the veins (hence also called interveinal necrosis)
- sulphur dioxide – larger bleached-looking areas
- nitrogen dioxide – irregular brown or white collapsed lesions on intercostal tissue and near the leaf edge
- hydrogen fluoride – brown necrosis at the tip or edge, working its way inward across the otherwise healthy tissue, with which there is a sharp division at a brown/red band
- ammonia – unnatural green appearance with tissue drying out.

9.1.3.2   Long-term response of agricultural yields to single gases

The economic importance of agricultural crops, together with the simplicity of working with single plant species compared to mixed natural systems, have made them the most heavily investigated plants. Research can be conveniently divided into NF/CF and additions experiments.

NF/CF experiments. Open-top chambers have been used in America and Europe to expose plants to air from which ambient pollutants have been removed by charcoal. Typically, plants from three treatments are compared – AA (an unchambered plot in ambient air), NF (chamber with non-filtered ambient air) and CF (chamber with charcoal-filtered air). Two factors are currently affecting the interpretation of such experiments. First, if the plants grow differently in the NF chambers compared to AA treatments, how should differences between CF and NF be interpreted? Second, with the small differences expected from low concentrations, how can we ever get enough statistical confidence from a practical number of replicates? In the US, only 7 of the 73 results from the National Crop Loss Assessment (NCLAN) programme showed a statistically significant yield reduction in NF compared to CF. It was concluded that the cumulative frequency of occurrence of intermediate ozone concentrations (50–87 ppb) was the best damage predictor, and this argument was supported with the observation that the highest concentrations tended to occur in the late afternoon when stomata were already closing.

Additions experiments. Experiments with laboratory chambers, open-top chambers (NCLAN and European OTC programme) and field release systems have shown changes in crop yield with increasing gas concentration for many species and varieties. Ideally, such an experiment involves the precise application of many different concentrations to different treatments so that response curves, called dose–response functions, can be fitted. One dangerous temptation with additions experiments is to apply unrealistically high concentrations in order to get a statistically significant response. This does not necessarily result in higher statistical confidence at low concentrations.

The dose–response function may take several forms: flat (e.g. barley and oats to ozone, Figure 9.5; a steady decline (e.g. wheat to ozone, Figure 9.6, or barley to SO_2, Figure 9.7); or a stimulation at low concentration followed by a decrease at higher concentrations (beans to ozone, Figure 9.8).

The common protocol of both the NCLAN and European OTC programmes involved varying the parameters of one family of curves to fit different experiments. The family chosen was the Weibull distribution, which gives the yield Y_i as a function of ozone concentration X_i:

$$Y_i = \alpha \exp\left[-\frac{X_i}{\omega} \right]^{\lambda}$$

where α = theoretical yield at zero ozone concentration, ω = ozone concentration giving a yield of $e^{-1}\alpha$ ($= 0.37\alpha$), λ = parameter controlling the shape of the

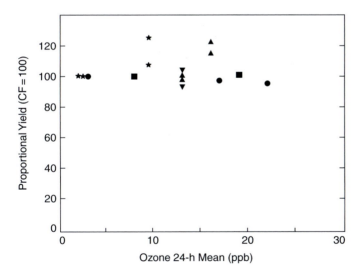

Figure 9.5 The responses of barley and oats yields to ozone in open-top chambers. ■, ▼ = oats; ●, ▲, * = barley. Yields are expressed as a percentage of the yield in charcoal-filtered air.
Source: Skarby, L. *et al.* (1993) 'Responses of cereals exposed to air pollutants in open-top chambers', In: H. J. Jager, M. H. Unsworth, L. de Temmerman and P. Mathy (eds) *Effects of Air Pollution on Agricultural Crops in Europe*, Air Pollution Research Report 46, CEC, Brussels.

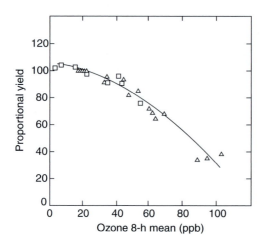

Figure 9.6 The response of spring wheat yield to ozone in open-top chambers. Yields are expressed as a percentage of the yield in charcoal-filtered air. Source: Skarby, L. *et al.* (1993) 'Responses of cereals exposed to air pollutants in open-top chambers', In: H. J. Jager, M. H. Unsworth, L. de Temmerman and P. Mathy (eds) *Effects of Air Pollution on Agricultural Crops in Europe*, Air Pollution Research Report 46, CEC, Brussels.

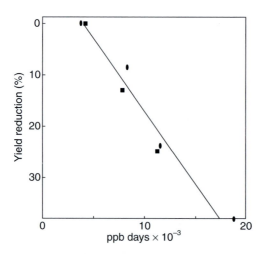

Figure 9.7 The response of winter barley yield to SO_2 in a field release experiment. Yield is expressed as a percentage reduction from that of unexposed control plants. Source: Baker, C. K., Colls, J. J., Fullwood, A. E. and Seaton, G. G. R. (1986) 'Depression of growth and yield in winter barley exposed to sulphur dioxide in the field', *New Phytologist* 104: 233–241.

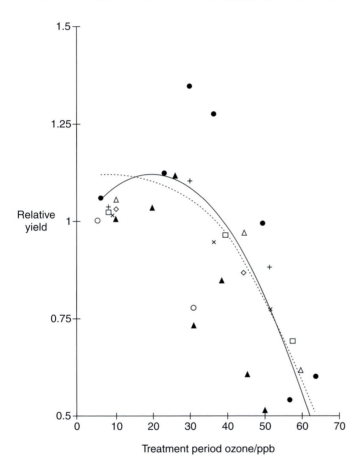

Figure 9.8 The response of green bean (*Phaseolus vulgaris* cv. Lit) yield to ozone. The yield is expressed as a proportion of the yield at an ozone concentration of 5 ppb.
Source: Colls, J. J. *et al.* (1993) 'The responses of beans exposed to air pollution in open-top chambers', In: H. J. Jager, M. H. Unsworth, L. de Temmerman and P. Mathy (eds) *Effects of Air Pollution on Agricultural Crops in Europe*, Air Pollution Research Report 46, CEC, Brussels, Belgium.

curve – usually in the range 1–5. The parameter λ can be adjusted to give curves that involve exponential decay, linear decrease, or a plateau followed by a sigmoidal decrease. Figure 9.9 shows the Weibull curves that were fitted by the NCLAN team to proportional yields from dicotyledenous agricultural crops. The simple Weibull distribution cannot give the stimulation and decline that has been observed in some experiments, but it can be modified to do so (by compounding it with a linear model).

A different approach has been used by a group in Germany to evaluate the combinations of ozone concentration (specified to be at the top of the plant canopy) and exposure period that might result in plant damage. They pooled results from published experiments and derived the relationships shown in Figure 9.10 for agricultural crops and wild plant species, and in Figure 9.11 for European coniferous and deciduous trees. In each case, the three shaded bands represent exposure–response regions in which direct ozone effects are unlikely, possible or likely, at the probability indicated.

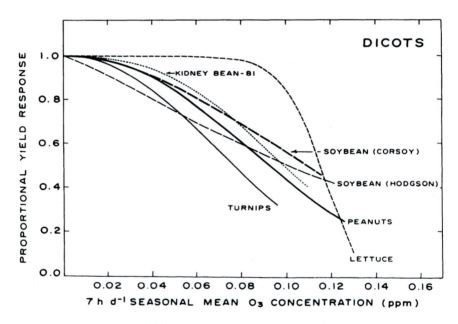

Figure 9.9 The range of Weibull curves fitted to experimental results for different dicotyledenous species during the NCLAN programme.
Source: Heagle, A. S., Kress, L. W., Temple, P. J., Kohut, R. J., Miller, J. E. and Heggestad, H. E. (1987) 'Factors influencing ozone dose–yield response relationships in open-top field chamber studies', In: W. W. Heck, O. C. Taylor and D. T. Tingey (eds) *Assessment of Crop Loss from Air Pollutants*, Elsevier, Barking, UK.

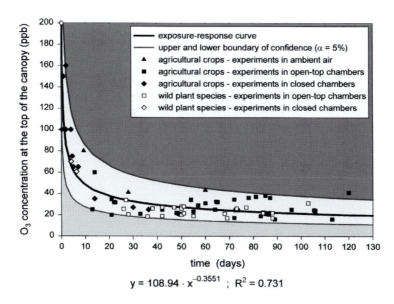

$$y = 108.94 \cdot x^{-0.3551} \; ; \; R^2 = 0.731$$

Figure 9.10 Ozone exposure-response for European crops and wild plant species.
Source: Grünhage, L., Krause, G. H. M., Köllner, B., Bender, J., Weigel, H.-J.,
Jäger, H.-J. and Guderian, R. (2001) 'A new flux-oriented concept to derive crit-
ical levels for ozone to protect vegetation', *Environmental Pollution* 111: 355–362.

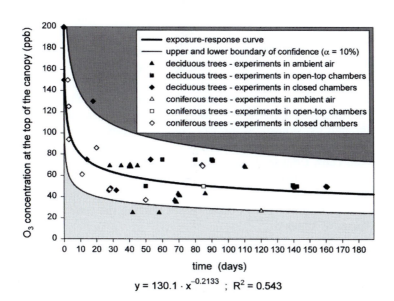

$$y = 130.1 \cdot x^{-0.2133} \; ; \; R^2 = 0.543$$

Figure 9.11 Ozone exposure-response for European trees.
Source: Grünhage, L., Krause, G. H. M., Köllner, B., Bender, J., Weigel, H.-J.,
Jäger, H.-J. and Guderian, R. (2001) 'A new flux-oriented concept to derive crit-
ical levels for ozone to protect vegetation', *Environmental Pollution* 111: 355–362.

9.1.3.3   Nutrient inputs

When atmospheric concentrations of SO_2 are high, plants can receive a significant proportion of their S input by deposition. There is evidence that in areas of the UK where there is limited sulphur available in the soil, some agricultural crops have benefitted from this atmospheric deposition. Conversely, as concentrations have declined, farmers in these areas have had to apply more S in fertiliser to compensate.

Plants in the UK are not thought to be significantly affected by the prevailing concentrations of NO or NO_2 *per se*. Nitrogen deposition, however, is another matter. Nitrogen is the key plant nutrient, and several hundred kg ha^{-1} may be added to agricultural crops to improve yield. Atmospheric inputs may be significant even on this scale. This is not likely to be important in areas where nitrogen availability is high. In many N-deficient areas these nitrogen inputs may affect the species composition, usually promoting grasses at the expense of shrubs, mosses and other understorey species that thrive in nitrogen-poor soils. This process has been most noticeable in parts of the Netherlands where high ammonia concentrations from intensive livestock production have led to atmospheric N deposition fluxes comparable to fertiliser inputs. High N deposition may also increase sensitivity to frost and droughts, and the severity of insect damage.

9.1.4   Pollutant entry into the plant

The simplest model of external leaf structure involves a waxy, relatively impermeable sheet of cuticle, punctuated at frequent intervals by stomata through which gas exchange occurs. The stomata, which are typically 10–20 μm long, are not simply open holes. They are protected by guard cells – pairs of sausage-shaped cells that open in response to turgor pressure. For most plants, the guard cells open during the daytime to allow gas exchange for photosynthesis while the sun is up, and close at night to conserve water. Gaseous pollutants are just as able to diffuse through the stomata as are CO_2 and water vapour. The single most important impact of air pollutants on plants is to impair the functioning of the stomatal guard cells. Conversely, it is unlikely that air pollutants will damage plants easily when the stomata are closed – at night or under drought stress, for instance. Although the area of the cuticle is far larger than the combined area of stomata, the rate of diffusion of pollutant gases through the cuticle is typically lower by a factor of 10 000.

The first site of action for ozone, as for other oxidative gaseous pollutants, is believed to be the cell plasma membrane. The integrity and full functioning of this membrane is particularly important for guard cells, because it controls water flow into and out of the cells, cell turgor and hence shape and stomatal aperture. If the guard cells do not function efficiently, there are several possible consequences: for well-watered plants, stomata will not open to their full capacity so that photosynthesis is reduced; for drought-stressed plants, they will not close as tightly so that water loss is increased; neither will they close as tightly when the plant is ozone-stressed, so that the ozone flux through the stomata is increased. Hence the ozone-exposed plant is in a lose-lose situation – it cannot photosynthesise as strongly when water is in good supply, and it cannot restrict water loss as effectively when water is scarce. Looked at from the opposite perspective, there are many experimental observations that stomata close in response to ozone exposure, potentially reducing gas exchange, photosynthesis and growth. There is a much smaller number of observations that suggest the opposite – that at low concentrations, ozone can stimulate stomatal opening.

Sulphur dioxide is converted to sulphite and bisulphite within the plant. These ions interfere with the fluxes into the guard cells of other materials such as potassium and calcium. When the relative humidity is high, these changes cause guard cells to become more turgid, so that they open wider and improve pollutant access. As with ozone, this opening may increase the transpirational water loss and lead to long-term water stress. At low humidities, stomatal apertures are reduced and the plant is protected from the pollutant.

9.1.5   Pollutants inside plants

Once pollutant gases have entered the stomata, they are in the highly aqueous environments of the substomatal cavity and the mesophyll. Their influence is strongly affected by their solubility in, and speed of reactions with, the water that makes up 90% of active plant tissue.

For SO_2, the sulphite and bisulphite from these initial processes are eventually oxidised to sulphate; during this sequence, reactive free radicals are formed that in turn promote disruptive effects such as oxidation of fatty acid double bonds. The reactions generate H^+ fluxes that disturb the pH – this change is buffered by proteins and amino acids. The rate constants of all these processes are not yet available, so quantitative response analysis is not possible. The overall effects of sulphur on plants depend on the exposure and other factors in the plants' sulphur environment. Since sulphur is a nutrient essential for growth, low concentrations of SO_2 may stimulate the growth of plants growing in sulphur-poor soils. On the other hand, high atmospheric concentrations offer no benefit if soil sulphur supplies are already adequate, and may only result in adverse effects due to the free radical formation.

As with sulphur, nitrogen is an element that is essential for plant growth, is usually accessed as nitrate via the roots from the soil, and is also taken up through the leaves when significant atmospheric concentrations of nitrogen oxides are present. Root nitrate is either reduced by nitrate reductase in the roots or is transported in the xylem flow to the leaves. Plants that are nitrogen-deficient are more likely to be at risk from atmospheric N deposition. Nitrogen oxides taken up through the stomata are thought to interfere with nitrate and nitrite reductase activity, although the exact damage mechanisms are not yet clear. It is thought that the main response is disruption of N metabolism by excess nitrite. Nitric oxide behaves quite differently to nitrogen dioxide, and these two gases should not be lumped together as NO_x for plant effects purposes. For example, NO is only sparingly soluble in aqueous fluids, while NO_2 reacts with water to increase its apparent solubility.

Plants respond to ozone by defensive or compensatory mechanisms such as stomatal closure, detoxification by chemical reaction, re-attainment of equilibrium by biochemical adjustment, repair of injured tissue. Many different responses have been observed, including:

- decreased photosynthesis
- decreased leaf area and leaf conductance
- decreased water use efficiency
- decreased flower and fruit number, and fruit yield
- decreased dry matter production and yield
- increased sensitivity to drought or nutrient stress
- increased or decreased disease susceptibility.

The type and degree of response depends on the intensity and timing of the ozone exposure, and on prevailing environmental factors, as well as on the species and cultivar. For example, a given ozone exposure will generally have a larger effect at high humidities than at low, because the stomata will be more open and gas conductance into the plant will be higher. Or a relatively low ozone flux into the plant may be more damaging late in the day when the defensive mechanisms have been weakened by previous exposure. Passive responses to ozone exposure involve the scavenging of ozone molecules by available plant antioxidants such as ascorbate and glutathione. Active defences involve the alteration of biochemical pathways or the production of additional antioxidants. There is only poor understanding of these processes in general. Ozone is highly reactive and is thought not to travel far from the stomata before interacting with other molecules and being transformed or decomposed. Hydrogen peroxide, some radicals and superoxide are known to be produced, but the proportions and the specific consequences for cell homeostasis are not yet understood. Neither do we yet understand rather simple damage phenomena, such as why visible ozone injury appears as spots of necrotic tissue, and the exact mechanism thereof. A popular recent hypothesis has been that ethylene, emitted from within the plant as a response to the ozone stress, reacts with the ozone within the plant to generate free radicals or hydroperoxides and disrupt cell biochemistry. Both ozone and sulphur dioxide generally decrease net photosynthesis and shoot:root ratio. Stomatal conductance may be either increased or decreased depending on other environmental conditions such as nutrient and water stress.

Plants also have long-term structural responses to pollutant stress, although these have not been investigated very intensively. For example, ozone exposure has been found to increase stomatal density on birch leaves – the birch responded to the reduced diffusive conductance of each stomata by increasing the number of stomata to maintain the overall rates of CO_2 diffusion and photosynthesis.

9.1.6   Exposure dynamics

There is a growing weight of evidence that we do not only need to consider the average concentration to which a plant is exposed, but also the concentration frequency distribution. This concept has only reluctantly been grasped by the effects community, because the extra level of variability makes it more difficult to compare one exposure with another, and to perform meaningful experiments, and to decide on practical guidelines for effects thresholds. It makes physiological sense, because the frequency distribution effectively defines the relative importance of the periods at low concentrations during which the plant has an opportunity to repair the damage done by episodes at high concentrations.

Exposure concentrations for ozone are normally described in terms of the treatment-period mean – the concentration averaged over the period of application – rather than the 24-h mean. This is a consequence of several factors: ozone concentrations at ground level follow a strong diurnal cycle as seen in Chapter 5, increasing after about 10 am to a maximum value at 2–3 pm before decreasing to the background value by about 5 pm; plant stomatal conductance follows a similar pattern; experimental ozone additions have frequently been made during the working day. Treatment means are therefore usually specified as 7-h or 8-h means. There is an increasing, and logical, trend to specify the mean as an average over daylight hours since that is the period over which the stomata will normally be open.

9.1.7   Premature senescence

It is now well established that one characteristic response to ozone stress is premature loss of leaves towards the end of the growing season. The higher the concentration, the lower the maximum number of leaves and the earlier in the season they die. This loss reduces the average green leaf area available for photosynthesis, and may therefore reduce the amount of plant growth.

9.1.8   Resistance

When a species grows in a region with high levels of a particular pollutant, natural selection may operate to promote plant genes that confer resistance to that pollutant and hence give a higher chance of long-term survival. We give two examples of experiments in which the tolerance of grasses to sulphur dioxide was investigated in contrasting ways. First, samples of grasses, including red fescue, were taken at different distances downwind of a smokeless fuel plant in Yorkshire. It was found that red fescue grown from seeds collected near the source was less sensitive to subsequent laboratory SO_2 exposure than that grown from seeds collected from further downwind (Figure 9.12). Second, grasses were sampled from lawns of different ages growing in the Temple district of central London. The red fescue from the oldest lawns (planted in about 1780, when

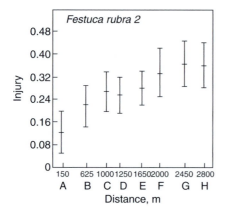

Figure 9.12 The injury to red fescue by SO$_2$ exposure in the laboratory, as a function of the distance from a point SO$_2$ source from which its seed was
harvested.
Source: Wilson, G. B. and Bell, J. N. B. (1986) 'Studies on the tolerance
to sulphur dioxide of grass populations in polluted areas. IV. The spatial
relationship between tolerance and a point source of pollution', *New
Phytologist* 102: 563–574.

pollution concentrations in central London were very high) was more tolerant to
laboratory SO$_2$ exposure than that from the newer lawns (the most recent one had
been planted in 1977) (Figure 9.13).

More recently, work at Newcastle has shown a corresponding response to
ozone. Seeds of the common Plantain (*Plantago major*) were gathered from 28
sites around the UK. As we have seen in Chapter 5, there is a strong gradient of
ozone concentration in the UK, increasing from north to south. The seeds were

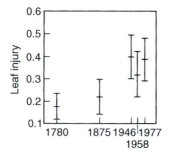

Figure 9.13 The injury to red fescue by SO$_2$ exposure in the laboratory, as a function
of the age of the plant population from which its seed was harvested.
Source: Wilson, G. B. and Bell, J. N. B. (1985) 'Studies on the tolerance
to sulphur dioxide of grass populations in polluted areas. III.
Investigations of the rate of development of tolerance', *New Phytologist*
100: 63–77.

germinated, exposed to ozone in laboratory chambers, and their relative growth rates (a measure of plant vigour) compared. The results, which are plotted as contours of ozone resistance in Figure 9.14, show that not only did ozone resistance also increase from north to south, but even reproduced the dip in ozone concentration found in the London urban region.

Figure 9.14 Contours of ozone resistance of plantain grown from seed. The resist-ance is the mean relative growth rate of ozone-treated plants as a percentage of the mean relative growth rate in clean air. The triangles mark the sites at which seed was collected.
Source: Reiling, K. and Davison, A. W. (1992) 'Spatial variations in ozone resistance of British populations of *Plantago major* L', *New Phytologist* 122: 699–708.

9.1.9   Trees

Attention has already been drawn to the special challenges posed by trees, because their long lives and large size make them difficult experimental subjects. Small effects are potentially much more serious than they are for annual plants such as most agricultural crops. For example, consider a tree that grows from being a 0.5 m seedling to its full height of 30 m in 100 years – equivalent to an average compound growth rate of 4.18% per year. Now suppose that air pollution stress reduces this growth rate by 1, 2, 5 or 10% per year. Table 9.1 below shows that these reductions all have appreciable effects on the final height, with the 10% reduction causing a significant 33% loss.

However, now consider the changes that would be measured in the course of the usual 3-year funded research programme, for a sapling that has an initial height of 3 m. (Table 9.2). Even the most severe growth reduction results in only a 41 mm difference in height over the 3 years, which would normally be extremely hard to detect because of the natural variability from tree to tree. The problem of detecting such tiny rates of change, in growth or any other parameter, remains unsolved.

Much of the work on tree responses has centred on the issue of 'acid rain', which is a popular term embracing the combination of dry and wet deposition of sulphur and nitrogen to vegetation. Although, there is now no doubt that acid deposition to fresh waters can destroy aquatic ecosystems, the cause and effect relationship for forests is far less clear-cut. The forest die-back that has been experienced in Germany and Scandinavia appears to be the result of many complex interacting factors, including drought, severe cold, disease and

Table 9.1 Effects of growth rate reductions on tree height

Reduction/%	Actual rate/% a^{-1}	Final height/m
0	4.18	60.0
1	4.14	57.8
2	4.10	55.5
5	3.97	49.1
10	3.76	40.1

Table 9.2 Reductions in tree height increment over a 3-year period

Reduction/%	Initial height/m	Ht after 1 year	Ht after 2 years	Ht after 3 years
0	3.000	3.125	3.256	3.392
1	3.000	3.124	3.254	3.388
2	3.000	3.123	3.251	3.384
5	3.000	3.119	3.242	3.371
10	3.000	3.113	3.230	3.351

management practice as well as air pollution. The roles of air pollutants are probably manifold

- direct effects of ozone and sulphur dioxide;
- interactive effects of ozone and sulphur dioxide on stomatal control, reducing photosynthesis and control over transpiration loss;
- nitrogen deposition stimulates new growth, which is then more susceptible to damage from severe frosts;
- ozone damages the leaf cuticle, so that acid rain can leach out cations such as calcium and magnesium;
- soil aluminium concentration is elevated by the lowered pH, which reduces the growth of the fine rootlets that are important for extracting nutrients from the soil.

For practical reasons, most experiments have been carried out for short periods on small juvenile trees and it has not been possible to reproduce the characteristic symptoms of forest dieback, such as storks' nests (adventitious shoots) and crown thinning (chronic defoliation of the upper parts of the tree). Although field surveys and recording of tree condition has been extensive in Europe and North America, much of the data has inevitably been semi-quantitative, using measures such as crown transparency which is obtained by visual assessment from the ground by a trained observer. Predictions of the demise of forests in Scandinavia and Germany have not been borne out, with continuing damage limited to relatively small areas such the black triangle in central Europe.

The Forestry Commission has operated a continuing survey of British tree health since 1986. Sitka and Norway spruce, Scots pine, oak and beech trees are surveyed at 80 sites, with a total 9600 trees. Results so far have not shown a clear dependence on air pollution. Other biotic (e.g. spruce aphid and winter moth attack) and abiotic (e.g. drought) stresses tend to obscure the geographical patterns that one might expect. Secondary effects may also be important. For example, air pollution might reduce the needle density in a coniferous tree, leading to a change in the insect or spider populations and hence in the populations of the birds that feed on them. Currently, there is not thought to be any hard evidence that either acid deposition or air pollution is resulting in chronic damage to UK forests. It remains possible that these anthropogenic stresses, especially that of increased nitrogen deposition, are enhancing the trees' sensitivity to the natural stresses, but that is an even more difficult case to prove.

Different kinds of damage have been reported for southern European forests. The ozone critical level for trees of 10 ppm · h (see Section 9.1.10) is exceeded over most of southern Europe, and visible damage is commonplace. There have been many reports on damage to coastal forests by polluted sea-spray. Strong onshore winds, without rain, blowing over seas that have been polluted with surfactants and/or oil, are the most damaging. As with the British forests, other damage symptoms may be confused with the effects of non-pollutant environmental stress and disease.

Effects of pollutant gases, especially ozone, have been studied in open-top chambers and field exposures. Again, the work has largely been carried out on seedlings or juvenile trees because the time taken and size needed make the use of mature trees impracticable. Unfortunately, it is not good practice to apply ozone exposures to already-mature trees, since they would have grown differently, and possibly adapted, had they been exposed to ozone throughout their lives. Many of the short-term responses reported above for crops have also been reported for trees, but it is uncertain to what extent these will become incorporated into the long-term response over periods of 50 or 100 years. Indeed, forests consist of communities of trees, and this aspect adds yet another layer of complexity and uncertainty. Since it is unlikely that really long-term experiments of sufficient quality will be carried out, predictions of long-term tree and forest responses will probably have to rely on scaling up and process modelling.

9.1.10   Critical loads and levels

In the 1985 Helsinki Protocol, many European countries agreed to try and achieve, by 1993, a 30% reduction from the 1980 sulphur emissions responsible for acid rain (the 30% club). This was a purely arbitrary target chosen for its political acceptability, without scientific basis. During the intervening period, it was recognised that a more rational basis for emission reduction would be needed for the future. The approach adopted has been that of critical loads and critical levels. According to this philosophy, which was first developed in Canada in the 1970s during investigations of lake acidification, the sensitivity of an ecosystem in any part of Europe to acid deposition or to gaseous pollutants will first be determined, and a critical load of wet deposition, or a critical level of gases, will be specified below which the ecosystem will not be significantly damaged. The critical loads are expressed in terms of acidity, sulphur or nitrogen fluxes (kg ha^{-1} a^{-1} of H$^+$, S or N) and the critical levels in terms of gaseous concentrations or doses (ppb, or ppb · h). Then modelling will be used to calculate the contribution to those loads and levels from different source regions of Europe. In cases for which the critical load or level is exceeded, emission reductions will be imposed to bring the region below the threshold of damage.

9.1.10.1   Critical loads

The definition of critical loads adopted by UNECE is 'a quantitative estimate of exposure to one or more pollutants below which significant harmful effects on - sensitive elements of the environment do not occur according to present knowledge'. The sensitivity of any particular ecosystem to pollutant inputs depends on physical, chemical, ecological, geological and hydrological factors. These can be mapped. Once a map of the spatial distribution of critical loads is available for a particular receptor–pollutant combination, comparison with current deposition loads (for example, of S and N from wet and dry deposition) generates a corresponding map of the areas of exceedance. A value of deposition above the critical load for a particular ecosystem is not, by definition, sustainable in the long term. The primary inputs to a calculation of critical loads are wet deposition, dry deposition of gases and particles and cloud droplet deposition. Reduced nitrogen species, such as gaseous ammonia and the ammonium in rain cloud and aerosols also contribute to acidification. The critical load of acidity for an ecosystem is determined by the net input of acidity, so that the atmospheric inputs of base cations – in particular, calcium and magnesium – must also be estimated. The dominant acidifying anion in precipitation is sulphate; areas of highest deposition are the high-rainfall areas of north-west England, west-central Scotland and north Wales. There is a clear gradient of decreasing wet deposition from north-west to south-east, with the low rainfall areas of East Anglia receiving 30–40% of their annual acidic inputs this way, compared to 80% for high-rainfall areas. The concentration of ions in cloud droplet water can be much higher than in rain water, and some areas of upland forest receive up to 30% of their acidic input by this route. In central and southern England, where rainfall amounts are lower and gaseous concentrations higher, dry deposition dominates.

The earliest critical loads were set to prevent soil chemical change (an increase in soil acidity or a decline in base saturation). It was then recognised that deposition of acidity, by whatever route, occurs to complex soil–plant–animal systems which may have many responses. The aim when setting critical loads for a given system was refined to determine the maximum load of a given pollutant which would not produce significant changes in the structure or function of the system. According to the formal procedure, a biological indicator is identified, the health of which is indicative of the whole soil–plant–animal system. A critical chemical value is then established for a given parameter above which the indicator will show an adverse response (e.g. a reduction in growth). The currently-favoured indicator is the health of the fine roots of trees, since these are believed to be sensitive to factors such as aluminium mobilisation that are associated with acidification. The favoured chemical value is the ratio of (calcium + magnesium) to aluminium in the soil solution. There are fundamental problems in defining critical loads – for example, should the critical load be set at the level that just causes a response in *current* (already damaged) ecosystems, or should it be set to a lower value that will allow the ecosystem to return to some previous condition? If so, what previous condition is the goal, and what acidity inputs are required to achieve it?

Once both total deposition and critical load maps of an area have been prepared, a critical load exceedance map can be made of the difference between the two values at each grid square. This map is the input to the emission/dispersion/deposition modelling process which is needed to identify which emission reductions will most efficiently reduce deposition to the exceedance grid squares and bring them below critical load. A high value of exceedance can be due either to a high value of deposition, or to a low critical load for that area. There are large possible variations – for example, the critical load for sulphur deposition can vary from 50 to 5000 mg S m^{-2} a^{-1}.

The principle of electroneutrality is that any change in the concentration of a negative ion (anion) contributed by acid deposition must be matched by corresponding changes in the concentrations of other anions or of positive ions (cations) so that overall charge neutrality is maintained in the surface water. Hence if sulphate and/or nitrate concentrations increase, some or all of the following changes will occur:

- decrease in bicarbonate anions, reducing the acid neutralising capacity
- increase in base cations (calcium, magnesium, sodium, potassium). This reduces acidification *per se*, but may deplete soil reserves and affect vegetation growth
- increase in hydrogen cation (decreased pH). Adverse effects on aquatic life
- increased aluminium cation. Adverse effects on aquatic life. This is the principal way in which fishes are affected by acidification.

An increase in the concentration of base cations in drainage waters due to acid deposition can work both ways. Removal of base cations from soils to balance sulphate or nitrate from acid deposition minimises the extent of surface water acidification. However, base cation reserves may be depleted from the soil over time if they are lost faster than they are replaced by deposition and weathering. The store of nutrients in the foliage, such as calcium, can be seriously depleted by acid deposition. This may alter the photosynthesis to respiration ratios and cause reductions in carbon fixation rates and growth rates.

The UNECE Second Sulphur Protocol, based on these critical load concepts, was signed in Oslo in July 1994, and further Protocols cover Nitrogen, Ozone and Volatile Organic Compounds. Modelling of the consequences of the Second Protocol indicated that there would only be a small recovery in the acid status of UK surface waters, and provided some of the arguments needed to develop the Gothenburg Protocol (see Chapter 13).

9.1.10.2 Critical levels

The corresponding definition of critical levels is 'the concentrations in the atmosphere above which direct adverse effects on receptors such as plants, ecosystems or materials may occur according to our present knowledge'.

International workshops were held in 1988, 1992 and 1993 to pool data on responses to ozone, sulphur dioxide and nitrogen oxides and decide on critical levels. Separate critical levels for crops and trees have been specified for ozone; they are based on the accumulated dose (expressed in ppb · h) over a threshold concentration of 40 ppb. Thus 1 h at a concentration of 55 ppb would add 15 ppb · h to the AOT40. AOT40s up to 20 000 ppb · h may be recorded in the areas of Europe with the highest ozone concentrations.

- for crops, 6000 ppb · h AOT40 for a 10% yield loss; 3000 ppb · h AOT40 for a 5% yield loss, accumulated during daylight hours (i.e. with solar radiation levels >50 W m^{-2}) during the 3-month period May to July

- for semi-natural vegetation, the same AOT40 is used as for crops
- for trees, 10 000 ppb · h AOT40, accumulated during daylight hours over a 6-month growing season
- serious plant damage can result from short episodes that do not cause the general critical level to be breached. The short-term critical level for ozone foliar injury involves an ozone exposure of more than 500 ppb · h during daylight hours within 5 consecutive days.

No critical level has been set for natural vegetation due to lack of sufficient decisive data.

It is important to note that, since damage for some species increases continuously with concentration rather than starting at some threshold, setting critical levels must involve some value judgement as to the level of acceptable damage. The 6000 ppb · hours critical level, for example, is expected to lead to a 10% reduction of crop yield below that of a control crop grown at background ozone concentrations. The critical level values are based on experiments with a small number of varieties of a small number of agricultural species – mainly wheat, barley and green beans. Although they are the best that can be done at present, they should only be regarded as crude and temporary indicators of plant responses to air pollutants.

There is a continuing debate about the best metric for predicting ozone impacts. The front runners at the moment are the AOT40 in Europe and the SUM60 in the US. SUM60 is the sum of all hourly mean concentrations ≥ 60 ppb. Unlike AOT40, the concentrations are added without subtracting the threshold, which makes it much less sensitive to threshold variability than the AOT40. The conclusion from the NCLAN programme mentioned above was that a 3-month, growing-season SUM60 of 26.4 ppm · h would protect 50% of the studied crops from yield losses of more than 10%. It is possible that more sophisticated metrics, weighted differently in different concentration ranges, will eventually be used. Future work may also lead to a redefinition of critical levels in terms of fluxes rather than concentrations. Peak monthly-average fluxes to wheat in Europe have been shown to be around 2 nmol m^{-2} s^{-1}, with cumulative fluxes over the growing season of around 7 mmol m^{-2}.

Ozone concentrations will eventually be reduced by the ongoing reductions in emissions of the ozone precursors (hydrocarbons and NO_x). It is now possible to predict the resulting changes in some detail, as shown in Figure 9.15. Although the sharp reductions in emissions of ozone precursors are expected to reduce AOT40 doses substantially, there will still be room for improvement, with significant areas of southern Europe exceeding the critical level in 2010. The main precursor sources for these exceedances are expected to be HGVs and evaporative emissions. In some areas, notably in parts of England, ozone concentrations were predicted to increase because there would not be so much NO around to react with the O_3.

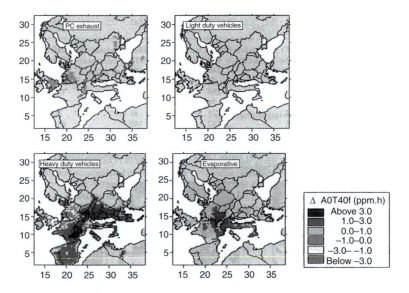

Figure 9.15 Modelled changes in ozone AOT40 in Europe, 1990–2010.
Source: Reis, S., Simpson, D., Friedrich, R., Jonson, J. E., Unger, S. and Obermeier, A. (2000) 'Road traffic emissions – predictions of future contributions to regional ozone levels in Europe', *Atmospheric Environment* 34: 4701–4710.

For sulphur dioxide, the long-term critical level has been set at 30 µg m^{-3} (11 ppb) for agricultural crops and 20 µg m^{-3} (8 ppb) for forests and natural ecosystems. It has also been proposed that in low-temperature regions (defined as being those for which the accumulated temperature does not exceed 1000 degree-days above 5 °C), the critical level should be set at 15 µg m^{-3} (6 ppb). This change recognises the field and laboratory evidence that SO$_2$ has a greater impact on plants in general, and trees in particular, at low temperatures. An additional critical level of 10 µg m^{-3} (as an annual mean) has been proposed for cyanobacterial lichens, which are especially sensitive to SO$_2$. This is a good example of the requirement that critical levels should be set low enough to encourage the recovery of damaged systems, rather than simply to maintain the *status quo*. Lichens have been eradicated from large areas of Europe, including central and north-west England, by centuries of SO$_2$ fumigation. Although we do not yet know whether they will recolonise, it is a reasonable assumption that the lower the SO$_2$ concentration, the faster the process will be. It should be noted that in practice, the only critical level that matters is the lowest one that applies to a grid square. Thus, although the critical level for forests might be 30 µg m^{-3}, all trees might be expected to support lichen, so that 10 µg m^{-3} will actually apply. Unless we decide that certain areas are forever to be lichen-free, 10 µg m^{-3} will have to be met everywhere.

For nitrogen oxides, the critical level for all plants has been set at an annual mean NO_x concentration of 30 $\mu g\ m^{-3}$ (16 ppb). The lack of discrimination between crops, trees and natural vegetation simply reflects the lack of experimental data on different plant communities. It is thought that natural mixed vegetation is likely to be more sensitive than managed monocultures, particularly for nitrogen deposition, because adequate nitrogen is usually applied to crops as fertiliser to increase yield.

Having defined critical levels, we are then in a position to compare them with measurements and hence to generate exceedance maps in just the same way as for critical loads. Figure 9.16 and Figure 9.17 show that, in the UK in 1994–1998, the ozone critical level for forests (10 000 ppb · h above 40 ppb over 6 months) was exceeded in the south of England only, whereas that for crops and

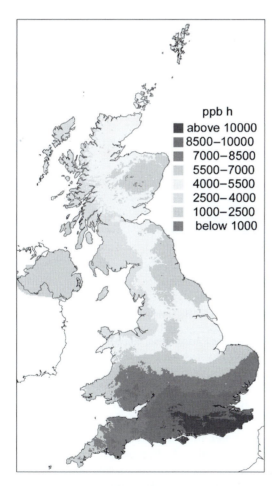

Figure 9.16 UK ozone AOT40 for forests, 1994–1998.
Source: National Expert Group on Transboundary Air Pollution (2001) *Transboundary Air Pollution: Acidification, Eutrophication and Ground-level Ozone in the UK*, Centre for Ecology and Hydrology, Edinburgh.

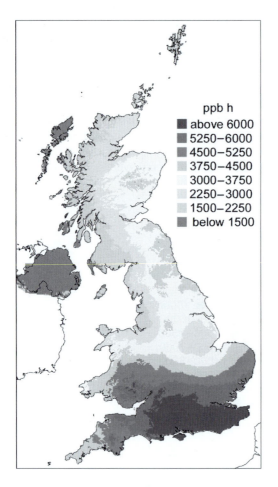

Figure 9.17 UK ozone AOT40 for crops and semi-natural vegetation 1994–1998.
Source: National Expert Group on Transboundary Air Pollution (2001)
*Transboundary Air Pollution: Acidification, Eutrophication and Ground-level
Ozone in the UK*, Centre for Ecology and Hydrology, Edinburgh.

semi-natural vegetation (3000 ppb · h above 40 ppb over 3 months) was exceeded
over most of the country. Average ozone concentrations are thought to be rising,
and this will probably increase the area in exceedance, although this depends to
some extent on the concentration variability. In Europe, data on ozone trends is
much more sparse than for the UK. The highest ozone concentrations are found in
the southern European and Adriatic countries. Peak AOT40s are currently up
to 14 000 ppb · h for annual plants (April–June) and 28 000 ppb · h for trees
(April–September). Both of these values are about three times the respective

UNECE critical levels. Even these values are probably underestimates at high altitude sites, because the ozone concentrations reported are for low altitude sites where the diurnal variation reduces the concentration in the evening, at night and in the early morning.

The critical level of sulphur dioxide for crops and trees is not now exceeded in the UK. However, the lowest lichen critical level is exceeded over a significant area of central England. Average SO_2 concentrations are falling, and will continue to do so while flue gas desulphurisation plants come on stream and coal continues to be displaced by gas. The database for rural NO_x is more sparse than for SO_2. Evidence from NO_2 measurements taken in 1991 suggests that the NO_x critical level would be exceeded over most of England. NO_x concentrations are expected to fall in the medium term as catalytic converters come into general use, then to rise again early next century as the vehicle population continues to increase.

9.1.11 Interactions

The majority of air pollutant effects experiments are carried out on plants that are experiencing no other stress – this is necessary in order to meet the usual scientific principle of only changing one variable at a time. In the real world, however, this condition is rarely met. Even crop plants, which represent the simplest plant monocultures and have their pests and diseases controlled for economic reasons, may experience drought stress. At the other extreme, a forest involves many plant and animal species, each of which, from the worms in the soil and fungi on the floor to the topmost leaf, may be affected directly or indirectly by air pollutants. There may be competition for resources between just two species (for example, a grass/clover sward) in which one species is more affected by a pollutant than the other. Alternatively, one stressor such as drought may affect the ability of a plant to respond to another stressor such as ozone.

9.2 VISUAL RANGE

Atmospheric particles reduce visibility by scattering and absorbing light. In Europe, this issue gets lower priority than many other environmental topics because there is no biological toxicity or clearly-definable cost associated with it; however, it results in a pervasive and apparent reduction in environmental quality. In the US, impairment of visibility has been a much more significant issue because of the large areas of pristine national park, and legislation has been designed specifically to protect and improve visual range. The original Clean Air Act made it a goal to remedy any existing impairment of visual range due to air pollution, but recognised that understanding of cause and effect was too limited at that time to bring in legislation.

9.2.1   Regional haze in the US

More recently, the Clinton–Gore administration announced in 1999 a major effort to improve air quality in national parks and wilderness areas. There is a natural variability of visual range with humidity and background aerosol; regional haze is visibility impairment (Table 9.3) caused by a multitude of sources and activities which emit fine particles and their precursors and which are located across broad geographic areas.

The value of atmospheric attenuation coefficient (σ) decreases (i.e. the visual range increases) with wavelength, with the value at 0.7 μm (red) typically being about half that at 0.4 μm (blue). Primary carbon and secondary sulphate particles have been identified as the main components of the aerosol responsible for light scattering. Sulphate has been associated with 45–90% of the light extinction coefficient on the 20% most impaired days. These particles or their precursors may be transported 1000 km from their sources. Hence all States are potentially responsible for visual range improvements, whether or not they contain Class I areas. The Regional Haze Rule calls for State and Federal Agencies to work together to improve visibility in 156 Class I areas, including Grand Canyon, Yosemite, the Great Smokies and Shenandoah. States must prepare regional implementation plans to show how they are going to meet the requirements, based on 3 years of $PM_{2.5}$ monitoring data. This monitoring period should have been completed by around 2003, and the first State Plans for haze control are due in 2003–2008, based on the designation of areas within the State as attainment or non-attainment areas. The States must set 'reasonable progress goals', expressed in deciviews (see below), to improve visual range on the most impaired days without allowing degradation on the clearest days. The eventual target is to achieve background natural visual range in a timescale of around 60 years. Wide-ranging measures might be needed to achieve the required targets, such as use of best available retrofit technology (BART) on particle and precursor sources, or emissions trading. This Regional Haze Rule recognises that the nationally-uniform particulate standards for $PM_{2.5}$ set for NAAQS would not necessarily restore the visual range in the Class I areas. In Class I areas in the Western US, current average visual ranges of 100–150 km are only 50–60% of their clean-air values.

Table 9.3 Variation of visual range and extinction coefficient

Descriptor	Visual range/km	Attenuation coefficient σ/km^{-1}
Haze	3	1.1
Light haze	8	0.5
Clear	15	0.28
Very clear	40	0.1
Exceptionally clear	60	0.07

The earliest haze measurements were made as coefficients of haze (COH). The COH is determined by drawing 300 m of air through a filter tape, and comparing the optical absorbance of the stain on the tape with standards. The COH = 100 × absorbance, and the critical value of six has been identified as a level that could cause adverse symptoms in sensitive individuals such as asthma sufferers. The US EPA uses deciviews to specify changes in visual range. The deciview is to visual range measurement what the decibel is to sound measurement. The definition is not very tight – 'a one deciview change in haziness is a small but noticeable change under most circumstances when viewing scenes in Class I areas.' The scale is derived by observing a large range of values of visual range for the same scenes and then comparing differences qualitatively to identify the smallest perceptible changes at different visual range values. It indicates the perceived change, and has the advantage that, for example, a three deciview change when the visual range is poor will mean as much to the observer as a three deciview change when the visual range is good. In contrast, an absolute change in visual range expressed in km would represent a huge change at poor visual range or a tiny one at large visual range.

These changes in visual range have also been related to the mass loading of particles in the air. From empirical analysis of extinction values, mass loadings and chemical analyses, a dry extinction coefficient has been developed for each aerosol component corresponding to the amount of light extinction (in units of Mm^{-1}, or inverse megametres) caused by 1 μg m^{-3} of that component. The dry value is then corrected for the effect of relative humidity.

We have already seen examples in this book of the size distributions of particles in the atmosphere from a multiplicity of natural and human sources. The optical properties are likely to be most strongly influenced by the surface area of the particles, rather than by their number or mass distributions. The key radius range for light scattering is around the wavelength of visible light, which covers the band 400–700 nm (0.4–0.7 μm). Although individual dust storms, which may be a result of wind erosion from poor land management practices, can produce very high particle loadings, the particles themselves are comparatively large in diameter, do not scatter light effectively per unit mass, and soon sediment out. In contrast, 'urban particles' are formed from gas-to-particle conversions (condensation of volatiles from combustion smoke, or photochemical formation of ammonium sulphate from ammonia and sulphur dioxide, for example), and are in the size range which not only accumulates in the atmosphere because Brownian diffusion and sedimentation are both ineffective, but also scatters light effectively.

9.2.2 Rayleigh scattering

Very small particles (<0.03 μm) scatter according to the Rayleigh mechanism that applies to molecules, so that the scattering coefficient is proportional to λ^{-4}. There are relatively few of these tiny particles in the atmosphere because they coagulate with each other very rapidly, so this strongly wavelength-dependent scattering mode is usually dominated by molecular scattering. Under some circumstances, however, such particles may be significant. The particles formed from the condensation of biogenic volatiles (e.g. pinenes from the conifers in the Smoky Mountains, US or from eucalyptus in the Blue Mountains, Australia) are around 0.1 μm in diameter; small enough to scatter blue light preferentially and create the eponymous blue haze. This wavelength dependence also accounts for the blue colour of cigarette smoke.

9.2.3   Mie scattering

At larger diameters, the scattering is described by Mie theory. The theory derives a scattering cross-section for each particle based on its diameter and refractive index and the wavelength of the light. The theory is very difficult to apply usefully, because we do not usually know the variation of refractive index with particle size distribution, and even if we did the calculations are extremely complex. Mie theory predicts a total scattering ratio, K_s, which is the ratio of the apparent scattering cross-section S to the geometrical cross-section.

$$K_s = S/\pi r^2$$

This factor K_s can then be incorporated into the Beer–Lambert law that describes the reduction of flux density through a uniform medium on a macroscopic scale:

$$I = I_0 \exp(-bL)$$

where I_0 is the initial flux density, I the flux density after the beam has passed through distance L of a medium having extinction coefficient b.

When scattering (the redirection of radiation without a change of wavelength) and absorption (the removal of energy followed by subsequent emission at a different, usually thermal, wavelength) are both present, b is the sum of a separate absorption coefficient (b_a) and a scattering coefficient (b_s), and we can write

$$b_s = \pi r^2 N K_s$$

where N is the number of scattering particles per unit volume. Since natural aerosols are always polydisperse (contain a range of sizes), any practical procedure involves the multiple applications of Beer–Lambert for different combinations of wavelength, particle size distribution and composition.

$$b_s = \pi \int_{r_{min}}^{r_{max}} r^2 \, n(r) K_s \, dr$$

Particles are highly effective light scattering materials, as can be seen from Table 9.4. The atmospheric extinction coefficient has been calculated for the 'air' molecules that make up most of the gas in the atmosphere, for typical concentrations of the common gaseous pollutants, and similarly for particles. The contributions of the gases depend both on their concentrations and on their scattering effectiveness in the visible wavelength range. Despite its overwhelming concentration, the air itself only makes a 6% contribution to the overall extinction coefficient. You really can see for ever on a clear day. In comparison, the tiny mass concentration of particles contributes nearly 90%. Note also that O_3 and NO_2 are much more effective than SO_2, which is one of the reasons for the severe effects of photochemical smog on visual range.

It should also be noted that it is the effectiveness of the particulate form of the material, rather than its opacity or refractive index, that is mainly responsible for the reduction in visual range. This can be seen most clearly by considering the effects of a specific concentration of water which is present either as vapour or as droplets. If 18 g m^{-3} water is present as vapour, the visual range would be over 200 km. If the same density of water is present as 10 μm droplets, the visual range drops to less than a metre.

9.2.4   Contrast reduction

An object can only be seen if there is an observable contrast between it and its background (i.e. a different flux density of light coming from the object as compared to that coming from the background). The concurrent processes of scattering and absorption *increase* the amount of light travelling from a dark object to the observer, and *reduce* the light coming from the background (or *vice versa*). The contrast between the object and the background is thereby reduced. When this contrast drops to about 0.01 the object is invisible for

Table 9.4 Calculated contributions of gases and particles to the atmospheric extinction coefficient[*]

	Material				
	Air	Particles	NO_2	SO_2	O_3
Concentration	1.3 kg m^{-3}	40 μg m^{-3}	50 μg m^{-3}	50 μg m^{-3}	200 μg m^{-3}
Extinction coefficient at 550 nm/Mm^{-1}	11.6	160	8.6	10^{-5}	1.0
Contribution to extinction/%	6.4	88.3	4.7	<0.05	0.6

Note

[*] The units for extinction coefficient are per million metres, or per 1000 km. An extinction coefficient of 10.0 Mm^{-1} means that the flux density will have been reduced by a factor of ten over a distance of 1 Mm.

practical purposes. Hence the visual range, commonly known as the visibility, is conventionally taken to be the distance in km at which the contrast between a white object and its dark background has fallen from 1 to 0.01. Even in a particle-free atmosphere, molecular scattering would limit the visual range to around 300 km.

Contrast reduction follows the Beer–Lambert law:

$$C = C_0 \exp(-bL)$$

where C_0 is the initial contrast and C the contrast as observed at distance L. If the object is black and white, then $C_0 = 1$ by definition. The critical contrast at which the object can no longer be discerned has been the subject of some discussion. The most careful measurements have given values of C/C_0 of less than 0.01. However, this means that the object will only be distinguished on 50% of occasions at that level of contrast. To be on the safe side, the threshold has been internationally accepted as 0.02, which leads to Koschmeider's formula

$$L = 3.91/b$$

where $3.91 = \ln(0.02)$. Thus for $b = 0.004 \text{ m}^{-1}$, which is typical of a photochemical haze, $L = 1$ km. This is a very crude approximation, which varies with the physical properties of the particles, with the initial darkness of the object, and between individual observers. Nevertheless, it has been adopted to define the international visibility code (Table 9.5).

The combined extinction coefficient of the atmosphere specified in Table 9.4 is 181 Mm^{-1}, which corresponds to a Koschmeider visual range of 22 km.

Table 9.5 International visibility code

International visibility code			
Code no	Description	L/km	b/km^{-1}
0	Dense fog	<0.05	>78.2
1	Thick fog	0.05–0.2	78.2–19.6
2	Moderate fog	0.2–0.5	19.6–7.8
3	Light fog	0.5–1.0	7.8–3.9
4	Thin fog	1–2	3.9–2.0
5	Haze	2–4	2.0–1.0
6	Light haze	4–10	1.0–0.39
7	Clear	10–20	0.39–0.20
8	Very clear	20–50	0.20–0.08
9	Exceptionally clear	>50	<0.08

9.2.5   Hygroscopic particles

Many atmospheric particles, such as the natural sea-salt nuclei on which cloud droplets form and the man-made salts of ammonium nitrate and ammonium sulphate, are hygroscopic (they absorb water vapour from the air). At a sufficiently high relative humidity they deliquesce (dissolve into their own water content) to form a highly saturated solution droplet. As the relative humidity increases further, each droplet takes up more water vapour, increases in diameter and scatters more light. Figure 9.18 gives examples of calculated and measured changes of extinction coefficient (on the left axis, with the corresponding visual range on the right axis) with humidity for particles of both ammonium sulphate (on the left) and ammonium nitrate (on the right).

At high relative humidities the visibility becomes a sensitive function, so that when such particles accumulate in stable air under a temperature inversion, rising atmospheric humidity can increase the effective size of the scattering centres and reduce the visible range, a characteristic feature of photochemical smogs. The difference between hygroscopic and non-hygroscopic particles is shown more clearly in Figure 9.19. Aerosol collected at the Grand Canyon was exposed to controlled relative humidities in the range 10–96%, and the scattering coefficient

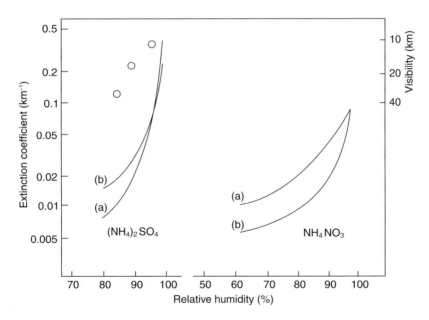

Figure 9.18 The variation of extinction coefficient (km^{-1}) with relative humidity for ammonium sulphate (left) and ammonium nitrate (right) aerosols. The salt concentrations are both 1 μg m^{-3}. The calculated curves (a) are for monodisperse aerosol; those marked (b) for polydisperse aerosol; and the open circles are experimental points for an ammonium sulphate concentration of 20 μg m^{-3}.

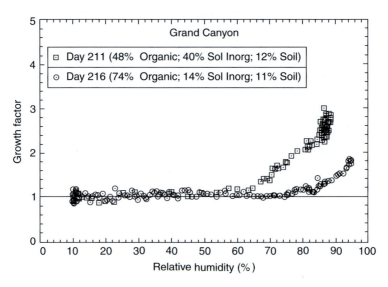

Figure 9.19 Scattering growth factors for aerosol at Grand Canyon.
Source: Day, D. E. and Malm, W. C. (2001) 'Aerosol light scattering measurements as a function of relative humidity: a comparison between measurements made at three different sites', *Atmospheric Environment* 35: 5169–5176.

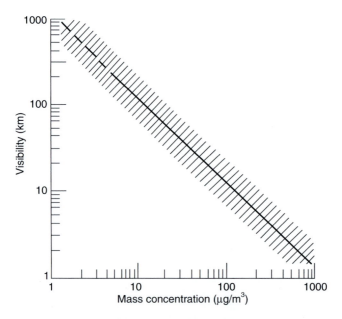

Figure 9.20 A general relationship between atmospheric visual range (km) and particle mass concentration, from US data.

determined. The change in scattering coefficient is expressed as the growth factor, which is the ratio of the scattering coefficient in the humid air to a reference scattering coefficient in dry air. On Day 216, when the proportion of soluble salts was low, the growth factor was quite insensitive to humidity. On Day 211, when soluble salts comprised 40% of the aerosol, the growth factor started to increase at a relative humidity of 60% and eventually rose by a factor of three.

An overall relationship between particle concentration and visual range, taken from the US National Acid Precipitation Assessment Program (NAPAP), is shown in Figure 9.20. The concentrations of total suspended particulate matter normally measured in the English Midlands (around 30 $\mu g\ m^{-3}$) would result in a visual range of around 20 km. The NAPAP data also showed clearly the relationship between increasing SO_2 emissions and the reduction in visual range

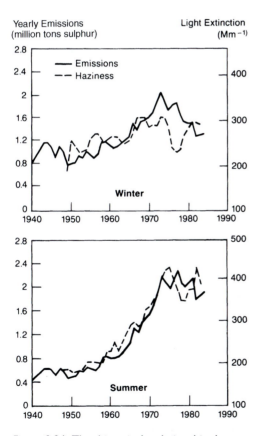

Figure 9.21 The historical relationship between sulphur emissions and atmospheric scattering for the south-eastern US, 1940–1990. Upper graph – winter; lower graph – summer.
Source: NAPAP (1990) *Integrated Assessment Report*, National Acid Precipitation Assessment Program, Washington, DC.

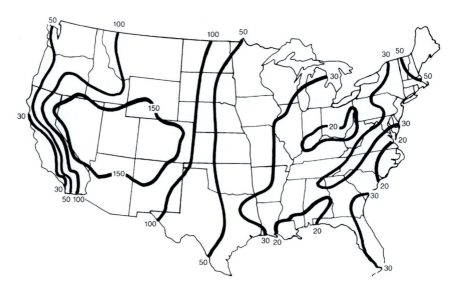

Figure 9.22 Contours of median visual range (km) for rural areas of the US.
　　　　Source: NAPAP (1990) *Integrated Assessment Report*, National Acid
　　　　Precipitation Assessment Program, Washington, DC.

due to sulphate aerosol (Figure 9.21), particularly during the summer months
(lower graph) when photochemical particle production rates are highest. Visual
range contours over the US (Figure 9.22) confirmed the expected pattern of high
range in less populated regions and owest ranges associated with the urban/in-
dustrial areas on the east and west coasts.

9.3   DAMAGE TO MATERIALS

Atmospheric corrosion is the consequence of the interaction between any material
such as metal, stone, glass, plastic or paint, and the gases and particles in the air.
The key mediating factor is water, which is present in thin films on most surfaces.
Even at 20% relative humidity there will be one or two equivalent monolayers
present, increasing to eight or ten at 80%. The physical distribution of these water
molecules depends strongly on the surface characteristics – whether it is
hydrophilic or hydrophobic, for example. Particles in the atmosphere cause
complications – hydrophilic or even hygroscopic particles may be deposited onto
a hydrophobic surface. Water molecules may bond in either molecular or dissociated form, with the latter resulting in the formation of surface hydroxyl groups.
Atmospheric gases such as CO_2, NO_2, SO_2, HCl, H_2S can all dissolve into the
water film. Aerosol particles supply sulphates, chlorides and nitrates.
Electrochemical reactions then occur at the interface between the solid substrate

and the water layer. Local microvariability means that some surface sites will support anode (e.g. metal dissolution, electron-generating) reactions, and others cathode (electron-consuming, reduction of oxygen to OH^-) reactions. For the cathodic sites, the required electrons are supplied by the easy diffusion of atmospheric oxygen through the thin water films. The composition of the aqueous layer changes dramatically and rapidly with the weather conditions, with orders of magnitude changes in ionic concentration, and hence several pH units, depending on the relative humidity. Eventually corrosion products, often hydroxides or oxyhydroxides, are precipitated from the ionic solution and cover the material surface. Corrosion rates are then affected by the transport of ions inward and outward across the layer, or by the transport of electrons from anodic to cathodic sites. We can expect corrosion to depend both on gaseous deposition velocity (supply of new gas to the water film) and on solubility (storage of material in the film). It may be very difficult to separate out the current effects from the historical ones. For example, even after a damaged stone surface has been cleaned, the effects of past pollution may make the stone more susceptible to current pollution.

9.3.1 Building materials

Attention in the UK has been mainly directed at historic monuments such as St Paul's (London), Lincoln and York cathedrals, which stand either in areas of high urban pollution or downwind from coal-fired power stations. On any stone building, we normally find a mixture of 'clean' surfaces that are stone-coloured, and blackened surfaces. The latter soiling may in itself be a serious environmental problem. The clean surfaces are exposed to falling rain and running rain water; damage is mainly due to rapid deposition of gaseous SO_2 to wet stone surfaces, rather than to the acid rain itself. The blackened surfaces are regions protected from the direct effects of the acid rain. Particles that are deposited to these surfaces are not washed off. Eventually the particles, together with dry-deposited gases such as SO_2, produce a crystalline gypsum lattice within the stonework interstices. The volume occupied by hydrated calcium sulphate is 1.7 times that of the carbonate that it replaces, so that pressure from the crystal growth breaks off small areas of stone, also exposing a more vulnerable surface. Furthermore, the loss of the sulphur results in the remaining solution becoming increasingly acidic, and dissolving the original carbonate stone more rapidly. All building stone is porous to a degree, and some is very porous. While the porosity of marble may be only 1%, that of limestone may be as high as 40%. Hence water, and any chemical species dissolved in it, has access to the interior of the stone as well as to the outside surface. The primary role of water in stone corrosion, as compared with metal corrosion, is to provide the medium into which the calcium carbonate can dissolve. The products are carbonates such as dolomite, sulphites, sulphates, nitrates and oxalates.

At St Paul's, lead dowels used in the original construction have been used to measure the long-term erosion rates of stonework. The dowels were originally flush with the stone surface, but have been left standing proud as the stonework has decayed. The average erosion rate over the period 1727–1982 was found to be 80 μm a^{-1} ($=$ 8 mm century^{-1}). In short-term measurements, the erosion rate was found to be higher (up to 220 μm a^{-1}) but this was an interpretation of a mass-loss measurement, and some of the mass could have been coming from below the surface. Of course, even clean rain is already weak carbonic acid (pH 5.6) that dissolves limestone, and building damage must be seen in this context. Portland stone, for example, has long been favoured as a building material because of its known resistance to pollution. In the National Materials Exposure Programme, test specimens of stone (Portland limestone, dolomitic sandstone and Monks Park limestone) and coated and bare metals (mild and galvanised steel, painted steel, aluminium and copper) were put out at 29 sites across the UK in the period 1987–1995 to map the effects, and to relate them to local SO_2 concentrations and acid deposition rates. The metal samples (75 $\times$ 50 $\times$ 3 mm) were mounted on sloping racks and exposed to direct rainfall. The stone samples (50 $\times$ 50 $\times$ 8 mm tablets) were either exposed to, or sheltered from, direct rainfall. Provisional results from these experimental programmes show that linear damage functions can account for up to 70% of the observed variation of surface erosion with environmental factors. Damage to stone is measured by weight change, because dissolution is not only a surface effect. At remote (unpolluted) sites with high rainfall, damage to exposed stone is dominated by dissolution losses, whereas sheltered stone at high SO_2 may show a weight gain due to gypsum formation. In direct comparisons, the erosion rates of stone samples in London were found to be three times those of identical samples in rural Hertfordshire.

9.3.2   Other materials

Sulphur dioxide reacts with many other materials, such as the zinc on galvanised steel (corrugated iron, electricity pylons). The former Central Electricity Generating Board was responsible for maintaining 20 000 galvanised steel pylons, and was therefore interested in the effects of pollution in shortening pylon life. Test squares of galvanised steel were put out all over the country; results showed a strong correlation between corrosion rate and known sources of SO_2, as well as with proximity to the sea. Another important effect is the expansion of metal as it corrodes. This process exerts huge pressures on the surrounding structure, which it can eventually split. Additional results from NAPAP have shown that corrosion rates in industrial US cities have declined in step with SO_2 concentrations, and that corrosion rates have been low in areas of low sulphur dioxide concentration. Paint consists of a substrate (natural or synthetic oil or resin) and a pigment (e.g. titanium dioxide, white lead, zinc oxide). Both the substrate and the pigment are subject to attack by SO_2, H_2S and O_3. SO_2 also attacks the paper in books and the leather on the binding of hardbacks, because metallic impurities in the paper and leather catalyse the conversion of SO_2 to sulphuric acid.

Ozone is a powerful oxidising agent, responsible for the decay of elastic materials by attacking the double bonds of compounds such as butadiene-styrene.

One of the factors responsible for the loss of elasticity of rubber bands is ozone in the air; in fact, the depth of cracks in rubber was used as an early O_3 measurement technique in California. Other effects are insidious and hard to control. For example, ozone bleaches the colours on paintings. Since paintings are expected to last for hundreds of years, even very low O_3 concentrations of around 1 ppb can supply a significant dose (= concentration $\times$ time) and threaten the appearance (and hence value) of works of art. Museums and galleries have to consider how to reduce the air concentration of ozone down to 1 ppb (compared to the ambient background of around 20 ppb and peak levels in Los Angeles of around 600 ppb). This can be achieved by charcoal filtration, but only if very tight controls are imposed on door opening and other routes for infiltration of ambient air. NO_2 has also had serious effects on colours of materials – in particular, the NO_2 generated by indoor gas fires discoloured nylon fabrics until various additives in the fabrics were changed by the manufacturers.

FURTHER READING

Agrawal, S. B. and Agrawal, M. (eds) (2000) *Environmental Pollution and Plant Responses*, CRC Press, Boca Raton, Florida.

Bussotti, F. and Ferretti, M. (1998) 'Air pollution, forest condition and forest decline in Southern Europe: an overview', *Environmental Pollution* 101: 49–65.

Critical Loads Advisory Group (1994) *Critical Loads of Acidity in the United Kingdom*, DoE/ITE, London, UK.

Heck, W. W., Taylor, O. C. and Tingey, D. T. (eds) (1988) *Assessment of Crop Loss from Air Pollutants*, Elsevier Applied Science, New York, USA.

Leygraf, C. and Graedel, T. (2000) *Atmospheric Corrosion*, Wiley, New York.

Singh, S. N. (ed.) (2000) *Trace Gas Emissions and Plants*, Kluwer Academic, Dordrecht.

Smith, W. H. (1990) *Air Pollution and Forests*, Springer-Verlag, New York, USA.

Treshow, M. and Anderson, F. K. (1989) *Plant Stress from Air Pollution*, Wiley, Chichester, UK.

United Kingdom Terrestrial Effects Review Group (1988) *The Effects of Acid Deposition on the Terrestrial Environment in the United Kingdom*, Department of the Environment, London, UK.

United Kingdom Terrestrial Effects Review Group (1993) *Air Pollution and Tree Health in the United Kingdom*, DoE/HMSO, London, UK.

Yunus, M. and Iqbal, M. (1997) *Plant Response to Air Pollution*, John Wiley & Sons, Chichester.

Chapter 10

Responses of humans and other animals

In this chapter on air pollution effects, we will look at the responses of animals, including people.

10.1   RESPONSES OF PEOPLE

As with plants, our understanding of human responses to air pollutants has come via a number of different routes. First, air pollution disasters where the effects have been clear-cut; second, epidemiological studies of affected populations; third, controlled experiments on individuals and animal models. A number of different types of adverse health effect have been attributed to air pollution, although the severity of response depends greatly on the type of pollution, the level of exposure and individual susceptibility. Typical health effects include

- reduced lung functioning
- irritation of the eyes, nose, mouth and throat
- asthma attacks
- respiratory symptoms such as coughing and wheezing
- restricted activity or reduced energy level
- increased use of medication
- increased hospital admissions
- increased respiratory disease such as bronchitis
- premature ('brought forward') death.

The WHO defines health as the state of complete physical, mental and social well-being, rather than the absence of disease or infirmity. By this definition, very few people could be classed as completely healthy. Hence health effects of pollutants (or anything else) become a matter of degree. The issue is further complicated by variations in public perception – for example, part of the reason for the steady increase in reported asthma cases is the increased public awareness that asthma exists and might be treatable.

10.1.1   Episodes and epidemiology

10.1.1.1   Severe combustion smogs

Ever since coal became the primary urban and industrial energy source, well before the industrial revolution got up to speed in the eighteenth century, people had suffered from exposure to high concentrations of toxic gases and particles. It is likely that significant numbers of people died from these exposures, but were not recorded as such. Matters came to a head in the middle of the twentieth century, when severe pollution incidents in the Meuse Valley, Belgium in 1930; Donora, Pennsylvania in 1948; and London, England in 1952 and 1962 marked the peak of acute air pollution problems due to emissions from fossil fuel combustion. In all these cases, emissions from industrial and domestic coal combustion were trapped by topography and/or atmospheric temperature inversions. Concentrations of SO_2 and particles increased to levels that seem amazingly high today – in the 1962 London episode, the daily average SO_2 was 3000–4000 $\mu g\ m^{-3}$ on three successive days, associated with peak particle loadings of 7000–8000 $\mu g\ m^{-3}$. Both morbidity (the illness rate) and mortality (the death rate) were clearly correlated with pollution, as is shown by the London data for 1962–1963 (Figure 10.1). Not only are the main spikes in early December coincident, but many of the minor features at other times within this period are apparently correlated. There were about 60 deaths that were statistically attributable to air pollution in the Meuse, 20 in Donora, and an extraordinary 4000 and 700 in London in 1952 and 1962 respectively. Mortality was highest among groups with existing respiratory or cardio-vascular disease. Morbidity was characterised by chest pain, coughing, shortness of breath, and eye, nose and throat irritation. Retrospective surveys have shown that similar but more minor incidents had occurred previously but not been recognised.

London had been known for its high incidence of fog for centuries. The severe air pollution fog, or smog (smog = smoke + fog) was due to a synergism between SO_2 and soot particles. On its own, SO_2 is not toxic at the concentrations experienced. For example, maximum urban concentrations were around 3–4 ppm. The statutory safe limit for occupational exposure to SO_2, in which one is supposed to be able to work for the whole of one's working life, is 5 ppm, and the concentration has to be around 100 ppm before it is lethal. The respiratory tract reduces exposure to toxic gases. Gaseous SO_2 has a fairly high solubility in the aqueous mucosal fluids, so that about 95% has been absorbed before the inhaled air reaches the lungs, and only 1% penetrates to the gas-exchange surfaces of the alveoli. However, another penetration route is available when soot particles and water vapour are also present. Water vapour and SO_2 are absorbed onto the soot particles while they are all present together in the ambient air, and trace metals such as vanadium in the particles catalyse the formation of sulphuric acid. These particles are then inhaled. This mechanism focussed attention on particle inhalation processes and led to the identification of the aerodynamic size classes such as PM_{10} and $PM_{2.5}$ discussed previously.

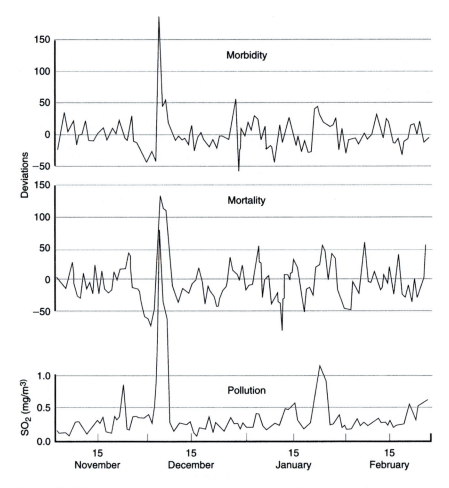

Figure 10.1 The variations of morbidity, mortality and SO₂ concentration during smog
episodes in London between November 1962 and February 1963.
Source: Quality of Urban Air Review Group (1993) *Diesel Vehicle Emissions and
Urban Air Quality*, Department of the Environment, London.

Compilation of data on morbidity, mortality and pollution concentrations from
a range of smog episodes (Figure 10.2) indicates the bands of joint occurrence of
particles and sulphur dioxide within which health effects of different severities
may be expected. There is a trade-off between the two pollutants, with higher levels of SO_2 being acceptable when the concentration of particles is lower, and *vice
versa*. It was concluded from the London smogs of 1952, 1956, 1962 and 1975
that a short-term significant response from chronic bronchitics to an air pollution
episode could not be detected when the daily mean concentration of SO_2 was
below 500 $\mu g\ m^{-3}$ while the smoke concentration was at 250 $\mu g\ m^{-3}$. This

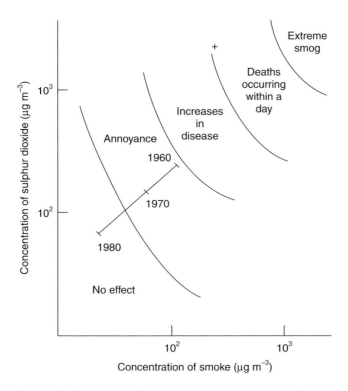

Figure 10.2 The health effects of combined exposure to SO_2 and smoke. The cross marks the worst day of the London smog in December 1952, while the straight line shows the trend in annual mean London concentrations between 1960 and 1980.

synergism has been incorporated into legislation: the EC, for example, has previously specified an annual mean of 120 μg m^{-3} of SO_2 unless the particle concentration is greater than 40 μg m^{-3}, in which case the upper limit for SO_2 is reduced to 80 μg m^{-3} (see Chapter 13).

10.1.1.2   Chronic low concentrations

The combustion smogs that were experienced in London in 1952 and 1962 were extreme examples of the responses of a large group of people to an air pollutant. In those and similar cases, the correlation between pollutant concentrations and morbidity or mortality was clearly identifiable. In the developed industrial countries, the reduction in air pollutant concentrations has almost eliminated these severe episodes. At the lower air pollution concentrations currently prevailing, effects are still probable, but relating them to a specific cause becomes much more difficult. The identification of the causes of a disease from a study of its occurrence is known as epidemiology. The results of such studies can be used to determine the effects of a particular pollutant, and to predict the benefits of any future reduction. Many different experimental designs have been used.

- In a case-control study, the previous pollutant exposures of diseased individuals are compared with those of non-diseased controls.
- In a cohort study, disease incidence in groups having different exposures is compared over a period.
- In a cross-sectional study, a snapshot is taken of disease incidence and pollution exposure.

The Six Cities study, for which some results are described in Section 10.1.1.3, is a good example of a cohort study. About 20 000 subjects (adults and children) in the six cities were identified between 1974 and 1976. The cities themselves were chosen to represent a gradient of pollutant concentrations. Key pollutants were measured. The subjects were tracked for 17 years, during which periodic measurements were made of lung function and respiratory symptoms. Data were also compiled on their morbidity and mortality. Much effort was put into reducing the effects of so-called confounding factors, which are the non-air pollutant factors that will vary between the populations and affect morbidity and mortality. Such factors include income, occupation, diet, exercise level, and smoking.

More recently, time-series analysis has been used to derive the relationships between concentrations and responses. Such analyses use sophisticated statistical methods such as Fourier analysis and autoregression to discriminate the signal (changes in morbidity or mortality due to pollutant concentration changes) from the noise (changes in morbidity or mortality due to other causes). For example, although high and low temperatures both increase morbidity, high and low O_3 have opposite effects. Effects of confounding variables are reduced in time series analysis, because these characteristics are relatively constant within the population from day to day. Seasonality (the recognised long-term correlation between time of year, pollutant concentrations and population responses) is the main confounding factor that must be taken into account.

There are many exposure metrics for which we need to know the response relationship. For a single pollutant such as ozone, there are acute responses to individual short term peaks (hourly means), or to sequences of such peaks (as would be experienced during a protracted ozone episode under unfavourable meteorological conditions). For such exposures the recovery time after each peak influences the severity of response to subsequent peaks. There are also chronic responses to living in an area that has high average ozone concentrations. In addition to these types of response for individual pollutants, there are many combinations of co-occurrence with either additive or antagonistic effects.

10.1.1.3   Photochemical smogs

As already described, long-term measurements of urban SO_2 and particle pollution have shown concentrations falling steadily during the second-half of the twentieth century. Acute pollution disasters of the type that occurred in London in 1952 have been averted, and short-term correlations between illness and pollution concentrations cannot now be seen. However, evidence is accumulating that chronic levels of air pollution are responsible for significantly reducing the lifespan of the exposed population. Furthermore, there has been a steady increase in the incidence of respiratory illnesses such as asthma, which is strongly associated in the public's mind with vehicle pollutant emissions and photochemical smog. In fact, there is not yet sufficient evidence to show that specific pollutants, or air pollution in general, either initiates or promotes asthma. Major research programmes are going on around the world to decipher the interrelationships between air quality and health.

In the Harvard Six-Cities study, the results from which were published in 1993, cohorts of young people were selected in the 1970s from six US cities. Their health records and mortality rates were studied throughout the following two decades, and the concentrations of various air pollutants were also measured. The health of the cohorts was expressed in terms of a relative mortality, after allowing for confounding factors such as age, body weight, income and tobacco consumption. Figure 10.3(a) shows that there was indeed a relationship between relative mortality (here expressed as the ratio of the mortality rate in the target cohort to the rate in the control) and PM_{10} concentration; an even stronger relationship (Figure 10.3(b)) was obtained when only the concentration of fine particles less than 2.5 μm was included.

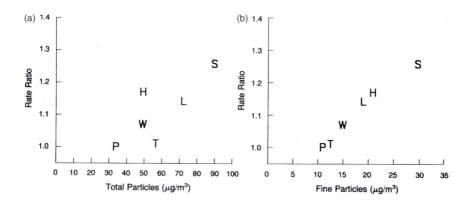

Figure 10.3 (a) The mortality ratio in six US cities as a function of total particle concentration. P is Portage, Wisconsin; T is Topeka, Kansas; W is Watertown, Massachusetts; L is St Louis; H is Harriman, Tennessee and S is Steubenville, Ohio. (b) The mortality ratio in six US cities as a function of $PM_{2.5}$ concentration.
Source: Dockery, D. W. *et al.* (1993) 'An association between air pollution and mortality in six US cities', *New England Journal of Medicine* 329: 1753–1759.

As mentioned above, this is the fraction that can penetrate right down into the alveolar region of the lungs. This study showed that particle concentrations similar to those measured in European cities increased the relative mortality by 26%, and reduced average lifespan by 2–3 years (compared with 15 years for smokers). The present view resulting from this and other work is that an increase of 10 μg m^{-3} in PM$_{10}$ increases hospital visits by 2–4% and total mortality by 1%, and that this gearing increases to 3.4% in the case of respiratory mortality. It has also been estimated that if the US had achieved the targets for sulphate concentrations set in the Clean Air Act Amendments of 1997, around 2500 premature deaths would have been saved that year in the Eastern US, with a monetary value of around $9 billion. The reduction would have had a correspondingly impressive effect on morbidity such as asthma, cardiac and respiratory illness. Attention has also been paid to ultrafine particles (less than 100 nm). These are always present in very large numbers (although their mass loading is usually small). Their small size makes them very reactive (large surface area per unit mass) and they may be able to penetrate through airway walls and into blood vessels.

Combustion smogs were identified by the co-occurrence of carbonaceous particles and sulphur dioxide gas. In a similar way, sulphur dioxide interacts with sulphate particles to make it hard to separate the health effects of one from the other. The emitted SO$_2$ is converted in the atmosphere to SO$_3$, which then combines with water vapour to create ultrafine droplets of H$_2$SO$_4$. This sulphuric acid aerosol is eventually neutralised by ammonia, via the strong acid ammonium bisulphate (NH$_4$HSO$_4$) to the nearly neutral salt ammonium sulphate (NH$_4$)$_2$SO$_4$. The resulting particles are typically 0.5–1.0 μm in size, putting them in the accumulation mode with long atmospheric residence times and high penetration into the respiratory system. Hence there is always a division of the sulphur between these solid, liquid and gaseous species. The exact partition depends on the history of the emissions in the airmass, critical factors being residence time, photochemical conditions and sources of ammonia (high over agricultural areas, medium over urban areas, low over water bodies). Principal responses to the sulphate aerosol component are bronchospasm in asthmatics and chronic bronchitis. Even though sulphur emissions and concentrations have been decreasing dramatically in the older industrialised nations during the last 30 years, measurable responses to sulphate particulate have been determined at concentrations as low as 0.04–0.8 μg m^{-3}.

10.1.1.4   Animal models

It is not usually possible to study the responses of people to air pollutants under controlled conditions. An animal model, in this sense, is an animal for which the behaviour or results can be applied with some confidence to humans. Such models have been used to understand aspects such as pollutant deposition and clearance, as well as biological action. The experimental animals (which have often been mice, rats, hamsters, guinea pigs, rabbits, dogs or primates) have usually been normally healthy. Certain research areas have favoured certain species. For example, rats have become the species of choice for work on carcinogenicity, whereas particle deposition studies have required animals such as pigs that have lungs more similar to those of humans. In addition to work with healthy animals, animals with pre-existing conditions have been used to model the enhanced susceptibility to air pollutants found in humans with equivalent pre-existing conditions. Such conditions include age, dietary deficiency, asthma, allergy, bronchitis and lung disease. There are many difficulties and caveats in the interpretation of results from animal trials. For example, sensitivity of mice to gaseous pollutants varies widely between strains, and pollutant deposition to small animal respiratory systems is very different to that of humans. Nevertheless, much of our current understanding of air pollutant toxicology is due to work with these animal models.

10.1.2   The respiratory system

The respiratory system is by far the most important route for entry of air pollutants into the body. The basic structure is shown in Figure 10.4. When the ribs and/or diaphragm are expanded, the increased volume (and hence reduced pressure) in the chest draws air in through the nose or mouth. The nasopharyngeal region is the complex of larger passages in the nose, sinuses and upper throat through which the air passes before going into the throat (trachea). In the tracheobronchial region, the trachea divides into two bronchi, each of which leads to one of the two lungs. The bronchi then branch repeatedly into bronchioles before ending, in the pulmonary region, in clusters of tiny sacs called alveoles (Figure 10.5). There are around 300 million alveoles in the lungs, each having an inflated diameter of about 0.25 mm. The total area of these alveoles is 50–100 m^2. The alveoles, where oxygen, water vapour and CO_2 are exchanged, effectively present a very thin membrane of large surface area to the blood supply.

All the internal surfaces of the respiratory tract are lined with various secretions. In many parts, the secretions are cleared continuously by cilia (motile hairs which wave in phase) to remove trapped material. There are about 200 cilia on the surface of each epithelial cell, they beat at about 10 Hz, and can move the surface fluid at speeds of up to 10 mm min^{-1}. This process is effective at removing dissolved gases and particles and undissolved particles. However, the very fine structure of the ciliated surface, with its large surface area, also makes it susceptible to the same gases. There are no cilia in the alveoles. The respiratory system is a beautiful example of a highly specialised multi-tasking design that has evolved to meet parallel requirements. The main task is the gas exchange in the alveolar region – 10–20 m^3 of air are drawn in per day by a healthy resting adult, substantially more during exercise. The air drawn in for this exchange is conditioned to constant temperature and humidity during its

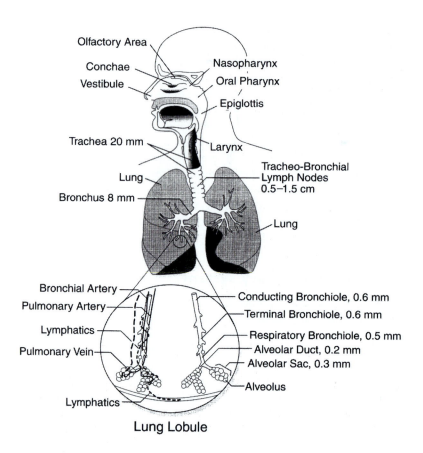

Figure 10.4 Structure of the respiratory tract.

passage into the lungs. The sensitive alveolar surfaces are protected from particles by tortuous passages, hairs and moist mucus linings. The moist linings also absorb some gases, and are rich in antioxidants such as ascorbic acid, uric acid and glutathione. Absorbed gases and trapped particles are removed by mucus flow. Hence, broadly speaking, penetration of gases depends on solubility in aqueous solution, and penetration of particles depends on aerodynamic diameter (Table 10.1).

The actual air exchange rate is itself variable. At rest, the tidal volume (the volume that flows in and out of the lung at each breath) might be only 0.5 l, exchanged every 5 s at 12 breaths min^{-1} to give a ventilation rate of 6 l min^{-1}. During heavy exercise these values could rise to 3 l every 2 s with a ventilation rate of 90 l min^{-1}. A value of 17.5 l min^{-1} is often used for gas and particle sampling.

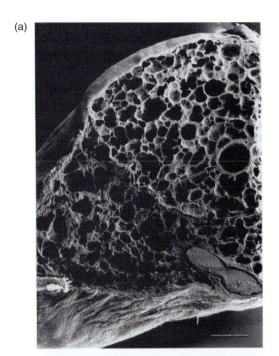

(a)

(b)

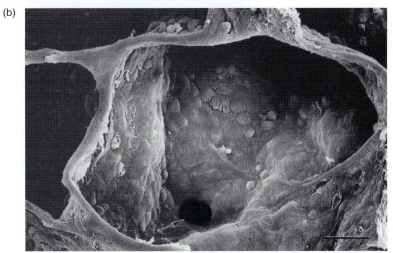

Figure 10.5 (a) Terminal bronchiole and associated alveole scale bar 200 μm; (b) an individual alveole scale bar 50 μm.
Source: Jeffery, P. K. (1999) 'Effects of cigarette smoke and air pollutants on
the lower respiratory tract', in: S. T. Holgate, J. M. Samet, H. S. Koren and
R. L. Maynard (eds) *Air Pollution and Health*, Academic Press, London.

Table 10.1 Dimensions and flows in the lungs

Region	Number	Diameter cm	Total cross section cm²	Cumulative volume cm³	Typical velocity cm s⁻¹	Reynolds no. Re
Nasal passages	2	0.65	0.6		550	2500
Trachea	1	1.7	2.3	70	145	1670
Main bronchi	2	1.3	2.7	80	125	1100
Divided bronchi	18	0.5	3.6	100	90	310
Bronchioles	500	0.2	16	150	20	28
Terminal bronchioles	1.2×10^4	0.07	47	200	7	3.4
Respiratory bronchioles	1.7×10^5	0.05	340	270	1	0.33
Alveolar ducts	8×10^5	0.08	4200	800	0.08	0.04
Alveoles	$3-5 \times 10^8$	0.015		3000		

Table 10.2 Air exchange rates during breathing

Activity level	Pulse rate/min⁻¹	Air flow rate/ l min⁻¹
Resting	70	6
Light exercise	80	10
Walking	110	20
Hard work	130–150	30–60

Flow in pipes tends to be turbulent above a Reynolds number of around 2000, so flow in the bronchi and below is laminar. The flow in the alveolar region is very slow and gentle (air speeds less than 1 mm s⁻¹), even during heavy breathing (Table 10.2).

10.1.3   Responses to particles

10.1.3.1   Particle deposition in the lung

In Chapter 2 we looked at particle deposition processes in general. One of the most important aspects of particle deposition is the way in which the construction of the human respiratory system interacts with the deposition processes to influence which particles are deposited in which regions. Research into this topic has been stimulated by problems in occupational health, cigarette smoking, and long-term exposures to low ambient aerosol concentrations. Studies have confirmed that particles are deposited in a fairly characteristic manner depending on the relationships between aerodynamic diameter, local airspeed, and residence time. The initially inhaled particles are typically less than 100 μm. Particles larger than 50 μm are unlikely to be drawn in through the nose and mouth, because their terminal speeds are greater than the air speed into the nostrils. If they are drawn in, their stopping distance is greater than the passage dimension and they will be deposited immediately, before entering the trachea. For example, a 30 μm particle with a stopping distance of 1.6 cm will probably be deposited in the nasal passages (passage diameters less than 1 cm).

Those particles greater than 10 μm aerodynamic diameter are deposited efficiently by inertial impaction and interception on hairs in the nasal passages and the walls of the nose, sinuses and throat. The respirable fraction (less than about 10 μm aerodynamic diameter) penetrate to the bronchi, and may penetrate into the lower respiratory tract. Larger particles have high inertial impaction efficiencies where the airspeeds are high and the passages tortuous. The smallest particles have high Brownian diffusion velocities to the walls where the airspeeds are low, the passages small and the residence time long. Sedimentation also acts, although efficiency is low because the small particles have low terminal velocities. The combined effect of these processes is the deposition curve of Figure 10.6; there is a diameter range centred on 0.5 μm where total deposition is a minimum. This is the diameter region where atmospheric mass loadings are a maximum, for just the same reasons. This sub-range of particles, around 0.1–1.0 μm in diameter, is too small to either impact

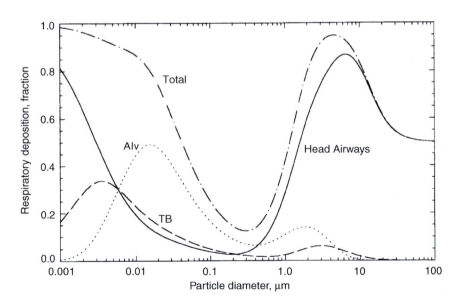

Figure 10.6 Deposition of particles in different regions of the respiratory tract.
Source: Hinds, W. C. (1999) *Aerosol Technology: Properties, Behaviour and Measurement of Airborne Particles*, Wiley, New York.

or diffuse to the walls efficiently, and may penetrate deep into the lung. For the same reasons, the particles might be breathed out again before they can be deposited.

Clearly the breathing pattern influences the balance between deposition modes. During periods of rest, with shallow breathing, there are low airspeeds during inhalation and exhalation, and a significant period between when the air is stationary. We would expect impaction losses to be reduced, and settling and diffusion losses to be increased. During heavy exercise, not only would the reverse hold, but the total volume of 'processed' air would be increased.

An additional effect of increased tidal volume will be mouth-breathing, which will bypass the particle cleaning properties of the nose and increase the penetration of coarse particles. The change from nose to mouth-breathing normally happens at a volume flow rate of 30–40 l min^{-1}.

If the inhaled particles are hygroscopic, as is the case with the most common photochemical ammonium sulphate and ammonium nitrate particles, the particles will absorb water vapour at the 100% humidity that prevails throughout the respiratory system. This will tend to increase their diameters and hence their deposition velocities, although the significance of this effect in humans is not known.

10.1.3.2   Standards for particle deposition

Since the different parts of the respiratory system sample particles of different aerodynamic diameters with different efficiencies, any measurement system that is used to evaluate, for example, the mass loading of particles that can penetrate down to the alveolar regions of the lung, must reproduce this efficiency variation. Several international groups have produced conventions that define the variation of sampling efficiency with diameter.

The International Standards Organisation (ISO 7708) has defined four main fractions, which all have exact mathematical descriptions:

- *Inhalable* – the mass fraction E_I of total airborne particles which is inhaled through the nose and/or mouth. This is an empirical relationship which has been determined from wind-tunnel measurements on full-scale mannequins. It describes the efficiency with which the nose and mouth extract particles from the airstream blowing past the face, and is given by

$$E_I = 0.5(1 + \exp[-0.06D]) + 10^{-5}U^{-2.75}\exp(0.05D)$$

where D is the aerodynamic diameter of the particle in μm and U is the windspeed in m/s (valid for $U < 10$). This fraction is potentially significant for understanding deposition and assimilation of coarse soluble particles in the nose and mouth.

The remaining three fractions all use cumulative lognormal curves to define the proportion of total airborne particles that penetrate to specific regions of the respiratory tract. In each case the curve is defined by the median aerodynamic diameter and geometric standard deviation given in Table 10.3.

Table 10.3 ISO sampling conventions

Convention	Median diameter/ μm	Geometric standard deviation
Thoracic	10	1.5
Respirable	4	1.5
'High-risk' respirable	2.5	1.5

- *Thoracic* – the mass fraction of inhaled particles that penetrates the respiratory system beyond the larynx. Potentially of significance for diseases such as bronchitis, asthma and other upper-airways illness.
- *Respirable* – the mass fraction of inhaled particles that penetrates to the unciliated regions of the lung. Potentially of significance for deep-lung diseases such as pneumoconioses.
- *'High-risk' respirable* – this is applicable to vulnerable groups such as the sick or infirm, or to children.

The conventions are also shown in Figure 10.7. It can be seen that the ISO thoracic convention corresponds closely to the PM_{10} sampling definition. The 50% cut diameter of the high risk respirable convention corresponds to that for $PM_{2.5}$, although the variation about that diameter is not as sharp. These are efficiencies, not size distributions. To find the size distribution of particles that penetrates to the thoracic region, the total airborne size distribution would be multiplied by the corresponding efficiency at each diameter across the distribution.

The only device that can accurately reproduce deposition in the lung is the lung itself. We can approximate the lung deposition by measuring the fraction of particles below certain aerodynamic diameters. PM_{10} is used as a surrogate for the thoracic convention, and $PM_{2.5}$ for the high-risk respirable convention. Penetration through to different parts of the respiratory tract can hence be measured by appropriate samplers. These are usually impactor or cyclone-based, with suitably-designed cut-offs. Porous polyurethane foam has also been used as a particle separator to allow the thoracic fraction to pass through on to a collection filter. Some devices are able to measure either thoracic or respirable fractions by adjusting the flowrate (and hence cut diameter). The penetration through such a device can often be reasonably described by a logistic model such as

$$\text{Penetration} = \frac{1}{1+(D_{\text{aer}}/k)^2}$$

where D_{aer} is the particle aerodynamic diameter, and k is a constant. For example, if $k = 10$ μm, there will be 50% penetration for 10 μm particles.

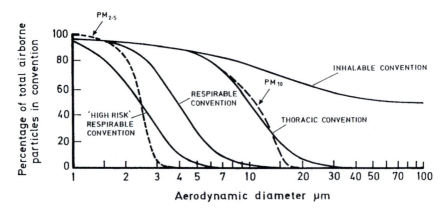

Figure 10.7 ISO particle deposition conventions.
Source: Harrison, R. M. (1999) 'Measurements of concentrations of air pollutants', In: S. T. Holgate, J. M. Samet, H. S. Koren and R. L. Mayhard (eds) Air Pollution and Health, Academic Press, London.

10.1.3.3   Particle clearance

Clearance of deposited particles is another essential function of the respiratory system. If it did not occur the system would rapidly clog up. Clearance is brought about in the following main ways

- dissolution and transfer into the blood
- macrophage consumption
- mucociliary transport to the larynx, then ingestion.

The layer of mucus which lines the breathing passages is moved towards the throat by the cilia and then ingested. Insoluble particles which have been deposited onto the epithelial surfaces of the lower respiratory tract are cleared by mucociliary action in a time scale of between a few hours and one day. The mucus speed is a few mm min^{-1} at the trachea, decreasing with each branching of the airways down to a few μm min^{-1}. Hence particles that penetrate deep into the lung have further to be cleared, at lower speeds, than those deposited closer to the trachea. After one day the clearance rate drops sharply. Particles which have penetrated into the alveolar region, which is not protected by the mucous lining, are cleared more slowly (on a time scale of months) by dissolution and by macrophages. The latter are mobile cells that move around the surface of the alveole and engulf any particles encountered – a process known as phagocytosis (Figure 10.8).

The macrophage then migrates to the terminal bronchiole and is removed from the lung by the mucociliary escalator. All these processes can be overloaded or bypassed under some conditions. The presence of particles stimulates mucus secretion by the goblet cells, although hypersecretion is characteristic of bron-

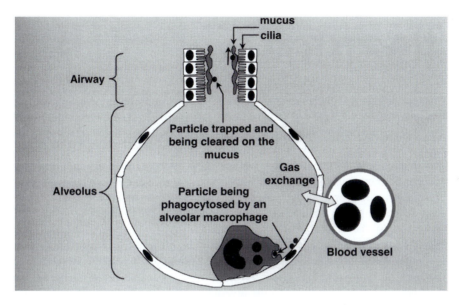

Figure 10.8 Particle removal mechanisms from the alveole.
Source: Stone, V. and Donaldson, K. (1998) 'Small particles – big problem',
Aerosol Society Newsletter 33: 12–15.

chitis. Chronic obstructive pulmonary disease (COPD) and smoking not only lead to hypersecretion of mucus by the goblet cells but also damage the mucociliary escalator. It is currently being suggested that the toxic effects of nanoparticles are a combination of overwhelming the macrophage defences by sheer numbers and the stimulation of oxidant species secretion with their large surface area.

Perhaps surprisingly, there is also evidence that inhaled particles can affect organs other than the lungs. For example, there is increased incidence of thrombosis (clots) being formed in either the heart or brain. The mechanism is uncertain, but is believed to involve the production of coagulating chemicals in the blood due to inflammation by particles. Particles may also carry transition metals into the alveolar region, where they catalyse the formation of free radicals and increase oxidative stress.

Most of the chemicals present in particles (such as ammonium salts and carbon) are not toxic in other ways – by ingestion, for example. Hence there is something of a mystery as to how the high toxicity of very low concentrations of PM_{10} is achieved. There are differing opinions as to whether it is the presence themselves of tiny particles in large numbers (regardless of their composition), or the ability of these nanoparticles to deliver specific chemicals directly to the highly effective exchange region of the lung, or a combination of these. There is also debate about whether it is the number or surface area of the particles that controls their effects. Certainly at very high concentrations found in occupational environments, defence mechanisms such as macrophage removal may be overloaded, but this is not relevant at low ambient concentrations.

10.1.3.4 Asbestos and other fibrous materials

The term asbestos includes several naturally-occurring fibrous inorganic minerals with remarkable thermal and structural properties, which have been widely used as a thermal insulating material in lagging, pipe insulation, fire doors and partitions, as an additive in the manufacture of roofing sheets, cement and gypsum, and in the manufacture of brake linings. The asbestos minerals fall into two groups: (1) chrysotile; and (2) the amphiboles. They are characterised by their crystallographic properties, which result in the formation of weakly-bound fibrils with an initial length of around 1 mm. The fibrils of all forms of asbestos, and especially of the amphiboles such as crocidolite (blue asbestos) break down lengthways so that the width of the fibre may be only 0.05–0.5 μm. This width gives them an aerodynamic diameter range of around 0.1–1 μm (the aerodynamic diameter of a long thin cylinder is around three times the cylinder diameter), so that they can easily penetrate to the alveolar regions of the lung. The fibres are deposited by the usual mechanisms of sedimentation, impaction, interception, electrostatic precipitation and diffusion. They are not cleared efficiently by the usual processes of dissolution and phagocytosis, especially if they are longer than about 10 μm, because the macrophages are 10–15 μm in diameter and cannot ingest long fibres. The attempts by alveolar macrophages to remove the fibres generates reactive oxygen species such as hydrogen peroxide which damage the alveolar epithelial lining and mesothelial lining. The regeneration of mesothelial cells in this environment can result in uncontrolled proliferation, which is the cancer known as malignant mesothelioma. Asbestosis, which is a fibrosis of the lung, may also result. Although asbestos has been replaced for most of its former uses, it can be released at any time – if old insulation is removed, for example. Asbestos health issues originally focused on the acute exposures (~1 fibre cm^{-3}) experienced by miners and factory workers, but it has since been realised that the chronic exposures of those living in asbestos-clad homes and other high ambient concentrations (~0.001 fibre cm^{-3}) pose a serious threat to population health. The death rate associated with asbestos exposure in industrialised countries, is believed to total tens of thousands of people per year. There is not thought to be a significant risk due to outdoor ambient concentrations.

Because the risk is associated with the presence of a fibre, rather than with the mass loading, the health guidelines are framed in terms of a number density at 0.1–0.2 fibres ml^{-1}. Average concentrations in rural air have been found to be around 0.00001 fibres ml^{-1}, rising by a factor of ten in urban air. There are also man-made vitreous fibres such as glass fibre. These materials are around 10 μm diameter and do not split lengthways, so that they do not penetrate beyond the thorax if inhaled and do not cause cancers.

10.1.4   Responses to gases

10.1.4.1   Sulphur dioxide

At low concentrations of gaseous SO_2 alone, the first symptom to develop is bronchoconstriction, which leads to wheezing and is apparent after 1 minute. Measurable increases in airway resistance of normally healthy people are found after 1 h of exposure to 1000 ppb (Table 10.4). Although, such concentrations were found in towns in the 1950s and 1960s, they would today only be associated with close proximity to a point source. Relief from exposure leads to recovery in about 30 min. As with other gaseous pollutants, responses increase in severity with the level of exercise, and tolerance develops during continuous exposure. People with respiratory complaints such as asthma or bronchitis are more at risk, responses having been measured at a minimum of around 100 ppb – this is lower than the peak UK concentrations of SO_2. Specific responses include changes to the tracheobronchial mucociliary clearance rate, alveolar macrophages and production of reactive oxygen species.

A research programme in Europe called Air Pollution and Health – a European Approach (APHEA) conducted a meta-analysis of air pollution and health data sets from 15 European cities. The data covered the general period 1976–1991, although the exact period varied between cities. APHEA found that an increase of 50 $\mu g\ m^{-3}$ in the 24-h average SO_2 concentration would lead to a 3% increase in mortality from all causes. Such an increase would also lead to an increase in hospital admissions due to chronic obstructive pulmonary disorder and a reduction in forced expiratory volume (FEV) of around 7%. There is currently no statistical basis on which these results could be applied to short pollution episodes.

10.1.4.2   Ozone

Fundamental ozone exposure is a direct function of the time sequence of airborne concentration inhaled by the individual, both outdoors and indoors. Since it is generated by outdoor photochemistry and stratospheric intrusion, and deposits efficiently to most surfaces, the indoor ozone concentration

Table 10.4 Health effects of sulphur dioxide

Concentration/ppm	Period	Effect
0.03–0.5	Continuous	Bronchitic patients affected
0.5–1.4	1 min	Odour detected
0.3–1.5	15 min	Eye irritation
1–5	30 min	Increased airway resistance sense of smell lost
5–20	>6 h	Reversible lung damage
>20	>6 h	Waterlogging of lung tissues

decays rapidly. The only likely indoor sources are photocopiers and electrostatic air cleaners. This concentration sequence is compounded by factors such as tidal volume – for example, a heavily exercising, heavily breathing person will extract the ozone (and any other air pollutants) from the air at a higher rate than a resting person. The removal efficiency of the respiratory system for ozone is around 90%. Unlike NO_2, O_3 is virtually insoluble in water, but is highly reactive. Hence it is deposited along the whole of the respiratory system. About 40% of inhaled ozone is absorbed in the nasopharyngeal region, leaving 60% that penetrates to the lower respiratory tract. At concentrations above about 200 ppb, the ozone is thought to target the fatty acids and proteins that comprise cell membranes, and hence to damage cells in the bronchiolar and alveolar epithelia, leading in the short term to changes in lung biochemistry and induced localised airway inflammation. The released pro-inflammatory mediators can cause inflammatory cells such as macrophages and leucocytes to adhere to the endothelial cells that line blood vessels close to the site of ozone attack. Chronic ozone exposure seems to be associated with an irreversible thickening or stiffening of the lung tissue structure. Evidence from exposure of laboratory animals to ozone was summarised by the UK Advisory Group on the Medical Aspects of Air Pollution Episodes in 1991. They concluded that it is unlikely that current UK levels of ozone exposure would cause irreversible changes in the lungs, or that summer pollution episodes would cause changes in lung biochemistry or localised airway inflammation. Getting at the truth can be difficult in these cases – for example, the highest O_3 concentrations in the UK occurred during the summer of 1976, and mortality and hospital admissions certainly increased. However, the episode was associated with unusually high air temperatures, which also result in increased mortality and hospital admissions. Unless a control is available for comparison, it is hard to determine the relative contributions of the two factors.

Experiments have also been performed on people in ozone exposure chambers, where the usual performance measure is FEV_1 (the volume of air in litres that can be breathed out by an experimental subject in 1 s). It has been shown that lung performance decreases with increasing ozone concentration (Figure 10.9), and that the more heavily the subject is exercising, the more she or he is affected. The results of these and similar experiments are summarised in Table 10.5.

Other measurements have shown that exposure on successive days shows acclimation, the response declining with repeated exposures. Because we breath in air pollutants, and asthma is an allergic response of the respiratory system, there is currently a widespread popular belief that ozone and photochemical smogs are affecting the health of the population in general, and of children and asthmatics in particular. The number of in-patient hospital treatments for asthma in England rose from around 40 000 in 1980 to over 100 000 in 1989, and a similar proportional increase in asthma incidence was reported by GPs over the same period.

However, there is no proven increased susceptibility to ambient concentrations of air pollutants for those with chronic bronchitis, cardiopulmonary disease or

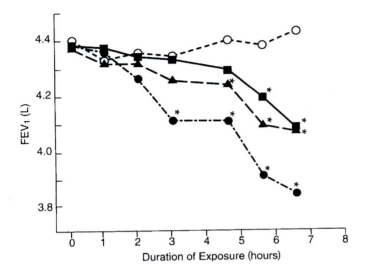

Figure 10.9 The effect of increasing concentrations of ozone on forced expiratory
volume after 1 s (FEV$_1$).
Source: Lippmann, M. (2000) 'Ozone', In: M. Lippmann (ed.) *Environmental Toxicants*, Wiley, New York.

Table 10.5 Health effects of ozone

	Ozone concentration (ppb) producing a 5% reduction in FEV$_1$ after various periods and levels of exercise			
	2 h moderate	2 h heavy	6.6 h moderate	2.5 h very heavy
Normal person	300	250	70	160
Most sensitive	150–100	120–80	–	–

asthma. An association with asthma is certainly suspected, but experimental and
epidemiological evidence of chronic health effects is lacking. Many other factors,
such as changes in lifestyle, exposure to new allergenic chemicals, and diet could
be involved.

10.1.4.3   Nitrogen oxides

We have seen in Section 1.4.3 that NO and NO$_2$ are present in variable proportions
in the outdoor atmosphere. While the initial emissions, principally from motor
vehicles and other combustion sources, are usually dominated by NO, the latter is

oxidised to NO_2 during photochemical reactions in the atmosphere, and ambient NO_2 concentrations may be higher than those of NO. There are also significant indoor emissions and exposures due to sources such as gas hobs. Of the two gases, NO_2 has much the higher toxicity to people; the occupational exposure limit for NO is 25 ppm, whereas that for NO_2 is 3 ppm. NO_2 has lower solubility than SO_2, so that a much higher proportion penetrates into the deep lung, and the main sites of action are the terminal bronchioles. As with other soluble gases, nasal breathing will protect better against NO_2 than oral breathing. NO_2 is a powerful oxidant, and it is thought that its main damage mechanism is destruction of cell membranes by lipid peroxidation. The body does have natural defences against this process due to antioxidant chemicals in the alveolar epithelium. High NO_2 concentrations can overwhelm the antioxidative defence capacity and allow penetration of the NO_2 through to the cell walls. There are also direct effects on the respiratory system, including increased airway resistance (due to inflammation), increased incidence of respiratory illness and damage to lung tissue.

Many of the studies over the past 30 years have been stimulated by concern over the indoor exposure due to gas cookers and other indoor combustion sources. Usually these have compared responses between houses with and without such sources. The consensus view is that there would be a 20% increase in the incidence of respiratory illness for children under the age of 12 if they live in a house with a gas cooker. Studies on adults have been less conclusive, showing a range of results from no effect to decreased lung function. Short-term exposure trials have shown that asthma sufferers may experience enhanced sensitivity after NO_2 exposure, and that those with normal respiratory function may experience increased airway resistance. For outdoor studies, it has often been difficult to separate NO_2 responses from those due to co-occurring pollutants, especially particulates. Short-duration exposure to NO_2 has also been shown to increase lung reactivity to other pollutants (Table 10.6). Different workers have linked NO_2 exposure to a wide range of disease symptoms, such as intrauterine, infant and cardiovascular death, hospital visits for asthma, and chronic obstructive pulmonary disease.

The low solubility of NO in aqueous media means that, when inhaled, it is more likely to penetrate deep into the lung and to diffuse through pulmonary

Table 10.6 Health effects of NO_2

NO_2 concentration range/ppm	Effects (short-term only)
0–0.2	No effects
0.1–0.2	Odour detected
0.2–1	Metabolic effects detected
1–2	Change of respiratory rate, increased likelihood of lung infection
2–5	Deterioration of lung tissue
>5	Severe damage to lung tissue, emphysema

tissue into the blood stream. There it binds to haemoglobin (Hb) to produce nitrosylhaemoglobin (NOHb), which is in turn oxidised to methaemoglobin (MetHb) in the presence of oxygen. However, exposures to NO concentrations as high as 10 ppm have shown no significant impairment of the blood's oxygen-transport capacity. In fact endogenous NO is an important component of the body's natural systems, such as nervous system signalling and regulation of pulmonary resistance.

10.1.4.4   Volatile organic compounds

Volatile organic compounds are important air pollutants for two distinct reasons. First, they are precursors required for the photochemical production of ozone. Second, they include species that are individually carcinogenic and mutagenic in their own right. The situation is complex, since there are several hundred chemical compounds involved and the most toxic (such as benzene) are not necessarily the most effective at ozone formation. It is usual to disregard methane from estimates of concentration, since although that gas is naturally present at ppm concentrations – dominating the other hydrocarbons – it is not toxic at those levels. It is, however, involved in ozone photochemistry. Typical compounds are propane, benzene, ethanol, methanol, ether, carbon tetrachloride and vinyl chloride; materials such as petrol, oil and resins may contain dozens of individual VOCs, and many more are produced during combustion processes.

10.1.4.5   Benzene

New data on urban hydrocarbon concentrations is now flowing from the UK DEFRA hydrocarbon network. Health effects have focused on benzene, for which a present air quality standard of 5 ppb has been suggested, with an eventual target of 1 ppb. Benzene makes up a maximum 5% of petrol. It is thought that benzene and 1,3-butadiene exposure may account for around 10% of leukaemia incidence across the UK. Ironically, it has recently been recognised that the benzene added to 'super plus' unleaded petrol to maintain vehicle performance may be a worse threat to human health than the lead that it displaced.

Around 70% of the 35 kt benzene emitted to atmosphere in the UK in 1995 was due to petrol (either by combustion or evaporation). Hence the air concentration is strongly influenced by traffic density and dispersion, and there is usually a gradient of increasing concentration from rural areas to urban centres. European surveys have found this gradient to range from annual averages of a few μg m^{-3} up to 10–20 μg m^{-3} in urban areas (1 ppb benzene = 3.25 μg m^{-3}). Particular microenvironments, such as petrol filling stations and road vehicles, may increase these concentrations significantly. Acute effects are found at concentrations above 10 ppm, higher than is ever experienced in ambient air. Responses to prolonged ambient exposures have been predicted *pro rata* from industrial exposures to higher concentrations. The main response is acute non-lymphocytic leukaemia. Hence there is a body of knowledge about responses to concentrations that are typically 1000 times those found in the ambient air, making it difficult to generalise or set limits. The WHO does not recommend a safe concentration for benzene because it is a carcinogen. Instead they give the geometric mean of the range of estimates of the excess lifetime risk of contracting leukaemia from a benzene concentration of 1 μg m^{-3}. This value is 6 cases per 100 000 population. Looked at another way, concentrations of 17, 1.7 and 0.17 μg m^{-3} would be expected to lead to 100, 10 and 1 case of cancer, respectively, in a population of one million.

10.1.4.6   Toluene

Although toluene occurs naturally in crude oil, natural gas and natural combustion products, there is now a much greater presence from human operations. The main industrial sources are catalytic reforming at petroleum refineries, styrene production, coke ovens and solvent evaporation from paint, rubber, printing, cosmetics and adhesives. It is one of the components of petrol and so is released by combustion and evaporation. Around 150 kt are emitted in the UK, of which 58% is from petrol exhaust and 35% from solvents.

As with benzene, there is documented evidence of the acute effects of exposure to high toluene concentrations from workplace environments. The main recorded effects are dysfunction of the central nervous system resulting in behavioural problems, memory loss and disturbance of the circadian rhythm. Annual average concentrations outdoors in urban areas have been found to be similar to those of benzene at around 1 ppb, but there is uncertainty about extending these findings to predict responses to prolonged low-level exposures. For non-cancer effects, the USEPA has recommended that the long-term concentration should not be greater than 100 ppb.

10.1.4.7   1,3-Butadiene

Urban ambient levels are similar to those of benzene and toluene at around 1 ppb. 1,3-butadiene is carcinogenic; hence a safe threshold concentration cannot be specified, and the hazard must be estimated in terms of the risk of contracting cancer through lifelong exposure to a reference concentration. There have been no epidemiological findings that have identified the risk due specifically to this chemical. Based on workplace observations at far higher concentrations and on experimental results from animal models, the UK set an air quality standard of 1 ppb.

10.1.4.8   Carbon monoxide

Carbon monoxide combines with blood haemoglobin 200 times more readily than oxygen. The resulting carboxyhaemoglobin (COHb) molecules can no longer transport oxygen from the lungs around the body, and hence the oxygen supply to the brain and other organs is reduced. The reaction is reversible, and exposure to clean air removes most of the gas from the body with a half-life of 3–4 h. The degree of CO absorption depends on its concentration in the air, the period of exposure and the activity of the individual. The WHO has calculated the relationship given in Table 10.7 between CO concentration and blood COHb for a lightly exercising person. The COHb values will be reduced by about a factor of two for a person resting, and increased by a similar factor by heavy exercise. The COHb concentration is around 0.5% when there is no CO in the atmosphere. Note that there is an equilibrium COHb concentration, so that longer exposure does not necessarily result in larger uptake. Also, the time to achieve equilibrium depends on air flow rate, so that equilibration that would take 4 h at rest would only take 1 h during hard exercise. Uptake rate depends on the difference between the applied concentration and the blood concentration. Hence the rate is maximum at the start, and falls away as the blood concentration increases. Don't jog in the city on bad air-pollution days.

Carbon monoxide is generated by combustion processes in which there is inadequate oxygen to produce CO_2. Three types of exposure represent the most severe situations. First, when an appliance such as a gas fire or central heating boiler has a blocked flue due to poor maintenance, or when open stoves are used for cooking. Not only do the combustion gases have a high CO concentration, but they cannot escape through the flue and come out into the building. Furthermore, CO is colourless and odorless which reduces its chances of detection. Around 1000 people in the US, and 50 in the UK, die each year by this cause. Clearly, indoor exposure is a hazard for the population at large. Second, cigarette smoke contains around 1% CO (10 000 ppm), so that heavy smokers may have 10% COHb. In addition, COHb crosses the placental barrier and results in a greater proportional change in the fetus than in the mother. Third, elevated outdoor ambient concentrations up to a few 10s of ppb CO may be created in urban centres by the restricted dispersion of motor vehicle exhaust gases. The highest COHb concentrations shown in Table 10.7 would be expected to result in a range of symptoms such as dizziness, nausea, headaches and measurably reduce mental performance. In the UK, CO concentrations in urban areas are monitored by the Automated Network, and show city-centre 8-h maximum concentrations of a few ppm. A blood COHb

Table 10.7 Health effects of carbon monoxide

Ambient CO		Carboxyhaemoglobin/%		
ppm	mg m^{-3}	After 1 h	After 8 h	At equilibrium
100	117	3.6	12.9	15
60	70	2.5	8.7	10
30	35	1.3	4.5	5
20	23	0.8	2.8	3.3
10	12	0.4	1.4	1.7

of around 1% might be achieved from an exposure to 30 ppm CO for 1 h or about 9 ppm for 8 h. Such an exposure, which is typical of those expected of someone outdoors in a severely polluted urban environment, will cause minor symptoms such as headache in sensitive members of the population. There is a lag in uptake of CO from the air and release back to the air, so calculating actual exposure from a time series of contacts with atmospheres at different concentrations is difficult. So far as long-term exposures are concerned, no epidemiological studies have shown the influence of chronic CO exposure on health outcomes.

10.1.5   Immunosuppression

It has been shown in many animal experiments that air pollutants such as ozone and nitrogen dioxide reduce the body's capacity to defend itself against harmful microorganisms such as bacteria. This appears to be due to a weakening of the pulmonary clearance mechanism – the bacteria remain in the lung rather than being removed from it. Similar responses have been found for particles.

10.1.6   Passive smoking

Passive smoking is the exposure to tobacco smoke created by others' smoking. Clearly it leads to inhalation of a similar range of compounds, albeit at lower concentrations, to those inhaled by the smoker. In some circumstances the total exposure may be significant, as for a bar worker who is in the bar all day. Mortality rates have been calculated from cohort studies, and it has been shown that around 46 000 deaths per year are attributable to passive smoking in the US, of which 3 000 are from lung cancer, 11 000 from other cancers, and 32 000 from heart disease.

10.1.7   Toxic organic micropollutants (TOMPs)

A huge range of other toxic chemicals is emitted into the atmosphere; many of them have been shown to be carcinogenic or mutagenic when given at relatively high concentrations to laboratory animals, and many can also be concentrated in terrestrial and aquatic food chains so that they achieve high concentrations in, for example, breast milk. The principal classes of chemical within the category are PCDDs, (or dioxins); PCDFs, (or furans); PCBs; and PAHs. Within each class there are many individual compounds with their own properties. Their atmospheric concentrations are normally extremely low in comparison to the air pollutants discussed above; while individual PAHs may be present at ng m^{-3} levels, individual PCDDs may only occur at pg m^{-3} or even fg m^{-3} levels. Nevertheless, there is increasing concern about their chronic long-term effects. In the US, these compounds are known as 'air toxics'. Since these compounds are carcinogenic, there is no threshold concentration. It is only possible to specify a risk of contracting cancer if exposed to that concentration for a lifetime.

Polychlorinated dibenzodioxins and dibenzofurans are released mainly from municipal and hospital waste incinerators, metallurgical processes and coal combustion. Their initial form is gaseous, but they are strongly absorbed onto particles (dust and fly ash) before emission. They are subsequently deposited onto soil, vegetation and water surfaces, and incorporated into food chains as well as being inhaled or ingested. With so many different chemicals to consider, it has become common to describe the effective toxicity of a range of these TOMPs with a single Toxic Equivalent (TEQ), which is the toxicity equivalent to the same weight or concentration of the most toxic TOMP, known as 2,3,7,8-TCDD. It has been estimated that only about 3% of daily TEQ intake comes directly from the air, compared with 16% from vegetables, 19% from meat and eggs, 36% from dairy products, and 26% from fish. The total average daily intake in European countries is about 120 pg TEQ day^{-1}. This represents about 2 pg per kg body weight (kgbw) per day for adults, a figure which may rise to ten for babies because of their intake of breast milk and dairy products. Experimental toxicological and epidemiological data indicate a lowest effect level of between 100 and 1000 pg kgbw^{-1} day^{-1}. The airborne dioxin concentration is around 5 pg m^{-3}; this is higher than would be expected from the source strengths and deposition velocity, so it is probably due largely to resuspension of particles from the ground.

PCBs, like CFCs, are a victim of their own success. Until the 1980s they were widely used in the electrical industry as an insulating fluid around power transformers and in capacitors, and as additives in industrial oils. However, the chemical stability that made them suitable for these applications has also resulted in very long lifetimes in the environment. Although most of the pollution has probably involved direct transfer to water or landfill sites, PCBs can only be destroyed by high temperature incineration, so that there is a risk of release to atmosphere if incineration conditions are inadequate. The relative occurrence of different PCB isomers has been found to be consistent with their relative volatilities. Recent UK measurements of eight common PCB congeners in four urban areas gave total atmospheric concentrations of a few ng m^{-3}. An elevation of the concentration in the summer was probably due to evaporation of the PCB from particulate reservoirs.

Polycyclic aromatic hydrocarbons are formed by incomplete combustion of fossil fuels, such as occurs in vehicle engines (especially diesel), furnaces, coke ovens, bonfires and barbecues. There are also natural sources such as volcanoes and forest fires. Eight of the possible PAHs are thought to be carcinogenic. Most attention has been paid to one or two particularly carcinogenic compounds such as benzo(a)pyrene. Measurements of 15 PAHs at UK urban sites have shown that total levels of 11–735 ng m^{-3} were dominated by the three lowest molecular weight compounds phenanthrene, fluorine and pyrene.

10.1.8   Quantification of health effects

10.1.8.1   PM$_{10}$ and SO$_2$

The UK Committee on the Health Effects of Air Pollutants (COMEAP) estimated the following effects due to PM$_{10}$ and SO$_2$ in Table 10.8. For comparison, the total

Table 10.8 Effects of PM_{10} and SO_2 on mortality and respiratory illness

Pollutant	Health outcome	No. of cases in urban GB
PM_{10}	Deaths brought forward	8100
	Hospital admissions (respiratory)	10 500
SO_2	Deaths brought forward	3500
	Hospital admissions (respiratory)	3500

deaths and respiratory hospital admissions in the equivalent areas of Britain are around 430 000 and 530 000 per year, respectively.

At present there is general uncertainty about which particle property or properties are mainly responsible for health effects, and hence what to measure and what standards to set in legislation. If is not $PM_{2.5}$ as a mass concentration, then it might be the sulphate fraction, or the number concentration of ultrafine particles (<0.50 μm), or whether the latter are acid-coated, or the mass concentration of soluble transition metals. The overall epidemiological evidence is that there is a linear relationship between some indicator of particle concentration and excess mortality – i.e., as for carcinogenic chemicals, that there is no threshold concentration above which effects are found and below which it is 'safe'.

As a result of this uncertainty, WHO–Europe has been unable to recommend a PM guideline. It has instead presented the PM_{10} concentration changes necessary to produce specific percentage changes in: (1) daily mortality; (2) hospital admissions for respiratory conditions; (3) bronchodilator use by asthmatics; (4) symptom exacerbation by asthmatics; and (5) peak expiratory flow. The ball is then placed firmly in the court of national and local authorities to set standards which create an acceptable risk for the population. The USEPA has been more definite, promulgating in 1997 revised standards for PM_{10}, and new standards for $PM_{2.5}$. The annual average $PM_{2.5}$ is not to exceed 15 μg m^{-3}, and the 24-h $PM_{2.5}$ is set at 65 μg m^{-3} taken as the 98th percentile value (the 8th highest in each year). It has been estimated that the value of the health benefits if the particle concentrations prevailing in 1995 are reduced to $PM_{2.5}$ equivalents of 15 μg m^{-3} would be \$32 billion (about £23 billion). If the $PM_{2.5}$ concentration were reduced to 12 μg m^{-3}, then this health benefits figure would rise to \$70 billion (about £50 billion).

10.1.8.2   Ozone

For acute exposures, the UK Advisory Group concluded that 200 ppb of ozone, which is below the UK record hourly mean and commonly achieved in the US, would cause eye, nose and throat irritation in more than 50% of the population at large; FEV_1 would be decreased by 25% for normal people being active outdoors, and by 50% for the most sensitive 10% of the population. Responses as large as these clearly have to be taken seriously. The compilation of many separate epidemiological studies has shown that relative mortality risk is increased by 4–5% when 1-h peak O_3 concentration increases by 100 ppb. The corresponding figure for morbidity (identified as hospital admissions due to respiratory illness) was 18%. For at-risk groups such as asthmatics, this can rise above 50%. Ironically, some of the most useful data on population responses to O_3 has been obtained from American summer camps, where healthy activities are enjoyed in rural high-altitude, but also high-O_3, locations. Figure 10.10 shows a prediction of the acute adverse health effects that could be avoided in New York by meeting the US EPA's standard of 80 ppb as an average daily maximum.

It is likely that there are chronic responses due to long-term exposure to low-O_3 concentrations, although it is even harder to separate this signal from the background noise. The UK Government has set the target for ozone concentration; the 97th percentile of the daily maximum 8-h running mean not to exceed 50 ppb (equivalent to exceedances on no more than 11 days per year). This health criterion was exceeded over almost all the UK in 1996, especially along the south coast of England. COMEAP considered the likely effects of O_3 on deaths and respiratory illness to be as given in Table 10.9. The European Environment Agency has calculated that 0.3% of EU hospital admissions are attributable to high ozone concentrations, with up to 0.7% in Belgium, Greece and France. For the UK this figure is probably less than 0.1%.

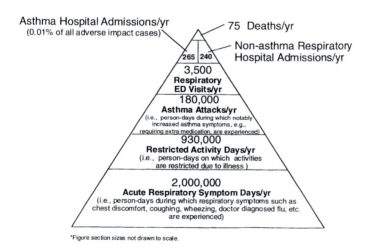

Figure 10.10 Health benefits to New Yorkers of meeting the ozone standard.
Source: Thurston, G. D. and Ito, K. (1999) 'Epidemiological studies of ozone exposure effects', In: S. T. Holgate, J. M. Samet, H. S. Koren and R. L. Maynard (eds) *Air Pollution and Health*, Academic Press, London.

Table 10.9 Effects of ozone on mortality and respiratory illness

Pollutant	Health outcome	No. of cases at specified threshold, per 10^6 population
O_3	Deaths brought forward	700, 50 ppb
		12 500, 0 ppb
	Hospital admissions	500, 50 ppb
		9900, 0 ppb

10.1.8.3   Benefits of achieving NAQS targets

Calculating the costs of pollution control is complex enough, but may seem straightforward in comparison to calculating the values of the benefits. Yet such a comparison is useful if the end-point for pollution control is to be judged. For example, we could continue to reduce the concentration of a given pollutant towards the natural background by insisting on tighter and tighter emission controls. The laws of diminishing returns would apply, and the cost per unit concentration (1 ppb, for example) by which the concentration was reduced would increase as the concentration was lowered. In order to find out the health benefits (purely in terms of morbidity and mortality) we need good understanding of the exposure–response relationship. Then we also need to convert those morbidity and mortality figures into financial equivalents.

The UK National Air Quality Strategy (NAQS) has determined targets for eight key air pollutants and a timetable for meeting them (see Chapter 13). These target values have been combined with previously-established dose–response relationships to predict the net health benefits of achieving the targets. The dose–response coefficients are given in Table 10.10. On the basis given, they are all remarkably similar, with coefficients close to 0.6% per 10 $\mu g\ m^{-3}$. The corresponding predictions of health benefits for an urban population of around 40 million people are given in Table 10.11. The figures imply that the total of deaths brought forward should be reduced by around 18 000 by the measures currently in place.

Table 10.10 Dose–response coefficients

Pollutant	Health outcome	Coefficient
PM_{10}	Deaths brought forward	+0.75% per 10 $\mu g\ m^{-3}$ (24-h mean)
	Respiratory hospital admissions	+0.8% per 10 $\mu g\ m^{-3}$ (24-h mean)
SO_2	Deaths brought forward	+0.6% per 10 $\mu g\ m^{-3}$ (24-h mean)
	Respiratory hospital admissions	+0.5% per 10 $\mu g\ m^{-3}$ (24-h mean)
O_3	Deaths brought forward	+0.6% per 10 $\mu g\ m^{-3}$ (8-h mean)
	Respiratory hospital admissions	+0.7% per 10 $\mu g\ m^{-3}$ (8-h mean)
NO_2	Respiratory hospital admissions	+0.5% per 10 $\mu g\ m^{-3}$ (24-h mean)

Table 10.11 Health benefits from achieving NAQS targets*

Pollutant	Deaths brought forward			Respiratory hospital admissions		
	In 1996 (1995 for O_3)	In 2005 (2010 for O_3)	Total reduction over period	In 1996 (1995 for O_3)	In 2005 (2010 for O_3)	Total reduction over period
PM_{10}	7390	6480	4860	6090	5345	4015
SO_2	3305	1370	9675	2125	880	6230
O_3	720	235	3890	615	200	3320
NO_2	Not predicted			5160	3505	8255

Is it worth it? One way of assessing effects in financial terms is by willingness to pay (WTP). The idea is to get people to estimate how much they are willing to pay for a safety improvement which will reduce their individual risk of death during the coming year by, say, 1 in 100 000 (equivalent to one fewer death in a group of 100 000 people). If this figure were £10, then the value of one life would be £10 × 100 000 = £1m. Another technique is to use quality-adjusted life years (QALYs). Quality of life is scored on a scale from one (full health) down to zero (death). Then the product of this score times the period in years for which it is experienced is the QALY.

Recently, DEFRA has reported on economic aspects of reducing particle concentrations, working from a base case in which total UK emissions of PM_{10} would decrease from 186 kt in 1999 to 98 kt in 2010. They found that the introduction of an 'illustrative package of additional measures' to further control mobile and stationary sources would reduce emissions in 2010 to 65 kt and PM_{10} concentrations in 2010 by 0.751 µg m^{-3} at a cost of around £1000m. The health benefits would amount to a gain of around 400 000 life-years over the period 2010–2110. Although this sounds a large figure, it is summed over a lot of people and a long time. Put another way, the gain corresponds to an increase in life expectancy of 2–3 days per person. The costs of such improvements can seem very high – the DEFRA report calculated that some measures, such as fitting particulate traps to diesel cars, would cost up to £1m per tonne of particulate emission saved.

10.1.8.4   WHO methodology

A detailed methodology for health impacts assessment has been developed by the WHO. Suppose a health effect such as respiratory disease can be causally related to exposure to a pollutant such as PM_{10}. Some cases of respiratory disease may be attributable to the pollutant, but others will not. The attributable proportion is called P_a.

$$P_a = \frac{\sum\{(R_R(c) - 1) \times p(c)\}}{\sum\{R_R(c) \times p(c)\}}$$

where $R_R(c)$ is the relative risk for the health effect in category c of exposure, and $p(c)$ is the proportion of the target population that lies in category c. The underlying (i.e. independent of pollution) frequency of the effect in the population is called I. Then the frequency I_E due to exposure to the pollutant is given by

$$I_E = P_a \times I$$

and the frequency of the effect in the population free from exposure is

$$I_{NE} = I - I_E = I \times (I - P_a).$$

For a population of N, the number of cases attributed to the exposure will be $N_E = N \times I_E$.

This response is often translated into an 'excess' number of cases, which is the number over and above the number that would have been anticipated without exposure to the pollutant.

The excess incidence $I_{excess}(c) = (R_R(c) - 1) \times p(c) \times I_{NE}$
and

$$N_{excess}(c) = I_{excess}(c) \times N$$

Example case study

Use one year of PM_{10} monitoring data to calculate the number of hospital admissions for respiratory diseases in a city of 400 000 inhabitants. First, form an exposure estimate for each of the 365 days of the year. This gives a frequency distribution of days that have a given category of pollution $p(c)$. The WHO Air Quality Guidelines for Europe give the relative risk for PM_{10} as 1.008 per 10 $\mu g\ m^{-3}$ increase in concentration, starting from a background concentration of 20 $\mu g\ m^{-3}$ at which the relative risk is one. Hence P_a can be calculated as 0.0276, which means that 2.76% of hospital admissions for respiratory illness can be attributed to PM_{10}. National statistics are then used to show that I = 126 per 10000 population per year, corresponding to 5040 for our city. Of this number, 139 (which is 2.76% of 5040) can be attributed to PM_{10}. Furthermore, the data in the last column of Table 10.12 show that only a small proportion of the cases is due to short exposures to very high concentrations,

Table 10.12 Example of hospital admissions due to PM_{10}

$PM_{10}/mg\ m^{-3}$	No of days in category c	Proportion of days in category c	Relative risk in category c	Excess cases due to category c
	(c)	p(c)	$R_R(c)$	$N_{excess}(c)$
<20	10	0.027	1.000	0
20–29	30	0.082	1.008	3
30–39	71	0.195	1.016	15
40–49	83	0.227	1.024	27
50–59	76	0.208	1.032	33
60–69	50	0.137	1.040	27
70–79	20	0.055	1.048	13
80–89	10	0.027	1.056	8
90–99	8	0.022	1.064	7
100–109	5	0.014	1.072	5
110–119	2	0.005	1.080	2
120–129	0	0.000	1.088	0
Total	365	1.000		139

and that benefits of reductions will only be felt if the whole pollution distribution is moved downwards.

10.1.9   Odour

Odour can be one of the most intractable and undefinable problems in air pollution. Some specific gases, such as ammonia and hydrogen sulphide, have specific and easily-defined odours. Yet even for those the odour threshold – the concentration above which they can be detected – can vary by an order of magnitude from one individual to another. Odour thresholds also vary widely from gas to gas, and give little idea of toxicity. Table 10.13 gives the odour thresholds of some common gases.

There are often complex mixtures of compounds for which there is no simple relationship between the responses. Detection of odour depends on molecules of the substance being received by the nose. Even materials usually thought of as inert, such as stainless steel, emit sufficient molecules to produce an odour. Although sophisticated analytical equipment, such as gas chromatography and mass spectroscopy, can shed light on the exact mix of compounds present in a sample, none of them can measure the smell of it. That is best left to the human nose. The response of the latter can be quantified through dynamic dilution olfactometry. A panel of 8–10 trained observers is exposed to a series of dilutions of an air sample, and record the odour concentration (the number of dilutions after which 50% of the panel can only just detect the odour). Electronic noses, based on rapid GC analysis of the components, are also increasingly being used.

Table 10.13 Odour thresholds of some gases

Gas	Threshold/ppb
Acetic acid	210
Butyric acid	0.5
Carbon disulphide	100
P-cresol	0.5
Dimethyl amine (fishy)	21
Dimethyl sulphide (bad cabbage)	1
Ethyl acrylate	0.1
Ethyl mercaptan	0.5
Formaldehyde	1000
Hydrogen cyanide (bitter almonds)	900
Hydrogen sulphide (bad eggs)	0.2
Methyl mercaptan (bad cabbage)	1
Nitrobenzene	4.7
Ozone (acrid)	100
Phenol	21
Phosgene	470
Pyridene	10
Styrene	47
Sulphur dioxide	470
Trimethylamine	0.2

10.2   EFFECTS ON OTHER ANIMALS

As compared to people, we know very little about the consequences of air pollution for the great majority of animals, whether they be mammals, insects or birds.

10.2.1   Fish and other aquatic species

One of the major environmental issues of the late twentieth century has been acid rain, meaning the wet and dry deposition of sulphur, nitrogen and acidity to terrestrial ecosystems. Combinations of observations, theory and experiments have confirmed that acidification of surface water bodies such as lakes is mainly due to acid deposition, exacerbated in some cases by natural processes and changes in catchment land-use and management. These changes certainly started around 150 years ago in industrialised parts of the world, and have been particularly severe where the location of the water body makes it sensitive to deposited acidity. For example, in Scandinavia, two factors combine to amplify the responses. First, much of the region is based on acidic granitic rock which does not buffer acid deposition. Second, the acid precipitation is stored as snow during the winter before being released during the short spring thaw. The combination of these two factors causes a large acid flush of reduced pH, to which salmonids are particularly sensitive, just during the fish breeding season.

With general S deposition, acidification of soils starts as base cations are stripped from the ion-exchange complex and base saturation decreases. The soil's capacity to adsorb SO_4 is eventually exhausted. If the rate of supply of base cations by primary weathering is less than the incoming flux of strong acid anions, the base cation store in the soil becomes depleted and there will be a net flux of hydrogen ions, acid anions and aluminium from the soil into the surface water. The rainfall pH is the general indicator of the severity of acid deposition – this can be seen to relate strongly to the strength of local emissions of acid gases (Figure 10.11).

Populations of freshwater animals such as brown trout have been declining in certain areas of both Europe (Wales, western Scotland, Norway and Sweden) and North-east North America. These declines are associated with parallel declines in the populations of animals such as mayfly larvae, beetle larvae and molluscs. Aquatic bird species such as the dipper, which feed on these benthic invertebrates, have also declined. For example, surveys on the river Irfon in Wales showed that the population of breeding dippers had declined from around 9–10 pairs per 10 km in the mid 1950s to around 1–2 pairs by 1982. The toxic effects on fish are believed to be due both to low pH by itself (in Canada, anglers use pH meters to find the best fishing) and to the enhanced solubility of aluminium from the stream bed and adjacent soil. The aluminium accumulates on specialised cells on the fish gills that are responsible for maintaining sodium balance. All freshwater animals have a higher salt concentration than the water they live in – if they excrete more salt than they absorb, they will die.

In contrast to S, N is a larger component of the terrestrial ecosystem. Hence most deposited N is immobilised in the soil, and some leaks to surface waters. This leakage will be increased by other factors such as local deforestation. If the N is deposited in the form of NH_4, it will first be assimilated by plants and microbes before being mineralised to NO_3, after which it can act as a strong acid anion.

Experimental studies have shown that the process of acidification is reversible, and many countries that experienced severe acidification of their freshwaters during the 1970s and 1980s are now adding lime, in the form of limestone or dolomite, to the lakes or their catchments rather than wait for natural restoration of the pH as acid deposition declines. The lime is either added from a boat as a fine powder mixed directly into the lake water, or dispersed as a powder across the surface by helicopter, or spread as larger lumps across the catchment. This is a really large-scale logistical operation. For example, the liming programme in Sweden involves 7500 lakes and 11 000 km of watercourses, into which 200 000 tonnes of limestone powder are dispersed each year. The typical dose to lakes is 10–30 g m^{-3}, and results in the pH recovering from 5.5 to 6.5.

The falling trends in emissions of S and N in the industrialised countries that were discussed in Chapter 1 will inevitably reduce S and N deposition to ecosystems, although this will not necessarily happen *pro rata*. Hence there is much interest

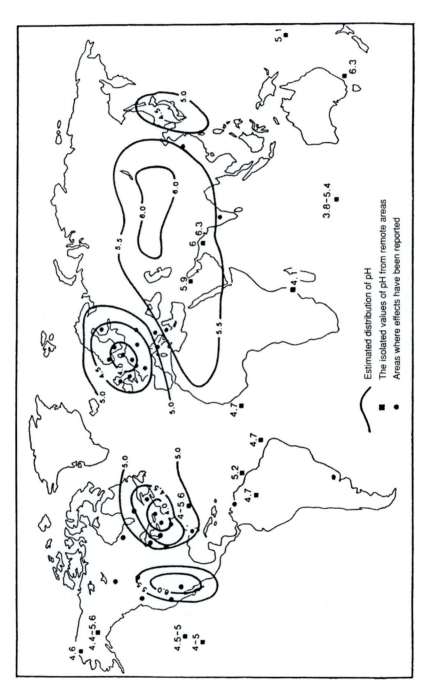

Figure 10.11 Global contours of precipitation acidity.
Source: Seinfeld, J. H. and Pandis, S. N. (2000) *Atmospheric Chemistry and Physics*, Wiley, New York.

in the extent and rate of recovery of affected soil and water ecosystems. Long-term measurements have been taken at many affected sites in the UK. These have generally shown a clear improvement in the chemical composition, marked by raised pH, raised acid neutralising capacity and lowered (non-sea salt) sulphate concentrations. However, biological improvements to indicator species such as aquatic invertebrates, fish, macrophytes, river birds and diatoms have not yet been identified. Modelling studies have indicated that it may take a further 50 years for this biological recovery to be achieved.

Although acid rain and acid deposition have been major air pollution issues in the past, they are now largely understood and under control. This environmental issue has effectively been transferred from the air to the soil and water, and we will not treat it here in any greater detail.

10.2.2   Industrial melanism

The peppered moth (*Biston betularia*) is found all over the UK. The species occurs mainly in two distinctive forms, one of which (the melanic form) is much darker than the other. In the natural environment, the paler form is better camouflaged from predators on bare or lichen-covered tree trunks and therefore predominates. In polluted environments, the trunks are blackened by soot and the pale lichen is killed by SO_2. Evolutionary pressures then favour the melanic form of the moth because it is better camouflaged, and it predominates. Transects between rural and urban environments have shown a smooth progression between the two forms. Recent surveys have shown the pale form to be recovering as soot levels decline. Similar responses have been found in spittel bugs and ladybirds in former Czechoslovakia, and with the peppered moth in Finland, Germany, Scandinavia and Russia.

10.2.3   Aphid stimulation

Aphids feed by drilling through the plant cuticle into the phloem and extracting the fluid. It has been shown that when plants are exposed to sulphur dioxide or nitrogen dioxide, aphids living on them grow and breed at a higher rate. Figure 10.12 shows the results of an experiment in which pea aphids were feeding on pea plants exposed to SO_2. The mean relative growth rate (MRGR, which is simply a measure of aphid vitality) increased steadily with SO_2 concentration until the latter reached 100 ppb. At higher concentrations, the health of the plant itself started to suffer, and consequently that of the aphids as well. It is thought that both S and N act as nutrients, improving (from the aphid's point of view) the quality of certain fractions of the phloem such as amino acids. Interactions such as these, which are only poorly understood as yet, have important implications for world food production.

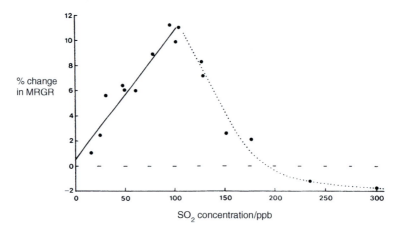

Figure 10.12 Percent change in mean relative growth rate (MRGR) of aphids feeding on plants exposed to sulphur dioxide at various concentrations. Source: Warrington, S. (1987) 'Relationship between SO_2 dose and growth of the Pea Aphid, *Acyrthosiphon pisum*, on peas', *Environmental Pollution* 43: 155–162.

FURTHER READING

Department of Health Advisory Group on the Medical Aspects of Air Pollution Episodes (1991) *Ozone*, HMSO, London, UK.

Department of Health Advisory Group on the Medical Aspects of Air Pollution Episodes (1992) *Sulphur Dioxide, Acid Aerosols and Particulates*, HMSO, London, UK.

Hester, R. E. and Harrison, R. M. (eds) (1998) *Air Pollution and Health*, Royal Society of Chemistry, Cambridge.

Holgate, S. T., Samet, J. M., Koren, H. S. and Maynard, R. L. (eds) (1999) *Air Pollution and Health*, Academic Press, London, UK.

Künzli, N. *et al.* (2000) 'Public-health impact of outdoor and traffic-related air pollution: a European assessment', *The Lancet* 356: 795–801.

Lippmann, M. (ed.) (2000) *Environmental Toxicants*, Wiley, New York.

Ostro, B. and Chestnut, L. (1998) 'Assessing the health benefits of reducing particulate matter air pollution in the United States', *Environmental Research* A 76: 94–106.

Parliamentary Office of Science and Technology (1994) *Breathing in Our Cities – Urban Air Pollution and Respiratory Health*, HMSO, London, UK.

Stedman, J. R., Linehan, E. and King, K. (2000) *Quantification of the Health Effects of Air Pollution in the UK for the Review of the National Air Quality Strategy*, The Department of the Environment, Transport and the Regions, London, UK.

United Kingdom Building Effects Review Group (1989) *The Effects of Acid Deposition on Buildings and Building Materials*, DoE/HMSO, London, UK.

World Health Organization (1999) *Monitoring Ambient Air Quality for Health Impact Assessment*, WHO Regional Office for Europe, Copenhagen.

Chapter 11

Greenhouse gases and climate change

People have been making impacts on the natural environment of the Earth on a steadily increasing scale for thousands of years. For most of that period, those impacts have been localised or temporary. As we neared the end of the twentieth century, however, we realised that we were entering an era in which our releases to atmosphere have the potential to change the whole environment of the planet. Although we have coped with air quality problems such as combustion smogs, photochemical smogs and depletion of the ozone layer by putting preventive measures in place when the symptoms got bad enough, climate change is in a different league. If you are young enough while you are reading this book to expect to live well into the second half of the twenty-first century, then you will probably experience the greatest level of environmental change, and the greatest efforts by humans to deal with it, that we have ever experienced. In this chapter and the one that follows, we will look at the physical causes, and at some of the possible consequences, of these changes.

The most authoritative attempts to look at the causes and consequences of climate change have been undertaken by the IPCC, which has been collating existing data and coordinating new studies on the topic for over a decade. Three Assessment Reports, each of which represents the efforts of hundreds of scientists around the world over periods of several years, and in which the whole topic of climate change is discussed in exhaustive detail, have been published since 1990. In addition, IPCC has published subsidiary reports on specialist topics such as aircraft emissions. Parts of this Chapter make extensive use of material from the First (FAR, 1990), Second (SAR, 1995) and Third (TAR, 2001) Assessment Reports.

The central piece of information that has focussed attention on climate change has been the global average temperature. We have a reasonably well-calibrated global data set going back to 1861, so that is the main starting date that has been used for these assessments of change (Figure 11.1). In fact, the SAR estimated that the temperature had increased by about 0.55 °C during the last 140 years, and the TAR raised this rate of increase to 0.6 °C during the twentieth century, with the 1990s being the warmest decade and 1998 the warmest year since the global data set started in 1861. However, the warming has been erratic, and there is so

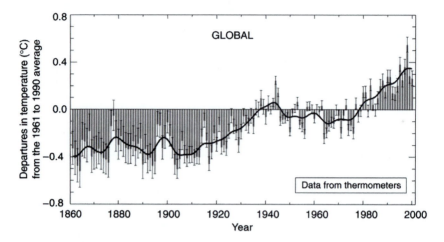

Figure 11.1 Anomalies in the annual average global surface temperature, 1860–2000, relative to the average temperature for the period 1961–1990.
Source: IPCC (2001) *Climate Change – Third Assessment Report*, Cambridge University Press, Cambridge, UK.

much noise on the air temperature signal that it was hard to show that the trend was not just due to natural variability. The average temperature appears to have been fairly steady from 1860 to 1910, to have increased between 1910 and 1940, to have declined slightly between 1940 and 1975, and to have resumed its increase since then.

11.1   OUR RADIATION ENVIRONMENT

All objects that have temperatures above absolute zero (0 K, or −273 °C) radiate electromagnetic energy. As the temperature increases, not only does the amount of radiated energy increase but the typical wavelength of emission falls. In everyday life, we experience radiation from objects having two quite distinct temperature ranges; first, incoming radiation from the sun, which has an effective surface temperature of 5800 K and a peak wavelength of 0.5 μm; second, radiation emitted by the surface of the Earth and all objects on it, which have temperatures of around 300 K and peak wavelengths of around 10 μm. The corresponding spectral bands are shown in Figure 11.2. The vertical scales have been adjusted to make them the same – it is only the relative positions along the wavelength scale that are relevant here. The solar radiation spectrum is almost completely confined to 0–3 μm, while the terrestrial spectrum occupies 3–30+ μm, so there is almost no overlap between them.

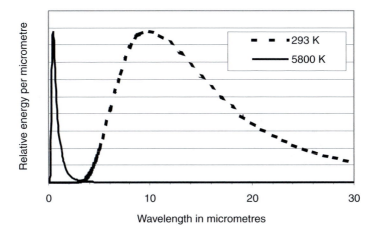

Figure 11.2 The solar and terrestrial spectral bands.

11.1.1 Variations in solar input

The Solar Constant is the name given to the global long-term average solar radiation flux density incident on the Earth from the Sun. It has a value of 1367 W m^{-2}; in fact it is not quite constant, having short-term variations over periods of months to years of plus or minus a few per cent due mainly to sunspots and the eccentricity of the Earth's orbit.

Sunspots are magnetic storms which create relatively cool areas of the Sun's surface, but which are surrounded by hotter rings called faculae. The number of sunspots varies on an 11-year cycle for reasons that are not understood, but have been observed by astronomers since telescopes were invented (Figure 11.3). The overall effect is to vary the solar constant by about 0.1% (the larger the number of sunspots, the higher the output), which in turn is thought to vary the Earth's temperature by 0.2% over the 11 years of the sunspot cycle.

Three quite regular variations occur in the orbit of the Earth about the Sun. First, the orbit is not exactly circular, but elliptical, and the eccentricity of the ellipse varies with a period of about 100 000 years. Although the total radiation incident on the Earth would remain constant, the maximum received at perihelion (when it is closest) and the minimum received at aphelion (when it is furthest away) would vary by ±10%. Second, the angle that the Earth's spin axis makes with the plane of the orbit (also known as the obliquity), which is currently 23.5°, varies between 21.6 and 24.5° over a period of 41 000 years. This again affects the seasonal variation in the distribution of radiation over the Earth's surface. Third, the time in the year when the Earth is at perihelion moves through the months of the year with a period of 23 000 years. These variations can all be summed independently to form a complicated time series. Celestial mechanics is a very exact science, and hence we can calculate the

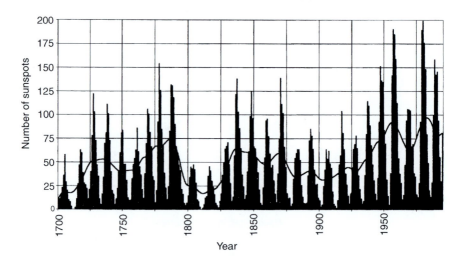

Figure 11.3 Variation in sunspot number, 1700–1990.
Source: Burroughs, W. J. (2001) *Climate Change – A Multidisciplinary Approach*, Cambridge University Press, Cambridge.

effects of these variations on the solar energy received at the Earth going back hundreds of thousands of years (Figure 11.4). The variations amount to about 0.2 MJ m^{-2} at the poles, about 10% of the summer sunshine received, and have been responsible for the quasi-regular ice ages. The regular cycles are called Milankovitch cycles, after the Serbian scientist who first recognised them.

We shall see later how the Milankovitch cycles show up in historical evidence about the Earth's climate, and can be used to explain the broad periodicity of the climate changes on the Earth solely in terms of the variation in received solar energy. However, the climate appears to respond in a more vigorous way than predicted by the magnitude of the solar energy change alone, and climate models indicate that some sort of positive feedback involving vegetation and greenhouse gases is also needed. It does get difficult to distinguish chickens from eggs in this analysis. Furthermore, it should be noted that the Milankovitch changes are mainly connected with the orientation of the Earth with respect to the Sun, not with the distance of the Earth from the Sun. Hence, while they affect the distribution of solar energy receipts over the surface of the Earth, they have much less effect on the overall total.

Apart from small amounts released internally by radioactive decay, the Earth receives all its initial energy supply as shortwave radiation from the Sun. The Earth reflects some of this energy, absorbs the rest, and then reradiates it into space in the longwave form. Since the Earth moves through the almost perfect vacuum of space, there can be no energy transfer by conduction or convection, and the sum of shortwave and longwave energy leaving the planet must balance

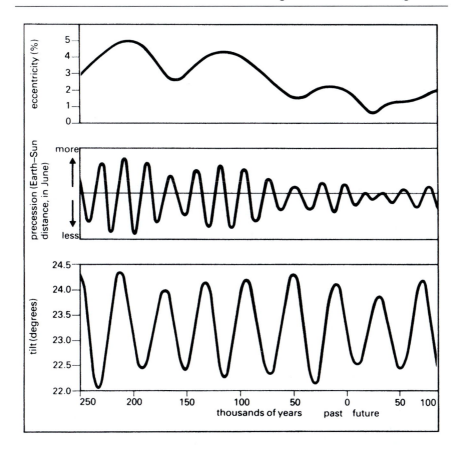

Figure 11.4 Changes in the Earth's orbital parameters.
Source: Burroughs, W. J. (2001) *Climate Change – A Multidisciplinary Approach*, Cambridge University Press, Cambridge.

the shortwave input in order to maintain thermal equilibrium. Figure 11.5 gives a stylised view of the way in which this input energy is partitioned during its flow through the system. In this figure, the Solar Constant of 1367 Wm^{-2} is represented by 100 units. Of these, for example, 8 are backscattered by the air, 17 reflected by clouds and 6 reflected from the Earth's surface. Of the remainder, 46 units are ultimately absorbed at the surface.

For the planet as a whole, the equilibrium radiative temperature (i.e. the temperature required to reradiate all the energy received from the sun) at the outside of the atmosphere is 255 K (-18 °C). If the atmosphere was perfectly transparent to all wavelengths of radiation, then it could neither absorb nor emit radiation, and the average temperature at the surface would also be close to 255 K. However, certain gases in the atmosphere, known as radiatively active gases (RAGs), or greenhouse gases (GHGs), absorb some of the upward-going longwave radiation after

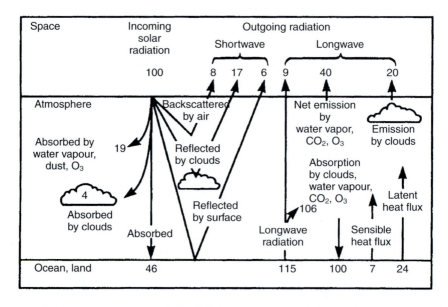

Figure 11.5 Main components of the global average energy balance. The initial input of shortwave solar energy is taken to be 100 units.
Source: Wuebbles, D. J. (1993) 'Global climate change due to radiatively active gases', In: C. N. Hewitt and W. T. Sturges (eds) *Global Atmospheric Chemical Change*, Chapman and Hall, London.

it has left the Earth's surface. The gases then emit longwave radiation in all directions. The downward-going portion adds to the shortwave radiation arriving at the surface of the Earth and raises the equilibrium temperature. Referring again to Figure 11.5, of 115 units of longwave energy emitted from the Earth's surface, 106 units are absorbed by clouds and GHGs in the atmosphere and 100 units eventually returned to the Earth. Thus, the atmosphere acts as an extremely effective longwave insulator. The overall process is known as radiative forcing, or colloquially as the Greenhouse Effect. The most important GHGs are water, carbon dioxide, methane, ozone and nitrous oxide.

For thousands of millions of years there has been a large radiative forcing of the surface energy balance due to GHGs in the atmosphere. In the early history of the Earth, the forcing was due partly to water vapour and mostly to a CO_2 concentration far higher than today's. It is thought that when the Earth was formed the power output from the Sun was only about 75% of its current value. In the absence of any greenhouse effect, the equilibrium radiative temperature would have been only 237 K ($-36\,°C$), the seas if any would have been frozen, and life could not have developed. The greenhouse effect was provided by CO_2, NH_3 and CH_4 concentrations that were far higher than any experienced today, equivalent in their radiative forcing to several hundred times the current CO_2 concentration. The power output from the Sun has increased during the 4 billion years of the Earth's lifetime, and during this period the CO_2 concentration has decreased, due largely to its sequestration in organic and later inorganic (e.g. calcium carbonate) forms by biological organisms as they developed. Hence, the surface temperature has remained within a relatively narrow band. The Gaia hypothesis is that this process is a consequence of the drive by living systems to optimise their survival environment. Today, water vapour is a completely natural, and by far the most important, GHG; its presence at concentrations of up to a few percent elevates the average temperature at the bottom of the atmosphere by 33 °C, from −18 °C to +15 °C, thereby making life on Earth possible. On Venus (90% of whose atmosphere consists of CO_2 at a much higher pressure than the Earth's) and Mars (80% CO_2), carbon dioxide is the dominant natural GHG, and elevates the surface temperatures by 523 and 10 °C respectively. On Earth, anthropogenic CO_2 and other GHGs such as methane, nitrous oxide, ozone and chlorofluorocarbons are now increasing in concentration to the point at which they may be appreciably raising the global average temperature and thereby also influencing rainfall patterns and sea level.

Each of the GHGs absorbs longwave radiation in characteristic regions of the spectrum shown on the top five graphs of Figure 11.6. These individual spectra are combined on the sixth graph. For example, water absorbs in the ranges

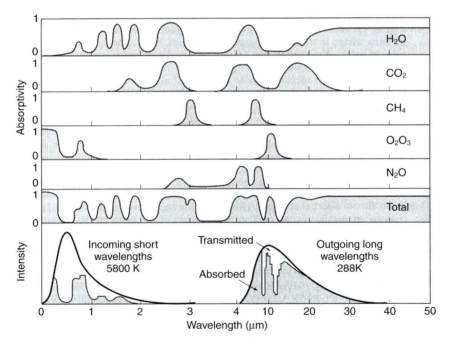

Figure 11.6 Absorption bands of the important GHGs.
Source: Masters, G. M. (1998) *Environmental Engineering and Science*, Prentice Hall, New Jersey.

2.5–3.5 μm, 5–7 μm, as well as a broad range above 13 μm. In the bottom graph of Figure 11.6, these absorption spectra have been combined within the solar and terrestrial spectra that we referred to in Figure 11.2. Note that the wavelength scale of the solar radiation spectrum has been expanded relative to that for terrestrial radiation. High absorption can have no influence if there is no radiation there to absorb, and although GHGs are effective absorbers in the 1–5 μm region, there is little longwave energy there owing to the separation of the black-body distributions. Between 5 and 20 μm all the GHGs have important absorption bands, and above 20 μm the atmosphere is opaque due to absorption tilt by water vapour. The region between 8 and 13 μm (the narrow spike centred on 10 μm in Figure 11.6) is known as the atmospheric window because its relative transparency allows the passage of radiation to the ground from space and *vice versa*. Water vapour absorption is confined to bands – for example, the high absorption between 5 and 8 μm is due to one such band. Water is not active in the window between 8 and 13 μm, although there is a narrow absorption band at 9.7 μm due to ozone. Beyond 14 μm, transmission is suppressed again by both water vapour and carbon dioxide.

11.2   THE ROLE OF GASES

11.2.1   Carbon dioxide

Carbon is at the heart not only of life, but also of energy production and climate change. The carbon content of past organisms has been stored in the form of oil, gas and coal. A rather small proportion of this carbon is accumulating in the atmosphere where it is helping the Earth to warm up faster than it has ever done previously. Although we have no direct evidence of CO_2 concentrations from millions of years ago, we have been able to derive them over a surprisingly long period. This history of the global atmospheric carbon dioxide concentration has been taken from ice-core measurements. When snow falls, air is trapped amongst the ice crystals. Subsequent snowfalls build up in layers, each of which is a record of the air and water quality prevailing at the time. Eventually, the snow layers become compressed into ice, and the air is trapped in bubbles within the ice. By drilling down up to 5 km in Antarctica and Greenland, it has been possible to reconstruct the history of fluctuations in CO_2, CH_4 and temperature (which is derived from $^{18}O/^{16}O$ isotopic ratios) going back over 400 000 years. Figure 11.7 shows such a record, taken from the 4000 m Vostok ice core in east Antarctica.

At the start of this record, carbon dioxide was close to its preindustrial concentration of 280 ppm. Over the next 400 000 years, the concentration fell four times to around 200 ppm before recovering rapidly (on a geological timescale) to its original value. There is a remarkable correlation between the CO_2, methane and temperature records, both for the long-term cycles and the more rapid fluctuations within them. Frequency analysis shows dominant periods of 100, 41, 23 and 19 000 years, which correlate with the Milankovitch cyclical changes in the Earth's orbital parameters, such as eccentricity, precession and tilt (Figure 11.4). These

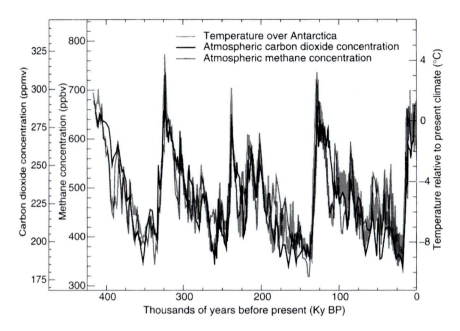

Figure 11.7 An example of the ice-core record of CO_2, methane and temperature changes over the last 400 000 years.
Source: IPCC (2001) *Climate Change – Third Assessment Report*, Cambridge University Press, Cambridge, UK.

changes affect the radiation budget directly, and hence temperature, plant life and CO_2 concentration. The simple explanation is that when the Earth is cooling, the surface waters of the oceans are cooler and therefore able to absorb more CO_2, so that the atmospheric concentration decreases, and *vice versa*. However, modelling shows that the orbital changes alone are not sufficient to cause the observed changes in climate, and hence some feedback or amplification mechanism must exist. There is no way of telling from this record alone whether it is CO_2 changes that are driving temperature changes, or *vice versa*, or indeed whether they are both being driven by an unmeasured third variable. Although there were major variations in CO_2 throughout the period, the concentration was never greater than 320 ppm. During the past 1000 years, the concentration has shot up at an unprecedented rate, from the previous typical peak level of 280 ppm to the current value of 370 ppm. This recent sharp rise in CO_2 concentration is due to the anthropogenic addition of CO_2 to the Earth's atmosphere by biomass burning and combustion of fossil fuels (Figure 11.8), strongly associated with population growth and industrialisation (Table 11.1). Figure 11.8 also demonstrates well the separation between the northern and southern hemispheres in this respect.

Although the total of 6.3 Gt is a huge absolute quantity in comparison to the fluxes of the sulphur and nitrogen pollutant gases that we have dealt with

CO$_2$ EMISSION FROM FOSSIL FUEL (1×10^{12} cm^{-2} s^{-1})

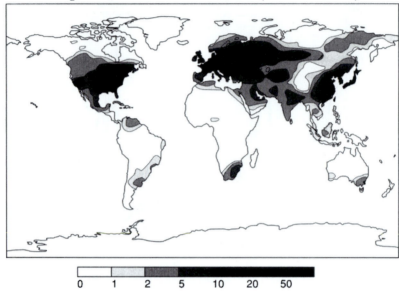

0   1   2   5   10   20   50

CO$_2$ EMISSION FROM DEFORESTATION (1×10^{12} cm^{-2} s^{-1})

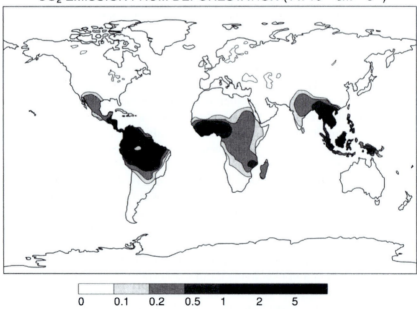

0   0.1   0.2   0.5   1   2   5

Figure 11.8 Global distribution of CO$_2$ emissions due to fossil fuel and deforestation. Source: Brasseur, G. *et al.* (1999) 'Trace gas exchanges in biogeochemical cycles', In: P. Brasseur, J. L. Orlando and G. S. Tyndall (eds) *Atmospheric Chemistry and Global Change*, Oxford University Press, New York, USA.

Table 11.1 Annual global budget of carbon dioxide

Component	Flux/Gt C a^{-1}
Emission to atmosphere from fossil fuel combustion and cement manufacture	6.3
Flux from atmosphere to ocean	−1.7
Flux from atmosphere to land	−1.4
Net atmospheric increase	3.2

previously (which are of order 0.1 Gt), it still represents only 3% of the 200 Gt C exchanged by natural processes such as photosynthesis and senescence every year, or just over 1% of the 560 Gt C that is locked up in terrestrial biomass such as forests, or 1% of the atmospheric stock of carbon (760 Gt C). A net increase of 2.12 Gt C corresponds to an increase of 1 ppm in the global average CO_2 concentration, so that the atmospheric concentration is increasing at about 1.5 ppm per year. A more detailed breakdown of CO_2 sources on a national basis is shown in Table 11.2. It is clear that emissions are spread quite evenly across all fuels and all activity sectors. It is this ubiquitous aspect of CO_2 emissions, as well as their sheer magnitude, that makes them hard to reduce.

Carbon is heavily involved in two natural cycles of very different timescales. The long-term cycle starts with the emission of CO_2 in volcanic gases, both as steady releases and violent explosions. Some of this CO_2 is dissolved into cloud droplets and eventually rained out. Clean rain is weak carbonic acid with a pH of 5.6 due to this effect. When the rainwater washes on igneous rocks, the resulting stream and river water contains bicarbonate ions. Eventually, this bicarbonate is precipitated as sedimentary rocks on the ocean floor, which are then in turn subducted at tectonic plate boundaries, where volcanic activity releases the CO_2. This cycle takes 100 million years, but without it the atmospheric CO_2 would eventually be depleted. The second cycle is a short term one which involves living organisms. On the plant side, the Earth's terrestrial and aquatic plants take up CO_2 from the atmosphere by photosynthesis and lock up the carbon as plant parts. The carbon is eventually returned to the atmosphere – slowly by respiration and rotting, or rapidly by forest fires. Animals burn food and release CO_2 as a byproduct. There is a large continuing exchange between the atmosphere and the oceans, in which the surface layers are saturated with CO_2 and alternately take up new CO_2 from the atmosphere and release it back again.

The recent ice-core record, together with direct measurements of atmospheric concentration taken at Mauna Loa in the Hawaiian Islands since 1958, are shown in Figure 11.9. The CO_2 concentration was steady at around 280 ppm until 1800, when it started to rise quickly. This figure also shows (inset) that the most recent ice-core values tie in well with the atmospheric measurements, which gives confidence to the whole ice-core record. About three-quarters of the increase in atmospheric CO_2 concentration is due to fossil fuel burning, with the remainder being due mainly to land-use change, especially deforestation. Superimposed on

Table 11.2 UK CO_2 emissions in 1970 and 1998

Source	Emissions (1970) Mt	Emissions (1998) Mt	Percentage of 1998 total
Combustion for energy production			
Electricity	57	41	27
Petroleum refining	5.6	5.7	4
Other combustion	8.2	5.8	4
Combustion in commercial and residential			
Residential	26.5	23.2	16
Commercial and			
agricultural	12.2	8.5	6
Combustion by industry			
Iron and steel production	15.8	6.3	4
Other industrial	32.5	16.9	11
Production processes	4.9	3.6	2
Extraction and distribution			
of fossil fuels	0.1	0.2	0
Road transport	16.3	31.5	21
Other transport and			
machinery	5.6	4.4	2
By fuel type			
Solid	90.3	34.2	24
Petroleum	70.7	52.3	35
Gas	20.3	57.0	38
Non-fuel	4.0	5.0	3
Total	185	149	100

the general increase measured at Mauna Loa is a strong annual variation due to seasonal changes in uptake by vegetation and oceanic phytoplankton (Figure 11.10).

Which fossil fuel is burned also makes a difference. The chemical reactions by which carbon (in coal), methane (in natural gas) and hydrocarbons (in oil) are oxidised all have different enthalpies and emit different masses of CO_2 for each unit of energy released. These amount to about 112, 50 and 70 kg kJ^{-1} respectively. The low CO_2 emission of gas has been another reason for it displacing coal in large scale energy operations such as electricity generation.

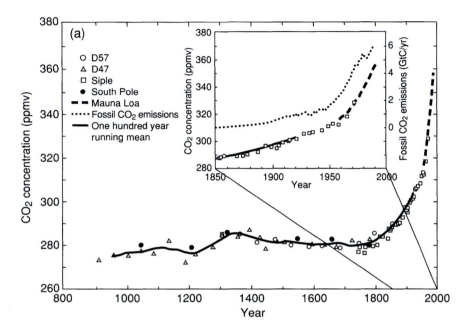

Figure 11.9 The time course of CO_2 concentration between 1000 and the present, compiled from ice cores and atmospheric measurements.
Source: IPCC (1994) *Radiative Forcing of Climate Change and An Evaluation of the IPCC IS92 Emissions Scenarios*, Cambridge University Press, Cambridge.

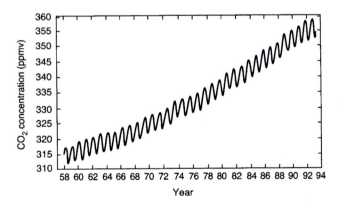

Figure 11.10 The background CO_2 concentration measured at Mauna Loa, Hawaii since 1958.
Source: IPCC (1994) *Radiative Forcing of Climate Change and An Evaluation of the IPCC IS92 Emissions Scenarios*, Cambridge University Press, Cambridge.

11.2.2 Methane

Methane is produced by a wide variety of anaerobic processes. The main natural source is bacterial decay in wetlands, while anthropogenic sources are paddy fields, the digestive tracts of ruminant animals such as cattle, and natural gas emissions from fossil fuel supplies. Methane is eventually oxidised to CO_2 (another GHG) in the atmosphere; since this is achieved via reaction with the OH radical, CH_4 is also involved in ozone photochemistry. If the oxidation occurs in the presence of high NO_x concentrations, then O_3 is produced; if NO_x is low, then O_3 is consumed (Table 11.3).

The atmospheric sink (due to reaction with OH in the troposphere, $CH_4 + OH \rightarrow CH_3 + H_2O$) is around 500 Mt a^{-1}, with a net atmospheric increase of 15 Mt a^{-1}. The long-term ice core record shows a very similar pattern of fluctuations to that of CO_2, ending up with a preindustrial concentration of 700–800 ppb. Since then the concentration has more than doubled to 1750 ppb (Figure 11.11).

However, the most recent data suggest that the rate of increase in CH_4 has decreased rather rapidly, from 20 ppb a^{-1} in 1975 to only a few ppb a^{-1} currently. This decrease is not understood at present, and may only be a temporary response to a short-term disturbance such as the eruption of Mt Pinatubo. A more detailed breakdown of methane emissions from the UK is given in Table 11.4.

Simple tables such as this conceal the huge amount of information retrieval which went into creating it. For example, the inventory for oil and gas production involved the identification of emissions from flare stacks, unignited vents, maintenance vents, gas turbine exhausts, fugitive emissions and drains. For each of these sources, a methodology was identified for estimating the emissions. In the case of flares, the rate of gas flow to flares, the percentage combustion efficiency and the proportion of methane in the flared gas were combined to calculate the total flow of uncombusted methane to the atmosphere. Where appropriate, direct measurements of emissions were taken in order to confirm the validity of the estimation procedure.

Table 11.3 Global methane emissions (Mt yr^{-1})

Natural		Human (biosphere)		Human (fossil fuels)	
Wetlands	100–200	Enteric fermentation (mainly cattle)	100	Natural gas	40
Termites	20	Rice paddies	60	Coal mining	20
Oceans	10	Biomass burning	40	Coal burning	25
Other	15	Wastes, landfill and sewage	90	Petroleum processing	15
Total	145–245	Total	290	Total	100

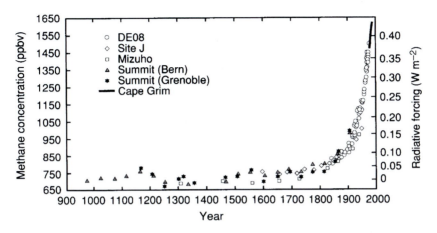

Figure 11.11 The time course of methane concentration between 1000 and the present, compiled from ice cores and atmospheric measurements.
Source: IPCC (1994) *Radiative Forcing of Climate Change and An Evaluation of the IPCC IS92 Emissions Scenarios*, Cambridge University Press, Cambridge.

The largest UK methane source is farm animals. Cattle and sheep are adapted to eat grass and other forage materials which have a high content of plant cell walls made up from structural carbohydrates. The rumen, which is the first stomach, acts as a fermentation chamber in which a microbial population of bacteria, fungi and protozoa break down the ingested carbohydrate into soluble digestible materials. The rumen bacteria generate methane as a waste product, which is lost by eructation (belching). One cow emits about 70 kg methane per year, one sheep about 8 kg, and one pig about 1.5 kg. Cattle and sheep emit about 70 and 20% respectively of the total animal emission. Each cow or sheep also creates around 15 or 1.2 t yr^{-1}, respectively, of excreta which eventually returns on to the land as dung, manure or slurry. A further 112 kt methane was emitted from anaerobic fermentation by methanogenic bacteria within these organic wastes.

Landfill gas emissions from the anaerobic decomposition of organic wastes are the next largest sources. Around 85% of UK municipal waste is disposed of to landfills. In properly designed, well-capped landfills the methane is an energy resource, and a substantial proportion of this figure could be recovered and used.

Coal and methane are always associated, and coal seams contain up to 20 m^3 methane t^{-1} from the 'firedamp' (which is 80–95% methane) absorbed in the coal. Hence, methane is released during mining and ventilated to atmosphere. It is also released from open-cast coal faces and from coal stocks. In the UK, the methane from coal mining used to dominate the total, but has declined with the coal industry.

Table 11.4 UK methane emissions in 1970 and 1998

Source	Emissions, (1970) kt	Emissions, (1998) kt	Percentage of 1998 total
Combustion for energy production			
Electricity	2	19	1
Petroleum refining	1	1	0
Other combustion	1	3	0
Combustion in commercial and residential			
Residential	327	35	1
Commercial and			
agricultural	2	4	0
Production processes	46	8	1
Extraction and distribution of fossil fuels	3	3	0
Coal mines	1540	263	10
Gas leakage	51	382	14
Offshore oil and gas	13	62	2
Road transport	16	19	1
Waste			
Landfill	764	774	29
Non-landfill	32	57	2
Land-use change			
Enteric fermentation	854	883	33
Animal wastes	115	112	4
Fuel emissions	388	84	3
Non-fuel emissions	3388	2553	97
Total	3788	2637	100

For gas distribution, the 382 kt represents about 1% of total gas throughput. The gas distribution system is divided into three distinct parts. There are 5 500 km of high pressure national transmission pipes, typically 1 m in diameter and carrying gas at pressures of 7–75 bar; 12 500 km of regional transmission system; and the regional distribution system of mains and services (e.g. a single pipe to one dwelling). Jointed mains are the main source of gas leaks, but others can be significant. It has even been estimated that 3 700 t methane are lost each year from the 1-s delay between turning on the gas and lighting the hob on a cooker. Domestic gas leaks are made more noticeable by the addition of an odoriser to gas at the terminal. The odoriser in general use is called Odorant BE, and consists of 72% diethyl sulphide, 22% tertiary butyl mercaptan and 6% ethyl mercaptan.

11.2.3 Nitrous oxide

Nitrous oxide (N_2O) is produced by a variety of biological nitrification and denitrification processes in water and in soil. Although not toxic at normal atmospheric concentrations, it is a radiatively active gas that contributes to the greenhouse effect and climate change. The soil component of the emission results from anaerobic microbial denitrification and nitrification. This is more intensive in agricultural areas where N is added to the soil, but wet and dry N deposition to other areas could be increasing N_2O emissions. Land-use changes such as forest clearance, drainage and cultivation stimulate emission. There are large spatial variations that make it difficult to quantify regional emissions; approximate global totals are given in Table 11.5.

The major removal mechanism is photolysis in the stratosphere, which accounts for 10 Mt a^{-1} (and for its long lifetime of 150 years), with the quantity in the atmosphere increasing by 3 Mt a^{-1}. Preindustrial concentrations of N_2O were about 285 ppb, have since increased to about 310 ppb, and are currently rising at 0.8 ppb a^{-1} (Figure 11.12). Natural emissions make up about 80% of the total. The component from cultivated soils is potentially the largest of the man-made sectors, although there is a lot of uncertainty attached to it. Corresponding calculations for the UK give emissions as shown in Table 11.6. Hence, around one half of the total is due to industrial emissions, and one quarter to fertilised land.

11.2.4 Halocarbons

Unlike carbon dioxide, methane and nitrous oxide, halocarbons are a recent invention and there is no natural source or background. There are at least 12 principal halocarbons that are a cause of concern for the Earth's radiation balance. They have been used for three main purposes: as the heat transfer medium in refrigerators and air conditioning systems; as the propellant in aerosol spray cans; and as the foaming agent in expanded polystyrene packaging and insulation. Two of the desirable properties that make them so suitable for these purposes are their chemical inertness and thermal stability, the very properties that also result in their

Table 11.5 Global sources of nitrous oxide

Source	Production/Mt N a^{-1}
Tropical forest soils	4.2
Temperate forest soils	4.0
Fertiliser application	1.5
Biomass burning	0.1
Fossil fuel combustion	0.6
Industrial sources	0.7
Oceans	2.0

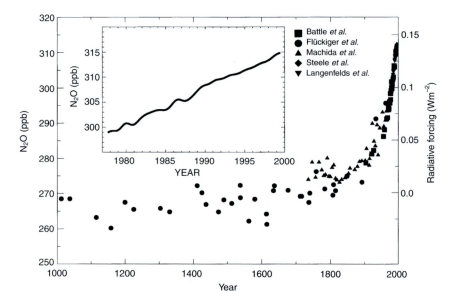

Figure 11.12 The time course of nitrous oxide concentrations over the past 1000 years compiled from ice cores and atmospheric measurements.
Source: IPCC 2001 *Climate Change – Third Assessment Report*, Cambridge University Press, Cambridge.

Table 11.6 UK N_2O emissions

Source	Emission/kt N a^{-1}
Fertilised land	22–36
Livestock waste	12
Forest and woodland	5
Semi-natural land	6
Road transport	3
Industry	49
Other fuel combustion	2
Total	100–115

having very long lifetimes in the atmosphere. The main consequent problem that has received attention is the chemical effect on stratospheric ozone (see Chapter 12), but there is also a physical effect. The halocarbon molecules absorb infrared radiation effectively in the waveband of the atmospheric window. They are hence strongly radiatively active on a per-molecule basis, and contribute a significant radiative forcing despite their low concentrations. One CFC molecule in the atmosphere is 5–10 000 times more effective at trapping longwave radiation than one molecule of CO_2. As a result, the tiny CFC concentrations account for about

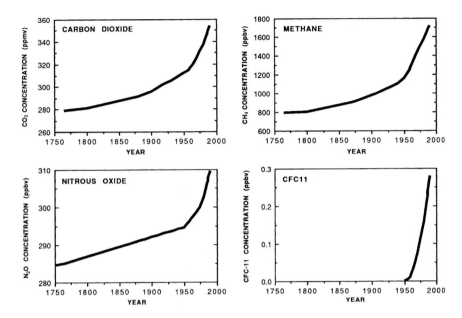

Figure 11.13 Summary graphs of the changes in atmospheric concentration of the four principal anthropogenic GHGs since 1750.
Source: IPCC (1990) *Climate Change – First Assessment Report*, Cambridge University Press, Cambridge.

20% of the radiative forcing in the tropics. The net effect is reduced at high latitudes because the CFCs have reduced the O_3, which is also a greenhouse gas. The two most important halocarbons are the chlorofluorocarbons CFC-11 and CFC-12, which have average concentrations of a few hundred ppt. Detailed information on the different halocarbons and their production rates is given in Chapter 12. Although their destructive effect on the ozone layer led to an international agreement (The Montreal Protocol) to phase out production, their long lifetimes mean that their effects will persist well into the twenty-first century.

In summary then, we have the changes in atmospheric concentration shown on Figure 11.13, with global industrialisation increasing the atmospheric concentrations of CO_2, CH_4 and N_2O at an accelerating rate since the eighteenth century, and CFCs coming in after the Second World War.

11.2.5   Contributions to radiative forcing

It is a major task to calculate the effect due to each GHG over the next century because it depends on the current and future release rates, on the lifetime of the gas in the atmosphere, and on the spectral absorptivity for radiation of the particular gas. These properties and parameters vary widely from gas to gas, as shown by Table 11.7 and Table 11.8.

In Table 11.7, the global warming potential (GWP) is a measure of the relative, globally-averaged warming effect due to the emission of unit mass of the gas. There is still uncertainty attached to this value, since although the direct warming by the gas can be calculated, indirect warming after its conversion to other gases is a far more complex process. The table shows that on a per-mass basis, the CFCs are thousands of times more effective than carbon dioxide as GHGs. However, they are emitted in far smaller quantities, so that their overall impact on radiation balance remains low. In Table 11.8, the relative importance of the different gases is shown in terms of the calculated increase in the radiative flux at the surface of the Earth between 1765 and 2000. Sixty per cent of the overall increase of 2.45 W m^{-2} has been due to carbon dioxide.

Note that despite the dominance of CO_2, methane also makes a significant contribution due to its higher GWP per kg. Although the contribution from CFCs is significant, it will decline over the next 50 years due to controls on CFC production designed to reduce damage to the ozone layer (see Chapter 12). In fact, the reduction in stratospheric ozone has caused a *negative* radiative forcing of -0.15 W m^2, so the decrease in one will tend to offset the increase in the

Table 11.7 Relevant properties of main GHGs

Gas	Lifetime in years	Relative radiative forcing per molecule	Relative radiative forcing per kg of gas	Global warming potential (relative to CO_2) over 100 years after 1 kg release
CO_2	120	1	1	1
CH_4	10	30	58	11
N_2O	150	160	206	290
CFC-11,12	100	21–25 000	4–5 000	5 000

Table 11.8 Effects of GHGs on global energy flux

Date	CO_2/ ppm	CH_4/ ppb	N_2O/ ppb	CFC-11/ ppt	CFC-12/ ppt	HCFC/ ppt
1765	279	790	285	0	0	0
2000	370	1750	310	280	500	320
Heat flux increase due to gas/W m^{-2}	1.48	0.48	0.15	0.07	0.17	0.10
% contribution	60	20	6		14	

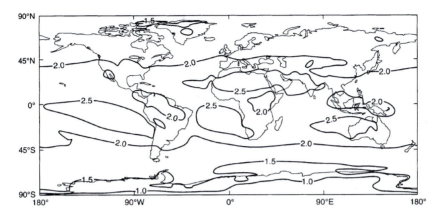

Figure 11.14 Global contours of the radiative forcing (W m^{-2}) due to GHGs since the start of the industrial revolution.
Source: IPCC (1994) *Radiative Forcing of Climate Change and An Evaluation of the IPCC IS92 Emissions Scenarios*, Cambridge University Press, Cambridge.

other in the medium-term future. On the other hand, *tropospheric* ozone has increased by about 36% since 1750, as we have seen previously in this book, increasing radiative forcing by about 0.35 W m^{-2}. The latter is an average figure, since tropospheric ozone is a local-scale short-lived gas which, unlike carbon dioxide, is not well mixed throughout the lower atmosphere. Looked at another way, if the CO_2 in the atmosphere were to be halved from its current value, the average radiative forcing would be reduced by around 4 W m^{-2}, and the global surface temperature would fall by 1–2 °C. If the CO_2 were to be removed altogether, outgoing radiation would increase by 25 W m^{-2} and temperature would fall by over 10 °C.

Perfluorocarbons (CF_4, C_2F_6) and sulphur hexafluoride (SF_6), which are also radiatively active, are used in some industrial processes and released to atmosphere in quite small quantities but have long lifetimes. The radiative forcing due to such gases is not significant at present.

There are considerable geographical variations in the global distribution of this forcing. Figure 11.14 shows that the calculated increase in forcing has been about 1 W m^{-2} at the poles, increasing to 2.5 W m^{-2} in the tropics. Note also that because the GHGs have fairly long lifetimes and are well-mixed into the atmosphere, the patterns of forcing reflect global differences in surface temperature and solar radiation rather than the source distributions of the GHGs themselves. This is a truly international problem, and emission reductions by one country will make little difference to that country's climate change.

11.3   THE ROLE OF AEROSOL

In addition to the gases described above, there is a second important factor that has also been modifying global energy fluxes – absorption and reflection of solar radiation by atmospheric aerosol. There are large natural sources of particles – volcanoes and forest fires, for example, which we would not have expected to increase or decrease on average during the last century. However, the increases in population and fossil fuel combustion that were responsible for the GHG emissions also produced particles – either directly, or indirectly via gas-to-particle conversions. If these particles remain in the troposphere, their lifetimes are only days or weeks. If they are injected into the stratosphere by a sufficiently violent process such as a volcanic eruption or a nuclear weapon test, they may remain there for years. Much of the aerosol mass consists of particles that are about the same size as the wavelength of visible light, so they interact strongly with it.

The exact effect of an aerosol layer depends on its optical depth and height in the atmosphere. In general, the aerosol will absorb both down-going solar radiation and up-going terrestrial radiation. Naturally, the greater the optical depth the greater the radiation exchange. If that radiation exchange occurs high in the atmosphere, there will tend to be warming at that height and cooling at the ground. Conversely, aerosol in the troposphere will tend to warm the troposphere. Aerosol radiative properties also depend on the size distribution of the particles – if the effective radius is greater than about 2 μm, then they act like a GHG and there will be net warming due to the absorption of longwave terrestrial radiation. If the effective radius is less than 2 μm, then reflection of shortwave solar radiation is more important, and there is net cooling. At the top of the aerosol cloud, warming by absorption of solar radiation in the near infrared dominates over the enhanced IR cooling from the aerosol particles. In the lower stratosphere, the atmosphere is warmed by absorption of upward longwave radiation from the troposphere and surface. In the troposphere, there are only small net radiative effects, because although the reduced supply of near IR means there is less warming due to absorption by water vapour, this is compensated for by the increased flux of longwave from the aerosol particles.

Table 11.9 gives a summary of the radiative forcing due to the main aerosol source categories, with the largest effect being a negative forcing due to sulphate from the sulphur released by fossil fuel combustion.

Figure 11.15 gives estimated global contours of the forcing due to sulphate concentrations in the troposphere, showing that over industrialised land areas of the

Table 11.9 Radiative forcing due to aerosols

Aerosol component	Forcing/W m^{-2}
Sulphate	−0.4
Biomass combustion	−0.2
Organic carbon from fossil fuel	−0.1
Black carbon from fossil fuel	+0.2

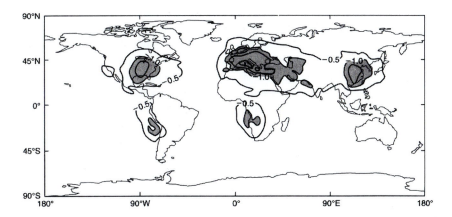

Figure 11.15 Global contours of the radiative forcing (W m^{-2}) due to sulphate aerosol increase since the start of the industrial revolution.
Source: IPCC (1994) *Radiative Forcing of Climate Change and An Evaluation of the IPCC IS92 Emissions Scenarios*, Cambridge University Press, Cambridge.

northern hemisphere the loadings, which are now up to 15 times their preindustrial level, account for *negative* forcing of up to 1.5 W m^{-2}. This is similar in magnitude to the positive forcing due to GHGs, although its geographical distribution is linked much more closely to the sulphate source areas.

11.3.1 Volcanoes and aerosol formation

Scientists have used large volcanic eruptions to test their models of climate change scenarios and to separate human from natural effects. An eruption is a perturbation to a system that is in quasi-equilibrium, and creates detectable oscillations in various measured climate change parameters. It also perturbs the ozone layer, as described in Chapter 12. There are two distinct families of volcano from this perspective.

Effusive volcanoes, of which Mt Etna in Sicily is a good example, emit continuously with low thermal energy and momentum. The emissions from effusive volcanoes do not have sufficient energy to penetrate the temperature inversion at the tropopause, so they remain in the troposphere. As a consequence, the emissions are involved in local small scale deposition processes on timescales of hours or days. Explosive volcanoes, on the other hand, have sufficient energy to inject large amounts of material into the stratosphere, where the inherent atmospheric stability gives long lifetimes of weeks to years and potentially global coverage. Ironically, the volcanic plume is not usually buoyant at source because of the high solid density, but becomes buoyant by entraining large volumes of ambient air which is heated by the hot solids. Hence, an injection into the stratosphere brings with it not only the volcanic material but tropospheric air with its own pollutant concentration profile.

On June 14, 1991, Mount Pinatubo in the Philippines erupted violently after 635 years of inactivity. The heaviest phase of the eruption lasted for 36 hours, and ash falls were observed in Thailand and Singapore. More than 19 Mt of SO_2 (2–3 months-worth of total global SO_2 emissions, or about 20-years of UK emissions) were injected up to an altitude of 20–30 km in the stratosphere, and eventually converted into 40–50 Mt of sulphuric acid aerosol. These aerosol are effective at scattering shortwave radiation (and at depleting stratospheric ozone – see Chapter 12). Data from the NASA earth radiation budget satellite (ERBS) stratospheric aerosol and gas experiment (SAGE) have shown that the optical depth of stratospheric aerosol increased by over two orders of magnitude between 40°N and 40°S during the first 5 months after the eruption, and that the total aerosol optical depth over the tropical Pacific increased by a factor of five. It was estimated that this increase in aerosol forcing would increase reflected radiation by up to 4.5 W m^{-2} in early 1992, decaying exponentially to a negligible amount by 1995, and that this change would in turn result in a short-term decrease in mean global surface temperature of up to 0.5 °C. Hence, natural events can certainly produce radiative and temperature changes of comparable magnitude to the human ones, but they are transient rather than incremental. Eruptions themselves may be transient or continuous in nature. If a volcanic plume rises 20 km into the atmosphere at a speed of 150 m s^{-1}, then it will take 2–3 min to become established. Measurements on many different eruptions have shown that plume height α (total heat content)$^{1/4}$ for a transient plume or (heat release rate)$^{1/4}$ for a continuous plume. Mt Pinatubo achieved a maximum height of 40 km. On a far larger scale, flood basalt eruptions have involved the ejection of over 1000 km^3 of solids within a few months, with an estimated plume height of 30 km. A caldera eruption such as Tambora or Krakatoa would eject *several thousand* km^3 in a day or so.

Two main measures of volcanic impact have been used. The volcanic explosivity index (VEI) ranks volcanoes according to the rate of energy release, giving a range for recent volcanoes of up to 7 for Tambora, Indonesia in 1815, and 6 for Krakatoa and Pinatubo. The dust veil index (DVI) is compounded from various observations of the severity of an eruption, and of the extent and effects of the resulting dust cloud. These might include historical reports of the eruption itself, optical phenomena such as obscuration of the sun or moon, radiation measurements, temperature measurements, or estimates of the volume of ejected material. The DVI takes on values of 3000, 1000 and 1000 for the same three volcanoes, respectively. The two measures are well correlated in general, although individual eruptions may not fit. For example, Mt St Helen's in 1980 was a powerful eruption with VEI = 5, but because it mainly came out sideways the stratospheric injection was low. Large eruptions (greater than 5 on the Volcanic Explosivity Index) overcome the temperature inversion at the stratopause and project material into the stratosphere, where it has long residence time. Smaller volcanic emissions are much more common, but material remains in the troposphere and has shorter lifetime.

It is not only the size of the eruption that matters, but where it happens on the Earth. A tropical volcano will generate larger tropospheric heating at the equator than at the poles, enhancing the pole-to-equator temperature gradient, especially in winter. Hence, because in winter in the northern hemisphere, advective forces dominate over convective, this increases the driving force on the Polar vortex (the jet stream) and perturbs the Arctic Oscillation. The actual effect on ground-level conditions depends on season and the state of El Niño (i.e. sea surface temperature). These interactions can lead to winter warming for temperate latitudes in the northern hemisphere, even due to a tropical volcanic eruption.

Volcanoes are responsible for around 10% of global atmospheric sulphate, but passive degassing from effusive volcanoes can raise this to 30% away from industrial sources. Etna, for example, is the largest continuous global source of SO_2 and releases around 5000 t day^{-1} at a concentration of about 1 ppm, equivalent to the emission from 10–20 2000 MW coal-fired power stations. During a more active phase, the flux rises to 25 000 t day^{-1}. Since the physical release height is 3300 m, and the effective release height due to buoyancy higher still, the emissions have a long residence time in the atmosphere in which to undergo gas-to-particle conversion. A contrast is seen with the Laki, Iceland eruption of 1783. This was a low-altitude event, which it is thought was effusive rather than eruptive. Estimates put the total sulphur emission at 122 Mt, equivalent to the total current annual human sulphur emission. Unusual climate events such as dry fogs and high temperatures were recorded across Europe, notably by Benjamin Franklin who was US ambassador to France at the time. Secondary effects of such eruptions include increased sulphur deposition to wetlands and a consequent increase in methane emissions.

For volcanic emissions, the initial dust and ash falls out in the troposphere, while the SO_2 rises in the plume. While the SO_2 is rising, it reacts with water vapour from both the plume and from entrained air to form sulphuric acid aerosol, which are small and very effective at scattering radiation. The processes are summarised in Figure 11.16.

Aerosol area densities of up to 40 μm^2 cm^{-3} were measured in the Pinatubo plume, compared to 1 μm^2 cm^{-3} present as background. These resulting changes in irradiance naturally change the temperature of the atmosphere as well. In Figure 11.17, the arrows on the time axis show major eruptions from El Chicon and Pinatubo. The global average monthly stratospheric temperature was increased by between 1 and 1.5 °C by the two eruptions.

As well as scattering solar radiation directly, these aerosol particles have indirect effects on the planetary albedo via cloud droplet formation. As seen previously, water droplets are unlikely to form spontaneously at realistic supersaturations because of the excess vapour pressure associated with the very small radius of curvature of the nascent droplet. Sulphate particles, which are hygroscopic and deliquesce to form 'starter' droplets at low supersaturations, are ideal nuclei for the formation of cloud drops. Any increase of such nuclei due to natural or man-made emissions will therefore have consequences for the

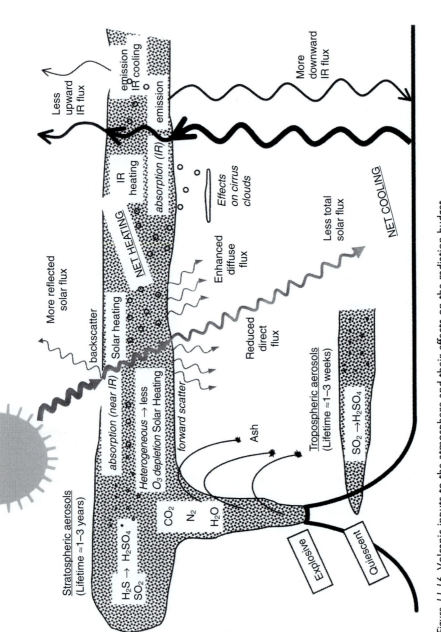

Figure 11.16 Volcanic inputs to the atmosphere and their effects on the radiation budget.
Source: Robock, A. (2000) 'Volcanic eruptions and climate', *Reviews of Geophysics* **38**: 191–219.

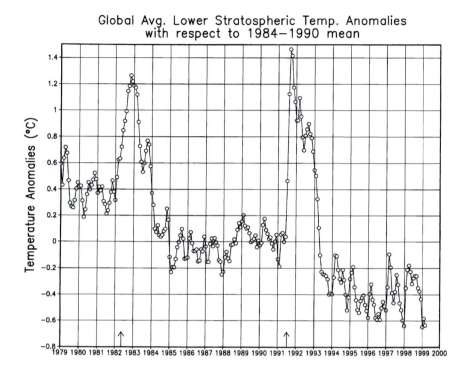

Figure 11.17 Stratospheric temperature anomalies due to volcanic eruptions.
Source: Robock, A. (2000) 'Volcanic eruptions and climate', *Reviews of Geophysics* 38: 191–219.

number density and diameter of cloud drops, and indirect effects on albedo. If a fixed amount of water vapour is available for cloud formation, then a larger number of smaller drops will give a higher reflectivity, changing the radiation balance in the same sense as did the aerosol particles themselves.

11.3.2   Modelling the radiative impact of aerosol

Radiation attenuation by aerosol particles is described by the Bouguer–Lambert Law

$$V_{m}(\lambda) = V_{0}(\lambda) \exp(-\tau_{atm} \, m)$$

where $V_{0}(\lambda)$ is the solar flux at zero airmass at wavelength λ, $V_{m}(\lambda)$ is the measured solar flux, τ_{atm} is the total optical depth of the atmosphere, m is the airmass factor, expressed as 1/cos(zenith angle).

Typically, the measured optical depth is the sum of several components, including those due to Rayleigh scattering and gaseous absorption; these are measured or calculated and the aerosol component obtained by difference. Measurements at many wavelengths enable the spectrum to be inverted and the particle size distribution to be calculated.

The size distributions of aerosol from volcanoes such as Pinatubo have been modelled by formulas such as:

$$n(r) = \frac{1}{r} \frac{N}{\sqrt{2\pi \ln \sigma}} \exp\left[-\frac{(\ln r - \ln r_0)^2}{2(\ln \sigma)^2} \right] \left[\frac{\text{number}}{\text{cm}^3 \ \mu\text{m}} \right]$$

where $n(r)$ is the number of aerosol particles per unit volume and radius, r is the aerosol radius, σ is the geometric standard deviation which characterises the width of the distribution, r_0 is the median radius, N is the total particle density, which varies in space and time. This is a unimodal log-normal size distribution, which in this case was sulphuric acid droplets. In this work, an experimental value r_{eff} was retrieved from SAGE data or from Upper Atmosphere Research Satellite (UARS) data.
Then

$$r_0 = r_{\text{eff}} \exp[-2.5 \, (\ln \sigma)^2]$$

The values used for Pinatubo aerosol were r_{eff} around 0.4–0.8 μm, σ around 1.2, peak aerosol extinction coefficients of around 0.015 km^{-1}.
The latter value E (m^{-1}) is found from

$$E(\lambda) = \int Q_{\text{ext}}\left\{ \frac{2\pi r}{\lambda}, m(\lambda) \right\} \pi r^2 \, n(r) \, dr \times 10^{-6}$$

where Q_{ext} is a dimensionless Mie refraction efficiency factor which is a function of complex refractive index $m(\lambda)$, wavelength λ (μm) and particle radius r (μm).

The overall optical depth is found by integrating these expressions down the atmospheric column. Typical peak values from Pinatubo were 0.3–0.45 at a wavelength of 0.55 μm. The Pinatubo scenario is that initially SO$_2$ gas (having low scattering) was injected into the stratosphere, and aerosol formation started. Aerosol extinction then increased for nearly 6 months. The largest particles, with r_{eff} ~0.5 μm, settle faster, which affects the radiation exchange differently at different altitudes. The highest, smallest particles (up at 10–20 mb) with r_{eff}~ 0.2–0.3 μm are more effective in scattering visible wavelengths than near IR. This in turn affects the radiative forcing. Total maximum radiative forcing was found to be ~−6 W m^{-2} (cooling) over large areas. An important additional feature is stratospheric warming due to absorption of near-IR by small particles near the top of the cloud.

11.4   GASES AND AEROSOL COMBINED

Figure 11.18 summarises the present balance between the gas and particle components of anthropogenic radiative forcing. Components above the horizontal line have positive forcing, increasing the tropospheric temperature, and *vice versa*. Although CO_2 has the largest effect, the other gases and factors may have equal significance and act to either increase or decrease atmospheric temperature.

When the GHG forcing contours of Figure 11.14 are combined with the sulphate aerosol forcing contours of Figure 11.15, we have the net forcing contours shown in Figure 11.19. Net forcing is positive everywhere except for very small areas in the industrial hearts of North America, Europe and Asia.

11.5   FUTURE SCENARIOS

The Inter-Governmental Panel on Climate Change has outlined a range of scenarios for the future pattern of GHG emissions up to 2100. These scenarios covered population ranges of 6.4–17.6 billion, annual economic growth rates between 1.2 and 3.0%, combined oil and gas consumptions between 15 and 25×10^{21} J, and a variety of national responses to international protocols on CO_2 and CFC reductions. For example, some individual countries, and groups such as the EU, are debating the imposition of a carbon tax as a financial incentive to reduce energy use. The IPCC scenarios are designed to show the possibilities – there is no implication that one sce-

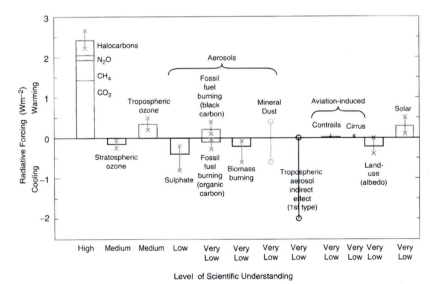

Figure 11.18 Contributions to radiative forcing from different factors.
Source: IPCC (2001) *Climate Change – Third Assessment Report*, Cambridge University Press, Cambridge, UK.

Net (sulphate aerosol + greenhouse gas) forcing

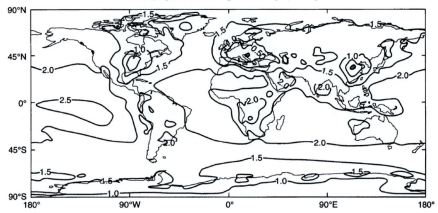

Figure 11.19 Global contours of the net radiative forcing (W m^{-2}) due to the combined effects of both GHGs and aerosol.
Source: IPCC (1994) *Radiative Forcing of Climate Change and An Evaluation of the IPCC IS92 Emissions Scenarios*, Cambridge University Press, Cambridge.

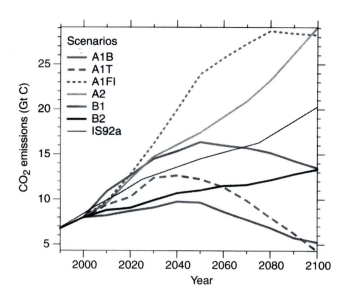

Figure 11.20 CO_2 emission scenarios.
Source: IPCC (2001) *Climate Change – Third Assessment Report*, Cambridge University Press, Cambridge, UK.

Table 11.10 IPCC GHG emission scenarios

Scenario	Predicted GHG Emission in 2100				
	CO_2 Gt C	CH_4 Tg	N_2O Tg N	CFCs kt	SO_x Tg S
1990	7.4	506	12.9	827	98
IS92c	4.6	546	13.7	3	77
IS92d	10.3	567	14.5	0	87
IS92b	19.1	917	16.9	0	164
IS92a	20.3	917	17	3	169
IS92f	26.6	1168	19.0	3	204
IS92e	35.8	1072	19.1	0	254

nario is more likely than another. Their range is shown in Figure 11.20. The range of resulting GHG emissions is huge and is shown in Table 11.10, with the 1990 values for comparison. IS92a is close to a 'business as usual' scenario in which the world neither goes mad for growth (overall, there is a slow reduction in the overall rate of growth) nor takes drastic measures to reduce energy use. Effects on temperature and other aspects of climate have been predicted by general circulation models (GCMs) – very complicated computer models of the physics of atmospheric and ocean movement and heat budgets. Although there are about five groups around the world running these models, they all tend to produce broadly similar predictions because the models make similar assumptions. The current state of the science is exemplified by the Hadley Centre climate model, which has a horizontal resolution of $2.5°$ latitude $\times 3.75°$ longitude, 20 layers in the ocean and 19 layers in the atmosphere. The models are first used to convert the GHG emission scenarios into concentration scenarios (Figure 11.21). The models predict that as atmospheric

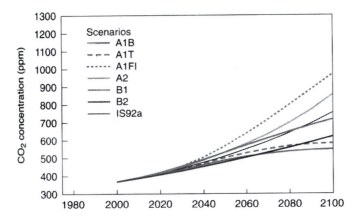

Figure 11.21 CO_2 concentration scenarios.
Source: IPCC (2001) *Climate Change – Third Assessment Report*, Cambridge University Press, Cambridge, UK.

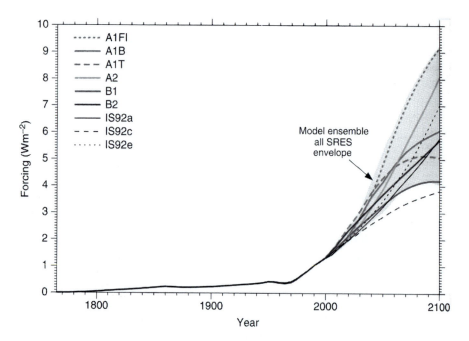

Figure 11.22 Anthropogenic forcing scenarios.
Source: IPCC (2001) *Climate Change – Third Assessment Report*, Cambridge University Press, Cambridge, UK.

CO_2 concentration increases, the take-up by sea and terrestrial surfaces will decrease, so that a higher proportion remains in the atmosphere as a greenhouse gas. The resulting atmospheric CO_2 concentration in 2100 will be between 500 and 970 ppm. The next step is to calculate the radiative forcing due to the GHGs, as seen in Figure 11.22. By 2100, the net forcing will have increased from its current value of 1.4 W m^{-2} to lie within a band from 4 to 9 W m^{-2}.

11.6   THE MAIN PREDICTIONS

11.6.1   Atmospheric temperature

The first useful outputs from the GCMs are predictions for global average temperature change. The IPCC scenarios give global average increases of between 1.4 and 6.0 °C by 2100 (Figure 11.23), with a most probable value of about 3 °C.

In addition, the temperature changes presented in Figure 11.23 are global averages, and there will be large regional variations, as shown in Figure 11.24 for the IPCC A2 scenario (which is somewhere in the middle of the distribution). The warming in nearly all land areas, especially in northern high-latitude

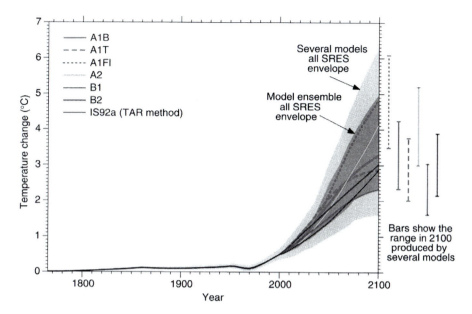

Figure 11.23 Temperature change scenarios.
Source: IPCC (2001) *Climate Change – Third Assessment Report*, Cambridge University Press, Cambridge, UK.

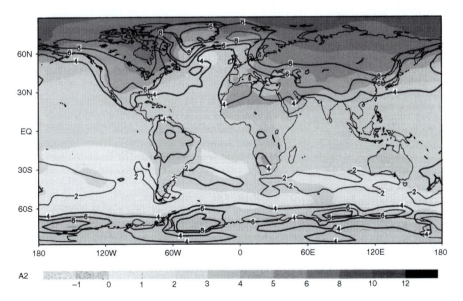

Figure 11.24 Variation of temperature change over the globe, expressed as the difference between the predicted temperature in 2071–2100 and the actual temperature in 1961–1990.
Source: IPCC (2001) *Climate Change – Third Assessment Report*, Cambridge University Press, Cambridge, UK.

winters, will be greater than the global average. Northern high latitudes will warm by 6–8 °C.

11.6.2   Model validation

Any model is only quantitatively useful if it has been validated against independent data. For the GCM predictions to be credible, it was important to validate the model on existing data such as the global temperature record of Figure 11.1. The model should not only predict the current warming, but also the general variations that have taken place during the period of measurements. Results from the Hadley Centre show that inclusion of the aerosol component discussed in Section 11.3 can explain the dip in global mean temperature between 1940 and 1970. As can be seen in Figure 11.25, the temperature anomaly based on GHGs alone showed too high a rate of temperature increase during the 1950s and 1960s. When the aerosol forcing term was included, the model correctly predicted the reduction in average temperature which was experienced during that period. It also predicted the subsequent increase which occurred when the sulphate aerosol was reduced by emission control while the GHG emissions continued unabated.

Changes from year to year are also apparent, and we are beginning to understand the factors that drive them. For example, look at the temperatures in 1990–1994 on Figure 11.1. After peaking in 1990, the global temperature fell sharply in 1991 and 1992, before resuming its upward trend again in 1992 and 1993. What looks at first like a random variation was probably due to the Mt Pinatubo emissions discussed in Section 11.3.1.

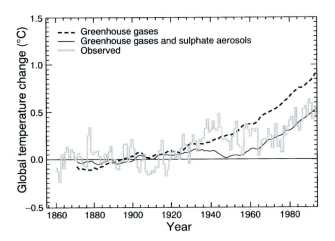

Figure 11.25 Modelled and measured surface temperature changes, 1860–1990. Source: Houghton, J. (1997) *Global Warming*, Cambridge University Press, Cambridge.

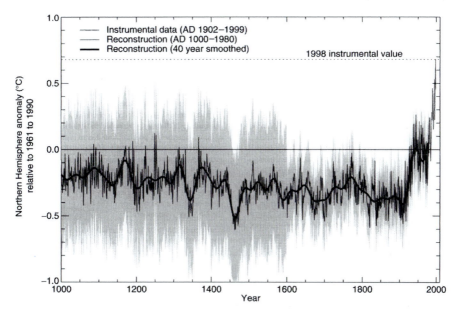

Figure 11.26 Variations in the Earth's surface temperature over the past 1000 years. Source: IPCC (2001) *Climate Change – Third Assessment Report*, Cambridge University Press, Cambridge, UK.

The TAR also presented an estimate of global temperature extended back 1000 years by the use of 'proxy' data. Of course, there were no weather stations and calibrated thermometers in use during most of that period. The temperature estimates have been made from observations of phenomena such as tree ring width, coral growth, ice cores and historical observations. The record shown in Figure 11.26 demonstrates that the gradual decline of global average temperature which had been taking place since AD 1000 was replaced in about 1900 by an increase at a much higher rate of just over 1 °C per century. This rate of increase is itself increasing, and is currently around 1.5 °C per century. Furthermore, the night time minimum has been increasing at about twice the rate of the daytime maximum, which lengthens the freeze-free season in mid- and high-latitude regions.

Looking back to Figure 11.25, it should be noted that we are discussing an *overall* increase of only around half a degree Celsius, and year-to-year variations of 0.01 °C. There is therefore some real uncertainty about the temperature database, and we have to be very careful to ensure that the data is free from systematic errors. For example, most air temperatures have been taken over the land, although most of the Earth's surface is water; most air temperatures have also been taken at sites that have been increasingly influenced by the geographical expansion of nearby urban heat sources. Owing to the small magnitude of the temperature changes involved, and the time-span of centuries, changes in methodology become paramount. For example, during most of the observation period, sea surface temperature has been measured by lowering a bucket on a rope from the ship's deck, and then recording the water temperature with a thermometer. Over the last 200 years, canvas buckets replaced wooden ones, giving an estimated 0.3 °C change in the recorded temperature because of their reduced thermal insulation. This change would correspond to about 70 years of global temperature change, making it vital to correct such deviations if the record is to be useful. More recently, the bucket method has been replaced by direct measurements of the temperature of the water entering the ship's cooling system. Nevertheless, over 60 million observations of sea surface temperatures taken from ships in these ways have contributed to the global temperature record. At the time of SAR, the IPCC belief was that the observed increase still lay within the range of natural climate variability. By the time that TAR was issued, IPCC had come to the conclusion that the changes that had occurred in global climate in the past 150 years were most likely to be due to human activities and that 'most of the observed warming over the last 50 years is likely to have been due to the increase in greenhouse gas concentration'. The very long-term ice core data provide an independent confirmation of the general relationship between GHGs and temperature, although it cannot be proven which is cause and which is effect.

11.6.3   Sea level

Two main factors control sea level. First, there is only so much water to share between sea and ice. Hence in cold periods of the Milankovitch cycle (ice ages), water will be locked up in the polar ice sheets and in glaciers and sea level will fall. Sea level 18 000 years ago was about 100 m lower than today, while 120 000 years ago it was 5 or 6 m higher. This factor predominates in long-term changes. Second, in the shorter term, the volume of a fixed mass of sea water varies with its temperature according to its coefficient of expansion, α. The value of α is not constant, since water has its maximum density at a temperature of 4 °C. At 5 °C, α is one part in 10 000 per degree Celsius, while at 25 °C it is three parts in 10 000. Hence, as the sea warms, its expansion accelerates.

In a warming world, sea level is predicted to rise due to glacier melt, ice sheet melt and thermal expansion of sea water. One of the initial reactions to predictions of increasing global temperatures over the next century was that the Antarctic and Greenland ice sheets might melt; they hold amounts of ice equivalent to global sea level increases of 65 and 7 m respectively. However, since saturated vapour pressure increases with temperature, the air will actually hold *more* water vapour at higher temperatures, more snow will fall, and more water will be locked up as snow and ice. Hence, temperature increases of a few degrees will *lower* sea level (from this cause), not raise it. This will remain true provided that the increasing temperature does not take the ground temperature above 0 °C, when all the snow would melt anyway.

Table 11.11 Predictions of sea-level rise

	Year			
	1990	2030	2060	2100
CO_2/ppm	354	470	600	850
Temp rise/°C		1.1	2.0	3.3
Sea-level rise/cm		18	38	65

Over the period of the global temperature record shown above, sea level is estimated to have risen at an average rate of 1–2 mm a^{-1}. Of more concern is the probable rise in the future. One set of responses predicted for average global temperature and sea-level rise is shown in Table 11.11 below. If the equivalent CO_2 concentration rises to 850 ppm by 2100, with a 3.3 °C temperature increase, sea level would rise by 0.65 m.

Increases of this magnitude would probably be contained in some low-lying areas such as the Netherlands, but threaten others such as Bangladesh, the Nile delta and the Maldives in the Indian Ocean, where the land is only 1–2 m above sea level. In many areas, the rise in sea level is already being compounded by subsidence of the land due to water extraction. The predicted range of sea-level rise given in TAR for 2100 is 0.11–0.77 m, with a central value of 0.48 m.

11.6.4 Rainfall

Global climate models predict that a warmer world will be a wetter world – a 4 °C increase in global mean temperature would increase precipitation by 8%, and there would also be some geographical redistribution. The IPCC estimated in TAR that precipitation had increased by 5–10% during the twentieth Century over most mid- to high-latitude continental areas in the Northern Hemisphere, and by 2–3% over tropical land areas. Overall, there is even less certainty about the rainfall changes than there is about temperature changes. For example, in the UK a predicted annual average temperature increase of 1.5–2.1 °C could be associated with rainfall changes of anywhere between ±15%. If precipitation is higher, then there would probably be an overall benefit from warmer temperatures and higher CO_2, whereas if it is lower the reverse would hold. The prediction of regional precipitation change is even more difficult than that of temperature – current GCMs cannot reliably model regional precipitation under present conditions, let alone those in the future. Because of these uncertainties, examination of the consequences of climate change can only be made via 'what if' scenarios, without any very clear idea as to which scenario will eventually turn out to be correct. A planners nightmare!

11.6.5   Effects in the UK

Temperature changes of around 1 °C may sound small, but they can have profound consequences for both natural and managed ecosystems. For comparison, a 'little ice age' occurred in the sixteenth century with temperatures about 1 °C cooler than now; when the Romans occupied Britain in the first centuries of the first millennium the temperature was about 1 °C warmer and grapes were grown all over England. We outline below some of the consequences that are currently thought to be most likely for the UK.

- Summers will be warmer, and summer precipitation about the same. Since evaporation will be higher, the soil will be drier. Winter temperatures and precipitation will both be higher.
- The summer/winter cycle of soil drying and wetting will become more pronounced, resulting in reduced soil organic matter content, changes in drainage pathways and an increase in swell/shrink factors (which affect building foundations on clay).
- The steady increase in atmospheric CO_2 concentration, which is not in dispute, combined with higher temperatures and an increase in the length of the growing season (for grasses and trees, about 15 days/°C temperature rise), will increase biomass production in both natural and managed ecosystems if water does not become a limiting factor. However, one of the responses to elevated CO_2 of many plant species is to partially close their stomata, thereby restricting the flux into the leaf. This results in increased plant dry matter without increasing the total water used, so that water use efficiency is increased. If woody species experience reduced stomatal conductance and evapotranspiration at elevated CO_2, then more water will be retained in the soil and woodland ecosystems will be affected. Many agricultural plant species yield about 10% more biomass when grown at the CO_2 concentrations anticipated in 2050. The response of economic yield may not be so clear cut – for example, higher temperatures reduce the grain-filling period for cereal crops, offsetting the benefit of higher CO_2.
- Some species (both animals and plants) will spread northwards and/or to higher altitude sites (the potential rate of northward movement is 200–300 km/°C, or around 100 km/decade). For plants, changes in viable latitude that become available due to temperature change may be restricted by photoperiod limitations.
- Many indigenous species will be lost from montane, coastal, wetland and peatland habitats. There will be increased pest and disease outbreaks – for example, milder winters with fewer frosts enable aphids to use the 'green bridge'.
- A number of low-lying areas such as East Anglia and the Thames estuary would be particularly vulnerable to sea-level rise. A possible increase in storm events would also increase wind damage and flooding.

- Both natural and man-made systems respond to extremes as well as to averages. A single frost or drought that is barely noticeable on the annual average temperature or rainfall total may cause damage from which ecosystems take years to recover. Small changes in mean temperature due to climate change will have disproportionate effects on the extremes. For example, an increase in mean winter temperature in the UK of 2.4 °C would increase the probability of a winter temperature exceeding 6.6 °C from 0.023 to 0.44, a factor of nearly 20. Of course, this calculation assumes that the variability will not itself change. Corresponding results for the distribution of summer temperatures show that summers like the one that occurred in 1976 (a drought year, when mean summer temperature was 2.4 °C above the average) may occur every third year by 2050. England has recently experienced the warmest November (1994), the driest summer (1995) and the wettest winter (2000/01) in around three centuries of records being kept – this may be coincidence, or it may be a harbinger of future change.

11.6.6   Effects in developing countries

Developing countries will be more adversely affected by climate change than industrialised economies. The IPCC projects a reduction in these countries' gross domestic products of 2–9% per year due to climate change, compared with 1–2% per year for more developed nations. Although food production has tripled in developing countries in the last 30 years, keeping up with population increase, crop production must increase by 40%, and that of meat by 60%, by 2020 to meet demand. The fertilisation effect of increasing CO_2 concentration will only increase crop production by 8–15%. This requirement is against a background of decreasing availability of suitable agricultural land, increased soil fertility depletion, increasing reliance on marginal lands. Less rainfall is predicted for much of Africa, which is already the one region in which increased food production has failed to match the population increase.

The hydrological cycle, which represents part of the coupling between atmosphere and oceans, drives the atmospheric energy budget and supports life. Global warming will affect this too. Worldwide, total water use is around 5000 km^3 per year, or about 10% of the available river and groundwater flow from the land to the sea. Use of water has increased by a factor of four in the last 50 years. Within regions, the ways in which societies have developed depends strongly on the availability of fresh water supplies. As the world warms, not only will the intensity of the hydrological cycle increase, but the distribution of rainfall will change so that some areas will have more than at present, but some less. Large communities will have to adjust to these changes (for example, by reducing use and increasing storage), over a relatively short period. Currently 7% of the World's population is at stress levels of fresh water availability (<2000 m^3 per person per year for all applications). By 2050, the figure is expected to be 70%.

The global N cycle is being altered, with industrial N fixation (140 Mt year^{-1}) now exceeding biological (100 Mt year^{-1}). Seventy per cent of anthropogenic N_2O emissions are from agriculture, and 8–16% of applied fertiliser N is emitted to atmosphere as N_2O. Demand for fertilisers in developing countries is projected to increase from 62 Mt in 1990 to 122 Mt in 2020. Again there are strong links between cultivation practices and pollution. If fertiliser N is applied at a time when there is a lot of runoff and plant growth is slow, then a high proportion will be lost to the water. If it is applied as urea or ammonia, then it will be lost to atmosphere. Plants are capable of recovering up to 60% of the applied N, but the proportion may be only 20% if the conditions are unfavourable. The introduction of high-yielding varieties as part of the 'green revolution' is estimated to have saved 170 million hectares of forest from being cleared in developing regions between 1970 and 1990. Assuming a typical carbon content of 100 t ha^{-1}, this represents a conservation of 17 Gt C, equivalent to 2–3 years of total carbon emissions. In fact, this figure would be offset somewhat by the carbon emitted making the fertiliser for the intensive agriculture.

The increased population will occupy more land, and hence the pressure will be on to gain the additional food production by intensification (more production per area). The main GHGs produced are then CH_4 and N_2O. However, there is also social pressure to increase production by extensification (more area). The main GHG is then CO_2. Extensification usually involves slash and burn forest clearances – these activities in Sumatra in 1997 were the cause of the long-lasting regional haze, poor air quality and loss of life.

There are great uncertainties about the interactions between CO_2 increase itself and the other related changes in climate factors such as soil water and temperature. Theoretical, growth-chamber and ecosystem studies have shown a greater proportional response to elevated CO_2 in dry years. If growth is water-limited, elevated CO_2 can result in greater water availability for longer in the growing season. Warmer temperatures will probably increase the response to CO_2 in warm-season grass land systems such as short grass steppe and tall-grass prairie. In other systems, however, responses may be more complex, possibly involving increased productivity at some times of the year and decreases at other times. The overall effect also depends on which component of the plant is economically important – cereal yields, for example, decrease at higher temperatures because of the reduced length of the fruiting season. There will also be changes in species composition. For rangelands, elevated CO_2 will increase woodland thickening, woody scrub invasion and grass suppression, tending to reduce nutritive value. If there are effects on the growth of pastures and forage plants, then there will inevitably be effects on the animals that consume them, although even less is understood about this at present. Overall, since the annual variability of agricultural output is high due to other factors, the additional perturbations due to climate change may not make much difference in this particular respect.

11.6.7   Biodiversity changes

Humans are having a huge impact on biodiversity as well as on the physical and chemical properties of the atmosphere. We are responsible for a large fraction of recent extinctions. The rate of species extinctions is running at 100–1000 times the ongoing background rate, and 20% of mammal species and 10% of bird species are currently threatened. The global change and terrestrial ecosystems (GCTE) project identified five main drivers that influence biodiversity – land use, climate, nitrogen deposition, biotic exchange and atmospheric CO_2 concentration. Of these five, three are primarily air pollution mediated. The balance between the five is very different depending on the situation – for example, GCTE predicted that climate change would predominate in the Arctic, because warming will be greatest in high latitudes whereas little land use change is expected. In tropical and southern temperate forests, on the other hand, land use change will dominate the other drivers. Overall, land use change was identified as the most important factor, with climate, N deposition and CO_2 concentration being about 65, 50 and 35% of that, respectively. However, understanding about interactions between the drivers is very limited, and the results are really more of an exploration of possibilities than a definite prediction.

11.7   FEEDBACKS

The processes that control the Earth's climate and generate the global average temperature of $+15°C$ that we take for granted are extremely complicated, and it is very hard to predict the consequences of change. The climate system contains many feedbacks, by which a change in one parameter changes a second parameter in such a way as to amplify the effect of the first change (positive feedback) or reduce it (negative feedback). Since the overall climate is determined by small differences between much larger terms, it is very sensitive to the sign and strength of these feedbacks. Consider a few examples:

Water vapour. An increase in global average temperature is predicted to increase the amount of evaporation from the oceans, and also the amount of water vapour in the atmosphere (because saturated vapour pressure increases with temperature). Water vapour is the most powerful GHG, so the additional water vapour will increase the longwave optical depth, trap more longwave radiation and increase the temperature further.

Clouds. About 30% of the solar radiation incident on the Earth is reflected back into space, more than half (17%) by clouds. This 17% amounts to about 228 W m^{-2}. Hence, if the increased water vapour in the atmosphere were to increase the cloud cover by 1% (i.e. increasing the albedo from 30% to 30.3%), that would decrease the radiant flux reaching the Earth's surface by 2 W m^{-2} and cancel out the radiative forcing due to GHGs. However, clouds also act as better absorbers

Table 11.12 Albedos of natural surfaces

Surface	Albedo/%
Ocean (calm)	2–4
Ocean (rough)	5–10
Forest (tall coniferous trees)	5–10
Forest (short broadleaved trees)	10–15
Grass (tall savannah-type)	10
Grass (short mown)	25
Soil	10–20
Desert sand	35–45
Ice (old and rough–new and smooth)	70–20
Fresh snow	80–95
Clouds – low	60–70
Clouds – middle	40–60
Clouds – high (cirrus)	18–24
Clouds – cumulus	65–75

of infrared radiation than the clear atmosphere, and hence can increase the positive forcing by around 100 W m^{-2}. The net effect of clouds on the radiation budget depends not only on the cloudiness, but on the detailed structure of the clouds and their distribution with height in the atmosphere. There is a great range of different cloud types, densities and altitudes. Some of them are made of water droplets, some of ice crystals of different shapes, and they all have different albedos (Table 11.12). In order to model such factors, we need a very sophisticated understanding of how aspects of cloud cover (amount, density, position in the atmosphere, droplet diameter distribution, ice crystal type) relate to other environmental factors such as water vapour density and air temperature. It turns out that low-level clouds tend to produce net cooling, whereas high-level clouds tend to produce net warming. Climate change models have only recently become sufficiently advanced that predictions of cloud cover are generated as output, rather than being specified as input.

Methane. Rising temperatures at high latitudes in the Northern Hemisphere are expected to melt regions that are currently permafrost. Very large quantities of methane are locked up in these permafrost regions as clathrate hydrates such as $CH_4 \cdot 6H_2O$, which are methane compounds trapped in the gaps between ice crystals. If the ice melts, the methane will be released. Since methane is a GHG, there would be a positive feedback, although the deposits are so deep down that release would take place over centuries rather than decades. In a more subtle example of a positive feedback, some of the methane diffuses up into the stratosphere and forms water molecules. The water molecules are not only radiatively active, but can form additional polar stratospheric cloud particles, enhance ozone depletion and raise tropospheric temperatures. Incidentally, there is also much interest in accessing the methane hydrates as the next supply of fossil fuels when the oil and gas wells have run dry.

Albedo. As well as the cloud albedo mentioned above, there are already strong seasonal and latitudinal variations caused by snow cover, which increases the mean albedo at 60 °N from 40% in July up to 65% in February. Although the albedo in north and south polar regions is higher still at 80%, the latter constitute <10% of the total surface area. The reflectivity of fresh ice and snow can be as high as 90%, compared to around 20% for bare soil and 25% for vegetation. If temperatures rise and significant areas of the Earth that are currently snow-covered become soil, the energy absorbed at the surface will increase, warming the atmosphere and melting more snow – another positive feedback. At a greater level of detail, any changes in vegetation will also have small effects on radiation balance, due to the range of albedos shown in Table 11.12.

Cloud nuclei. Cloud condensation nuclei (CCN) are a significant factor in the Earth's albedo, and an increase of 30% in CCN would reduce radiative forcing by about 1 W m^{-2}. Some CCN are produced by the conversion of gaseous dimethyl sulphide (DMS) to sulphate particles. Dimethyl sulphide in turn is produced by oceanic phytoplankton that will breed faster in warmer water. Hence, higher temperatures could produce more DMS and more cloud cover and less radiation penetrating to the ground and lower temperatures – a negative feedback. Other CCN are produced from sulphate particles formed in the atmosphere from gaseous SO_2 produced by the burning of fossil fuel. The more CO_2 that is generated to give a warming, the more sulphate aerosol is generated to give a cooling.

Ocean circulation. The atmosphere is coupled to the oceans by processes such as evaporation and heat transfer, and anything that affects the atmosphere must affect the oceans in due course. The changes tend to be slower, because the mass and heat capacity of the oceans is so huge compared to that of the atmosphere. For example, the top 3 m of the oceans hold as much heat as the whole atmosphere. As with cloud cover, small percentage changes in ocean heat transfer may have large effects on heat balance. Changes in rainfall and evaporation will affect salinity, which will affect the global scale currents that continually redistribute heat energy in the oceans.

There may be interactions that are much more powerful, and much harder to predict, than any of those mentioned above. The Earth has many complex systems that we take for granted on a day-to-day basis; climate change may disturb these systems in profound ways. For example, the global thermohaline conveyor belt moves huge quantities of energy around the planet in the form of slow-moving currents of water having different salinities and temperatures. It is estimated that 10^{12} kW is transported from the southern to the northern hemisphere by the Atlantic Ocean. Part of that flow, the Gulf Stream, brings an amount of energy equivalent to about 30 000 times the entire UK electricity generating capacity. This keeps the UK several degrees Celsius warmer than it would be otherwise. There is a possibility that the flow is unstable, and that perturbations in the climate system could switch the flow quite quickly (on a geological timescale) into another pattern.

11.8   GLOBAL RESPONSES

Initial responses to the threat of climate change were made on a rather piecemeal basis – for example, the EU decided in 1990 on the objective to stabilise EU CO_2 emissions at 1990 levels by the year 2000. These efforts were coordinated on a global basis by the United Nations Framework Convention on Climate Change (UNFCCC), which was agreed in Rio de Janeiro in 1992. At the Third Conference of Parties (COP3), held in Kyoto in 1997, different binding GHG emission targets were set for different countries. The Kyoto Protocol aims to reduce aggregated emissions of six GHGs (CO_2, CH_4, N_2O, HFCs, PFCs, SF_6) by at least 5% below 1990 levels before a 'commitment' period 2008–2012. Within this Protocol, the EU accepted a target reduction of 8%, distributed among the member states according to a burden-sharing agreement which allowed the less developed nations to increase their emissions (by up to a maximum of 27% for Portugal) while the more developed ones cut theirs (by a maximum of 21% for Germany).

In fact, substantially greater reductions, of around 60–70%, will be needed if the CO_2 concentration is to be stabilised even at 550 ppm. During the 1990s, world CO_2 emissions from fossil fuel use increased by 8%, and those from industrialised nations by 10%. In order to meet Kyoto commitments, OECD countries would have to reduce their CO_2 emissions to 9644 Mt a^{-1} by 2010, compared to 1997 emissions of 12 236 Mt a^{-1} and predicted 2010 emissions of 13 427 Mt a^{-1}. If energy consumption was not reduced, this would involve generating 35–40% of their energy requirements from non-fossil fuel sources. Nuclear power is not really seen as an option, and is predicted to decrease from the current 6.6% to 4.4% by 2020. Global total emissions of CO_2 due to fuel combustion increased steadily throughout the 1980s and 1990s. However, the countries involved in the Protocol have not yet agreed to it. Indeed, significant events occurred during the writing of this book. In December 2000, COP6 at the Hague ended acrimoniously without agreement on how the Kyoto target would be achieved. In March 2001, the new American President Bush withdrew from his predecessor's commitment to control CO_2 emissions. With the US contributing around one quarter of total global emissions, this was a serious setback to progress. In July 2001, the EU committed to achieve their own Kyoto objectives, independently of the US.

11.8.1   European responses

There are wide differences in the approach of different countries to reducing emissions, and in their overall success. For example, although the EU as a whole reduced GHG emissions by 2.5% from 4.15 Gg to 4.05 Gg between 1990 and 1998, emissions from Spain increased by 19% while those from Germany decreased by 16%. Overall, the EU is not far off the linear trend necessary to meet the GHG target by 2008–2012. However, CO_2 emissions, which make up around 80% of the GHG total, are almost stable. The constancy of the CO_2 emissions is not as bad as it might seem, since overall output as measured by real GDP increased by 17%, and energy consumption by 8%, over the period 1990–2000. Hence, energy production became less carbon-intensive, due to switching from coal to gas, renewables and nuclear power. Also, the EU needs to achieve its 8% cut overall, so it does not matter if some countries do not manage it provided that others compensate. During the same period, the population grew by 3%, so that the CO_2 emission per person fell from 9.1 to 8.9 tonnes. In terms of production sector, 32% of total CO_2 production was from energy industries, 24% from transport and 20% from small combustion plant. In terms of fuel use, 42% was from oil, 21% from gas, 16% from solid fuel, 15% from nuclear and 6% from renewables. The EU has a target to produce 10% of its energy from renewables by 2010, so the last figure shows that there is some way to go.

11.8.2   UK responses

For the six gases controlled under the Kyoto Protocol, the baseline 1990 UK emissions are shown in Table 11.13. Under the EU burden-sharing agreement, the UK agreed to reduce GHG emissions by 12.5% below 1990 levels by 2008–2012. The UK Government independently increased this target to a 20% reduction in order to emphasise how committed it was to solving the problem. By 2000, the UK was ahead of the linear reduction trend required to meet the EU target by about 6.5%. All GHGs were reduced over this period, with a sharp methane reduction being partly due to the decline of the UK coal mining industry and a small reduction in N_2O being due to the increase in transport N_2O emissions.

The change in energy consumption pattern in the UK has been broadly similar to that in the EU15. Solid fuel consumption fell by 40% between 1990 and 1998, while gas consumption rose by 60%. Twenty-six per cent of electricity was from nuclear stations in 1998, and this reduced the UK carbon emissions by around 10%. Only around 2.5% of UK electricity is from renewables. The Government

Table 11.13 Baseline UK GHG emissions

Gas	Emission/Mt C_E a^{-1*}
CO_2	168.0
CH_4	20.8
N_2O	18.0
HFC	3.1
PFC	0.62
SF_6	0.20

Note
* MtC_E = million tonnes of carbon equivalent. The emissions of each gas are weighted according to their global warming potential.

is expecting to increase this to 10% by 2010, reducing carbon emissions by 5 Mt pa. The Non Fossil Fuel Obligation (NFFO) requires electricity suppliers to generate specific amounts of electricity from non-fossil fuel sources. In the UK these are mainly biofuels, waste incineration, landfill gas, wind and hydro. A Fossil Fuel Levy (FFL), currently 0.3%, is charged to consumers to reimburse the suppliers for the higher cost of these renewables. The levy funds the development of new renewable schemes.

The Royal Commission on Environmental Pollution (RCEP) recommended that global CO_2 concentration should not be allowed to rise higher than 550 ppm, compared to its current value of 370 ppm. In order to 'pull its weight' in achieving this, the UK should cut CO_2 emissions by 60% by 2050, and by 80% by 2100. The RCEP supported the principle of contraction and convergence, in which per capita energy consumption would end up the same for all people. Developed countries would have to cut their emissions, developing nations would be able to increase their emissions in order to raise living standards. Since GHG emissions are mixed into the global atmosphere from wherever they are released, there might be a role for international trading in emission quotas, by which nations that found it costly or difficult to reduce emissions could buy quota from nations that found it easy or cheap to reduce emissions below their quota. A general carbon tax was recommended based on carbon emitted per unit energy supplied.

Emission reductions on such a scale would have great implications for our way of life. If overall energy use remains the same, we would need to replace a number of fossil fuel power stations with nuclear or add CO_2 storage to existing fossil fuel plants, and build 200 offshore windfarms, 7500 small wave power generators, thousands of CHP generators fuelled by energy crops, and a tidal barrage across the Severn. Nuclear is a poor option at the moment because current stations will cease operating by 2020, and new ones are unlikely until a safe long-term storage or disposal method for radioactive waste has been devised. Some of the carbon can be locked up in vegetation, though this is not sufficient, because fossil fuel combustion is a highly concentrated energy release, and it simply takes too much vegetated land area to absorb the CO_2 again. Similarly, renewable energy sources are very diffuse compared to fossil fuels. Typical wind turbines, for example, generate 0.5–1.0 MW of electricity, compared to 2000 MW by a major power station. Hence, it would take 2000–4000 such turbines to replace the electricity output from a single fossil-fuel power station. There are schemes for concentrating the CO_2 in the waste gases and pumping it either into permeable rock strata or into the deep ocean, but neither of these options has been tested on a large scale and the environmental and ecological consequences are uncertain. Such methods should be seen as short-term technical fixes which simply transfer the problem without solving it. The better option is to reduce energy usage by many different measures, which would have the added spin-off of reducing pollutant emissions. UK energy consumption has been rising at an average of 0.5% per year since 1965;

this sounds small, but if it continues at this rate it will increase by a further 30% by 2050. The fastest rate of expansion has been in the transport sector, while domestic use has remained about constant and industrial use has declined. The latter decline is partly due to substitution of gas for coal, a process which has largely been completed. Energy costs have declined in real terms over the same period, so there has been no incentive to reduce consumption. Domestic energy efficiency, especially in the housing stock built in the nineteenth and early twentieth centuries, is very poor. Combined heat and power schemes, in which the waste heat from electricity generation is used to provide hot water for industrial and domestic space heating, increases thermal efficiency from below 40% to around 80%. The RCEP anticipated that such CHP schemes would be linked up into heat networks and that the energy supply would be largely energy crops and electrically powered heat pumps.

FURTHER READING

Burroughs, W. J. (2001) *Climate Change – A Multidisciplinary Approach*, Cambridge University Press, Cambridge.

European Environment Agency (2000) *European Community and Member States Greenhouse Gas Emission Trends 1990–1998*, EEA, Copenhagen.

Houghton, J. (1997) *Global Warming – The Complete Briefing*, Cambridge University Press, Cambridge, UK.

Hewitt, C. N. and Sturges, W. T. (1995) *Global Atmospheric Chemical Change*, Chapman and Hall, London, UK.

IPCC (1990) *Climate Change – The IPCC Scientific Assessment*, J. T. Houghton, G. J. Jenkins and J. J. Ephraums (eds), Cambridge University Press, Cambridge, UK.

IPCC (1992) *Climate Change 1992 – The Supplementary Report*, J. T. Houghton, B. A. Callendar and S. K. Varney (eds), Cambridge University Press, Cambridge, UK.

IPCC (1994) *Climate Change 1994 – Radiative Forcing of Climate Change*, J. T. Houghton, L. G. Meira Filho, J. Bruce, H. Lee, B. A. Callender, E. Haites, N. Harris and K. Maskell (eds), Cambridge University Press, Cambridge, UK.

IPCC (1997) *Stabilization of Atmospheric Greenhouse Gases: Physical, Biological and Socio-economic Implications*, IPCC, Geneva.

IPCC (2001) *Climate Change – Third Assessment Report*, Cambridge University Press, Cambridge.

Royal Commission on Environmental Pollution (2000) *Energy – The Changing Climate*, The Stationery Office, London, UK.

United Kingdom Climate Change Impacts Review Group (1991) *The Potential Effects of Climate Change in the United Kingdom*, DoE/HMSO, London, UK.

Chapter 12

Ozone depletion and ultraviolet radiation

Only about 10% of atmospheric ozone occurs in the troposphere, either naturally or as a pollutant. The remaining 90% occurs in the stratosphere, where ozone concentrations of up to several ppm serve a vital biological role in absorbing high-energy photons of ultraviolet radiation from the sun. In this book so far, we have been largely concerned with the troposphere because that is where the emissions, transformations, deposition and effects mainly occur. The stratosphere, though, is not some remote, disconnected and inert structure where nothing happens that has any influence. The stratosphere is an important reaction chamber for a number of processes that depend on either very cold conditions, very high solar UV fluxes, or both. In 1985, it was discovered that the ozone concentration above the continent of Antarctica was decreasing; this was found to be due to chemical reactions with chlorine derived from stable chemicals known as chlorofluorocarbons. The reduction in ozone concentration has now been observed globally, and has potentially severe implications for life on the surface.

12.1   OZONE IN THE STRATOSPHERE

In Chapter 1, Figure 1.1 showed the average vertical temperature structure of the atmosphere. In the troposphere, which extends from ground level up to around 10 km, the temperature decreases with height, reaching just above 200 K at the tropopause. From there up to the stratopause at 50 km, the temperature increases again, although at a gentler rate, to around 270 K. This increase in temperature is partly due to the absorption, within a layer of the atmosphere containing a higher proportion of ozone (O_3) molecules, of solar UV radiation with wavelengths between 200 and 300 nm. The layer of ozone is formed at around 25 km by the action of solar UV on atmospheric oxygen molecules. UV photons having a wavelength of less than 242 nm have sufficient energy to split the oxygen molecules into two oxygen atoms.

$$O_2 + h\nu \rightarrow 2O \qquad (\lambda < 242 \text{ nm})$$

These then combine with other oxygen molecules, in the presence of a third impartial molecule M such as nitrogen, to give

$$O + O_2 + M \rightarrow O_3 + M$$

Local ozone concentrations are usually expressed as partial pressures (Pa), volume mixing ratios (ppb or ppm) or more rarely as number densities (molecules cm^{-3}). The global average atmospheric concentration is around 0.5 ppm at 15 km, rising to 8 ppm at 35 km and falling back to 3 ppm at 45 km. An alternative measure of ozone amount is the column depth, which integrates the ozone density over the height of the atmosphere. This value, which is the physical depth that all the ozone would occupy if it were a pure gas at standard temperature and pressure (STP, which is 0 °C and 1 atmosphere) is measured in Dobson Units (DU) after the scientist who devised the method. One Dobson Unit is 0.01 mm of ozone (sometimes expressed as milli-atmosphere-centimetres), and typical column depths are in the range 300–400 DU. It is remarkable that a layer of the pure gas that is effectively only a few mm deep turns out to have played such an important role in the development of life on the Earth. The most common ground-based method of measuring total column ozone is the Dobson spectrophotometer, which determines the differences in atmospheric attenuation of solar UV radiation across spectral lines centred on 320 nm. Satellite measurements are also being made of both total column ozone and vertical variation, using ultraviolet radiation backscattered from the atmosphere.

Ozone production is at a maximum in the tropical stratosphere. The Earth's meridional circulation pattern then transports the ozone towards the Poles, with a preference for the winter pole. Hence, the latitudinal ozone distribution (Figure 12.1) shows that total column amounts are higher towards

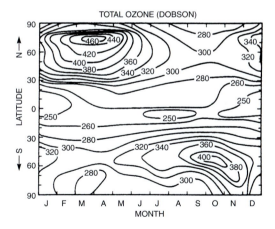

Figure 12.1 Contours to show the average variation of total column ozone with latitude and season over the globe.
Source: Peter, T. (1994) 'The stratospheric ozone layer – an overview', *Environmental Pollution* 83: 69–79.

the poles than in equatorial regions, and that the peak values are reached in March/April in the northern hemisphere and September/October in the southern hemisphere.

There are several routes by which ozone is naturally depleted. First, by photo dissociation and recombination with atomic oxygen

$$O_3 + h\nu \rightarrow O + O_2 \quad (\lambda < 1140 \text{ nm})$$
$$O + O_3 \rightarrow 2O_2$$

If the O_3 absorbs a higher-energy photon with a wavelength of less than 310 μm, then it can dissociate into O_2 and an excited oxygen atom such as $O(^1D)$. The latter energetic atom will then participate in reactions with other atmospheric species. The formation of $O(^1D)$ is doubly dependent on wavelength, because the quantum yield increases sharply in just the same spectral range as the solar radiation flux is decreasing sharply (Figure 12.2). Second, by more complex chains of catalytic reactions involving NO, HO and Cl. These are pairs of reactions of the form

$$H + O_3 \rightarrow OH + O_2$$
$$OH + O \rightarrow H + O_2$$

net: $O + O_3 \rightarrow O_2 + O_2$

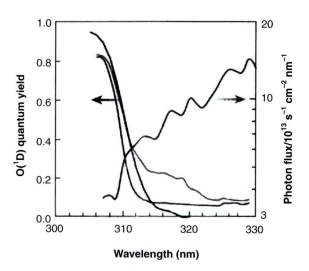

Figure 12.2 Quantum yield for O(^{1}D) formation.
Source: Ravishankara, A. R., Hancock, G., Kawasaki, M. and Matsumi, Y. (1998) 'Photochemistry of ozone: Surprises and recent lessons', *Science* 280: 60–61.

and

$$NO + O_3 \rightarrow NO_2 + O_2$$
$$NO_2 + O \rightarrow NO + O_2$$

net: $O + O_3 \rightarrow O_2 + O_2$

When many such species are participating in ozone destruction, there will be interactions between species that may increase or decrease the overall rate.

In the 'clean' environment, the ozone concentration takes on a stable value at any particular altitude, depending on the local balance between production and destruction. At very high altitudes, there is plenty of UV radiation but little oxygen for it to act on, whereas at low altitudes there is plenty of oxygen but little UV radiation. The interaction of these two competing processes results in an optimum height for O_3 formation at about 30 km. Hence, the characteristic vertical profile shown in Figure 12.3 is established. The left-hand graph is the number of O_3 molecules per cm^3, which will naturally tend to decrease anyway with height because the total pressure and density decrease. In the right-hand graph of this figure, the concentration has been expressed as a volume mixing ratio, which improves the definition of the peak concentration at around 30 km. Until the mid-1970s, the ozone concentration over Antarctica had a similar vertical profile (the right-hand curve on Figure 12.4), with the partial pressure typically increasing from a few mPa at 8 km to 15 mPa at 18 km, and decreasing again to a few mPa at 30 km.

Figure 12.4 also shows the startling change in this profile that now occurs in the Austral Spring (in October, when the sun first shines after the Austral winter). The ozone concentration declines throughout the whole profile, decreasing close to zero where it used to be at its maximum. Although this abrupt seasonal change was only noticed in the mid-1980s, it soon became clear that the trend had existed for

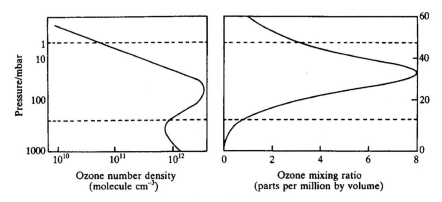

Figure 12.3 Vertical ozone profiles.
Source: Stratospheric Ozone Review Group (1988) *Stratospheric Ozone 1988*, HMSO, London.

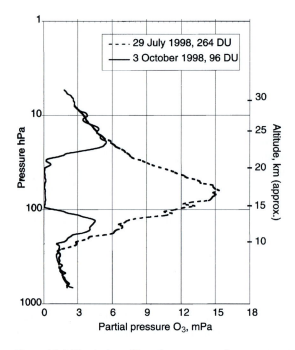

Figure 12.4 Vertical profiles of ozone partial pressure over the South Pole. The nor-
mal profile (dashed line) is replaced during the Austral Spring by the
severely depleted profile shown by the full line.
Source: World Meteorological Organisation (1999) 'WMO Global
Ozone Research and Monitoring Report No 44', *Scientific Assessment of
Ozone Depletion 1998*, WMO, Geneva.

some time. Retrospective analysis of data from all sources showed that gradual
decreases had started in the 1960s and accelerated dramatically in the last decade
(Figure 12.5).

To some extent, these decreases had been concealed by other effects such
as instrument calibration changes, the annual cycle of the seasons, the quasi-
biennial oscillation of the stratospheric winds, and the 11-year cycle of sunspots. However, in 1993 Antarctic column ozone fell from its winter value of
350 DU to just 90 DU, the lowest-ever recorded value.

The ozone 'hole' appeared in the Antarctic, rather than elsewhere, because
of distinctive atmospheric circulations in that region; these are shown in
Figure 12.6. There is a specific meteorology during the winter and spring that sets
up a cold isolated airmass (the polar vortex) within which O_3 cannot be replen-
ished by flows from lower-latitude regions. The vortex develops a core of very
cold air, within which polar stratospheric cloud particles form, on the surfaces of
which many of the important ozone-destroying reactions occur. However, the

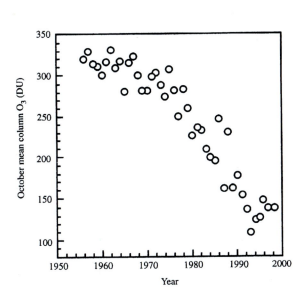

Figure 12.5 The time course of October column ozone over Halley, Antarctica between 1957 and 1998.
Source: Updated from Stratospheric Ozone Review Group (1996) *Stratospheric Ozone 1996*, Department of the Environment, London.

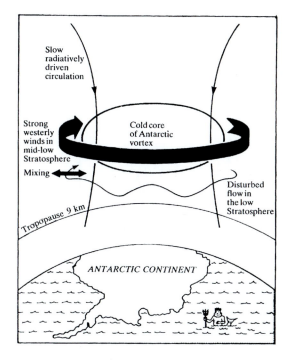

Figure 12.6 The Antarctic winter vortex.
Source: Wayne, R. P. (2000) *Chemistry of Atmospheres*, Oxford University Press, Oxford.

problem is not confined to the Antarctic, and less severe reductions are now being measured over the Arctic and elsewhere.

Major European and American experimental programmes, called THESEO and SOLVE respectively, are now shedding light on ozone destruction in the Northern hemisphere. Between them, these programmes in 1999/2000 involved contributions from over 500 scientists, 30 instrumented research balloons, 600 ozone sondes, 6 aircraft and 30 ground stations, as well as various satellites. It has been shown that a polar vortex system similar to, but weaker than, that in the Antarctic exists in the Northern hemisphere. In the winter of 1999–2000, over 60% of the O_3 at an altitude of 18 km was destroyed. After this springtime ozone destruction, ozone-depleted air moves out into lower latitude regions and reduces the column ozone. Figure 12.7 gives the total column ozone at Arosa, Switzerland, between 1930 and 1997; there is a clear downward trend after 1960.

In March 2000, mean column ozone amounts over Europe were down 15% on their 1976 values. In the United States, column ozone averaged over a network of six measuring sites decreased to record low values during the summer of 1993, with a minimum of around 280 DU compared with an historical April peak of 365 DU. Average measurements from networks of Dobson spectrophotometers in Europe and North America show reductions of around 20 DU between 1960 and 1988 (Figure 12.8). It is not known to what extent the latter reductions are due to the movement of ozone-depleted air from the Poles, rather than to ozone destruction *in situ*. Overall, column ozone between latitudes 66N and 66S has been declining by 3% per decade (4% over Europe, less over the

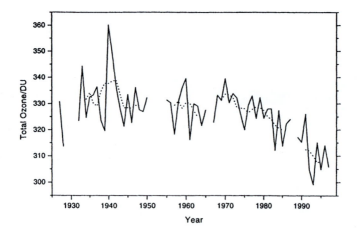

Figure 12.7 The time course of average column ozone over Arosa, Switzerland between 1930 and 1997.
Source: Proceedings, Schonbein International Ozone Symposium, Basel Switzerland (1999).

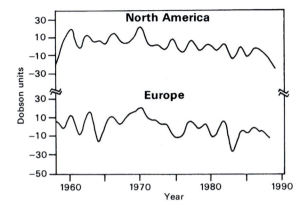

Figure 12.8 The time courses of change in average column ozone above North America and Europe between 1958 and 1988.
Source: Stratospheric Ozone Review Group (1990) *Stratospheric Ozone 1990*, HMSO, London.

tropics). In 1992 and 1993, the column ozone over Halley Bay, Antarctica reached record lows of 111 and 90 DU respectively, and the 'hole' (defined as the area enclosed by the 200 DU contour) had extended to a record area of 23×10^6 km^2 in 1994.

12.2   DESTRUCTIVE CHEMISTRY

So, why is this recent ozone reduction occurring? In the 1960s and 1970s there was speculation by atmospheric scientists that supersonic transports (SSTs) would affect stratospheric chemistry. It was expected that eventually several hundred SSTs would be flying at altitudes of 17–20 km – well into the stratosphere – and that their emissions, especially of the catalytically active NO, might have serious consequences. Original predictions of the impact were very large – up to 10% reduction in global average ozone column. Although these were subsequently scaled down by improved understanding of stratospheric chemistry, they became irrelevant in any case when the SST fleet did not materialise. However, they did establish an interest in the processes in the stratosphere which could lead to ozone destruction by other routes. Combination with a free oxygen atom is not the only route for stratospheric ozone destruction. Ozone can also be lost by catalytic reactions with a variety of other substances, the most important groups being free radical species in the nitrogen, hydrogen, bromine and chlorine families such as OH and NO.

In general

$$O_3 + X \rightarrow O_2 + OX$$
$$OX + O \rightarrow O_2 + X$$
_____

net: $O + O_3 \rightarrow 2O_2$

where X is the catalytic compound, which destroys two odd oxygens – both an O (which could later have formed an O_3) and an O_3.

specifically: $Cl + O_3 \rightarrow ClO + O_2$

$$Cl + O_3 \rightarrow ClO + O_2$$
$$ClO + ClO + M \rightarrow Cl_2O_2 + M$$
$$Cl_2O_2 + h\nu \rightarrow 2Cl + O_2$$
_____

net: $2O_3 + h\nu \rightarrow 3O_2$

Although these ozone consumers are present at extremely low concentrations, they are conserved through the reaction. Hence, one chlorine molecule can be involved in the catalytic destruction of tens of thousands of ozone molecules. When ozone depletion first became a matter of international concern, the reactions given above were hypothetical. Figure 12.9 shows the concentrations of both O_3 and ClO measured from an aeroplane as it flew towards Antarctica in the spring of 1987, as the ozone hole was developing. As O_3 concentration declined, there was a symmetrical increase in ClO concentration, powerful though circumstantial evidence for the mechanism given above. This destruction cycle is thought to be responsible for 70% of the ozone loss in Antarctica.

In addition, more complicated and much faster heterogeneous reactions take place on the surfaces of polar stratospheric cloud particles (PSCs) – frozen mixtures of nitric acid and water that form at temperatures below 195 K ($-78\ °C$) and make up the nacreous or mother-of-pearl clouds frequently observed above Antarctica. In particular, chlorine nitrate and hydrogen chloride attach to the particle surfaces, whence

$$ClONO_2 + HCl \rightarrow Cl_2 + HNO_3$$

The chlorine released by this and other related reactions is volatilised to take part in ozone destruction. The nitric acid remains bound to the particle surface, and hence depletes NO_2 (denoxification) which is then not available to neutralise ClO. Hence, the level of active chlorine atoms is maintained.

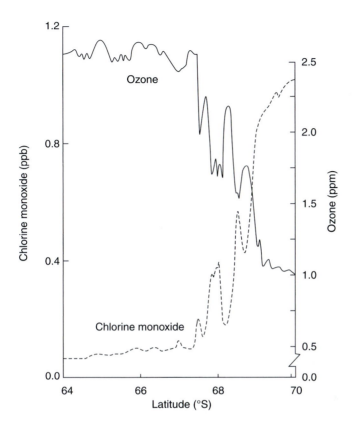

Figure 12.9 Variation of ozone and chlorine monoxide concentrations with latitude, between 64 and 70°S.

The most effective type of PSC particle is formed at the very low Antarctic temperatures, but not at Arctic temperatures which are some 10–15 °C warmer. It appears that stratospheric sulphuric acid aerosol particles may offer surface conditions which can accelerate the destruction of ozone in much the same way as the polar clouds. This has been particularly apparent after major injections of sulphate particles and SO_2 (which eventually forms particles) into the stratosphere by volcanoes such as Mt Pinatubo in 1991. A second ozone destruction cycle is believed to be important under these slightly warmer conditions:

$$ClO + BrO \rightarrow Br + Cl + O_2$$
$$Cl + O_3 \rightarrow ClO + O_2$$
$$Br + O_3 \rightarrow BrO + O_2$$

$$\text{net:} \quad 2O_3 \rightarrow 3O_2$$

Modelling suggests that one result of the ozone loss will be a cooling of the lower stratosphere, and this has been supported by temperature soundings that show a decline in the 15 km temperature in November. There is a possible positive feedback mechanism involved – ozone loss cools the air, more PSC particles are formed, more ozone is destroyed. It is also becoming apparent that PSC formation is not restricted to the polar vortices, but can occur at temperate latitudes in the core of the stratospheric jet stream. It has become clear from the Antarctic aerosol investigations that heterogeneous processes involving particles of ice, soot, sulphuric acid and sulphates are an important component of stratospheric chemistry everywhere.

12.2.1   The role of chlorofluorocarbons

The ClO chemistry described above depends on the arrival in the stratosphere of a supply of Cl atoms. There has always been a natural supply, in the form of methyl chloride (CH_3Cl) from marine biological emissions, forest fires and volcanoes. The main post-war disturbance to the stability of the ozone layer came about through the release at ground level of chlorofluorocarbons (CFCs), artificial chemicals which have been used specifically as heat transfer fluids in refrigerators, as propellants in aerosol sprays, as blowing agents in expanded polystyrene packaging and insulation, and as cleansing fluids for electronic circuits. The molecules are very stable and unreactive (that was why they were developed in the first place), and so have very long lifetimes in the atmosphere. A proportion of the released molecules has been diffusing upwards into the stratosphere, a process with a time constant of tens of years because of the stable lapse rate. When sufficiently high in the stratosphere, they are broken down by the UV and release chlorine atoms to take part in the above reaction sequence. Other molecules, carrying halons such as bromine instead of chlorine, are also effective ozone scavengers.

The most prevalent chlorofluorocarbons are listed in Table 12.1. CFC nomenclature specifies the first digit as the number of carbon atoms minus one (omitted if zero), the second digit as the number of hydrogen atoms plus one, and the final digit as the number of fluorine atoms. Thus, the CFC-114 molecule has two carbon atoms, no hydrogen atoms and four fluorine atoms (as well as two chlorine atoms).

Although there are substantial natural sources of chlorine, the combined effect of these synthetic molecules has been to increase the global average concentration of chlorine atoms from the natural background of 0.6 ppb to a current value of 3.8 ppb. The Antarctic ozone hole is believed to have become noticeable when global average chlorine concentration rose above 2 ppb. In 1990, the different

Table 12.1 CFCs involved in ozone destruction

Name	Chemical formula
CFC-11	Trichlorofluoromethane $CFCl_3$
CFC-12	Dichlorodifluoromethane, CF_2Cl_2
CFC-113	Trichlorotrifluoromethane $C_2F_3Cl_3$
CFC-114	Dichlorotetrafluoroethane $C_2Cl_2F_4$
CFC-115	Chloropentafluoroethane C_2ClF_5
Halon-1211	Bromochlorodifluoromethane CF_2ClBr
Halon-1301	Bromotrifluoromethane CF_3Br

halocarbons made up the total in the proportions shown in Table 12.2. It is clear from the column on the right that, if left uncontrolled, the equilibrium total chlorine concentration resulting from 1990 release rates would have risen to around 13 ppb, more than three times the prevailing concentration, with severe consequences for stratospheric ozone.

The discovery of ozone depletion in the Antarctic in 1985, and the recognition of the potentially severe consequences for the UV climate, resulted in a fast piece of international pollution control legislation. The Vienna Convention for the Protection of the Ozone Layer was established in 1985, and the Montreal protocol was signed in 1987 and strengthened in London (1990), Copenhagen (1992), Vienna (1995) and Montreal (1997). A timetable was set out for production of the major CFCs, and other ozone-depleting substances, to be severely reduced, with a different timetable for developed and developing countries. Although the Montreal Protocol originally identified five CFCs and three halons for reduction, by 1999 no fewer than 95 individual substances were controlled. The agreed control measures resulted in a rapid reduction of halocarbon production, as summarised in Figure 12.10.

Table 12.2 Concentrations and release rates of halocarbons

Halocarbon	Concentration ppt Cl	Release rate ppt Cl a^{-1}	Steady state concentration ppt
CFC-11	858	56	3346
CFC-12	962	53	6349
CFC-113	170	19	1683
CCl_4	428	15	728
HCFC-22	103	12	175
CH_3CCl_3	479	96	623
Others	16		
CH_3Cl	600		
Total	3624	250	12904

On the other hand, Figure 12.11 demonstrates very clearly what would have happened to the equivalent effective stratospheric chlorine (EESC) concentration if the Montreal Protocol had not been agreed, and why the follow-up agreements were needed as well. WMO estimated that if the Montreal Protocol had not been put into effect, the concentration of ozone-depleting substances in 2050 would have reached 17 ppb (five times the current value), there would have been 50–70% ozone depletion at mid latitudes, and surface UVB radiation would have been doubled in northern mid-latitudes and quadrupled in southern mid-latitudes. With the emission reductions specified in the Protocol, the chlorine loading of the stratosphere would peak in the late 1990s, and by 2050 will have fallen back to the 2 ppb concentration at which ozone destruction was first observed in the 1970s. Furthermore, some of the global stock of CFCs that is presently contained in refrigerators and air conditioners will continue to be released, despite improved efforts to recover and destroy it. In order to reduce the global average chlorine concentration below 2 ppb by 2050, we will have to have achieved an 85% reduction (from 1990 values) in the release of all long-lived halocarbons by the mid-1990s and zero

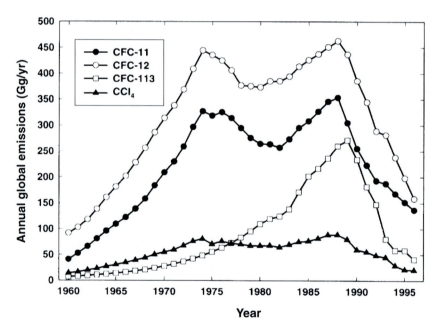

Figure 12.10 Effect of international legislation on global emissions of long-lived hydrocarbons.
Source: World Meteorological Organisation (1999) 'WMO Global Ozone Research and Monitoring Report No 44', *Scientific Assessment of Ozone Depletion 1998*, WMO, Geneva.

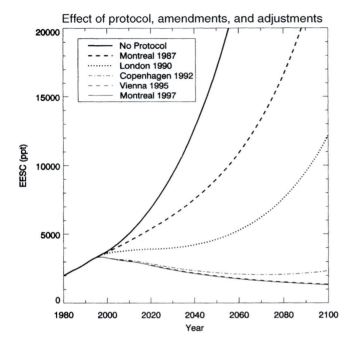

Figure 12.11 The impact of the Montreal Protocol on stratospheric concentrations of chlorine-equivalent gases.
Source: World Meteorological Organisation (1999) 'WMO Global Ozone Research and Monitoring Report No 44', *Scientific Assessment of Ozone Depletion 1998*, WMO, Geneva.

release by 2000; all release of short-lived halocarbons would also have to end by 2030.

12.2.2   Ozone depletion potentials

Other substances have been developed to replace CFCs as refrigerants and aerosol propellants, and naturally there is continued concern as to their environmental impact. Hydrochlorofluorocarbons (HCFCs) and hydrofluorocarbons (HFCs) contain at least one hydrogen atom in each molecule, and are oxidised in the troposphere during reactions with OH radicals:

$$OH + CHCl_xF_y \rightarrow H_2O + CCl_xF_y \rightarrow products$$

This is thought to result in a shorter overall lifetime, although the chain of atmospheric chemistry and the reaction rates remain uncertain. In order to compare the harmful effects of different CFCs and their proposed substitutes, the concept of ozone depletion potential (ODP) has been established. The ODP for a given halocarbon is defined by

$$ODP = \frac{\text{ozone depletion due to compound}}{\text{ozone depletion due to CFC-11}}$$

where the ozone depletion is the equilibrium change in the globally averaged total column ozone per unit mass emission rate. Calculation of ODPs relies on models of atmospheric transport and chemistry, and values therefore vary according to the sophistication of the model used. Relative values should nevertheless give a useful guide to the threat from different compounds (Table 12.3).

These steady-state ODPs assume that equilibrium concentrations have been achieved in the atmosphere – a process which itself has a timescale of a hundred years or so. Since emissions at the Earth's surface are changing with a timescale of tens of years, the ODPs will seriously underestimate the potential impact of many halocarbons. An alternative approach has been to calculate the total chlorine loading, defined as the instantaneous tropospheric mixing ratio of the chlorine atoms in all the halocarbons. When combined with atmospheric transport models, the time and altitude dependencies of chlorine atom concentration can be derived. Such calculations have shown a time lag of 5–10 years between the onsets of maximum chlorine loading in the troposphere and of maximum inorganic chlorine release in the upper stratosphere, implying that ozone depletion will get worse before it gets better.

Table 12.3 Ozone depletion potentials of different halocarbons

Halocarbon	ODP	Lifetime/years
CFC-11	1.00 (by definition)	60
12	0.90	120
113	0.85	90
114	0.60	200
115	0.37	400
HCFC-22	0.05	15.3
123	0.017	1.6
124	0.020	6.6
HFC-125	0	28.1
134a	0	15.5
HCFC-141b	0.095	7.8
142b	0.05	19.1
HFC-143a	0	41.0
152a	0	1.7
Carbon tetrachloride (CCl_4)	1.1	50
Methyl chloroform (CH_3CCl_3)	0.14	6.3

12.2.3   Methyl bromide

Atom for atom, stratospheric bromine is about 50 times more effective than chlorine at destroying ozone in the stratosphere; it is thought that bromine reactions have been responsible for 20% of the loss in the Antarctic ozone hole, and for an even higher proportion in the Arctic and at lower latitudes. The principal bromine carrier is methyl bromide (CH_3Br), but much less is known about sources and cycles for this gas than for CFCs. The sources are both natural (oceanic emissions, biomass burning) and human (soil fumigation, leaded petrol combustion) amounting to some 100 kt a^{-1}. These emissions, and a lifetime of 2 years, give an average global concentration of 10 ppt. The short lifetime means that, if international measures are taken to reduce the anthropogenic component, improvements will be far more rapid than in the case of CFCs.

12.3   THE CURRENT SITUATION

The World Meteorological Organisation held a meeting in 1998 to review the position on ozone depletion. They concluded that the tropospheric concentration of ozone-depleting substances had indeed peaked in around 1994, when the chlorine concentration was 3.7 ppb, and that the stratospheric concentration would peak in around 2000 (due to the lag time taken by the gases to diffuse up through the troposphere). The peak was so flat, and the random variability so great, however, that it would take a further 20 years to identify whether the ozone layer had started to recover. Figure 12.12 shows the dramatic impact of

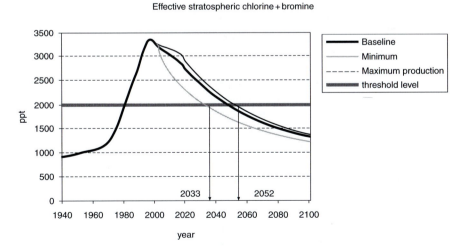

Figure 12.12 Historical and predicted concentrations of chlorine and bromine.
Source: European Environment Agency (1999) *Annual Topic Update on Air Quality*, EEA, Copenhagen.

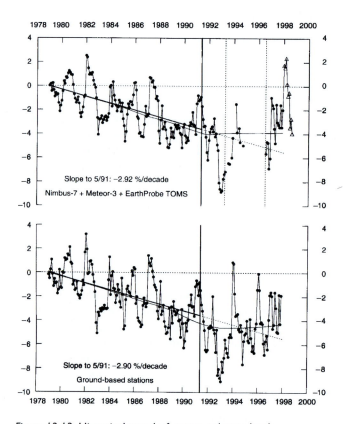

Figure 12.13 Historical trend of ozone column depth.
Source: Proceedings, Schonbein International Ozone Symposium, Basel Switzerland (1999).

the international controls on the stratospheric concentration of chlorine + bromine. After the peak in 2000, the concentration will decline less quickly than it rose, falling below 2000 ppt between 2033 and 2052 (depending on which scenario for stopping the emission of ozone-depleting substances actually happens). In fact, it appears that the data from the last few years of the twentieth century do indicate a recovery of column depth. Figure 12.13 shows that both satellite and ground-based measurements gave a linear decline in mid-latitude (25–60°N) ozone with time between 1979 and about 1993. In 1993, there was a large perturbation caused by the eruption of Mt Pinatubo. From 1996 the ozone column depth has been stable or slightly increasing. The September/October 2000 ozone hole (<220 DU) was the largest yet. With an area of 27 million sq km, it was the size of North America and twice the size of Antarctica. However, although the area of the hole was the largest, the minimum value of 98 DU was not as low as the 1993 record minimum. These observations are extended in Figure 12.14, which shows modelled future OCDs based on the halocarbon

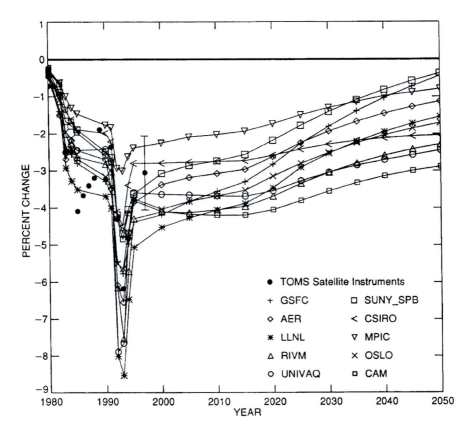

Figure 12.14 Changes in OCD, 1980–2050.
Source: World Meteorological Organisation (1999) 'WMO Global Ozone Research and Monitoring Report No 44', *Scientific Assessment of Ozone Depletion 1998*, WMO, Geneva.

trends from Figure 12.12. Although there is a significant difference between the predictions of the different models used, they agree as to the overall trend. It will be at least 2050 before the OCD recovers to the pre-1979 value, and could be much later. Note also that the model outputs are generally supported by the measurements from the TOMS satellite, although frustratingly the scatter is too large to distinguish between models.

Measurements can be misleading if there is not enough understanding to interpret them correctly. For example, an ozone 'mini-hole' was observed over the UK and North Sea in November 1999. Column depths were below 200 DU compared to an expected 290 DU. The cause was found to be poleward transport of subtropical air containing low ozone concentrations, rather than chemical depletion of the ozone layer.

12.4 OZONE AND ULTRAVIOLET

Ozone and ultraviolet radiation are inseparable. It takes UV photons to form ozone from oxygen and UV photons to release the chlorine from CFCs to destroy ozone. The ozone layer absorbs UV, and it is the potential increase of ground-level UV that is seen as the main environmental hazard of ozone depletion. Known effects of this increased UV in humans are sunburn, melanoma and cataracts; there may be corresponding effects that are not yet understood in plants and other animals.

The waveband of the electromagnetic spectrum that is visible to the human eye and photosynthetically active for plants lies between 400 nm (the short wavelength blue end) and 700 nm (the long wavelength red end). Wavelengths below 400 nm in the ultraviolet are invisible to the unaided human eye. The UV region is conventionally subdivided into UVA (320–400 nm), UVB (280–320 nm) and UVC (200–280 nm). Some workers divide UVA at 315 nm rather than 320, or at 290 nm rather than 280. Natural UV is supplied exclusively by the sun, which radiates energy as a black body with an effective surface temperature of about 6000 K. Outside the Earth's atmosphere, the total solar radiation flux density of 1367 W m^{-2} is made up as shown in Table 12.4.

12.4.1 Penetration of UV through the atmosphere

The solar spectrum is modified as it passes through the atmosphere by two main processes. First, Rayleigh scattering by molecules of oxygen and nitrogen is more effective the shorter the wavelength (by a factor of 16 at 200 nm compared to 400 nm). Second, the stratospheric ozone layer absorbs very effectively over a band centred on 255 nm. Figure 12.15 shows the absorption cross sections for both ozone and oxygen in the UV and visible wavelength range. The oxygen peak is centred on 150 nm where there is almost no radiation, whereas the ozone peak occurs well over the shortwave edge of the solar spectrum. This strong absorption and scattering in the UV has a profound effect on that end of the spectrum, as seen in Figure 12.16.

Table 12.4 Flux densities in different wavebands of the solar spectrum

Waveband	Flux density/ W m^{-2} (%)
Visible and infrared	1254 (91.7)
UVA	85.7 (6.3)
UVB	21.1 (1.5)
UVC	6.4 (0.5)
Total	1367 (100)

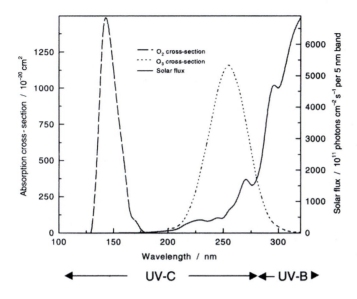

Figure 12.15 Absorption cross sections of ozone and oxygen.
Source: van Loon, G. W. and Duffy, S. J. (2000) *Environmental Chemistry – A Global Perspective*, Oxford University Press, Oxford.

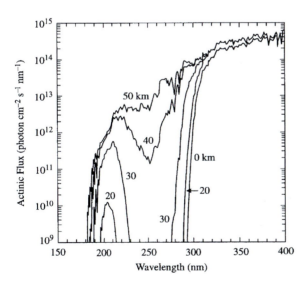

Figure 12.16 Variation of UV radiation with altitude.
Source: Jacob, D. J. (1999) *Introduction to Atmospheric Chemistry*, Princeton University Press, Princeton, New Jersey.

Close to the top of the Earth's atmosphere (50 km), the spectrum is essentially extra-terrestrial. By 40 km O_3 absorption has already made a significant difference, reducing the flux most around 250 nm because that corresponds to the centre of the O_3 absorption band. At 20 km above the surface, the photon flux is almost the same as the surface value, except for a residual peak at 200 nm. Note, however, that although the ground level and 20-km spectra look very similar around 300 nm, at any *individual* wavelength the logarithmic scale gives an order of magnitude difference in the fluxes.

UVC (wavelengths below 280 nm), which is small to start with, is effectively eliminated. Although UVB is attenuated, the UV source flux increases towards the blue and the ozone absorption band weakens towards 300 nm – both these influences combine to give a significant UVB flux which is very sensitive to small shifts in ozone absorption. The general relationship is shown in Figure 12.17, where measurements from six stations around the world confirm that the percentage increase in surface UVB is around twice the reduction in OCD.

The increase in UV becomes more marked at shorter wavelengths, as seen in Figure 12.18. The UV irradiance was measured on two days having OCDs of 177 and 410 DU, and is presented both as the measured spectral distributions and as the ratio of the values on the two days.

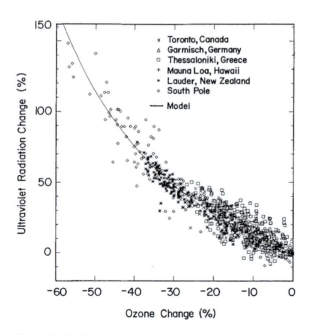

Figure 12.17 Measured and modelled relationship between OCD and UVB.
Source: World Meteorological Organisation (1999) 'WMO Global Ozone Research and Monitoring Report No 44', *Scientific Assessment of Ozone Depletion 1998*, WMO, Geneva.

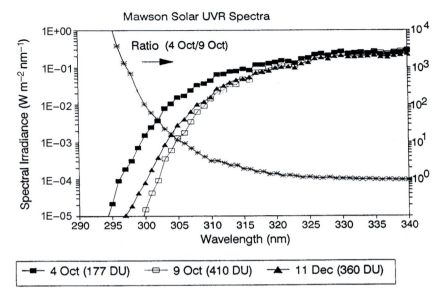

Figure 12.18 Effect of ozone column depth on spectral composition.
Source: Roy, C. R., Gies, H. P. and Tomlinson, D. W. (1994) 'Effects of ozone depletion on the ultraviolet radiation environment at the Australian stations in Antarctica', In: *Ultraviolet Radiation in Antarctica: Measurements and Biological Effects*, Antarctic Research Series 62: 1–15.

The decline in ozone caused what looks at first sight to be a rather moderate increase in the UV penetration below wavelengths of 320 nm. However, when this change is plotted as the ratio of the flux density at that wavelength to the flux density at 340 nm (chosen because it would not be affected by ozone changes), a dramatic increase is seen, up to a factor of 10 000 at 295 nm. Put another way, any change in OCD is amplified, with the amplification factor increasing as wavelength falls. Figure 12.19 shows the variation of this radiation amplification factor with wavelength for a 1% reduction in OCD at a range of solar zenith angles (at a zenith angle of zero the sun is overhead). At the shortest wavelengths, the irradiance is increased by a factor of three or four.

The optical amplification factor (OAF) is the percentage increase in carcinogenically effective irradiance caused by a 1% decrease in column ozone. The factor is different for different effects, but is usually in the range 1.6–2.0. These changes are not confined to the remote Antarctic. In 1989, a patch of the ozone hole drifted up over Melbourne in December (Australian summer), and the column ozone was reduced by 50%.

Some molecules in biological organisms, notably DNA, are sensitive to these short wavelengths. In addition, the total or global flux involves a direct component (that travels straight to the receptor) and a diffuse component (that

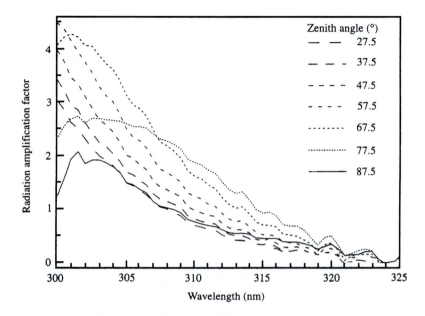

Figure 12.19 Variation of radiation amplification factor with wavelength for a 1% decrease in OCD.
Source: Wayne, R. P. (2000) *Chemistry of Atmospheres*, Oxford University Press, Oxford.

comes to the receptor after scattering in the atmosphere). When the angle of the sun above the horizontal is low, the path length through the atmosphere is long and the ratio of scattered diffuse radiation to direct radiation is high. The significance of diffuse/direct ratios depends on the geometry of the receptor, but typically UVB in the lower troposphere is 50% diffuse. The proportion of total UV to total solar radiation, which is around 8% outside the atmosphere, is decreased to 5%, 3% and 1% at zenith angles of 0, 60 and 80°, respectively. This effect protects the Arctic and Antarctic regions from high UV fluxes because of the high zenith angles of the solar radiation. Even with ozone depletion, UV fluxes in the Antarctic are less than those at the equator. Cloud transmissivity is often higher in the UV than in the visible, and surface reflectivities are different (Table 12.5).

Table 12.5 UV albedos

Surface	UV albedo %
Grass	2
Concrete	10
Sand	20
Snow	60–90

12.4.2   Penetration of UV into water

Aquatic and marine ecologies depend on solar radiation just as do terrestrial ones, and UV changes will impact on them too. The first attenuation takes place when the light strikes the water surface. If the water surface is calm, then direct reflectance increases from a few percent when the sun is overhead to more than 60% at grazing incidence. When diffuse radiation from a clear sky is included, the maximum percentage reflected drops to 40. When diffuse only is considered (as under a cloudy sky), about 6% is reflected. When the sea is rough, which is usually linked to increased windspeed, the effect is small for high solar elevations but tends to reduce the percentage reflected at low solar elevations.

As the transmitted light penetrates the water column, it is both scattered and absorbed by the water itself, by suspended particles and by dissolved organic material. The overall effect of these processes is to attenuate the flux density according to the Beer–Lambert law with an attenuation coefficient K_d (m^{-1}). If the variation of flux with depth is measured, then 10% of the initial flux will remain at a depth of $2.3/K_d$. K_d varies with wavelength, so that usually UV will be much more rapidly attenuated than PAR or IR, especially in coastal waters that are high in dissolved organic carbon. The 90% attenuation depth of 305 nm UV, for example, might range from 1 m in the rather murky Baltic to 5 m in the North Sea and 15–20 m in the clean ocean. The value of K_d for pure water is about 0.01 m^{-1}. An interesting interaction here is that one of the effects of acidification is to increase water clarity, thereby reducing the attenuation of UV in the water column.

12.4.3   Absorption of UVB

The energy E associated with individual photons of radiation varies inversely with their wavelength λ according to Planck's formula $E = hc/\lambda$, where h is Planck's constant and c is the speed of light. The shorter the wavelength, therefore, the more the energy per photon and the more damage can potentially be caused. The effects of UV, however, can be much more sensitive to wavelength than indicated by that relationship. The variation of effectiveness of the radiation with wavelength can be described by an action spectrum, usually obtained from laboratory experiments. Any tissue molecules capable of absorbing electromagnetic energy are called chromophores. Absorption of the photon puts the molecule into a high-energy unstable state, initiating a series of biochemical and cellular changes. Important UV chromophores include DNA, proteins, urocyanic acid and melanins. Ideally, the action spectrum will correspond to the absorption spectrum of the chromophores; this is modified for different parts of the body by site-specific factors such as the thickness of overlying tissue. Figure 12.20 shows the action spectra for erythema (see Section 12.4.4.1) and for damage to the DNA molecule. The effectiveness increases by a factor of over 10 000 as the wavelength is reduced from 320 to 280 nm – the upper and lower limits of the UVB range.

Also shown on Figure 12.20 is the solar spectrum in the UVA and UVB (the thick solid line). The combination of the sharp changes in both these spectra makes the biological effectiveness extremely sensitive to small changes in UV radiation, and therefore in ozone amount. Thus, it is very important when predicting health effects to have sound knowledge of the spectral distribution as well as the bandwidth flux density. This is shown in another way by Figure 12.21. Here, the dashed curve represents the per cent change in spectral irradiance when the OCD is reduced by 1%, from 300 to 297 DU. The corresponding absolute change is shown by the dotted curve (which is read from the right scale). When the irradiance is weighted by the erythemal action spectrum, the solid curve, which shows a peak in erythemal effectiveness at 305 nm, results.

The biological amplification factor (BAF) is the power to which the UV dose D is raised to predict the incidence of cancer.

i.e.   incidence $=$ Constant D^{BAF}

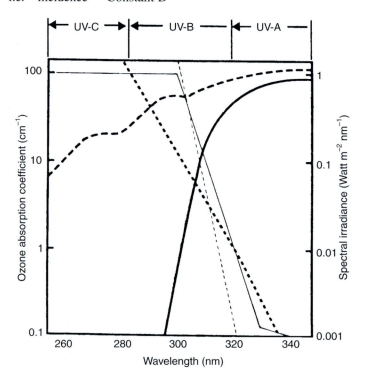

Figure 12.20 The action spectrum of DNA sensitivity to UV radiation, together with the UV spectrum of solar radiation. Note that both vertical scales are logarithmic.
Source: Webb, A. R. (2000) 'Ozone depletion and changes in environmental UV-B radiation', In: R. E. Hester and R. M. Harrison (eds) *Causes and Environmental Implications of Increased UV-B Radiation*, Royal Society of Chemistry, Cambridge.

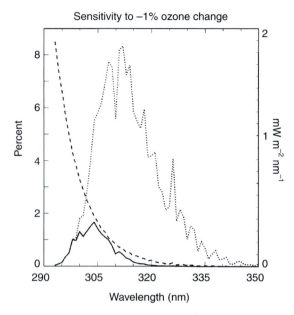

Figure 12.21 Changes in spectral irradiance and erythemal effect for a 1% reduction in OCD. The dashed curve shows the percent increase in the spectral irradiance (left scale); the dotted curve shows the absolute change in spectral irradiance (right scale); and the solid curve shows the absolute change in spectral irradiance weighted by the erythemal action spectrum (right scale) Source: World Meteorological Organisation (1999) 'WMO Global Ozone Research and Monitoring Report No 44', *Scientific Assessment of Ozone Depletion 1998*, WMO, Geneva.

Values of BAF are different for different types of skin cancer, lying in the range 1.4–2.5 for fair-skinned people. The overall biological response to a change in the ozone column density is thus the product of both the optical and biological amplification factors.

12.4.4   Effects of UV on people

The main responses that have received attention for humans are those involving damage to the skin (sunburn, non-melanoma skin cancer and melanoma), eyes (cataracts) and the immune system.

12.4.4.1   Skin responses

Coco Chanel is reputed to have started the tanning fashion after coming back tanned from a yachting holiday in the 1930s, and the number of holidays abroad taken by UK residents increased from 5 to 20 million between 1970 and 2000.

Tanning is the induced production of melanin by epidermal melanocytes, as a consequence of UV-induced damage to DNA. Being tanned confers a photoprotection equivalent to a Sun Protection Factor of 2 or 3. Ironically, sunscreen users are more likely to burn than shade seekers; they probably feel that any application of sunscreen will give them protection for any length of time, despite the fact that the average application is only one quarter of the recommended 2 mg cm^{-2}. There are some medical benefits: vitamin D is produced, but this only needs about 10 min exposure per day; a feeling of well-being is created; psoriasis and eczema are reduced.

Acute responses

The most well-known acute responses are sunburn and pigmentation. Both UVA (320–400 μm) and UVB (280–320 μm) develop similar visual responses in human skin, although UVA typically requires a flux density that is several orders of magnitude higher. UVB activity is confined to the outer 0.03 mm of epidermis, while UVA penetrates three times as far – through the epidermis and into the dermis. About 1% of UVA penetrates into the subcutaneous tissue. UVB stimulates melanin (brown pigment) synthesis; the tanning effect becomes visible 3 days after exposure, continues for some days after, and provides some protection against subsequent UVB exposure. The longer wavelength UVA is more immediate in action but previous exposure provides little subsequent protection. Over-exposure induces erythema (the characteristic red inflammation due to dilation of small blood vessels just below the epidermis), weeping and blistering, followed by shedding of skin layers and pigmentation. Interestingly, some drugs, for example those used for heart complaints, photosensitise the skin so that it becomes very sunburned very quickly. Contact with giant hogweed has the same effect.

The melanin pigment molecules that characterise naturally dark skin absorb UV and prevent its penetration into the skin (Figure 12.22). Fair skin, on the other hand, transmits effectively in the visible and UVA regions, and retains a higher transmissivity well into the UVB region, before cutting off sharply below 290 nm. The total sunburning potential of a given radiation environment can be summarised by the erythemally effective radiation (EER, with units of mW m^{-2}), which is the integral of the energy in the solar spectrum at each wavelength weighted by the erythemal action spectrum. When skin is exposed to EER, a dose accumulates which eventually causes just-perceptible reddening of the skin. This standard erythemal dose (SED) has been set at 200 J m^{-2} for the most sensitive skins, based on exposure to monochromatic radiation at the maximum spectral efficiency for erythema (around 300 nm). Maximum daily UV doses under clear summer skies are about 70 SED in the tropics, 60 SED at mid-latitudes and 45 SED at temperate (UK) latitudes. It takes an exposure of about 1.5 SED to just redden the skin (erythema) of an unacclimatised, sun-sensitive person who never tans (skin type I); 2 SED for those who burn easily without tanning much (Type II) and 3 SED for those who burn and tan (Type III). In the British population,

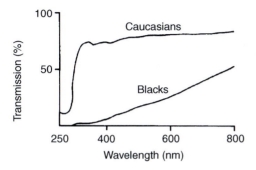

Figure 12.22 Spectral transmissivity of skin tissue from Caucasian and Black subjects. Source: Jones, R. R. (1989) 'Consequences for human health of stratospheric ozone depletion', In: R. R. Jones and T. Wigley (eds) *Ozone Depletion: Health and Environmental Consequences*, John Wiley, Chichester.

about 11, 30 and 31% occupy those types, respectively. Northern European indoor workers get an annual dose of around 200 SED, which is only about 5% of the available dose. However, they are liable to receive 30% of the dose from one fortnight's holiday. Some of the solar UVB is direct (it reaches the surface having passed through the atmosphere in a straight line) and some is diffuse (it reaches the surface indirectly, after scattering in the atmosphere, reflection from clouds or surfaces, transmission through intervening surfaces). The direct component is greatest when the sun is at its highest altitude because the depth of the atmosphere is then least. The diffuse component is increased at low solar angles because the atmospheric path length (and hence scattering) is greatest. Hence calculation of the total UVB is complex. Table 12.6 shows the maximum values for erythemally-effective clear-sky UVB irradiance for the northern hemisphere.

The effects of clouds are highly variable, depending on their type and coverage. For radiation in general, a formula has been used that cloud cover of C tenths reduces the solar flux density to a fraction $F = 1 - 0.06C$. However, greater reductions in UVB than are suggested by this formula have been observed in practice.

The UK Met Office now issues forecasts of a UV index based on the predicted EER, divided by 25 to bring it into line with indices used elsewhere – the index

Table 12.6 Maximum EER values in the northern hemisphere/mW m^{-2}

Latitude, degrees	January	March	June	December
0	229	260	207	221
30	79	157	215	56
60	4	21	105	2
90	0	0	16	0

is typically in the range 4–8 at noon on a cloud-free summer day in the UK, increasing to 10 or 11 in the tropics. A UV index of 6 corresponds to sunburn times of 20 min and 2 h for the most and least sensitive skin types, respectively. Other factors naturally have to be considered – cloud and aerosol will reduce exposure, while height in the atmosphere will increase it. At solar noon, UV at 300 nm is ten times greater than it is at 9 am or 3 pm.

Chronic responses

Chronic responses to UV include skin photoageing and skin cancer. First, long-term exposure to high UV fluxes leaves skin dry, wrinkled and leathery, a common end result for lifelong sunbathers. This is mainly a cosmetic problem, but the change in skin texture can lead to chronic cracking and infection. Another chronic response is dilated blood vessels, giving the skin an inflamed appearance. Solar keratoses are small scaly lesions, which are considered to be premalignant and which may progress into squamous cell carcinoma, although this is rare. More importantly, DNA is one of the strongest UVA chromophores and genotoxic responses are to be expected.

The carcinomas are divided into melanomas and non-melanomas. For our purposes, a melanoma is a malignant tumour that arises from melanocytes – the cells of the skin that produce melanin. It is the least treatable, and hence the most important, type of skin cancer. There are around 1 300 deaths from melanoma in England and Wales each year, over twice as many as from all other skin cancers combined, although the incidence is only one-tenth. Typically 40% of patients with melanoma die within 5 years, as compared to 1% for non-melanoma skin cancers. Predisposing factors are: fair skin, particularly when associated with freckles; blond or red hair; intense, short-term exposure of previously untanned skin; and a large number of naevi (moles). There is some evidence that a single episode of severe sunburn in infancy or childhood may trigger a melanoma in later life. If the melanoma is treated early, it greatly increases the chance of cure.

In 1995, there were 5 400 cases of malignant melanoma, which was the 12th most common cancer, in the UK. In the context of this chapter, it should be stressed that the melanoma increase due to UV reduction has never been detected among the general increase due to sunbathing and other social changes. However, it has certainly been shown that the closer to the equator that people live, and the higher is their exposure to solar UV, the greater their risk of melanoma (Figure 12.23). The rate of occurrence is around 40 per 100 000 at the equator, falling to 10 in temperate latitudes, and with a 6:4 ratio of women to men.

Although there is a strong relationship between the site on the body where the melanoma occurs and whether that site is sometimes exposed to the sun, there is no clear corresponding relationship with exposure itself. This highly behaviour-dependant response has made it hard to predict reliably the consequences of ozone depletion.

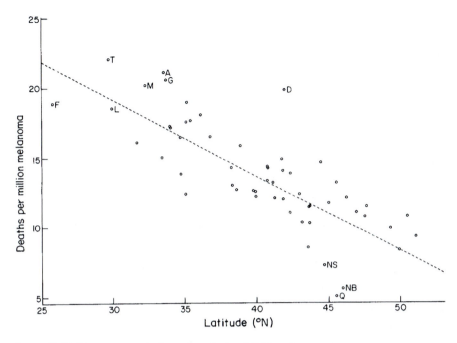

Figure 12.23 Variation with latitude of the US/Canadian melanoma mortality rate, 1950–1967.
Source: Elwood, J. M. (1989) Epidemiology of melanoma. Jones, R. R. (1989) 'Consequences for human health of stratospheric ozone depletion', In: R. R. Jones and T. Wigley (eds) *Ozone Depletion: Health and Environmental Consequences*, John Wiley, Chichester.

Non-melanoma cancers are divided into basal cell carcinomas (BCCs, which account for about 75% of cases) and squamous cell carcinomas (SCCs). Basal cell carcinomas are raised translucent nodules, which develop slowly on the face, neck and head. Squamous cell carcinomas appear as persistent red crusted lesions on sun-exposed skin. Chronic over-exposure increases the incidence of both melanoma and non-melanoma skin cancers. Skin carcinomas are already the most common human malignancies, with 40 000 new lesions annually in the UK and 600 000 in the United States. Eighty per cent of such cancers arise on areas of the body that are naturally exposed to the sun, and Figure 12.24 confirms that the incidence of non-melanoma skin cancer is a strong function of UV exposure wherever you live.

Multifactorial analysis has shown that age and UV exposure are the two most important factors in the epidemiology of the carcinomas, with the risk being expressed as a power law of the form

$$\text{Risk} = (\text{annual UV dose})^a \, (\text{age})^b$$

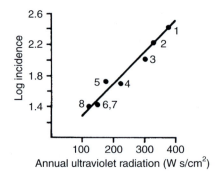

Figure 12.24 Variation with solar UV irradiance of incidence rate for non-melanoma skin cancers.
Source: Jones, R. R. (1989) 'Consequences for human health of stratospheric ozone depletion', In: R. R. Jones and T. Wigley (eds) *Ozone Depletion: Health and Environmental Consequences*, John Wiley, Chichester.

a and *b* are typically 1.7 and 3.2 for basal cell carcinoma, and 2.3 and 5.1 for squamous cell carcinoma.

Action spectra for SCC have been derived from experiments with mice, although the applicability of the results to people has been questioned. Not only is there doubt about the mouse as a model, but we do not understand the dependance of response on the details of the spectrum.

Global UV monitoring networks with reliable long-term calibrations have not been in existence long enough to identify whether the UV changes described above are actually occurring, particularly at low latitudes. The data flow is increasing rapidly, with around 250 monitoring stations now in operation. However, it has been estimated that 10 years data will be needed to detect a trend of 5% per decade amongst the variations due to local pollution sources, cloud and haze cover, and instrument calibration. The most popular instruments in use have been Robertson–Berger meters, which give an integrated measurement of erythemally-effective radiation over the whole 280–400 nm waveband and are hence unsuitable for detailed analysis of trends in UVA and UVB. In a typical instrumental set-up, the incident UV passes through a black glass filter to absorb visible and IR radiation. The remaining radiation is wavelength-shifted into the 400–600 nm spectral region by a magnesium-tungstate phosphor, and detected by a vacuum photodiode with a bialkali photocathode. The relative spectral response of the instrument is close to the spectral efficacy curve for erythemal response of human skin. Personal dosimeters are also available, based on the absorption of UV by polysulphone film. Again this can be produced to match closely the human skin erythemal response. Wavelength-resolved fluxes are really required, but the instruments are much more complex and intercomparisons less reliable. As discussed elsewhere, the ozone layer depletion should have peaked in 2000, and concentrations will be restored over the next 50 years. With this scenario, it has been estimated that today's children living in Britain will experience an increased risk of non-melanoma skin cancer, due to ozone depletion, of 4–10%. Globally, there will be a peak increase in the incidence halfway through this century of 7 per 100 000, corresponding to 4 200 cases in the UK.

12.5   CLOTHING PROTECTION FROM UV

Clothing provides a complex and variable level of protection from UV radiation. The level of protection depends on fabric structure (fibre content and thickness, and the knit or weave) and colour. Individual fabrics can be tested and given a Clothing Protection Factor (CPF) analogous to the Sun Protection Factor (SPF) given to sunscreens. The test firstly involves spectral measurement of the transmissivity (the fraction of the incident radiation that is transmitted through the fabric) over the wavelength range 290–400 nm. The solar irradiance at each wavelength is then multiplied first by the transmissivity and second by the erythemal activity, and the results integrated across the wavelength range to find the erythemally effective UV radiation that penetrates the fabric. The CPF is then the ratio of the erythemally effective UV incident on the fabric to that transmitted through the fabric. Hence, a high value of CPF indicates a high level of protection. Clothing Protection Factor is not constant for any one fabric, since it may change with stretch or wetness. Clothing with a CPF of 30 or more confers a high degree of protection. If the CPF is less than 10, then a hazardous level of UV exposure will be experienced if the wearer is outdoors all day in strong sunlight. In some countries, CPF is indicated on clothing for sale if the garment is likely to be worn under UV-hazardous conditions – T shirts, for example. The CPFs of some clothing are surprisingly low. For example, measurements on a white football shirt gave a CPF of 5–10 (equivalent to using a sunscreen with a protection factor of 5–10).

12.5.1   Eye responses

Potential UV damage to any area of the body is initially a question of exposure. It has been calculated that because the front of the eye is mainly vertical, eyelid exposure (i.e. the flux density actually available at the eye, as compared to the irradiance on a horizontal surface) is about 20% of the corresponding irradiance falling on the top of the head. Under strong sun conditions, radiation reaching the eye itself is normally further reduced to 5% by the squint reflex. The cornea absorbs UV strongly, with over 90% of UV <300 nm being absorbed. The young lens transmits UVA well (75%), although this falls away to <20% for adults. On the retina, the focussing action of the lens produces an enhancement of flux density by many orders of magnitude. UV can cause two main types of damage to the eye – at the cornea, and at the lens. There is a smaller prevalence of a condition called pterigium – small benign growths on the white of the eye.

Although the cornea has high transmissivity to radiation in the visible band, it does absorb UV. In principle, the cornea can develop photokeratitis in just the same manner as skin develops erythema, but rarely does so because of natural exposure avoidance. The most common example is snow-blindness, when total UV exposure is elevated by a combination of high direct values from cloud-free skies and high diffuse values by reflection from the snow. Sensitivity is very high, being typically 30–40 J m^{-2} compared to the 200 J m^{-2} for skin erythema. There is an issue here for pilots and mountaineers, because UV flux at 10 000 m is 100 times that at the ground.

A cataract is a progressive opacity of the eye lens, which occurs at an increasing frequency with age. Production of cataracts is accelerated by UV radiation (Figure 12.25), and there is an observable dependency on latitude, as shown in Table 12.7.

The USEPA has quantified the relationship between UVB flux and cataracts, estimating that a 30% increase in UVB (which would result from a 15% decrease in total column ozone) would increase cataract prevalence by around 14%,

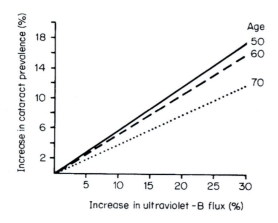

Figure 12.25 Sensitivity of cataract prevalence in different age groups to proportionate increase in UVB flux.
Source: Jones, R. R. (1989) 'Consequences for human health of stratospheric ozone depletion', in: R. R. Jones and T. Wigley (eds) *Ozone Depletion: Health and Environmental Consequences*, John Wiley, Chichester.

Table 12.7 Variation of cataract occurrence with latitude

Location	Cataract occurrence/%
Rochester, New York (42°N)	9
Tampa, Florida (28°N)	20
Manila (15°N)	43

depending on age. Calculations using similar action spectrum ideas to those that we used above for skin lesions have predicted that a 10% reduction in column ozone over the US would increase the number of cataract cases by about 600 000 in the current population, and by about 4 million in the cohort born between 1986 and 2029. In their worst-case scenario for growth in the use of ozone-depleting substances, the same cohort would experience an additional 10 million cataracts.

12.5.2   Immune responses

The skin is a major site of entry for foreign antigens into the body, and plays an active role in immunological defence. UV affects the immune system by interaction with the skin. Many skin cells, such as the keratinocytes, secrete cytokines which mediate the control of immune and inflammatory responses. A number of studies has shown that exposure to UV suppresses responses of the immune system and resistance to viral and bacterial infections. For example, activation of the herpes virus, leading to cold sores, commonly occurs at the start of a summer holiday due to the immunosuppressive effect of UV on the skin.

12.5.3   Effects of UV radiation on terrestrial biota

It is thought that early in the Earth's existence the Sun produced much higher UV than today even though its total output was lower. Since there was no oxygen, and hence no ozone layer, the UV levels at the Earth's surface were too great for life to develop. One of the reasons why early life developed underwater was that the water shielded the UV. As oxygen and ozone increased in step, UV levels were reduced to tolerable amounts by about 500 million years ago when the first terrestrial plants appeared. Even so, UV levels were almost certainly higher then than they are now, which raises the possibility that some ancient and more primitive plant species such as algae and lichen may show greater UV tolerance than more recent evolutionary developments. Some developments in plants involved the manufacture of phenolic compounds such as tannins, flavonoids and lignin, which in addition to their other plant functions were UV-screening. In addition, plants have defence mechanisms such as DNA repair, radical scavenging and polyamine stabilisation of injured membranes.

Many experimental studies of plant responses to elevated UVB have now been published. Contrary to expectations, these have not generally shown a reduction in primary productivity. However, we must stress again here that the results of such an experiment are only as good as the experiment itself. It is always a major, and in many ways insuperable, problem to change only the UVB radiation that is the variable of interest. If the plants are grown under controlled conditions (in a laboratory growth chamber, for example) then the temperature, humidity, soil structure, windspeed and other key variables will all be different to those outdoors. In particular, the spectral composition of the lighting will be different, even after strenuous and expensive efforts, using contributions from a range of sources, have been made in order to simulate solar daylight. For example, Figure 12.26 shows that a QTH (Quartz envelope, Tungsten element, Halogen gas fill) or Xenon bulb, which both cross the natural UV spectrum at 305 nm, under-represent the PAR and exceed the shorter wavelength UV by an increasing proportion. When this is translated into an action spectrum (Figure 12.27) it is clear that if we simply go outdoors and shine lights on vegetation we cannot be certain of getting valid responses. In this case, the weighted irradiance from the artificial sources is 10 000 times higher than that from natural light at 290 nm.

Instead of supplementing the UV with artificial sources, filtering can be used to deplete it. Figure 12.28 shows the transmission spectra of four 'transparent' materials, characterised by the wavelength at which the transmittance is 50% – 220 nm for quartz, 305 nm for Pyrex, 320 nm for mylar and 350 nm for Plexiglas. Cellulose diacetate (300 nm) and polycarbonate (390 nm) filters are also used. Even within the UVB band, a supplementary source will have a different distribution of flux with wavelength to that in sunlight, and a weighting function will be used to correct for this. Some studies, particularly of secondary responses such as plant pathogens, have used sources with such high UVC components that they are not representative of conditions at the surface. If the experiments are done outdoors, the naturalness of many factors can be improved, although the spectral problem will remain whether it is increased by supplementation or reduced by

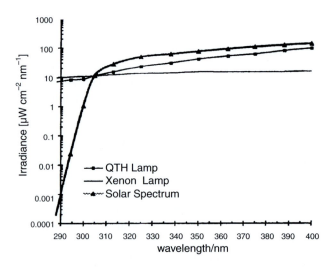

Figure 12.26 Spectra of UV sources and solar radiation.
Source: Diaz, S. B., Morrow, J. H. and Booth, C. R. (2000) 'UV physics and optics', In: S. de Mora, S. Demers and M. Vernet (eds) *The Effects of UV Radiation in the Marine Environment*, Cambridge University Press, Cambridge, UK.

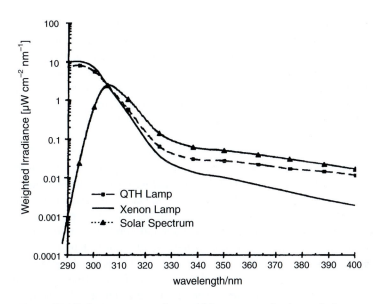

Figure 12.27 Action spectra due to UV sources and solar radiation.
Source: Diaz, S. B., Morrow, J. H. and Booth, C. R. (2000) 'UV physics and optics', in: S. de Mora, S. Demers and M. Vernet (eds) *The Effects of UV Radiation in the Marine Environment*, Cambridge University Press, Cambridge, UK.

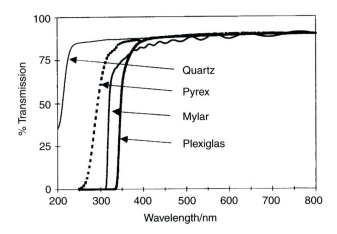

Figure 12.28 Transmission curves of four materials used to deplete UV radiation.
Source: Vernet, M., Brody, E. A., Holm-Hansen, O. and Mitchell, B. G. (1994) 'The response of Antarctic phytoplankton to ultraviolet radiation: absorption, photosynthesis and taxonomic composition', In: *Ultraviolet Radiation in Antarctica: Measurements and Biological Effects*, Antarctic Research Series 62: 1–15.

filtering. In addition, the variability of the conditions and the growth will increase, so that responses have to be correspondingly larger for statistical confidence to be maintained. For trees, the detailed optical properties of individual leaves (needles for conifers) and communities is important. Only 10–20% of the incident radiation penetrates down to the lowest foliage in typical forests, so the exposure of the youngest and oldest leaves may vary by a factor of ten. As with air pollution studies, there have been few controlled experiments on mature trees – for example, most of our knowledge on conifer responses has been acquired from experiments on 1-year old seedlings.

The greatest reductions of column ozone and associated increases in UV have occurred in Antarctica, which is not known for its high plant density. Over the globe as a whole, we would not expect the total column ozone changes that have occurred so far (a few percent over 20 years) to be reflected in observable changes in plant communities. Hence, we have to rely on the results of relatively short experiments conducted in middle latitudes on a small number of mainly agricultural species to predict the future consequences of ozone depletion. A few hundred species have been screened for UV tolerance. Of these, about two thirds have proved sensitive to some degree, and there is a wide variation of sensitivity between cultivars of the same species. There are large differences in the quantitative sensitivity of field-grown plants, although the qualitative responses are typically: reduced photosynthetic activity, reduced leaf and stem growth, and reduced dry weight. For some crops, there are also important quality changes – for example, the protein and oil content of soybean is reduced. As with tropospheric air pollution, it is impractical to investigate the responses of mature trees owing to their size and longevity. Plant diseases such as fungi have been amplified by enhanced UVB, because sporulation is enhanced. On the other hand, diseases such as *Septoria* (leaf blotch) have been reduced because the UVB damages spores and reduces infection. In much of the experimental work, the validity of the results is governed by the degree to which the imposed UV spectrum matches the natural spectrum, because of the extreme sensitivity of the action spectrum to small changes in wavelength.

We know very little about the responses of non-human animals to elevated UV exposures. Clearly there are possible carcinoma and eye implications. Observations were reported of increased eye defects in rabbits living in the southern parts of South America (i.e. close to the Antarctic ozone hole), but these have not been substantiated.

12.5.4   Effects on marine organisms

Although the UVB waveband represents only 0.8% of the total energy reaching the surface of the Earth, it is responsible for almost half the photochemical effects on marine organisms. UVB penetrates the water column to significant depths – 20–40 m – and may have a serious impact on aquatic life. This issue has been particularly identified with the seas around Antarctica, where the increase in UVB is greatest below the ozone hole. Furthermore, the protection afforded by low solar elevations in reducing UV flux densities has led to the evolution of UV-intolerant ecologies which are especially liable to disturbance by enhanced UV. Increased UV may have consequences for food web dynamics, starting with bacteria and algae. The UV will increase the rate at which substrates are made available photochemically, promoting growth rates in areas where carbon is in short supply. Photosynthetic organisms such as phytoplankton naturally populate the upper water layers where the flux of solar radiation, and hence of UV, is strongest. UV cannot only damage photosynthesis, but also carbon allocation, photosynthetic pigmentation, nutrient uptake and cell motility. Some organisms can limit the damage by UV avoidance – for example, by moving to a greater depth below the surface where the UV is weaker. The overall response to enhanced UV is to reduce primary productivity. The base of the food chain is then weakened, so that krill and whales are in turn affected. There are also direct effects on animals – zooplankton DNA is affected, fish eggs are vulnerable and fish can get sunburned. Results are now being published for the biological weighting functions for damage to individual species. These generally show the critical influence of wavelengths in the 280–320 nm region.

Depletion of the ozone layer has spurred interest in marine photochemistry, which is largely driven by UVB. The key factor here is the relationship of the photon energies (which decrease as wavelength increases) to the bond energies of the relevant molecules in natural systems. Bonds in the energy ranges 300–370 kJ mol^{-1} (H$-$C$=$C, for example), 370–425 kJ mol^{-1} (S$-$S, for example) and 425–600 kJ mol^{-1} (O$=$O or H$-$OH, for example) can be photolysed by UVA, UVB and UVC respectively.

Pure water and the inorganic salts that seawater contains absorb only weakly in the UV. The absorption is mainly due to dissolved organic matter (DOM), which is present in the range 1–4 mg C l^{-1} in surface seawater. This represents a major pool of reduced carbon at the surface of the Earth. Typically less than half of the DOM is common biomolecules such as amino acids and polysaccharides, the remainder being complex humic substances, which are not well characterised but are responsible for absorbing UVB photons and triggering the resulting photochemical responses. Among the immediate photoproducts, hydrogen peroxide, hydroxyl radical, carbonyl sulphide and formaldehyde have been determined. These react rapidly to generate a wide range of stable low molecular weight organic end products, which have important consequences for carbon cycling in the oceans. Some of the species, for example CO, CO_2, CS_2 and methyl iodide, can diffuse into the atmosphere and affect atmospheric chemistry. We do not know yet how these relationships are being impacted by the increase in UVB. Experimental investigations of these processes are hampered not only by the variability of the chromophoric DOM, but also by the strong wavelength dependence of the photochemistry which makes accurate replication of the natural spectrum very difficult.

FURTHER READING

de Mora, S., Demers, S. and Vernet, M. (eds) (2000) *The Effects of UV Radiation in the Marine Environment*, Cambridge University Press, Cambridge.

Hester, R. E. and Harrison, R. M. (eds) (2000) *Causes and Environmental Implications of Increased UV-B Radiation*, Royal Society of Chemistry, Cambridge.

International Ozone Association (1999) *Proceedings, Schonbein International Ozone Symposium, Basel, Switzerland*, IOA, Paris.

Krupa, S. (1998) *Elevated Ultraviolet (UV)-B Radiation and Agriculture*, Landes Bioscience & Springer, Austin, Texas.

Lumsden, P. (1997) *Plants and UV-B: Responses to Environmental Change*, Cambridge University Press, Cambridge.

Rozema, J. (1999) *Stratospheric Ozone Depletion: the Effects of Enhanced UVB Radiation on Terrestrial Ecosystems*, Backhuys, Leiden.

Russell Jones, R. and Wigley, T. (eds) (1989) *Ozone Depletion: Health and Environmental Consequences*, Wiley Chichester, UK.

United Kingdom Stratospheric Ozone Review Group (1990) *Stratospheric Ozone 1990*, DoE/HMSO, London, UK.

United Kingdom Stratospheric Ozone Review Group (1993) *Stratospheric Ozone 1993*, DoE/HMSO, London, UK.

United Kingdom Stratospheric Ozone Review Group (1997) *Stratospheric Ozone 1997*, DoE/HMSO, London, UK.

World Meteorological Organisation (1998) *WMO Global Ozone Research and Monitoring Report No 44. Scientific Assessment of Ozone Depletion*, WMO, Geneva.

Chapter 13

Standards and legislation

We have now come full circle. Having seen how air pollutants are produced and dispersed, and the effects that result from their presence in the air and deposition to receptors, we will review in this Chapter the national and international limits that have been set. Air pollution standards are emission or concentration or deposition rates that are designed to protect people and the environment from harm. If there were precise concentrations at which such harm was observed to occur, then setting the standards would be straightforward. Unfortunately life is never that simple.

Inputs to the standard-setting process come from two main sources:

- observations of the short-term or long-term impact of environmental pollutants; actual incidents that result in human morbidity or mortality, or quantifiable damage to a receptor; epidemiological studies of populations or other targets; and
- specific experiments under controlled conditions. These might involve experiments on laboratory animals that then have to be interpreted for humans, or studies on a small number of plant species that have to represent the whole gamut of possible plants.

Increasingly, the powerful computer models that have been developed to understand the relationships between the many interacting variables are also being used to determine standards. Emission limits of sulphur and nitrogen are calculated from dispersion, deposition and critical load models; emission limits of ozone precursors are calculated from dispersion and critical level models; and reductions in CFC release rates set from models of ozone destruction in the stratosphere.

The Earth's atmosphere is one of the Global Commons – those regions, which also include Antarctica, the ocean floor and outer space, over which no one nation has individual control. There is clear distinction between 'airspace', which is the 3-D volume extending from the surface of a nation up to the top of the atmosphere, and the air that flows through it. Although each nation has sole responsibility for its own airspace, it has joint responsibility for the air within it.

A variety of different philosophies has been used to control air pollution. They fall into three general categories. Historically, the philosophy in the UK has been based on emission standards, the assumption being that if we set appropriate maximum emission concentrations then ambient concentration targets can be achieved. In Europe and the USA, on the other hand, legislation has been framed in terms of air quality standards (AQS) – potential new emitters have to show that their emissions will not cause AQS to be exceeded. On a larger scale, the critical loads and levels approach (see Chapter 9) is a more recent concept that relates the maximum deposition fluxes of gaseous and liquid pollutants to their effects on local ecosystems. The concept depends on sophisticated dispersion modelling to calculate the maximum emissions that can be accepted if critical loads and levels at a given receptor site are not to be exceeded.

There is a common thread to these ideas. It is assumed that exposure to a pollutant stresses an organism, and that the organism's natural repair processes can make good any damage provided that the exposure does not rise above some threshold value. This threshold value, sometimes divided by some arbitrary 'safety' factor such as 2 or 10, can then be set as the standard. Although this idea is convenient for both scientists and legislators, it seems increasingly likely that organisms do not respond in such a straightforward fashion, but show some response, however small, at all concentrations. Whether such responses are economically and socially significant then becomes a matter of comparing the risk from that pollutant against the other risks to which the population is exposed.

Very often receptors that are superficially similar (whether they are people, animals, plants or materials) display a very wide range of responses between sensitive and resistant. Standards will usually be set to protect the most sensitive receptor. It is often not possible to determine whether short-term experimental responses result in long-term health effects. In addition, the responses depend not only on the concentration but on the period over which it occurs. A statistical indicator, such as the 98th percentile of daily means, then has to be determined and agreed. In some cases (for example with particles and SO_2), there may be synergism between pollutants so that standards for one cannot be considered without regard to the other. For carcinogenic or genotoxic materials, it is thought that just one molecule can initiate a cancer, so that there is no safe limit. Instead, the standard is expressed in terms of a risk of cancer formation per population per unit time at a given concentration.

Different organisations have independently evaluated the evidence on different pollutants and arrived at somewhat different standards. The World Health Organisation (WHO), European Union (EU), United Nations Economic Commission for Europe (UNECE) and the United States Environmental Protection Agency (USEPA) have been the most influential groups. Various terms such as standards, guidelines and limit values are used by different organisations. European Union limit values are mandatory, and must be met by the member states, while guide values only give guidance. Standards (or EU Directives) can contain both limit and guide values. World Health Organisation Guideline values have no statutory basis, but are widely used by governments to assess the impact of pollutants for which there are no national standards.

13.1   UK LEGISLATION

13.1.1   Background

In the 1980s, it became recognised that control of air pollution usually had implications for water quality and solid waste management, and that one body should regulate across the media boundaries. That body is the Environment Agency. The relevant legislation is the Pollution Prevention and Control Act (PPCA) 1999, which brought into force in the UK the EU Directive 96/91 on integrated pollution prevention and control (IPPC). Integrated pollution prevention and control applies to around 6000 larger installations, while around 20 000 smaller ones come under Local Authority control. The IPPC philosophy is to target all emissions into all environmental media, so that it embraces not only air soil and water, but waste generation and energy efficiency as well. New installations have been subject to IPPC regulation since 1999, while existing ones will have to be brought within IPPC by 2007. Each process operator must apply to the appropriate regulator (Environment Agency or Local Authority in England and Wales, Scottish Environmental Protection Agency in Scotland) for a permit to operate. If granted, the permit will include a set of emission limit values (ELVs) for the relevant pollutants; identify Best Available Technologies to ensure the necessary level of environmental protection; and specify the monitoring that must be done to confirm compliance. IPPC targets toxic pollutants rather than climate change – for example, carbon dioxide is not on the list of pollutants that should be considered, although it would be affected by controls on energy efficiency.

In the PPCA, best available techniques not entailing excessive cost (BATNEEC) must be employed. This philosophy (with the T standing for Technology rather than Techniques) was first used in EC Directive 84/360/EEC, which limited emissions from large industrial plant. The BAT part covers both the process itself – the technology – and the operating practices. Thus, it includes factors such as staff numbers, working methods and training. The NEEC part implies a level of judgement as to whether the extra cost of a higher emission standard is warranted, or indeed supportable by the financial strength of the organisation. There are around 80 Process Guidance Notes, which set out not only the BATNEEC for specific processes, but give emission limits, the timetable for bringing old plant up to the control standard required of new plant, and details of any relevant environmental quality standards or EU legislation. These Process Guidance Notes are available for a wide range of processes, from animal carcass incineration to tobacco processing, from bitumen processing to vegetable matter drying. Clearly, the words such as 'best' and 'excessive' used in BATNEEC are to some degree judgmental, and have to be interpreted with caution. Emission limits are specified as a mass concentration of the pollutant per unit volume of gas. For particles in general, the limits are designed to make the emission invisible – for low-volume emissions of coarse particles 0.46 g m^{-3} might be adequate, reducing to 0.115 g m^{-3} for large volume flow rates carrying fine material (the arbitrary-looking values are derived from old Imperial units – they were originally specified as round numbers of grains per cubic foot). In individual cases, for example when there is an unusually large volume flow rate, a lower limit such as 0.070 g m^{-3} might be set. For toxic materials, whether particles or gases, the limits are based on dispersion calculations designed to reduce the ground-level concentration to some small fraction of the occupational exposure limit for that material, or the prevailing air quality limit value, as appropriate. Emissions to atmosphere may also be considered within a wider regulatory process involving planning permission from local authorities, environmental impact assessment, environmental management systems and environmental auditing.

There remains an important division in UK law between large-scale processes that come under central (Environment Agency) control, and smaller ones that remain under the control of local authorities. For example, a major iron and steel works using 100-t electric arc furnaces would have to meet EA standards for emissions, whereas small furnaces (<7 t), of which there might be many around a manufacturing centre such as Birmingham, are regulated by the local authority.

Although industrial emissions have been controlled by national legislation since 1863, no systematic action was taken against domestic emissions until much later. Following investigation by the Beaver Committee of the pollution smog in London in December 1952, the Clean Air Act was passed in 1956. This was extended in 1968, and both Acts were consolidated into the Clean Air Act 1993. The main provisions of the Act are to control the emission of dark and black smoke from chimneys; to control the emission of smoke, grit, dust and fumes from boilers and processes not regulated by the EA; to control the heights of new industrial chimneys; and to require the use of smokeless fuel in domestic grates in urban areas. Emission of dark smoke is generally prohibited. However, it is recognised in the Act that such emissions are technically unavoidable when lighting up small boilers from cold, and three standards of emission limit must be met. For example, an aggregate of 10 min in any period of 8 h is permitted from a chimney serving a single furnace, provided that no dark smoke emission involves a continuous period of more than 4 min or an aggregate of more than 2 min of black smoke in any 30-min period. In addition to the above emission regulations, certain other forms of air pollution control fall within the Clean Air Act – the lead content of petrol, sulphur content of gas oil and fuel oil, and cable burning.

Increasingly, UK legislation is having to march in step with overarching EC Directives. The EEC had historically taken an approach to air pollution based on air quality standards rather than emission limits – EU legislation on air quality, such as 80/779/EEC on SO_2 and suspended particulates, is implemented in the UK through Air Quality Standards Regulations. The Regulations are identical in most respects to their EU counterparts, except that local provisions (called derogations) are incorporated. For example, a number of districts in the UK were allowed additional time to meet the required suspended particulate levels, provided that they submitted detailed plans in advance of how this was to be achieved.

13.1.2   The National Air Quality Strategy (NAQS)

This important national policy document was published in early 1997, adopted by the new Labour Government when it took office in May 1997, and reviewed in 1998. The outcome of these deliberations was a revised NAQS which was published in January 2000. The NAQS worked in a timeframe up to 2005, and set out both the *standards* for air pollutants and *objectives* to be achieved within a specified time, after allowing for costs and benefits as well as feasibility and practicality. These are summarised in Table 13.1. All the objectives, except those for O_3 and ecosystem protection, were given statutory force by the Air Quality Regulations 2000. The values largely reflect the conclusions of

Table 13.1 NAQS objectives

Pollutant	NAQS objective and timetable
Benzene	Running annual mean of 16.25 μg m^{-3} (5 ppb) by end 2003, and 1 ppb indicative value by end 2005
1,3-butadiene	Running annual mean of 2.25 μg m^{-3} (1 ppb) by end 2003
Carbon monoxide	Running 8-h mean of 11.6 mg m^{-3} (10 ppm) by end 2003
Lead	Annual mean of 0.5 μg m^{-3} by end 2004, and 0.25 μg m^{-3} by end 2008
NO$_2$	1-h mean of 200 μg m^{-3} (105 ppb) by end 2005 (maximum of 18 exceedances per year)
	Annual mean of 40 μg m^{-3} (21 ppb) by end 2005
	Annual national mean of 30 μg m^{-3} (16 ppb) by end 2000 (vegetation and ecosystem protection)
O$_3$	Running 8-h mean of 100 μg m^{-3} (50 ppb) by end 2005. Expressed as 97th percentile of daily maximum 8-h running mean
Particles (PM$_{10}$)	Annual mean 40 μg m^{-3}, running 24-h mean 50 μg m^{-3} (maximum 28 exceedances per year). It is expected that this will be tightened to maximum 7 exceedances per year, to be achieved by 2010.
Sulphur dioxide	1-h mean of 350 μg m^{-3} (132 ppb) by end 2004 (maximum 24 exceedances per year)
	24-h mean of 125 μg m^{-3} (47 ppb) by end 2004 (maximum 3 exceedances per year)
	15-min mean of 266 μg m^{-3} (100 ppb) by end 2005 (maximum 35 exceedances per year)
	Annual mean and winter average of 20 μg m^{-3} (8 ppb) by end 2000 for vegetation and ecosystem protection

detailed previous reviews by the expert panel on air quality standards (EPAQS).

But the NAQS goes much further than that, because it lays out the ways in which NAQS is influenced by, and interacts with, other legislation and programmes. These include public transport provision, reduction of traffic congestion, bicycle tracks, road user charging and workplace parking levies. Some Local Authorities are implementing Low Emission Zones to limit access by bad emitters. Such a control system might be based on the Euro standards (Stages 1–4) achieved by vehicles. Some minor tax incentives have also been brought in to favour smaller engine sizes. The review document predicts the exceedances in 2005 for each of the targeted pollutants, and estimates the costs and benefits of reducing or eliminating such exceedances. The cost of meeting the NO_2 and PM_{10} objectives, assuming that this could be achieved by reducing vehicle emissions alone, came out at around £100 M per annum. For ozone, the situation is much more complex because much UK O_3 is transboundary and can only be reduced by Europe-wide control measures. The UK share of the cost of bringing O_3 concentrations down to the Strategy objectives could be as high as £10 billion per annum. The benefits analysis, which was largely based on a 1998 COMEAP report, showed that rather small additional benefits would acrue compared to those already expected from existing or planned reductions. For example, the additional reduction in deaths brought forward due to PM_{10} was predicted to be 20–30, compared to nearly 5 000 from measures already in place.

13.1.2.1   Additional measures

The NAQS also considers what additional measures might be needed in the longer term to bring the concentration targets below the objectives and meet the standards. The problem pollutants were identified as:

- NO_2, which is unlikely to meet the objectives in dense urban areas or at busy roadsides. The main mechanism for further improvement was thought to be local traffic management.
- O_3, which is certainly the most intractable pollutant. The NAQS objectives are currently being exceeded across most of the UK, and computer models of O_3 formation indicate that large reductions in precursor emissions, of the order of 50% or more across the whole of Europe, will be needed to meet the standards.

In 1997, the NAQS estimated that reductions in vehicle particle emissions of around 65% from 1995 levels would be needed to achieve the objective in urban areas. However, the increased stringency of the objective means that even complete elimination of particle emissions from vehicles in urban areas would be unlikely to achieve the objective.

13.1.3   Local Air Quality Management

Part IV of the Environment Act 1995 also sets up the legal framework for Local Air Quality Management (LAQM). Local Authorities (LAs) must conduct a Review and Assessment of air quality in their area, in the light of the objectives set in the NAQS. The DETR has provided extensive guidance to help LAs carry out their responsibilities. The necessary first stage of such a review is an initial screening of all pollution sources (mainly road transport and industrial) that could have a significant impact within the LA's area, and a collation of all relevant existing data. It was expected that some LAs would be able to complete this first stage simply by accessing the national database. For other LAs, additional basic monitoring and/or modelling would be needed. Second and third stages may be required if the local air quality is falling short of the objectives. Second and third stage reviews involve increasingly intensive consideration of locations that are at risk of not meeting the NAQS objectives for any of the prescribed pollutants, together with more sophisticated modelling and monitoring. If the third stage review shows that prescribed air quality objectives are not likely to be met by the appropriate dates, the Authority must designate an air quality management area (AQMA), and work out an action plan which will achieve the air quality objectives. Such a plan would take into consideration the relevant local circumstances (meteorology, topography, densities of people and vehicles, specialised sources etc.) in a cost-effective manner. There are many elements that the action plan might include, such as: low-emission zones in the city centre, roadside testing of vehicle emissions, increased park and ride provision, increased city-centre parking charges, workplace parking levies, traffic management improvements, increased bus lanes. By itself, each of these changes would only have a small impact on air pollution, but collectively they will not only reduce concentrations but, perhaps more significantly, change the public attitude towards pollutant emissions.

13.1.3.1   Rushcliffe review and assessment

We will cite Rushcliffe Borough Council (RBC) in the south-east of Nottinghamshire as an example of the LAQM process in action. Rushcliffe covers 157 square miles and has a population of around 105 000. The main town, West Bridgford, is adjacent to the city of Nottingham but separated from it by the River Trent. A comprehensive emissions inventory was compiled for Nottinghamshire by the combined Nottinghamshire Boroughs, and the results of this were used by RBC. The following categories of sources were included

- Processes authorised under Parts A or B of the Environment Protection Act 1990
- Combustion processes related to Part B processes
- Boiler plant >0.4 MW
- Petrol stations
- Railway and road traffic.

There are four Part A (large source) processes in the Borough: a 2000 MW electricity generating station, a cement plant, an animal carcass incinerator and a crude oil processor. There are 33 Part B (smaller source) processes such as waste oil burning at garages, petrol unloading at filling stations, plasterboard manufacture and vehicle paint spraying. Rushcliffe Borough Council published a Draft Consultation Report for Stages One and Two of their Review and Assessment in March 2000. Together with other LAs in Nottinghamshire, RBC had engaged an environmental consultancy to model current and future concentrations of seven of the eight key air pollutants (ozone not included). Where necessary, additional monitoring was undertaken to verify the predictions for PM_{10} and NO_2.

- For benzene, 1,3 butadiene, carbon monoxide and lead, it was not necessary to proceed beyond the first stage review because the concentrations were already well below the objectives set out in the NAQS.
- For NO_2, the modelling identified existing areas of exceedance along some of the more heavily trafficked roads within the Borough (typically having an annual average daily traffic flow of 35 000–50 000 vehicles per day). However, the study also predicted that these exceedances would cease before the target date of end 2005 because of the continuing increase in the use of catalytic converters, and that third stage review was unnecessary.
- For PM_{10}, modelled concentrations exceeded the objective along some major roads in West Bridgford. Although these exceedances were predicted to cease before end 2004, the predicted concentrations were still quite close to the objective. Hence, it would be necessary to proceed to Stage 3. For Stage 3, short-term monitoring of PM_{10} concentrations was carried out which confirmed the calculated concentrations. The projected modelled concentrations for 2004 would not exceed the national objective, and it was decided unnecessary to declare an AQMA in West Bridgford. The national objective is that a daily average concentration of 50 $\mu g\ m^{-3}$ should not be exceeded more than 35 times per year. The urban centre of Nottingham came quite close to this with 31 exceedances in 1997. The measurements made by RBC exceeded 50 $\mu g\ m^{-3}$ on three occasions in a 3-month period.
- For SO_2, Stage 1 indicated that one specific oil-fired boiler might cause local exceedance. This was remodelled with new data and it was decided that Stage 3 was not necessary.

An important aspect of the Environment Act 1995 is the need for local consultation. Rushcliffe Borough Council met this need in two ways. First, it is a member of the Nottinghamshire Air Quality Steering Group, which was established to promote the coordination of consultations between the LAs in Nottinghamshire and other organisations involved in air quality management, such as health authorities, universities and environmental, community and business groups. Second, a consultation exercise was carried out across the county through the Nottinghamshire Pollution Working Group. Local businesses, MPs and representatives of the local community were all consulted.

13.2   EU AIR QUALITY LEGISLATION

The EU has consistently put the improvement of air quality high on its priorities. Legislation has been framed in terms of two categories: limit values, and guide values. Limit values are concentrations that must not be exceeded in the territory of the Member States during specified periods, based on WHO guidelines on human health effects. Guide values are set as long-term precautions for health and the environment.

Table 13.2 EU Air Quality Directive concentrations

Pollutant	Target or limit value	Timetable
O_3	Daily mean of 120 $\mu g\ m^{-3}$ not to be exceeded on more than 25 days per year. Population to be warned if 1-h mean exceeds 240 $\mu g\ m^{-3}$. Countermeasures to be taken if this level is exceeded for more than 3 h.	1 Jan 2010
SO_2	350 $\mu g\ m^{-3}$ hourly limit value for health protection. No more than 24 exceedances per calendar year.	To be met by 1 Jan 2005
	125 $\mu g\ m^{-3}$ daily limit value for health protection. No more than 3 exceedances per calendar year.	1 Jan 2005
	20 $\mu g\ m^{-3}$ for ecosystem protection, as an annual and winter average.	2 years after Directive enters force
NO_2	200 $\mu g\ m^{-3}$ hourly limit value for health protection; no more than 8 exceedances per calendar year.	1 Jan 2010
	40 $\mu g\ m^{-3}$ annual limit value for health protection.	1 Jan 2010
NO_x	30 $\mu g\ m^{-3}$ annual limit value for ecosystem protection.	2 years after Directive enters force
PM_{10}	*Stage 1* 50 $\mu g\ m^{-3}$ as a 24-h average for health protection; no more than 25 exceedances per calendar year.	1 Jan 2005
	30 $\mu g\ m^{-3}$ limit value as an annual mean for health protection.	1 Jan 2005
	Stage 2 50 $\mu g\ m^{-3}$ as a 24-h average for health protection; no more than 7 exceedances per calendar year.	1 Jan 2010
	20 $\mu g\ m^{-3}$ limit value as an annual mean for health protection.	1 Jan 2010
Lead	0.5 $\mu g\ m^{-3}$ as an annual mean for health protection.	1 Jan 2005
Benzene	10 $\mu g\ m^{-3}$ as annual mean	1 Jan 2005
	5 $\mu g\ m^{-3}$ as an annual mean	1 Jan 2010

The central piece of legislation is the Air Quality Framework Directive (96/62/EC) on air quality assessment and management. Twelve air pollutants are identified for which limit values will be required, and provision is made for a sequence of Daughter Directives which address these specifically. The 12 air pollutants are those that have been examined by EPAQS (except for 1,3-butadiene), plus cadmium, arsenic, nickel and mercury. The first Daughter Directive, on SO_2, NO_2, particulate matter and lead, was adopted in April 1999 and entered into force in July 1999 (99/30/EC). Member States must implement the Directive within 2 years. The second Daughter Directive (2000/69/EC) covered CO (to be met by 2005) and benzene (2010). The third Daughter Directive will require ozone standards corresponding to those set by the WHO to be met by 2010. Proposals are in hand for the remaining pollutants. The Framework Directive envisages two types of air quality objective. The usual one will be the limit value, defined as

> a level fixed on the basis of scientific knowledge with the aim of avoiding, preventing or reducing harmful effects on human health and/or the environment as a whole, to be attained within a given period and not to be exceeded once attained

For ozone, in recognition of the transboundary regional nature of this gas, target values can be set instead of limit values. Target values must be met by States if possible. If they cannot be met, the EC will evaluate whether additional measures are necessary at EU level (Table 13.2).

13.3   UNECE

The United Nations Economic Commission for Europe is an organisation of 34 countries that includes the USA , the EU and many non-EU countries in Europe. United Nations Economic Commission for Europe has been at the forefront in identifying environmental issues and working out measures to ameliorate them. The measures are formulated in terms of Protocols and Conventions. Although these are not legally binding, countries that ratify them are under a strong moral obligation to achieve the identified pollution control targets.

- The first such agreement was the Convention on Long Range Transboundary Air Pollution (CLRTAP), agreed at Geneva in 1979, which committed countries to take a BATNEEC approach to curb the long range transport and deposition of sulphur and nitrogen.
- The 1984 European Monitoring and Evaluation Protocol set up the machinery and funding for a network of measurement sites across Europe to provide data on all substances that are the subjects of UNECE agreements.
- The 1985 Helsinki Sulphur Protocol requires ratifying countries to reduce national S emissions or their transboundary fluxes by 30% from a 1980 base

by 1993. This set up the '30% club', which was not joined by either the EU as a whole or by the UK in particular. The UK objected to the choice of 1980 as the base date, since an industrial recession had caused its own sulphur emissions to be particularly low that year. The Protocol was redrawn in terms of critical loads in 1994. The Second Sulphur Protocol came into force in August 1998. The UK agreed to target a reduction of at least 80% in its national annual SO_2 emissions, compared to 1980.

- The 1987 Montreal Protocol on substances which deplete the ozone layer has been described elsewhere.
- The 1988 Sofia Nitrogen Oxides Protocol required ratifying countries to restore their NO_x emissions or transboundary fluxes to a 1987 base by 1994. The target was met comfortably by the EU15.
- The 1991 Geneva VOC Protocol requires a 30% reduction in VOC emissions from a 1988 base by 1999, and subsequently to adopt a further range of measures to reduce VOC emissions.
- The Aarhus Heavy Metal and POP Protocol (1998). Emissions of heavy metals – lead, cadmium and mercury – to be reduced below 1990 levels. The main sources of these materials are leaded petrol, battery production and disposal, and the chlor-alkali industry. It was also required to ban certain POP emissions (mainly pesticides) and to reduce emissions and/or use of others such as dioxins, furans and PAH. This would be achieved mainly by upgrading municipal waste incinerators.
- The December 1999 Gothenburg Protocol requires the following emission reductions for the EU as a whole, and for Britain in particular, by 2010, compared to 1990 levels (Table 13.3).

These successive Protocols have had a profound effect on EU sulphur emissions since the 1970s, as seen in Figure 13.1. So far as the UK is concerned, achieving the Protocol will reduce sulphur emission in 2010 to a value below that which prevailed in 1860.

Correspondingly, the critical load exceedances in Europe have fallen by 76% for S since 1985 and by 65% for acidifying N. The S exceedances will fall to a negligible area when the targets for the Gothenburg Protocol are achieved. However, eutrophying N critical loads will fall more slowly and substantial exceedances will

Table 13.3 Annual emission reductions required by the Gothenburg Protocol

Pollutant	UK 1990/kt	UK 2010/kt	UK change, %	EU 1990/ kt	EU 2010/kt	EU change %
SO_2	2500	625	75	38040	13990	63
NO_x	2315	1181	49	23330	13846	41
NH_3	349	297	15	7653	6280	18
VOCs	2791	1200	57	23964	13590	43

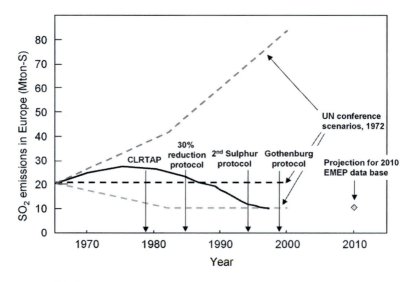

Figure 13.1 The effect of CLRTAP on European sulphur emissions.
Source: National Expert Group on Transboundary Air Pollution (2001)
*Transboundary Air Pollution Acidification, Eutrophication and Ground-level
Ozone in the UK*, Centre for Ecology and Hydrology, Edinburgh. After
P. Grennfelt.

still be observed. It has been estimated that once the Protocol is implemented, the
area in Europe with excessive levels of acidification will shrink from 93 million
hectares in 1990 to 15 million hectares and that with excessive levels of eutrophication will fall from 165 million hectares in 1990 to 108 million hectares. The number of days with excessive ozone levels will be halved. Consequently, it is estimated
that life-years lost as a result of the chronic effects of ozone exposure will be about
2 300 000 lower in 2010 than in 1990, and there will be approximately 47 500
fewer premature deaths resulting from ozone and particulate matter in the air. The
exposure of vegetation to excessive ozone levels will be 44% down on 1990.

13.4 WORLD HEALTH ORGANISATION

The WHO has played a key role in assessing the responses of people to pollutant
concentrations. In 1987, it published a summary of air quality guidelines for 28
substances, based on human health criteria. For 18 toxic chemicals, WHO gave a
concentration below which there was believed to be no risk to human health. For
7 of the 9 carcinogenic chemicals, for which there is no safe low concentration, an
estimate was given of the risk of cancer for a lifetime exposure to a concentration
of 1 μg m^{-3}. These guidelines were revised in 1999 (see Tables 13.4–13.6).

Hence, if the PM_{10} concentration experienced by one group is $10 \mu g \ m^{-3}$ higher than that experienced by another, then, other things being equal, we would expect a 0.8% increase in hospital admissions and a 0.74% increase in mortality.

13.4.1   Exceedances

Having established monitoring programmes and legislated concentration guidelines, the natural next step is to work out whether the guidelines are being exceeded and if so, how badly. For example, the EU ozone threshold for the protection of human health ($110 \mu g \ m^{-3}$ as an 8-h mean) is being exceeded for about 330 million people (most of the population of the EU15) at least once per year; and the threshold value for the protection of vegetation is being exceeded 'widely and frequently'.

Table 13.4 WHO guideline values for toxic chemicals (partial list only)

Substance	Time-weighted average concentration	Averaging time
Cadmium	$1–5 \ ng \ m^{-3}$	1 year (rural areas)
	$5 \ ng \ m^{-3}$	1 year (urban areas)
Carbon disulphide	$100 \ \mu g \ m^{-3}$	24 h
Carbon monoxide	$100 \ mg \ m^{-3}$	15 min
	$60 \ mg \ m^{-3}$	30 min
	$30 \ mg \ m^{-3}$	1 h
	$10 \ mg \ m^{-3}$	8 h
1,2 Dichloroethane	$0.7 \ mg \ m^{-3}$	24 h
Dichloromethane	$3 \ mg \ m^{-3}$	24 h
Diesel exhaust	$2.3–5.6 \ \mu g \ m^{-3}$	1 year
Fluorides	$1 \ \mu g \ m^{-3}$	1 year
Formaldehyde	$100 \ \mu g \ m^{-3}$	30 min
Hydrogen sulphide	$150 \ \mu g \ m^{-3}$	24 h
	$7 \ \mu g \ m^{-3}$	30 min
Lead	$0.5 \ \mu g \ m^{-3}$	1 year
Manganese	$0.15 \ \mu g \ m^{-3}$	1 year
Mercury	$1 \ \mu g \ m^{-3}$	1 year
Nitrogen dioxide	$200 \ \mu g \ m^{-3}$	1 h
	$40 \ \mu g \ m^{-3}$	24 h
Ozone	$150–200 \ \mu g \ m^{-3}$	1 h
	$120 \ \mu g \ m^{-3}$	8 h
Styrene	$260 \ \mu g \ m^{-3}$	1 week
Sulphur dioxide	$500 \ \mu g \ m^{-3}$	10 min
	$125 \ \mu g \ m^{-3}$	24 h
	$50 \ \mu g \ m^{-3}$	1 year
Tetrachloroethylene	$0.25 \ mg \ m^{-3}$	24 h
Toluene	$0.26 \ mg \ m^{-3}$	1 week
Vanadium	$1 \ \mu g \ m^{-3}$	24 h

Table 13.5 WHO guideline values for carcinogenic chemicals

Substance	Unit risk* $[\mu g\ m^{-3}]^{-1}$/site of tumour
Acrylonitrile	2×10^{-5}/lung
Arsenic	1.5×10^{-3}/lung
Benzene	6×10^{-6}/blood (leukaemia)
Benzo[a]pyrene	8.7×10^{-2}/lung
Chromium (VI)	4×10^{-2}/lung
Diesel exhaust	4×10^{-5}/lung
Nickel	3.8×10^{-4}/lung
PAH (expressed as BaP)	8.7×10^{-2}/lung
Trichloroethylene	4.3×10^{-7}
Vinyl chloride	1×10^{-6}/liver and other sites
Asbestos (concentration of 500 fibres m^{-3},	10^{-6}–10^{-5} (lung cancer)
rather than $1\ \mu g\ m^{-3}$)	10^{-5}–10^{-4} (mesothelioma)

Note

* Unit risk is defined as 'the additional lifetime cancer risk occurring in a hypothetical population in which all individuals are exposed continuously from birth throughout their lifetimes to a concentration of $1\ \mu g\ m^{-3}$ of the agent in the air they breathe'.

Table 13.6 WHO guideline values for PM

Health outcome	Relative risk*
Use of bronchodilator	1.0305
Cough	1.0356
Lower respiratory symptoms	1.0324
Change in peak expiratory flow rate	− 0.13%
Respiratory hospital admissions	1.0080
Death	1.0074 (PM$_{10}$)
	1.015 (PM$_{2.5}$)

Note

* Relative risk associated with a $10\ \mu g\ m^{-3}$ increase in the 24-h average concentration of PM$_{10}$ (or PM$_{2.5}$).

The WHO ozone guideline for human health is $120\ \mu g\ m^{-3}$ maximum 8-h concentration. In April–September 1995, for example, this was exceeded on 40–50 days in Central Europe and northern Italy, decreasing to a few days only in northern UK and Scandinavia. A similar conclusion applies as with the EU guideline, that most of the population of the EU15 is exposed to concentrations in excess of the threshold for some of the time. A complication is caused by the fact that health effects are not related to concentration in a black and white fashion, and that a small part of the population may experience symptoms at concentrations around the guideline value. For example, 8-h exposure to an O_3

concentration of 120 $\mu g \, m^{-3}$ is expected to reduce the FEV_1 of 10% of the population (the most active, healthy and sensitive 10%) by 5%. This level was exceeded for around 10% of person days in the EU15 in March–October 1995. 160 $\mu g \, m^{-3}$, which was exceeded for 1.6% of person-days, is expected to reduce FEV_1 by 10%. In all, it has been calculated that ozone exposure would have resulted in 3000 hospitalisation cases in the EU15 in 1995.

Overall, it is evident that air quality legislation is by no means fixed. There is a dynamic, iterative series of international comparisons under way in which new studies of pollutant effects are being incorporated year by year. Two phases have been seen. In the first phase, during the 1970s and 1980s, a higher set of limits and guidelines was identified which would reduce acute effects during short-term episodes. Subsequently, epidemiology has been used more effectively to specify a lower set of values which will protect populations from chronic effects due to long-term low-level exposures. One consequence of this sequence is that emission reductions are always chasing ever-lower ambient targets. The major part of the work has probably now been done, for it is becoming clear that future fine tuning will yield diminishing returns in reduced morbidity and mortality.

13.5   EU INDUSTRIAL EMISSION LEGISLATION

Acid deposition has been a major environmental concern in the EU since the 1970s, and legislation has been passed to reduce deposition levels. The Large Combustion Plant Directive (LCPD, 88/609/EEC) is designed to reduce emissions of sulphur and nitrogen oxides from industrial plant (>50 MW thermal, mainly power stations) burning fossil fuel. The UK committed to reduce SO_2 emissions from existing plant by 20, 40 and 60% of the 1980 value by the years 1993, 1998 and 2003 respectively. NO_x emissions had also to be reduced by 15 and 30% of the 1980 value by 1993 and 1998. These reductions are effectively national totals – it is up to individual countries how they meet the targets overall. A National Emissions Plan was drawn up for the UK that shared out the need for reductions between PowerGen and National Power – the major electricity generators at that time. The National Emissions Plan specifies a maximum sulphur emission per year for each major source, negotiated with the source operator. In the case of a coal-fired power station, the limit will correspond to a certain combination of sulphur concentration and load factor (the proportion of the time for which the station is used). If the operator wants to increase the load factor, then a corresponding reduction in sulphur emissions (probably by the use of lower-sulphur imported coal) will be needed. Currently, amendments are proposed to the LCPD to further reduce emissions, saving an estimated 100 kt SO_2, 4000 kt NO_2 and 100 kt particles by 2010.

13.6   EU VEHICLE EMISSIONS

13.6.1   Emission testing cycles

Tailpipe emission standards specify the maximum amount of pollutants allowed in exhaust gases discharged from the engine. Regulated emissions include:

- Diesel particulate matter (PM), measured by gravimetric methods. Sometimes diesel smoke opacity measured by optical methods is also regulated.
- Nitrogen oxides (NO_x), composed of nitric oxide (NO) and nitrogen dioxide (NO_2). Other oxides of nitrogen which may be present in exhaust gases, such as N_2O, are not regulated.
- Hydrocarbons (HC), regulated either as total hydrocarbon emissions (THC) or as non-methane hydrocarbons (NMHC). One combined limit for HC + NO_x is sometimes used instead of two separate limits.
- Carbon monoxide (CO).

The permitted pollutant emissions must be specified in a standard way, and new models must be tested in the same standard way to find out whether they meet the specification. Emissions are therefore measured over an engine or vehicle test cycle which is an important part of every emission standard. Regulatory test procedures are necessary to verify and ensure compliance with the various standards. These test cycles are intended to create repeatable emission measurement conditions and, at the same time, simulate a real driving condition of a given application.

Since it is clearly impractical to instrument all new models and drive the test cycle on real roads, emission test cycles are used which are a sequence of speed and load conditions performed on an engine or chassis dynamometer. The former will usually be more appropriate if the engine is a general-purpose power source that might be used in a wide range of applications (not necessarily vehicles). Emissions measured on vehicle (chassis) dynamometers (instrumented rolling roads) are usually expressed in grams of pollutant per unit of travelled distance, e.g. g km^{-1} or g mi^{-1}. Emissions measured according to an engine dynamometer test cycle are expressed in grams of pollutant per unit of mechanical energy delivered by the engine, typically g$(kWh)^{-1}$ or g$(bhph)^{-1}$. Depending on the character of speed and load changes, cycles can be divided into steady-state cycles and transient cycles. Steady-state cycles are sequences of constant engine speed and load modes. Emissions are analysed for each test mode. Then, the overall emission result is calculated as a (time-weighted) average from all test modes. In a transient cycle, the vehicle (or engine) follows a prescribed driving pattern which includes accelerations, decelerations and changes of speed and load. The final test results can be obtained either by analysis of exhaust gas samples collected over the duration of the cycle or by electronic integration measurements from fast response, continuous emission monitors. Regulatory authorities in different countries have not been consistent in adopting emission test procedures and at least 25 types of cycle are in use. Since exhaust emissions depend on the engine speed and load conditions, specific engine emissions which have been measured on different test cycles may not be comparable.

13.6.1.1   The ECE15 + EUDC test cycle

The ECE + EUDC cycle, also known as the MVEG-A cycle, is an urban driving cycle for light-duty vehicles (such as cars and small vans), devised to represent city driving conditions and performed on a chassis dynamometer. It is characterised by low vehicle speed, low engine load, and low exhaust gas temperature. The vehicle is taken through four cycles of steady acceleration, cruise at constant speed, and steady deceleration (see top graph on Figure 13.2). Emissions are sampled during the cycle according to the 'Constant Volume Sampling' technique, analysed, and expressed in g km^{-1} for each of the pollutants.

The above urban driving cycle represents a Type I test, as defined by the original ECE 15 emissions procedure. The Type II test is a warmed-up-idle tailpipe CO test conducted immediately after the fourth cycle of the Type I test. The Type III test is a two-mode (idle and 50 km h^{-1}) chassis dynamometer procedure for crankcase emission determination. The EUDC (extra urban driving cycle) segment has been added after the fourth ECE cycle to account for more aggressive, high speed driving modes, and has a maximum speed of 120 km h^{-1}. There has been a lot of debate about the exact details of the driving cycles, because these affect emissions and therefore the costs for the vehicle manufacturer in meeting the targets. In fact, the legislated driving cycles are quite simplified compared to the real thing, as can be seen in the lower graph on Figure 13.2. This is a problem because the more transient the driving cycle, the worse the engine combustion control and the worse the catalyst efficiency, as we saw in Chapter 3. Regulatory driving cycles remain generally less transient than real ones.

13.6.2   EU emission legislation

EU legislation against vehicular air pollution divides vehicles into cars (<2.5 t), light duty (<3.5 t) and heavy-duty vehicles. EU standards for petrol-engined vehicles are divided by engine capacity categories into small (less than 1.4 l), medium (1.4–2 l), and large (over 2 l). The first Directive on vehicles was approved in 1970, being an adaptation of UNECE Regulation No 15. Since then there have been seven major Directives. The most significant early Directives were 83/351/EEC (maximum values of CO, HC and NO_x), 88/76/EEC (reduced maximum values for the two larger engine capacities) and 89/458/EEC (reduced maximum values for the smallest engine capacity). These Directives were soon superseded by the Consolidated Directive 91/441/EEC, which implemented tighter controls for all passenger cars (regardless of engine capacity, but excluding those having direct injection diesel engines) weighing less than 2500 kg. Since then there have been

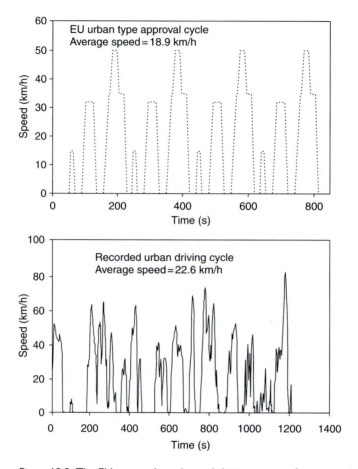

Figure 13.2 The EU test cycle and a real driving pattern for comparison.
Source: European Commission (1999) *MEET – Methodology for Calculating Transport Emissions and Energy Consumption.* EC, Brussels.

four successively tighter waves of EU legislation, known as Euro I, II, III and IV respectively (Table 13.7).

From 1 January 1993, all new petrol cars were required to meet emission standards that imply the use of three-way catalysts. The standards specified that emissions be less than 2.72 g CO, 0.97 g (HC + NO$_x$) and 0.14 g particles km^{-1} when driven over a 7 km test track that included conditions equivalent to urban and motorway driving. The cars also had to be able to pass the test after having been driven 30 000 km. The tighter limits set by Euro II are given in Table 13.8.

Table 13.7 Timetable for Euro I–IV

Standard	Directive	Vehicles affected	Date of Introduction
Euro I	91/444/EEC	Passenger cars	31 Dec 1992
	93/59/EEC	LGVs	1 Oct 1994
	91/542/EEC	HGVs	1 Oct 1993
Euro II	94/12/EC	Passenger cars	1 Jan 1997
	96/69/EC	LGVs	1 Oct 1997
	91/542/EEC	HGVs	1 Oct 1996
Euro III	98/69/EC	Passenger cars and LGVs	1 Jan 2001
	common position	HGVs	1 Jan 2001
Euro IV	98/69/EC	Passenger cars and LGVs	1 Jan 2006
	common position	HGVs	1 Jan 2006

Table 13.8 EU car exhaust emission limits for 1997

Engine Type	Emission limit/g km^{-1}		
	CO	HC + NO$_x$	Particles
Petrol	2.2	0.5	
Diesel			
Indirect injection	1.0	0.7	0.08
Direct injection	1.0	0.9	0.10

An integrated approach to air pollution problems has been followed in the European Commission's Auto-Oil Programmes. Auto-Oil I was completed in 1996 (COM(96)248), while Auto-Oil II is ongoing. The programmes have calculated the emission reductions which would be needed from road transport in order to comply with the Commission's air quality objectives by 2010, and identified cost-effective packages of measures for introduction in 2000 that would achieve those targets. The Directives introduced as a result of Auto-Oil I will result in reductions in emissions per vehicle of around 70% as compared to prevailing standards. The stringent new emission standards of Euro III (Table 13.9) will be applied to cars and light vans from 1 January 2001. Auto-Oil II has shown that measures currently in place will reduce vehicle emissions of toxic pollutants and ozone precursors to below around 15% of their 1995 values by 2020.

These measures will be further tightened by Euro IV standards from 1 January 2006. There will be a parallel tightening of fuel quality specifications which will apply to both petrol and diesel. The sale of leaded petrol has been banned since 1 January 2000. This last change has been accompanied by the provision at filling stations of lead replacement petrol (LRP), in which sodium and other additives are used to maintain the lubricant properties of lead.

Table 13.9 EURO III Limits (Directive 98/69/EC) for cars/g km⁻¹

Year	CO		HC		NO$_x$		HC + NO$_x$		PM	
	Petrol	*Diesel*	*Petrol*	*Diesel*	*Petrol*	*Diesel*	*Petrol*	*Diesel*	*Petrol*	*Diesel*
2000	2.3	0.64	0.20	–	0.15	0.50	–	0.56	–	0.05
2005	1.0	0.50	0.10	–	0.08	0.25	–	0.30	–	0.025

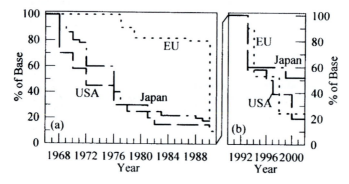

Figure 13.3 Relative vehicle emission changes due to European legislation.
Source: Eastwood, P. (2000) *Critical Topics in Exhaust Gas Aftertreatment*,
Research Studies Press Ltd, Baldock, UK.

The long-term consequence for relative emissions of the measures enforced in these various European programmes is shown in Figure 13.3. For CO, HC and NO$_x$ the baseline has been taken as a pre-Euro I petrol car. For PM, it has been taken as a pre-Euro I diesel car. These translate into quite spectacular absolute emission reductions, as seen in Figure 13.4. The reductions are most marked for petrol-powered passenger cars and vans, but significant improvements to diesel vehicle emissions have followed as well. Any computer model that is designed to predict vehicle emissions has to take all this complexity into account – for example, there have been seven levels of emission control for petrol cars since 1970, and cars built to meet all those different emission standards will be on the road today.

The particle limits for diesel-engined cars will replace current limits under 88/436/EEC, under which emissions are limited to 1.1 g per 7 km test cycle. Additional EU Directives are designed to reduce VOC emissions from the storage and distribution of petrol by 90% over a 10-year period (Com(92) 277); to ensure that catalytic converters meet a CO emission standard of 0.5% when idling and 0.3% at 2 000 rpm (92/55/EEC); to ensure that non-catalysed exhausts meet a CO emission standard of 3.5%; and to restrict the benzene content of petrol to 5% (85/210/EEC).

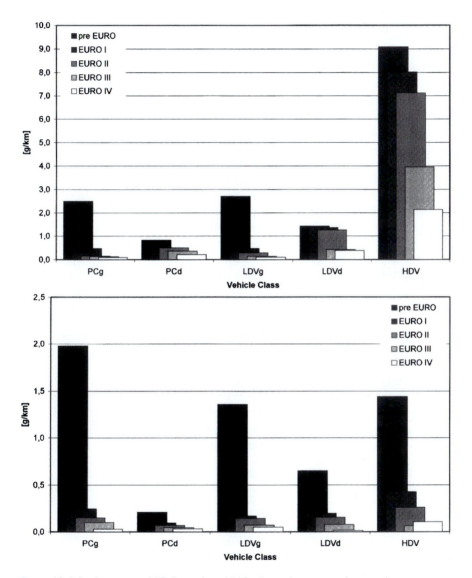

Figure 13.4 Reductions in HC (lower) and NO$_x$ (upper) emission factors due to successive European legislation.

Source: Reiss, S., Simpson, D., Friedrich, R., Jonson, J. E., Unger, S. and Obermeier, A. (2000) 'Road traffic emissions – predictions of future contributions to regional ozone levels in Europe', *Atmospheric Environment* **34**: 4701–4710.

13.7   US LEGISLATION

13.7.1   Background

The organisation responsible for air quality in the USA is the environmental protection agency (EPA). The Clean Air Act (CAA) was passed in America in 1963, and was air quality-driven, rather than emissions-driven. Procedures were set up for setting standards for the main air pollutants. The CAA was first amended in 1970, and the new standards promulgated in 1971. National Ambient Air Quality Standards (NAAQS) were established, plus control measures designed to achieve them. There were two types of NAAQS: (1) primary (designed to protect against adverse health effects in sensitive population groups, with a safety margin); and (2) secondary (the same or lower, designed to protect against welfare effects such as vegetation damage and visibility reduction). The original intention was to meet primary standards everywhere in the US by 1977, and secondary standards 'within a reasonable time' (Table 13.10).

These standards for the most widely-prevailing and ubiquitous pollutants are known as criteria, and hence the pollutants are referred to as criteria pollutants. The standards are set centrally (by the Federal Government), but the individual American States are responsible for achieving them. Each State must develop a State Implementation Plan (SIP) describing the control strategies that the State will use to bring non-attainment areas into compliance with NAAQS for the criteria pollutants. The EPA in turn approves these plans. The CAA is adaptable, in the sense that health justifications for all NAAQS had to be re-evaluated every

Table 13.10 US ambient air quality standards

Pollutant	Concentration limit/μg m^{-3} (ppb)	Averaging Time
Carbon monoxide	10 000 (9 000)	8 h[a]
	40 000 (35 000)	1 h[a]
Ozone	235 (120)	1 h[a]
	157 (80)	8 h
Non-methane HC	160 (240)	3 h[a]
Nitrogen oxides	100 (53)	1 year
Sulphur oxides	80 (30) – primary	1 year
	365 (140) – primary	1 day[a]
	1300 (500) – secondary	3 h[a]
Particulate matter <10 μm	50 – primary	1 year[c]
	150 – primary	1 day[c]
Particulate matter <2.5 μm	15	1 year
	65	1 day
Lead	1.5 μg m^{-3}	3 months

Notes
Primary and secondary standards are the same unless shown otherwise.
a Not to be exceeded more than once per year.
b Geometric mean.
c Not to be exceeded more than three times in 3 years.

5 years, and modified to reflect new scientific understanding of effects. There is a complex process of updating whereby new scientific findings on health effects are used to generate an air quality criteria document (AQCD), which leads to proposed decisions and then final decisions. At all stages there are opportunities for review and comment by non-EPA scientists, interested parties and the public at large, so that the process of revision can take some years. For example, the particles standard was originally defined (in 1971) in terms of TSP; when it became clear that health effects were largely due to the PM_{10} fraction, the standard was redefined in 1987 in those terms. Subsequently, $PM_{2.5}$ has come to the fore. A periodic review of the health effects of PM led to an AQCD in 1996. In 1997, new NAAQSs for PM were promulgated which incorporated the standards for $PM_{2.5}$ shown above. In the course of the consultation process on this one standard, over 50 000 written and oral comments had been received on the proposed revisions. The same revision process was applied to O_3 over the same period, resulting in the addition of the maximum 8-h average. It has been estimated that 113 million Americans were living in areas that failed to meet the new O_3 standard. This process of rolling re-evaluation sometimes results in changes, but not always. The original standards for SO_2 and NO_2 have remained unchanged, despite detailed re-evaluations.

Despite the flexibility and continuous updating of this approach, air pollution has proved to be a more intractable problem than expected, and the control measures have failed to achieve the intended air quality standards. A major revision and expansion of the CAA – the Clean Air Act Amendments (CAAA) was passed in 1990. It was nearly 800 pages long, in place of the original Act's 50 pages. The CAAA includes provision for operating permits for stationary sources, and tradeable emission credits – if an emitter reduces emissions below the target standard or ahead of the target timetable, emission credits are earned that can be applied to future emissions or sold to another emitter. These market-based initiatives were characteristic of the new legislation. Non-attainment areas are identified that fail to meet NAAQS for CO, O_3 or PM_{10}, in categories called marginal, moderate, serious or severe. Los Angeles merits the special category of extreme. The size of source regarded as major depends on how badly the NAAQSs in that area are exceeded. Exceeding regions will be required to make steady progress towards meeting targets. States must reduce overall emissions from all sources by 15% per year for the first 6 years. Vapour recovery must be installed at petrol stations in exceedance areas, and vehicle inspection and maintenance programmes operated. If State Implementation Plans are inadequate, then the USEPA can impose Federal Implementation Plans.

Reformulation of petrol can be used to reduce volatility, reduce sulphur (to increase catalyst efficiency and reduce sulphate emissions), or add oxygenated blend stocks (to reduce the proportion of high-octane aromatic HCs such as benzene that might otherwise be required). The most popular additive has been methyl tertiary butyl ether (MTBE), although other ethers and alcohols (especially ethanol) have been used. Oxygenated fuels bring additional oxygen to the combustion reactions, reducing the formation of CO but increasing NO_x; on the other hand, the flame temperature is reduced which lowers NO_x. However, they also raise the Reid Vapour Pressure – the RVP of ethanol is about 7 kPa higher than for standard petrol – which is also the subject of controls in order to reduce evaporation losses. Vehicle evaporative emissions can account for 30–50% of total HC emissions, and increasing the RVP from 62 to 82 kPa can double the evaporative emissions. If the vehicle has evaporative emission controls, although the mass emissions are lower, the proportional increase may be greater – about five times, if the fuel tank carbon canister saturates. Reduced fuel RVP also has an immediate benefit for HC emissions without having to wait for the vehicle stock to be replaced. The CAAA requires RVP to be reduced from 62 kPa to 56 kPa in northern (cooler) US cities and to 50 kPa in southern (warmer) ones. There are complex wider issues here. For example, ethanol is produced from wheat, and the wheat is grown on large areas of high-intensity agricultural land. The emissions from the fossil fuels used in making the fertiliser and other agricultural inputs must be offset against the benefits of reduced vehicle CO emissions.

13.7.2   Hazardous Air Pollutants (HAPs)

There are dozens of air pollutants that normally occur at ppb or ppt concentrations in ambient air and which are known to be harmful. More than half of these pollutants are known or suspected to be carcinogenic or mutagenic. Others have respiratory, neurological, immune or reproductive effects, especially for more sensitive populations such as children or frail adults. Many are also known to have adverse effects on plants and animals. These are materials such as heavy metals, chloroform and polycyclic organics which are not mainstream criteria pollutants but which can reasonably be expected to cause a health hazard. These compounds, or subsets of them, are variously known as air toxics, hazardous air pollutants (HAPs) or toxic organic micropollutants (TOMPs). Diesel exhaust particles, although not listed as HAP, contains many of them and is therefore treated as such. The USEPA currently identifies 188 HAPs, including 17 that represent groups of chemicals such as PAH rather than discrete compounds. Maximum Achievable Control Technology (MACT) must be used on any source emitting more than 10 t annually of any one HAP, or 25 t total. The US EPA is committed to reducing air toxics emissions by 75% below their 1993 level. By the end of a 10-year upgrading period (in 2002) it was predicted that total annual HAP emissions would be reduced by at least 1 million tons. The emission controls are accompanied by a complex procedure for assessing the residual risk from regulated emissions.

Thirty-three of the listed HAPs (Table 13.11) have been identified as posing the greatest threat to public health in the majority of urban areas, and 30 of these are emitted from smaller industrial sources known as 'area' sources; such sources are characteristic of urban areas in general, rather than particular process plants or similar operations. The potential sources are small and diverse, such as

Table 13.11 Hazardous air pollutants for primary control

Chemical name	Chemical name
Acetaldehyde	Formaldehyde
Acrolein	Hexachlorobenzene
Acrylonitrile	Hydrazine
Arsenic compounds	Lead
Benzene	
Beryllium compounds	Manganese
1,3 Butadiene	
Cadmium compounds	Mercury compounds
Carbon Tetrachloride	Methylene Chloride
Chloroform	Nickel
Chromium	Polychlorinated biphenyls (PCBs)
Coke oven emissions	Polycyclic organic matter (POM)
1,2 dibromoethane	Quinoline
1,2 dichloropropane	Tetrachloroethane
1,3 dichloropropene	Tetrachloroethylene
Dioxins	Trichloroethylene
Ethylene dichloride	Vinyl Chloride
Ethylene oxide	

wood burning stoves, residential space heaters, landfills, filling stations and vehicles. The subset of 33 was chosen by means of a ranking system based on toxicity, risk and concentration (measured and modelled). These area sources will be controlled by an Integrated Urban Air Toxics Strategy.

13.7.3   NAPAP

Another aspect of US pollution control is the National Acid Precipitation Assessment Program (NAPAP). NAPAP, and its successor NADP (National Atmospheric Deposition Program) is responsible for research and monitoring of acid rain, and reporting the results to Congress on a biennial basis. Title IV of the CAAA requires the reduction of acid rain precursors – essentially SO_2 and NO_x from power stations. The Acid Deposition Control Program is being implemented in two phases and has two major goals:

1   Reduction of total SO_2 emissions by 10 million tons per year (below 1980 levels) by 2010. Initial enforcement of a limit of 2.5 lb SO_2 per million BTU, followed by reduction to 1.2 lb in 2000.
2   Reduction of NO_x emissions from coal-fired boilers that will contribute to an overall target reduction of 2 million tons (below 1980 levels) by 2000.

Phase 1 of the program started in 1995 for SO_2 and in 1996 for NO_x, and involved a total of 445 power stations, mainly in the eastern US. Phase 2 started in 2000 and applies to over 2000 stations across the country. One distinctive feature of the programme is that emission allowances are tradeable. Each allowance permits a station to emit 1 t of SO_2 during or after a specified year, known as the vintage year. Once allocated, the allowances are fully marketable. They can be banked if it is thought that emission control will be more expensive in the future, or sold to another generator. By the end of 1996, over 3100 transactions involving over 51 million allowances had occurred. About 3.8 Mt of SO_2 was being removed, half by new FGD scrubbers and half by burning low-sulphur coal. By 1995, total anthropogenic SO_2 emissions had dropped to 18.3 Mt from 26 Mt in 1980.

These measures represent a complicated attempt to gain control over major environmental problems that have beset America for the last 50 years and have defied previous attempts to clean up the air.

13.7.4   US vehicle emissions

13.7.4.1   US drive cycles

The US has a set of drive cycles for emission testing, known as Federal Test Procedures (FTPs) that correspond to those used in the EU. The 1975 FTP was the outcome of the progressive development in the US of a dynamometer drive cycle designed to suit American roads, vehicles and driving habits. Development started in the 1950s, and was based on a driving cycle through Los Angeles, California. FTP 75 has the characteristics shown in Table 13.12.

During each of the 3 phases, the vehicle is taken through a number of exactly specified acceleration/cruise/deceleration cycles. The FTP heavy-duty transient cycle consists of four phases: the first is a New York Non Freeway (NYNF) phase typical of light urban traffic with frequent stops and starts; the second is Los Angeles Non Freeway (LANF) phase typical of crowded urban traffic with few stops; the third is a Los Angeles Freeway (LAFY) phase simulating crowded expressway traffic in Los Angeles; and the fourth repeats the first NYNF phase.

Table 13.12 The FTP 75 emissions drive cycle

Cycle length	11.115 miles
Cycle duration	1875 s plus 600 s hot soak
Phase 1 (cold start)	0–505 s
Phase 2 (hot stabilised)	505–1370 s
Hot soak	600 s
Phase 3 (hot start)	0–505 s
Average speed	34.1 km hr^{-1}
Maximum speed	91.2 km hr^{-1}
Number of hills	23
Number of modes	112

It comprises a cold start after overnight parking, followed by idling, acceleration and deceleration phases, and a wide variety of different speeds and loads sequenced to simulate the running of the vehicle that corresponds to the engine being tested. There are few stabilized running conditions, and the average load factor is about 20–25% of the maximum horsepower available at a given speed. The cycle is carried out twice and the repetition is made with a warm start after a stop of 1200 s (20 min) on completion of the first cycle. The equivalent average speed is about 30 km h^{-1} and the equivalent distance travelled is 10.3 km for a running time of 1200 s.

13.7.4.2   US cars and LDVs

Federal Standards

The US was the first country to mandate the use of catalytic converters, following recognition of the causes of the photochemical smogs in the 1960s. The first oxidation catalysts were fitted in that decade, and emission standards steadily increased through the following three decades, with all new cars required to have a TWC after 1981. Two sets of standards, Tier 1 and Tier 2, were defined for light-duty vehicles in the CAAA. The Tier 1 regulations were published as a final rule on June 5, 1991 and fully implemented in 1997. The Tier 2 standards were adopted on December 21, 1999, to be phased-in starting in 2004.

TIER I STANDARDS

Tier 1 light-duty standards apply to all new light duty vehicles (LDV), such as passenger cars, light duty trucks, sport utility vehicles (SUV), minivans and pick-up trucks. The LDV category includes all vehicles of less than 8 500 lb gross vehicle weight rating (GVWR, i.e. vehicle weight plus rated cargo capacity). Light duty vehicles are further divided into the following sub-categories:

- passenger cars
- light light-duty trucks (LLDT), below 6000 lbs GVWR
- heavy light-duty trucks (HLDT), above 6000 lbs GVWR.

The standards apply to a full vehicle useful life of 100 000 miles (effective 1996). The regulation also defines an intermediate standard to be met over a 50 000 miles period. The difference between diesel and petrol car standards is a more relaxed NO$_x$ limit for diesels, which applies to vehicles through the 2003 model year.

Car and light truck emissions are measured over the FTP75 test and expressed in g mi^{-1}. In addition to the FTP75 test, a supplemental Federal Test Procedure (SFTP) will be phased-in between 2000 and 2004. The SFTP includes additional test cycles to measure emissions during aggressive highway driving (US06 cycle), and also to measure urban driving emissions while the vehicle's air conditioning system is operating (SC03 cycle) (Table 13.13).

Table 13.13 EPA Tier 1 Emission standards for passenger cars and light-duty trucks, FTP75 g mi^{-1}

Category	100 000 miles/10 years					
	THC	NMHC	CO	NO$_x$ diesel	NO$_x$ petrol	PM
Passenger cars	–	0.31	4.2	1.25	0.6	0.10
<3 750 lbs	0.80	0.31	4.2	1.25	0.6	0.10
>3 750 lbs	0.80	0.40	5.5	0.97	0.97	0.10
Light duty truck <5 750 lbs	0.80	0.46	6.4	0.98	0.98	0.10
Light duty truck >5 750 lbs	0.80	0.56	7.3	1.53	1.53	0.12

TIER 2 STANDARDS

The Tier 2 standards bring significant emission reductions relative to Tier 1. In addition to more stringent numerical emission limits, the regulation introduces a number of important changes that make the standard more stringent for larger vehicles. Under the Tier 2 standard, the same emission standards apply to all vehicle weight categories, i.e. cars, minivans, light duty trucks, and SUVs have the same emission limit. Since light-duty emission standards are expressed in grams of pollutants per mile, large engines (such as those used in light trucks or SUVs) will have to utilise more advanced emission control technologies than smaller engines in order to meet the standard.

In Tier 2, the applicability of light-duty emission standards has been extended to cover some of the heavier vehicle categories. The Tier 1 standards applied to vehicles up to 8 500 lbs GWVR. The Tier 2 standard applies to all vehicles that were covered by Tier 1 and, additionally, to 'medium-duty passenger vehicles' (MDPV). The MDPV is a new class of vehicles that are rated between 8 500 and 10 000 GVWR and are used for personal transportation. This category includes primarily larger SUVs and passenger vans. Engines in commercial vehicles above 8 500 lbs GVWR, such as cargo vans or light trucks, will continue to certify to heavy-duty engine emission standards. The same emission limits apply to all engines regardless of the fuel they use. That is, vehicles fuelled by petrol, diesel, or alternative fuels all must meet the same standards.

The Tier 2 exhaust standards are structured into 8 certification levels of different stringency, called 'certification bins', and an average fleet standard for NO$_x$ emissions. Vehicle manufacturers will have a choice to certify particular vehicles to any of the 8 bins. At the same time, the average NO$_x$ emissions of the entire vehicle fleet sold by each manufacturer will have to meet the average NO$_x$ standard of 0.07 g mi^{-1}.

The Tier 2 standards will be phased-in between 2004 and 2009. For new passenger cars and light LDTs, Tier 2 standards will phase in beginning in 2004, with the standards to be fully phased in by 2007. For heavy LDTs and MDPVs, the Tier 2

Table 13.14 Tier 2 emission standards, FTP 75, g mi^{-1}

Bin#	120 000 miles				
	NMOG	CO	NO$_x^*$	PM	HCHO
8	0.125	4.2	0.20	0.02	0.018
7	(0.156) 0.090	4.2	0.15	0.02	0.018
6	0.090	4.2	0.10	0.01	0.018
5	0.090	4.2	0.07	0.01	0.018
4	0.070	2.1	0.04	0.01	0.011
3	0.055	2.1	0.03	0.01	0.011
2	0.010	2.1	0.02	0.01	0.004
1	0.000	0.0	0.00	0.00	0.000

Note
* average manufacturer fleet NO$_x$ standard is 0.07 g/mile.

standards will be phased in beginning in 2008, with full compliance in 2009. During the phase-in period from 2004 to 2007, all passenger cars and light LDTs not certified to the primary Tier 2 standards will have to meet an interim average standard of 0.30 g mi^{-1} NO$_x$, equivalent to the current NLEV standards for LDVs. During the period 2004–2008, heavy LDTs and MDPVs not certified to the final Tier 2 standards will phase in to an interim programme with an average standard of 0.20 g mi^{-1} NO$_x$, with those not covered by the phase-in meeting a per-vehicle standard (i.e. an emissions 'cap') of 0.60 g mi^{-1} NO$_x$ (for HLDTs) and 0.09 g mi^{-1} NO$_x$ (for MDPVs).

The emission standards for all pollutants (certification bins) are shown in Table 13.14. The vehicle 'full useful life' period has been extended to 120 000 miles. The EPA bins cover California LEV II emission categories, to make certification to the federal and California standards easier for vehicle manufacturers. The Tier 2 regulation also brings new requirements for fuel quality. Cleaner fuels will be required by advanced emission after-treatment devices (e.g. catalysts) that are needed to meet the regulations.

13.7.4.3   Other measures

The US has also legislated to increase fuel efficiency through the Corporate Average Fuel Economy (CAFE) standards designed to raise average fuel economy to 27.5 mpg by 1985. This was achieved for cars, but the increasing popularity of light trucks (4 × 4 s, or sport utility vehicles) which had only to achieve 20.2 mpg, rather undermined the improvement. There are also standards for evaporative emissions, which cover losses in the categories of diurnal (due to temperature changes while at rest), hot soak (evaporation driven by a hot engine block after a trip), running, resting and refuelling. These emission are measured by a standard procedure called sealed housing evaporative determination (SHED), for which a full test takes 5 days.

13.7.4.4   National LEV programme

On December 16, 1997, EPA finalised the regulations for the National Low Emission Vehicle (NLEV) programme. The NLEV is a voluntary programme, that came into effect through an agreement by the north-eastern States and the car manufacturers. It provides more stringent emission standards for the transitional period before the Tier 2 standards are introduced.

Starting in the north-eastern States in model year 1999 and nationally in model year 2001, new cars and light-duty trucks have to meet exhaust standards that are more stringent than EPA can legally mandate prior to model year 2004. However, after the NLEV programme was agreed upon, these standards are enforceable in the same manner as any other federal new motor vehicle programme.

The National LEV programme harmonises the federal and California motor vehicle standards and provides emission reductions that are basically equivalent to the California Low Emission Vehicle programme. The programme is phased-in through schedules that require car manufacturers to certify a percentage of their vehicle fleets to increasingly cleaner standards (TLEV, LEV, ULEV). The National LEV program extends only to lighter vehicles and does not include the Heavy LDT (HLDT, GVW >6 000 lbs) vehicle category.

The program also requires most refiners and importers to meet a corporate average petrol sulphur standard of 120 ppm and a cap of 300 ppm beginning in 2004. By 2006, the average standard will be reduced to 30 ppm with 80 ppm sulphur cap.

13.7.5   Comparison of EU and US vehicle emission standards

The EU and US are proceeding down similar courses of increasingly stringent emission standards, attempting to offset the ever-increasing number of vehicles and the distances driven in them (Table 13.15).

Table 13.15 Comparison of EU and US vehicle emission standards

Year	Emission Factor/g km^{-1}							
	CO		HC		NO$_x$		TSP	
	US	EU	US	EU	US	EU	US	EU
1980	6.3	16–36	0.3	1.5–2.4	1.4	2.1–3.4		
1990	3.06	14	0.3	Combined with NO$_x$	0.7	4.7–6.9		
1994	3.06	2.72	0.18		0.3	0.97	0.06	0.14
1996	3.06	2.2	0.18		0.3	0.5	0.08	
2004	1.53		0.09		0.13			

13.7.6 Vehicle emissions in California

California has had the world's worst air quality problems due to motor vehicles for half a century, and is taking draconian measures – beyond those required by the CAAA – to try and control them. The California Air Resources Board (CARB) has set low-emission vehicle standards for passenger cars and light-duty trucks. They are designed to reduce precursor emissions and O_3 formation, and to allow for and encourage the introduction of alternative fuels such as ethanol, methanol, compressed natural gas, or liquified petroleum gas. The current California emission standards are expressed through the following emission categories:

- Tier 1
- Transitional Low Emission Vehicles (TLEV)
- Low Emission Vehicles (LEV)
- Ultra Low Emission Vehicles (ULEV)
- Super Ultra Low Emission Vehicles (SULEV)
- Zero Emission Vehicles (ZEV, assumed to be electric).

Car manufacturers are required to produce a percentage of vehicles certified to increasingly more stringent emission categories. The phase-in schedules, which are fairly complex, are generally based on vehicle fleet emission averages. The exact schedules can be found in the EPA web site. After 2003, Tier 1 and TLEV standards will be eliminated as available emission categories. California Air Resources Board also stipulated that 10% of new cars sold in California in and after 2003 would be ZEVs. The initiative has spawned a host of new developments in ZEV technology, together with advances in hybrid vehicles and alternative fuels. In fact, the ZEV mandate proved to be too optimistic, and was softened in 1998. The major car manufacturers are allowed to substitute 6% of the 10% with 'almost-ZEVs' which achieve extremely low emissions. Since totally ZEVs can only be battery-powered at present, this change has allowed the use of alternative fuels, hybrid vehicles and fuel cells. The main limitation of the battery powered vehicles produced so far is their short range – typically 100 miles.

The same standards for gaseous pollutants apply to diesel- and petrol-fueled vehicles. Particulate matter standards apply to diesel vehicles only. Emissions are measured over the FTP 75 test and are expressed in g mi^{-1}. The additional SFTP procedure will be phased-in in California between 2001 and 2005.

The Californian emission values incorporate the concept of reactivity-weighted VOC mass emissions, together with NO_x reductions. The reactivity adjustment factor (RAF) for any new fuel is defined as the mass of O_3 formed from one gram of emissions from the new fuel, taken as a proportion of the mass of O_3 formed from one gram of emissions from standard-grade petrol. Hence, the allowed NMVOC for a particular fuel is found from the value above divided by the appropriate RAF. For example, the typical fuel for a TLEV would be 85%

methanol, 15% petrol, with a RAF of 0.41 and permitted NMVOC emissions of $0.125/0.41 = 0.30$ g mi^{-1}. Accurate assessment of the RAF for a particular fuel requires quantitative speciation of at least 50, and probably 170, combustion products.

The CARB estimates that 27 000 t of PM are emitted each year in California from around 1.25 million diesel engines, and plans to reduce this by 90%. The control programme involves 12 measures, including: the use of PM traps on all new and most existing diesel engines (this covers private as well as commercial, although private diesel vehicles are less common in the US than in Europe); low-sulphur diesel fuel; in-use emission testing; and broader use of alternative fuels in place of diesel.

13.7.6.1   Low Emission Vehicle II (LEV II) standards

On November 5, 1998 the CARB adopted the LEV II emission standards which will extend from the year 2004 until 2010. Under the LEV II regulation, the light-duty truck and medium-duty vehicle categories of below 8 500 lbs gross weight are reclassified and will have to meet passenger car requirements. As a result, most pick-up trucks and sport utility vehicles will be required to meet the passenger car emission standards. The reclassification will be phased in by the year 2007. Medium duty vehicles above 8 500 lbs gross weight (old MDV4 and MDV5) will still certify to the medium-duty vehicle standard.

Under the LEV II standard, NO$_x$ and PM standards for all emission categories are significantly tightened. The same standards apply to both petrol and diesel vehicles. Light-duty LEVs and ULEVs will certify to a 0.05 g mi^{-1} NOx standard, to be phased-in starting with the 2004 model year. A full useful life PM standard of 0.010 g mi^{-1} is introduced for light-duty diesel vehicles and trucks less than 8 500 lbs gross weight certifying to LEV, ULEV, and SULEV standards. The TLEV emission category has been eliminated in the final regulatory text. It is, therefore, believed that emission certification of light duty diesel vehicles in California will be possible only if advanced emission control technologies, such as particulate traps and NO$_x$ catalysts, are developed.

13.7.7   US locomotives

In the United States, emissions from diesel-powered locomotives are also regulated. On December 17, 1997 the EPA adopted emission standards for oxides of nitrogen (NO$_x$), hydrocarbons (HC), carbon monoxide (CO), particulate matter (PM) and smoke for newly manufactured and remanufactured railroad locomotives and locomotive engines. The ruling, which takes effect in the year 2000, applies to locomotives originally manufactured from 1973, any time they are manufactured or remanufactured. Electric locomotives, historic steam-powered locomotives, and locomotives originally manufactured before 1973 are not regulated.

Three separate sets of emission standards have been adopted, with applicability of the standards dependent on the date a locomotive is first manufactured. The first set of standards (Tier 0) applies to locomotives and locomotive engines originally manufactured from 1973 through 2001, any time they are manufactured or remanufactured. The second set of standards (Tier 1) applies to locomotives and locomotive engines originally manufactured from 2002 through 2004. These locomotives and locomotive engines will be required to meet the Tier 1 standards at the time of the manufacture and each subsequent remanufacture. The final set of standards (Tier 2) applies to locomotives and locomotive engines originally manufactured in 2005 and later. Tier 2 locomotives and locomotive engines will be required to meet the applicable standards at the time of original manufacture and each subsequent remanufacture.

The emissions are measured over two steady-state test cycles which represent two different types of service including the *line-haul* and *switch* locomotives. The duty cycles include different weighting factors for each of the 8 throttle notch modes, which are used to operate locomotive engines at different power levels, as well as for idle and dynamic brake modes. The switch operation involves much time in idle and low power notches, whereas the line-haul operation is characterised by a much higher percentage of time in the high power notches, especially notch 8.

A dual cycle approach has been adopted in the regulation, i.e. all locomotives are required to comply with both the line-haul and switch duty cycle standards, regardless of intended usage. The emission standards and current locomotive emission levels are listed in Table 13.16. The smoke opacity standards are listed in Table 13.17. The regulations contain several other provisions, including a production line testing (PLT) programme, in-use compliance emission testing, as well as averaging, banking and trading (ABT) of emissions.

13.7.8   National parks and wilderness areas

In the US there are many forests and national parks, where controlled burning is a routine management requirement. Yet such burning unavoidably creates emissions of many air pollutants. The US Clean Air Act designated 156 Natural Parks, Wilderness Areas and wildlife protection zones as Class I areas. Many of these are household names such as Grand Canyon, Great Smokies, Yellowstone, Yosemite and the Everglades. The managers of such areas have an 'affirmative responsibility' to protect the air quality related values (AQRVs) from adverse air pollution impacts. Air Quality Related Values are those features or properties of a Class I area which can be changed by air pollution. If the area concerned is adjacent to a non-attainment area for NAAQS, the proposed burn must include prediction of the impact on local air quality. If any process or activity is proposed, then it must be shown that it will not result in significant deterioration of air quality.

Table 13.16 Emission standards for locomotives/g (bhp h)$^{-1}$

Duty Cycle	HC*	CO	NO$_x$	PM
Tier 0 (1973–2001)				
Line-haul	1.0	5.0	9.5	0.60
Switch	2.1	8.0	14.0	0.72
Tier 1 (2002–2004)				
Line-haul	0.55	2.2	7.4	0.45
Switch	1.2	2.5	11.0	0.54
Tier 2 (2005 and later)				
Line-haul	0.3	1.5	5.5	0.20
Switch	0.6	2.4	8.1	0.24
Current Estimated Locomotive Emission Rates (1997)				
Line-haul	0.5	1.5	13.5	0.34
Switch	1.1	2.4	19.8	0.41

Note
* HC standard is in the form of THC for diesel engines.

Table 13.17 Smoke standards for locomotives/% opacity – normalised

	Steady-state	*30-s peak*	*3-s peak*
Tier 0	30	40	50
Tier 1	25	40	50
Tier 2	20	40	50

13.8   AIR POLLUTION INDICES

While scientists, administrators and legislators may anguish over the relation-ships between the concentrations of a particular pollutant and their effects, non-expert members of the public need more immediate information to tell them how clean the 'air' is, whether it is safe and to warn them if it is not. Various air quality indices have been developed to meet this need. The general approach is to avoid the complexities of units such as ppb and μg m^{-3} by relating effects to a standard scale such as 1–100, and to describe the air quality in everyday terms such as good or moderate or hazardous.

13.8.1 Air quality banding

The first approach along these lines in the UK was the Air Pollution Information System, which is designed to inform the public about air quality without using specific concentrations or units. The main components are three threshold concentrations (standard, information and alert) for each of five pollutants, together with four corresponding bands (low, moderate, high and very high). The thresholds were chosen for their significance to health effects, and these bandings have been widely used, for example during TV weather forecasts, to alert people both to prevailing and predicted conditions (Table 13.18).

13.8.2 Headline Indicator

Another approach that has been investigated in the UK is that of the Headline Indicator. This has been suggested as useful not only for communicating with the public, but also to feed in to assessments of overall environmental quality and sustainability. After considering various options, DETR (now Department for Environment, Food and the Regions) adopted the concept of Days Exceeding as the criterion for a headline indicator. Correct terminology is vital in this subject – for example, Days Exceeding is a different concept, and generates quite different values, to Daily Exceedances. Days Exceeding is the sum of days with an exceedence of any pollutant, maximum of 1 exceedence per day. It was recommended that the Standard threshold be used for counting exceedances, since this corresponds to the Standard concentration set in NAQS, and to the concentration below which health effects are unlikely to be found. Thus, if one

Table 13.18 UK air quality bandings

	Carbon monoxide	Nitrogen dioxide	Ozone	PM$_{10}$	Sulphur dioxide
Time averaging period	8-h running mean	1-h mean	1-h mean	24-h running mean	15-min mean
Unit	ppm	ppb	ppb	μg m^{-3}	ppb
Low pollution	<10	<150	<50[1]	<50	<100
Standard threshold	**10**	**150**	**50**	**50**	**100**
Moderate Pollution	10–14	150–299	50–89[2]	50–74	100–199
Information threshold	**15**	**300**	**90**	**75**	**200**
High pollution	15–19	300–399	90–179	75–99	200–399
Alert threshold	**20**	**400**	**180**	**100**	**400**
Very high pollution	>20	>400	>180	>100	>400

Notes
1 Running 8-h mean.
2 Running 8-h mean or 1-h mean.

monitoring site recorded CO concentrations above Standard on 5 days, SO_2 concentrations above standard on 6 days, and PM_{10} concentrations above standard on 8 days during 1 year, the Days Exceeding figure for that site would be 19. If the days overlap, however, they should only be counted once.

13.8.3 Pollutant Standards Index

In America, the Pollutant Standards Index (PSI) equates the scale value 100 to the shortest-term NAAQS concentration, which is called moderate. The index value then rises, approximately *pro rata* with the actual concentration, to values of 200 (alert), 300 (warning), 400 (emergency) and 500 (significant harm). At each level, the general health effects are summarised and appropriate precautions indicated. Of course, one danger of this approach is that equal index values for quite different pollutants will be perceived as equally hazardous to health. The PSI is used by metropolitan areas to report an overall assessment of each day's air quality to the public. The PSI is calculated by first working out a subindex, on the same 0–500 scale, for each of five pollutants – 1 h O_3, 8 h CO, 24 h PM_{10}, 24 h SO_2 and 1 h NO_2. Then, the highest of the five subindices is used as the PSI. Hence, the PSI method does not attempt to evaluate the overall impact of combined pollutants, about which even less is understood than the impacts of the individual pollutants. Trends in potential health impacts of air pollution have been analysed in terms of the number of days on which PSI values of 100 were exceeded.

FURTHER READING

Bell, S. (2000) *Ball & Bell on Environmental Law: The Law and Policy Relating to the Protection of the Environment*, 5th edn, Blackstone Press.

Bennett, G. (ed) (1992) *Air Pollution Control in the European Community*, Graham and Trotman, London, UK.

Biermann, F. (1995) *Saving the Atmosphere: International Law, Developing Countries and Air Pollution*, P. Lang.

Department for Environment, Food & Rural Affairs (2001) *Air Quality Strategy: An Economic Analysis to Inform the Review of the Objectives for Particles*, DEFRA, London.

Department of the Environment, Transport and the Regions (1999) *The Air Quality Strategy for England, Scotland, Wales and Northern Ireland: A Consultation Document*, DETR, London, UK.

European Environment Agency (1999) *Annual Topic Update on Air Quality*, EEA, Copenhagen.

Garbutt, J. H. (2000) *Environmental Law: A Practical Handbook*, 3rd edn, Bembridge Palladian.

Hughes, D. (1998) *Air Pollution: Law and Legislation*, Parpworth, Joan Upson.

Jans, J. H. (2000) *European Environmental Law*, 2nd rev. edn, Europa Law Publishing, Groningen.

Kiss, A. and Shelton, D. (1992) *Manual of European Environmental Law*, Grotius, Cambridge, UK.

Krämer, L. (2000) *E.C. Environmental Law,* 4th edn, Sweet and Maxwell, London.

Leeson, J. D. (1995) *Environmental Law*, Pitman, London, UK.

McEldowney, J. F. (2001) *Environmental Law and Regulation*.

Mumma, A. (1995) *Environmental Law: Meeting UK and EC Requirements*, McGraw-Hill, Maidenhead, UK.

UK Department of the Environment Expert Panel on Air Quality Standards. *Benzene* (1994); *Ozone* (1994); *Carbon Monoxide* (1994); *1,3 Butadiene* (1994); *Particles* (1995); *Sulphur dioxide* (1995); *Nitrogen dioxide* (1996); *Lead* (1998); *PAH* (1999), Department of the Environment, London, UK.

World Health Organisation (1987) *Air Quality Guidelines for Europe*, WHO, Copenhagen, Denmark; also see *web sites*.

Conclusion

History has taught us that curing environmental problems is like mending a rusty bucket. Even while we are fixing one leak, another one is starting. The first wave of air pollution problems was due to stationary combustion of fossil fuels, and we fixed that in the developed nations with tall chimneys, desulphurisation and natural gas. In the developing nations, however, the problem is far from fixed, since those countries are concentrating on raising their standards of living rather than environmental protection. The second wave was due to motor vehicles in urban areas, and we fixed that with better engines, cleaner fuels and catalytic converters. Again, there is still a lot of potential growth of air pollution from this cause in developing nations. The third wave is climate change, which we have seen is due to aspects of our culture which are much more fundamental than the pollutants that we have dealt with previously. I believe that the only way to fix the problem is to develop a less energy-intensive society, and that will take a coherent effort by the majority of both developed and developing nations. We did not get a proper grip on fossil fuel pollutants until there had been environmental disasters which killed thousands of people in a few days. Even after that, extended studies of the long-term consequences of exposures to low concentrations have shown that thousands more people are still having their lives shortened by air pollution. In the same way, we will probably have to wait some decades before sufficiently serious events take place, due to climate change, to motivate a serious response. Although the pace of build-up in greenhouse gases is very rapid on a geological time-scale, it is rather leisurely by comparison with a human lifetime. Hence it is not inconceivable that we will stave off the consequences by a series of increasingly desperate measures, such as irrigation schemes, flood defences and the burial of carbon dioxide down mine shafts. There will be an ongoing struggle between those determined to go for growth and those seeking to restrain them. The outcome of that struggle will affect us all in due course.

Appendix: useful websites

Many of these sites cover several of the topics in this book, so I have not divided them between Chapters. Many of them are also entry sites to large organisations having dozens of links, not only within their own site but to other sites.

http://www.nsca.org.uk – UK National Society for Clean Air and Environmental Protection. An independent British society which is involved with the interpretation and application of Government legislation, particularly with Local Authorities.

http://www.aeat.co.uk/netcen/airqual – UK National Environmental Technology Centre, which processes the data from the Automatic Network of air pollution analysers and houses the National Atmospheric Emissions Inventory.

http://www.rsk.co.uk/ukefd/siteidx.htm – UK data base of emission factors for stationary and mobile, point and area sources.

http://www.defra.gov.uk – UK Department for Environment, Food and Rural Affairs.

http://www.dtlr.gov.uk – UK Department for Transport, Local Government and the Regions.

http://www.doh.gov.uk/comeap – UK Department of Health Committee on the Medical Effects of Air Pollution. COMEAP has published several authoritative reviews on health impacts of gases and particles.

http://www.met-office.gov.uk/research/hadleycentre – The Hadley Centre at the Meteorological Office is the main UK Government Research Centre for modelling climate change.

http://www.concawe.be – The oil producers environmental organisation, which has been involved in aspects of air pollution such as emission inventories and dispersion modelling.

http://europa.eu.int/comm/environment/air – Huge EU site, to which this is the entry for air quality material.

http://europa.eu.int/comm/environment/autooil – Home page for the EU AutoOil programme on the impact of vehicle emissions on air quality.

http://www.eea.eu.int – European Environment Agency.

http://etc-acc.eionet.eu.int – The European Topic Centre on Air and Climate Change, which was established by the EEA in 2001. The remit of ETC/ACC is to coordinate measurement, monitoring and assessment of data on climate change and air quality, informing the EEA for programmes such as CLRTAP and CORINAIR.

http://www.oar.noaa.gov/organization/napap.html – Home page of the US National Acid Precipitation Assessment Program.

http://nadp.sws.uiuc.edu – US National Atmospheric Deposition Program, which is the successor to NAPAP. Data sets and contour plots on deposition of acidity, ions, metals in the US since 1994.

http://www.fe.doe.gov – US Department of Energy.

http://gcmd.gsfc.nasa.gov – The Global Change Master Directory, run by NASA, specialist searchable reports on research projects.

http://www.oar.noaa.gov/atmosphere – US National Ocean and Atmosphere Administration.

http://www.epa.gov – US Environmental Protection Agency. A huge site covering climate change as well as all aspects of pollution.

http://www.arb.ca.gov/homepage.htm – The California Air Resources Board, which has had to be proactive and ahead of US Federal standards because of the severe air quality problems in California.

http://www.eia.doe.gov – Official energy statistics from the US Government.

http://www.ipcc.ch – Intergovernmental Panel on Climate Change. The coordinating body for international assessments by hundreds of scientists on the causes and consequences of climate change.

http://www.unfccc.int – United Nations Framework Convention on Climate Change. Driving the global response on climate change.

http://www.who.int/home-page – The World Health Organisation is actively involved in health effects of air pollution, including publication of the Air Quality Guidelines on threshold values for individual pollutants.

http://www.wmo.ch/index-en.html – The World Meteorological Organisation carries out fundamental reviews of stratospheric ozone depletion, and tropospheric air pollution.

http://www.knmi.nl/gome_fd – Near-real time data from the GOME satellite of ozone column depths and concentrations.

www.unece.org/env/lrtap – United Nations Economic Commission for Europe. Has taken a leading role in air quality.

http://www1.oecd.org/env – The Organisation for Economic Cooperation and Development, a global community of 30 nations which has been instrumental in identifying environmental problems.

http://www.iiasa.ac.at – The International Institute for Applied Systems Analysis (IIASA) in Austria. Has developed the RAINS model for regional forecasting of air pollution, and published reports on topics such as particulate control.

http://www.unep.org – United Nations Environmental Programme.

Index